Dielectric Ceramic Materials

Related titles published by the American Ceramic Society:

Advances in Dielectric Ceramic Materials
Edited by K.M. Nair and A.S. Bhalla
© 1998, ISBN 1-57498-033-5

Hybrid Microelectronic Materials
Edited by K.M. Nair and V.N. Shukla
© 1996, ISBN 1-57498-013-0

Ferroic Materials: Design, Preparation, and Characteristics
Edited by A.S. Bhalla, K.M. Nair, I.K. Lloyd, H. Yanagida, and D.A. Payne
© 1994, ISBN 0-944904-77-7

Grain Boundaries and Interfacial Phenomena in Electronic Ceramics
Edited by Lionel M. Levinson and Shin-ichi Hirano
© 1994, ISBN 0-944904-73-4

Dielectric Ceramics: Processing, Properties, and Applications
Edited by K.M. Nair, J.P. Guha, and A. Okamoto
© 1993, ISBN 0-944904-60-2

For information on ordering titles published by The American Ceramic Society, or to request a publications catalog, please contact our Customer Service Department at 614-794-5890 (phone), 614-794-5892 (fax),<customersrvc@acers.org> (e-mail), or write to Customer Service Department, 735 Ceramic Place, Westerville, OH 43081, USA.

Visit our on-line book catalog at <www.acers.org>.

Ceramic
Transactions
Volume 100

Dielectric Ceramic Materials

Edited by

K.M. Nair
E.I. duPont de Nemours & Company, Inc.

A.S. Bhalla
The Pennsylvania State University

Published by
The American Ceramic Society
735 Ceramic Place
Westerville, Ohio 43081

Proceedings of the International Symposium on Dielectric Ceramics, held at the 100th Annual Meeting of The American Ceramic Society in Cincinnati, Ohio, May 3-6, 1998.

COVER PHOTO: "SEM micrograph of Pure and Dy-doped BaTiO$_3$ Powders — Dy:BaTiO$_3$, 1500°C, Bar = 10 μm," is courtesy of Ersin E. Ören and A. Cüneyt Tas, and appears as figure 4 in their paper "Hydrothermal Synthesis of Pure and Dy:BaTiO$_3$ Powders at 90°C and Their Sintering Behavior," which begins on page 95.

Library of Congress Cataloging-in-Publication Data
A CIP record for this book is available from the Library of Congress.

For information on ordering titles published by The American Ceramic Society, or to request a publications catalog, please call 614-794-5890.

1 2 3 4–02 01 00 99

ISSN 1042-1122
ISBN 978-1-57498-066-0

Contents

viii

Preface

The "electronic revolution" of this century will take an expanded role in the living standards of the human beings when we enter the twenty-first century. The initial areas of progress will be realized in automotive, telecommunications, and medical electronics. The development of multilayer microelectronics that are mainly based on ceramics and ceramic-based systems has increased the reliability, cost-effectiveness, and the number of diversified applications of electronic circuits. Further improvements in these areas will come from the development of new materials, composite materials, and from the use of mixed technologies.

Materials societies like The American Ceramic Society will have to play a key role in the future development of new materials, processes, and technologies. For the last 15 years, The American Ceramic Society has organized several international symposia covering many aspects of the advanced electronic materials systems and has brought together leading researchers and practitioners of electronics. The proceedings of these conferences have been published in the *Ceramic Transactions*, a leading up-to-date international materials book series.

This volume contains a collection of selected papers that were presented at the International Symposium on Dielectric Ceramics during the 100th Annual Meeting of The American Ceramic Society held in Cincinnati, Ohio, May 3–6, 1998. The major topics of the symposium were fundamental and historical perspectives of dielectric materials; advanced aspects of powder preparation, characterization, and properties; materials for thick and thin films; materials for low- and high-frequency applications; processing–microstructure–property relationships; and potential areas of applications. More than 20 invited and 25 contributed papers were peer-reviewed and are included in this volume.

We, the editors, acknowledge and appreciate the contributions of the speakers, conference session chairs, manuscript reviewers, and Society officials for making this endeavor a successful one.

K.M. Nair
A.S. Bhalla

BARIUM TITANATE - PAST, PRESENT AND FUTURE

Alan Rae, Mike Chu and Vladimir Ganine
TAM Ceramics Inc.
4511 Hyde Park Blvd.
Niagara Falls
NY14305-0067, USA

ABSTRACT

Barium titanate's initial commercialization as a piezo, thermistor and capacitor material has driven its strong growth over the past 50 years. Now the trusty perovskite faces opportunities and challenges in its markets, as competing materials and technologies race to meet the better-faster-smaller-cheaper needs of the electronics industry. The paper reviews market and technology trends and attempts to extrapolate to future demand and technologies.

THE HISTORY AND DEVELOPMENT OF BARIUM TITANATE

Ferroelectric materials can be polarized by an applied electric field. This polarization has a dramatic effect on charge storage, physical properties and thermal properties. The first studies of ferroelectric crystals were by Pierre and Jacques Curie around 1880 who measured surface charges on a variety of materials under applied mechanical stress.

Barium titanate ceramic as a ferroelectric was discovered independently in the USA, Japan and Russia in 1943 and single crystals were first produced in 1947.
Although in single crystals there are 3 phase transitions:

Rhombohedral => -90^0C => Orthorhombic => 50C => Tetragonal => 120^0C => Cubic

the most interesting property change in ceramics occurs around the 120^0C transition, commonly referred to as the "Curie Point". At the Curie Point there is the highest dielectric constant, above the Curie point there is a sharp decrease in electrical conductivity.

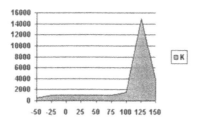

Figure 1 - Dielectric Constant (K) vs. Temperature - schematic

Useful dielectric formulations need to shift this peak and flatten it so that electrical properties either "switch" to a technologically useful temperature (as in PTC thermistors) or are uniform over a specified range.

Here we are helped by the compliant structure of barium titanate which has the so-called Perovskite structure, named after the mineral form of calcium titanate. These minerals have a formula approximating to ABO3 where metal A is in an octahedral site and element B is in a tetrahedral site.

With charge compensation substitution for barium and titanium is possible in the structure with resulting distortion and transformation of the character of the Curie peak. Typical A site substitutes would be Ca, Mg, Sr, Pb, Y and selected rare earths, and B site substitutes include Zr, Sn and Nb.

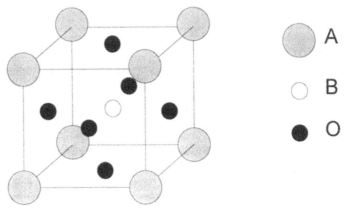

Figure 2. The Perovskite Structure - schematic

Transition temperatures for isomorphous compounds range from -100^{0}C (Ba-SrTiO3) through 490^{0}C (PbTiO3) and many complex solid solutions are possible.

Other additives are used to enhance sintering as transient fluxes, or as glassy or crystalline phases which may affect physical, chemical and electrical properties.

TAM Ceramics originally started production of barium titanates and additives for capacitors and piezos in the 1940's and until the 1960's was the sole producer of these materials.

BARIUM TITANATE IN MULTILAYER CERAMIC CAPACITORS

The statistics are staggering. Nearly a trillion capacitors are produced a year and the vast majority are multilayer capacitors based on barium titanate. The selling price can be as low as 3000 for $1 for a typical 0603 capacitor! Layer thickness can be as low as 2 microns and the number of layers can be as high as 500. The smallest parts, EIA 0201 are approximately 0.5 x 0.25 x 0.25 mm in size.

A typical watch contains 2-4 capacitors, a video camera or cell phone 250, a laptop computer 400 and an automobile over 1000. When you look around your office or home you start to realize just how many devices you own have significant inexpensive computing capacity - phones, pagers, bread makers, temperature controllers, remote controls - the list is very extensive.

Why has the multilayer ceramic capacitor (commonly abbreviated to MLCC or MLC) been so successful? A combination of:
- Performance
- Size
- Surface mount tolerance
- Manufacturability
- Cost

What are the competitors to MLCCs?
- Tantalum capacitors
- Integrated capacitance (on-chip)
- Embedded capacitance (in-board)
- Capacitor arrays

MULTILAYER CERAMIC CAPACITOR CONSTRUCTION

Approximately a billion capacitors are produced per day, 70% of them being ceramic capacitors, mostly surface mount multilayer capacitors. 75% of ceramic capacitors are based on barium titanate. Although higher dielectric constants are possible with lead magnesium niobates, barium titanate-based systems give the most compromise of performance, cost and environmental friendliness.

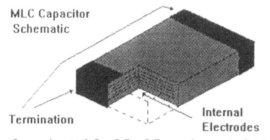

MLC Capacitor
Schematic

Termination   Internal
               Electrodes

Case size to 1.0 × 0.5 × 0.5 mm; layers to 2 microns

Figure 3. Multilayer ceramic capacitor schematic.

MLCCs are mainly used in bypassing, decoupling and tuning circuits in computers, communications and consumer electronics. Especially in consumer electronics, the pressure is to smaller and smaller capacitors with a growing emphasis on 0402 capacitors for size and weight-sensitive applications such as the palmtop computer on which this paper was written.

Dielectric Ceramic Materials   3

CERAMIC PIEZOELECTRIC APPLICATIONS AND DEVICE CONSTRUCTION

Although barium titanate was the first ceramic material used in sonar applications it has been supplanted in many areas by lead zirconate titanate (PZT) for sonar and audio applications because of PZT's greater coupling coefficient which means more movement for the same applied voltage. Barium titanate is still used in a number of applications by choice including commercial fish finder sonar.

The commercial market is now growing at double-digit rates for piezos in non-traditional applications such as knock sensors and acceleration sensors in automobiles and vibration sensors in washing machines and many industrial uses involving flow or level control.

Piezo parts can be simple disks or complex curved shapes designed for a particular acoustic profile. They are commonly formed by dry pressing. Large arrays can be produced by assembling panels of piezo elements or using ceramic-polymer composites.

PTC THERMISTOR APPLICATIONS AND DEVICE CONSTRUCTION

PTC thermistors are used both to detect temperature and to control current. The Curie temperature is manipulated by the use of additives to give the best combination of switch temperature and conductivity profile. Typical applications include TV degausser, the largest volume application, fluorescent light ballasts, motor starters and self-regulating heaters including those available to the consumer market designed to avoid overheating by limiting temperature if air flow is impeded.

Typical shapes include disks (similar to disk capacitors) and simple shapes which may be attached to heat sinks or perforated to allow better heat transfer. Components are typically produced by dry pressing or extrusion.

ELECTROLUMINESCENT APPLICATIONS AND DEVICE CONSTRUCTION

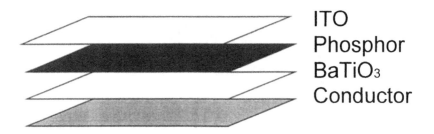

ITO
Phosphor
BaTiO$_3$
Conductor

Figure 4. Electroluminescent panel construction – schematic.

A thick film layer of barium titanate is applied to the electrode as a backing for the phosphor and is covered with a transparent indium-tin oxide (ITO) electrode. These panels are increasingly used in watches, phones, auto entry keypads, and aircraft instruments. In addition, there is growing interest in their use for advertising and any displays where a cold flexible light is required. Panels greater than 1000 cm2 are readily available.

BARIUM TITANATE MANUFACTURE

There are essentially four ways of making barium titanate:

Solid state

$BaCO_3 + TiO_2 => BaTiO_3 + CO_2$

This requires heating at about 1100^0C. The technique was developed in the 1940's and limitations include the ability to control stoichiometry, crystallinity, purity and particle size.

Oxalate

$BaCl2 + TiCl4 + oxalic\ acid\ (C_2O_4H_2) => BaTi\ oxalate\ hydrate + HCl$

This requires heating around 1000^0C to optimize the properties required. The process was developed in the 1960's and 1970's to give higher purity products for PTC thermistors. Oxalate synthesis is widely used commercially to prepare metal oxides such as rare earths in a pure form for electronics applications. The precipitation and calcination are highly controllable, giving enhanced stoichiometry control, controlled crystallinity, controlled purity and controllable particle size.

Hydrothermal

$Ba(OH)_2 + TiO_2 => BaTiO_3 + H_2O$

This technique requires temperatures much less than the above techniques, around 200^0C, and several atmospheres of pressure. A low temperature calcination around 600^0C may be required to remove hydroxyl groups and improve crystallinity. Even at the original precipitation temperature barium titanate is identifiable by x-ray diffraction. Interestingly, this is not a new technique - the first patent on this process, USP 2,216,655 dates from 1940! Hydrothermal barium titanate is characterized by uniform particle size and high reactivity but its use thus far has been limited because of economics and processing challenges.

Alcoxide

$Ba(OH)_2 + Ti\ alcoxide => BaTiO_3 + alcohol$

Typically Ti isopropoxide is used. pH change gives a gel, which is coated on a substrate by a technique such as spin coating, then decomposed by heating to barium titanate. This technique shows promise for forming layers below 2 microns but still has significant commercial and technical barriers to its success.

Conventional wisdom says that COG dielectrics with a low but very stable dielectric constant are produced by solid state routes from materials such as neodymium titanate. X7R and Z5U dielectrics with higher dielectric constant but higher variability with temperature are based on high-purity precipitated or solid state barium titanates. Most commercial parts are X7R or Y5V parts. The powders are partially reacted and react further during MLC sintering. Reaction is normally not fully complete, either to restrict grain size or deliberately develop an inhomogeneous structure such as the X7R core-shell.

Conventional powders are typically 1 to 1.5 microns in size, with fired grain sizes ranging from less than a micron in COG to more than 3 microns in Y5V.

MINIATURIZATION OF CERAMIC CAPACITORS

Consumers demand better, smaller, faster and cheaper performance from each generation of electronic device. Passive components occupy a relatively large proportion of the circuit board area and there is tremendous pressure to reduce component size.

It used to be so simple! If you looked at capacitance, the pecking order was quite simple. In increasing order of capacitance value, designers would specify:

$$Film < Ceramic < Tantalum < Aluminum.$$

It's not so simple now. Because of the dramatic increase in ceramic volumetric efficiency and the constant electronics pressure for better-smaller-faster-cheaper parts, the market dynamics are changing. First of all, Aluminum capacitors are not well suited to high frequencies and Tantalums and lead relaxor MLCCs are taking share from them. In turn, base metal MLCCs are nibbling at the lower end of the Tantalum market and in response Tantalums are being downsized and made more surface mount friendly as their volumetric efficiency increases!

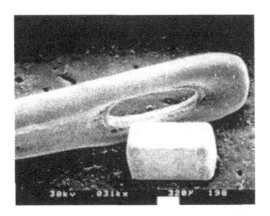

Figure 5. An 0402 capacitor and the eye of a needle.

The 0402 capacitor above is 2 mm x 1 mm x 1 mm and is used in a cell phone, palmtop computer or digital camera. How is it possible to engineer useful capacitance in such a small package?

ENGINEERING USEFUL SMALL CAPACITORS

The key proportional equation governing volumetric efficiency in capacitors is:

$$C_V \propto nK_0K/t^2$$

Where:

C_V is capacitance per unit volume;

N is the number of layers;

K_0 is the permittivity of free space, a constant;

T is the layer thickness.

The two key variables are dielectric constant and layer thickness. You will note that volumetric capacity is proportional to dielectric constant – but is proportional to the inverse square of layer thickness.

Dielectric constant is controlled by the chemical composition and microstructure of the barium titanate formulation. There are tradeoffs, though with other electrical, physical, chemical and environmental factors.

Layer thickness is controlled by the ceramic casting process – typically tape or waterfall type casting on a substrate or previous layer, with thickness down to 2 microns.

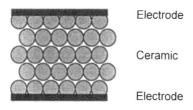

Electrode

Ceramic

Electrode

Figure 6. Schematic showing 5 grains per ceramic layer.

The above diagram shows an important factor relating fired grain size to layer thickness. It is a generally held belief that at least 5 grains are necessary to separate electrode layers in a multilayer capacitor. A lower number of grains yields a more direct grain boundary path between electrodes and a breakdown of electrical properties. For a 5 micron layer part therefore, 1 micron fired barium titanate grains are vital. As will be described elsewhere in this paper, this can dramatically influence the choice of formulation and processing. There is not always a good correlation between unfired and fired grain size; the fired grain size depends on raw materials, formulation and thermal history.

THIN LAYER IMPLICATIONS – ELECTRODE METALS

The fastest growing electrode material is Ni (often referred to as base metal technology, BME or TME). Other technologies such as lead relaxors, composites and regular low fires are not standing still though; a year ago the magic layer count was 100 with base metal owning the territory above 100 layers. Now processing details have been worked out, advanced low fire parts with up to 500 layers have been produced.

BASE METAL ELECTRODE SYSTEMS

In order to be compatible with traditional barium titanate based systems with acceptable electrical and physical properties, co-fired electrodes in the 13000C range were developed. Typically in Japan these were 100% Pd, elsewhere in the world they were 70%Pd-30%Ag. Because of the concern in Japan about premature failure due to Ag, precious metal inks have stayed predominantly 100% Pd there. In the rest of the world, however most "high-fire" (13000C) capacitors have used 70% Pd. Another class, "low-fire" dielectrics are similar to high fire dielectrics but have frits and fluxes added to make a compromise between the 11000Cfiring temperature and dielectric properties. Lead relaxor dielectrics fire at around 10000C but typically also use 30% Pd.

The real problem with Pd is that it is a commodity metal with only two sources of supply, South Africa and Russia. There are competing applications such as auto catalysis and even dentistry and as a result the price can fluctuate. Although it has hovered in the mid $100s per Troy Ounce for the past couple of years, recent supply issues in Russia pushed spot prices over $300, higher than gold! Even in the mid $100s the value of precious metals in capacitor electrodes was 150% of the cost of the functional ceramic in a MLCC.

While this price fluctuation has gone on, the price of a given MLCC has been dropping at between 15% and 30% per annum. This means that MLC makers find themselves in a rather unpleasant squeeze. The only way out is to either reduce the cost of existing parts by reducing the cost, thickness or number of electrode layers, or to develop new products with either smaller case sizes or larger capacitance values to allow competition with Aluminum or Tantalum capacitors

Ni electrode systems have a particular edge in that base metal parts made with this technology have lower ESR and higher reliability than traditional Tantalum capacitors.

Pb alloy injection

Lead alloy injection has been successfully used for many years to produce high reliability parts. The technology, originally patented by TAM in the 1970's, involves making MLCCs with a fugitive ink which is burned out during the firing cycle. The ceramic firing is thus independent of the electrode formation, which occurs after sintering when parts cool and low-melting alloy is injected in such a way as to fill the electrode patterns by capillary action.

Co-fired Ni base metal

Ni systems have been used in the USA and Japan for many years. The real breakthroughs however have occurred in the last 5 years when new high quality Ni powders became available, the powder technology developed to form extremely thin layers and X7R formulations were developed and reliability was dramatically increased. Coupled with the new ability to make very high layer count parts, the Ni base metal parts have taken significant market share from precious metal electrode parts.

The barriers to entry for a new base metal MLCC manufacturer have been dramatic. Until recently all the formulation technology has been proprietary to a very few in-house makers, the technology has been heavily patented worldwide, and processing equipment was unavailable. Furthermore, high quality Ni powder was very scarce. Now however, formulations and processing technology, materials and equipment are all becoming available but it still will require a major monetary and technical effort for any MLCC maker to begin to compete in this field.

Dielectric Ceramic Materials

A variant of the Ni base metal system is the NiO system developed at TAM. Although Ni powders are less expensive than Pd they are by no means cheap. With the right formulations and processing, NiO pre-electrodes can be co-fired with the ceramic in the MLCC and then subsequently reduced. Although originally proposed as a low-cost start-up route (existing sintering furnaces can be used) the technology was too novel to be applied as a first step. Instead the technology is showing real promise as a way for customers already in the base metal business to make a dramatic reduction in their electrode costs and an increase in their throughput.

Co-fired Cu base metal

The real benefit of Cu is in its electrical properties, especially for high frequencies. Although Cu end terminations are applicable to Ni base metal parts, Cu electrodes now seem to have a market niche developing for high frequency COG parts and for film capacitor replacement.

THIN LAYER IMPLICATIONS - CERAMIC POWDERS AND MLCC PROCESSING

Thin layers with different metal systems and different processing conditions abound. Particle size control, surface condition and interactions with additives need to be tightly controlled. And that is before we consider compatibility with inks, binders, terminations, plating systems or build design!

Powder formulation

The right distribution of additives is critical. This can be achieved by co-precipitation, super-precipitation, pre-calcination as well as more traditional co-milling and mixing.

The choice of raw material now runs much more to finer particle sizes, much less than 1 micron, finer additive systems and an overall package which is resistant to dissolution, unmixing or undesirable mixing in the next step, slip preparation.

Base metal system formulation involves a careful preparation of reduction-resistant phases.

Slip processing

Slips for ultra thin layers must be uniformly dispersed, stable and appropriate for the casting technology chosen. Cooperation between the slip technologist and the ceramist is vital here to ensure that additives do not selectively settle or become chemically changed to the extent that either the slip or ceramic properties are affected.

The jury at this stage is evenly divided on water-based or solvent-based systems, and tape or wet lay-up systems. Each permutation has its champions. Each process has its unique challenges as MLCC makers "push the envelope".

Firing

This is key to base metal production success. Carefully controlled burn-out, firing and annealing brings a large capital cost penalty.

Termination

Termination is also a critical issue especially with base metal systems. Interactions with electrodes and ceramic that never used to be a problem with simpler parts have become more complex.

Although Ag terminations suitably plated are the most common types, Cu and/ or Ni for base metal and in some cases Ag-Pd terminations are used.

The aggressiveness of plating solutions can cause problems with low fire materials if porosity is present or grain boundary etching takes place.

Electrical property requirements

The game is changing here all the time. The EIA is changing DF and other specifications to allow thinner lower-voltage parts. One of the key issues is the standardization of HALT (Highly Accelerated Life Testing). There isn't any standardization and each MLCC maker and user seems to have their own set of conditions. HALT tests are sometimes used rigidly as go / no go conditions to approve MLCC performance rather than being used for their original purpose, as a tool to investigate failure mechanisms.

CHOOSING THE MOST APPROPRIATE MLCC SYSTEM

This really depends on the MLCC house and their customers' preferences.

Performance

High volumetric efficiency - relaxor, base metal, composite.
High strength, toughness and thermal shock - any BaTiO3 based system.

Cost and Flexibility

In high volumes - base metal
In moderate volumes - low fire, relaxor, composite.

CHALLENGES TO - AND RESPONSES FROM - BARIUM TITANATE MARKETS

We saw how barium titanate was replaced by PZT in piezos. Which technologies are threatening barium titanate in its current applications?

MLCC challenges

There are two challenges – real estate and cost.

On real estate, up to 50% of the wiring board's area is devoted to passive components and their connections. This could be solved by on-chip integration, integrated passives within the board and passive device networks.

All of the above represent significant threats but the fact is, MLCCs are extraordinarily cheap and flexible. Usually passives are designed in last in order to get the circuit to work (that's how an engineer friend of mine described it) and integrated systems find it tough to cope with that. In addition, grouping capacitors together can increase conductor length and inductance, which is becoming a real issue in high frequency devices.

Indirect cost is the really interesting challenge. Although you can buy a capacitor for less than half a cent it will cost you three cents at least to place it in its circuit! This gives a strong incentive to combine components in a designer-friendly way.

Dielectric Ceramic Materials

Figure 7. Four 0603 MLCC capacitors replaced by one 1206 array.

There is great interest currently in using a practical halfway step - using capacitor arrays rather than discretes. Board spacing is reduced dramatically by replacing 4 0603 capacitors with one 1206 capacitor.

Will tantalum capacitors fight back strongly? We think not. Although outstanding engineering work allows the production of 0402 surface mount capacitors, they are still polarized and still have higher equivalent series resistance than their ceramic counterparts. Aluminum capacitors cannot compete on lifetime or frequency capability, and film capacitors are thermally challenged.

Piezo challenges
There may be more opportunities for barium titanate here as lead based materials become more widely challenged in all areas of electronics.

Thermistor challenges
It is difficult to see a direct replacement for barium titanate as a material – the replacement would be of the entire thermistor heating or sensor system by an alternate technology rather than a materials substitution.

Electroluminescent panel challenges
It is difficult to see whether a substitution would be desirable or economic in the future.

THE PROGNOSIS FOR BARIUM TITANATE?
It's here to stay!

BIBLIOGRAPHY
General
A good basic primer in ferroelectrics is:
"Ceramic Materials for Electronics" ed. R. Buchanan, Marcel Dekker 1986.

References to much of the early work can be found in:
"Ferroelectric Crystals" by Jona @ Shirane, MacMillan, 1962. www.macmillanusa.com

Technical Information
Proceedings of Capacitor and Resistor Technical Symposia (CARTS), annually, available from Components Technology Institute Inc., Huntsville AL. www.mindpring.com/~ctiinc/CARTS.htm
Proceedings of the Electroceramic Device Manufacturing Workshop, April 1997, available from The Particulate Materials Center, Pennsylvania State University. http://pmcenter.cerse.psu.edu
Proceedings of the Eighth US-Japan Seminar on Dielectric and Piezoelectric Ceramics, sponsored by The Office of Naval Research, administrated by the Materials Research Laboratory, The Pennsylvania State University. www.mri.psu.edu

Market Information
Regularly published by Paumanok Publications Inc.
www.cicada.com/priv/paumanok

Internet
Running a search on barium titanate on regular search engines is not particularly useful! There is a huge volume of data that is really tricky to sort through but there are some real gems out there!

For a listing of companies in passive components with links to their web sites contact:
www.yahoo.com.sg/Business_and -Economy/Companies/
Electronics/Passive_Component_Manufacturer/index.htm
Capacitors
There are many good capacitor company sites including:
www.avxcorp.com
www.johanson-caps.com
www.kemet.com
www.murata.com
www.netcom.com/~dlongden/philips.html
www.syfer.com
t-yuden.com
...and many more companies also!
On the information end:
http://leonardo.eeug.caltech.edu/~eel4/capacitor_guide.html
is a useful guide to identifying capacitors and their properties.

Thermistors
These two sites have a wealth of useful information:
www.wecc.com/ptceng.html
http://ceramite.com

Piezos
A history of piezo development is at:
www.piezo.demon.co.uk/histry.htm
and an outstanding technical briefing is at:
www.stco-stettner/com/englisch/home.htm

Electroluminescent Displays
This area has some of the most fascinating sites - to see how barium titanate is used, visit:
www.electroluminescent.thomasregister.com/olc/electroluminescent/novalite.htm
and to see applications visit:
www.us.net/quantex/el.htm

Finally, don't forget to visit us at:
www.tamceramics.com

ACKNOWLEDGEMENTS

The authors thank their colleagues for support and encouragement and TAM Ceramics Inc. for permission to publish this paper.

Dielectric Ceramic Materials

FIFTY YEARS OF INVESTIGATIONS AND DEVELOPMENTS TO CREATE FERRO-PIEZOCERAMIC MATERIALS.

A. Ya. Dantziger, L.A. Reznitchenko, O.N. Rasumovskaya, V. P. Sakhnenko, A.P. Naumov
Institute of Physics of Rostov State University, Pr. Stachky, 194, Rostov-na-Donu, 344090, Russia. Technology Compilations. Inc., 6 Forest Park Dr., Farmington, CT 06032

HISTORICAL REVIEW

The modern history of the development of ferro-piezoelectric ceramic materials (FPCM) began approximately fifty years ago. During this time, many compositions were investigated yet, only a small number of these compositions have been used in any applications. FPCM are based on barium titanate, lead-titanate-zirconate (PZT), lead niobate and sodium niobate. During the first period of development of industrial materials, the period from 1950 - 60's, most investigators chose a purely empirical way coming, mainly, to the selection of all the possible atoms - modificators and their combinations. As a base for modification, the solid solutions of the PZT system were used most often that was caused by their high piezoelectric parameters, wide isomorphism, the presence of the morphotropic region in this system - a region of the structural phase transition between the different ferroelectric phases [1,2] accompanied by the extreme electrophysical parameters. Piezoelectric ceramic materials for various purposes obtained in the 1960's were based on the PZT system; some of them haven't lost their efficiency up to now. Among them the PZT - type materials are considered to be best [3,4].

Transition in late 1960's - at the beginning of 1970's from the PZT system to the ternary systems on its basis enabled one to increase the parameters of ferroelectric solid solutions and to improve their ability to sinter [5]. The following modifications of solid solutions with different oxides have led to the essential improvement of their characteristics and the creation of industrial materials of PCM type (Japan) [6]. In Russia the contribution to development of novel FPCM was made in these years by the researchers of Rostov State University [7,8], proceeded in early 1970's to a study of multi component (quaternary and 5-component systems based on PZT:

$$PbTiO_3 - PbZrO_3 - \sum_n PbB'_{1-\alpha} B''_{\alpha}O_3 \qquad (n=2,3)$$

where

B' is 5 and 6 valent cations, B" is 1, 2 and 3 valent cations, α is 1/2, 1/3, 1/4, depending on B' and B" valencies.

As a number of components increases, the regions of materials with optimal combinations of parameters broaden, a variety of properties of solid solutions increases and the most important electrophysical parameters become higher. It makes evident the considerable advantages of multi component systems over the simpler ones constituting of them [7].

The development of FPCM of new generation was impossible without solving the triplex problem:

- fundamental - the establishment of crystallochemical conditions of existence of complex multi component oxides of different structural types; the determination of criteria of stability of polar states in them; elucidation of the origin and piezoelectric properties; the development of scientific bases of the physics of real polycrystals;

- search - development of the methodology of directional search and the algorithm of prediction of the FPCM properties of different chemical composition belonging to the different types of crystal structure (perovskites, pseudoilmenites, lagered perovskite-like), their computer modeling with the characteristics for various purposes, designing of practically important ferro-piezoelectric substances on an atomic level;
- applied - development of formulation and technologies of small-tonnage production, data bases on developed ferro-piezoelectric materials; fabrication of ferro-piezoelectric materials in the form of synthesized powders, high - density ceramics and piezoelectric elements of different standard sizes.

The development of the techniques of determination of phase states and regularities of their origination, and also alternation with variation of composition and external influences basing on the general symmetry and thermodynamical principals became a base for solving all these problems. The main types of morphotropic boundaries in the compounds of oxygen-octahedral types and in particular, in the oxides with perovskite structure were determined theoretically.

The introduced idea of the unstrung cation-anion bonds and the built crystallographical theory enabled one to predict the lattice parameters of possible compounds and solid solutions. This in turn, became a basis for development of the physicochemical theory of solid solutions based on the oxides with perovskite-type structure that enabled us within the shortest possible time
- to establish the correlation relationships between chemical composition, atomic structure, super atomic structure, physical properties and application of FPCM;
- to synthesize and completely investigate more than 20 thousand novel compositions;
- to develop the methods of their preparation with optimal characteristics;
- to create about 200 novel FPCM and techniques of their preparation, having the deference in the form of author's certificate and patents, for practically, all the known piezoelectric fields.
Description of different groups of FPCM and their comparison with the industrial analogs are given below.

MATERIALS

Group I. Materials stable towards electric and mechanical action

PCR-8 is ferrohard, while PCR-12, PCR-22 and PCR-6 are of moderate ferrohardness. They all are intended for devices working at high levels of the resonance excitation, such as piezotransformers, piezomotors, ultrasonic radiators. The PCR-8 parameters are given for two methods of sintering, namely, hot-pressing and conventional (*) ones. Their closeness indicates good reproducibility of PCR-8. Comparison with analogs (for subgroups of various ferrohardness) PZT-8, PZT-4 (Vernitron, USA) has shown that, while possessing close values of dielectric losses, the PCR materials are superior as to the piezoelectric parameters and mechanical quality. The Russian analogs ZTSS-3, ZTBS-3 are, by contrast, characterized by much higher losses (both dielectric and mechanical), while their piezoelectric parameters are rather close. Most interesting is the comparison with regard to complex parameters that include both losses and piezoelectric coefficients. For example, in regard to the $K^2_{31}Q_m$, value proportional to the efficiency coefficient of piezotransformers, PCR-8 is better by about one order of magnitude than PZT-4. The same may be stated in regard to $K^2_{31}Q_m\varepsilon^T_{33}/\varepsilon_0$ for PCR-22. The maximal efficiency coefficient of the low-voltage piezotransformers manufactured from PCR materials approaches 99%. As regards the value of $K_{33}tg\delta$ (E=100 kV/m) which is proportional to an acoustic radiator efficiency, the PCR materials are better than the analogs mentioned (about

Dielectric Ceramic Materials

twice better compared to the Russian analogs). The maximum of this value is observed in PCR-8 prepared by conventional methods.

The parameters of the materials of this group are presented in Table I There are T_c - Curie temperature; $\varepsilon^T_{33}/\varepsilon_0$ - relative dielectric permittivity; K_i - electromechanical coupling factor; d_{ij} - piezomodules; g_{ij} - piezoelectric constants; $tg\delta$ - dielectric loss tangent; Q_m - mechanical quality; V_R - speed of radial vibrations; T_W - operating temperature; $\delta f_0/f_t$ - relative deviation of resonance frequency in operating temperature range; $C^{D,E}_{ij}$ - elastic stiffnesses in this and mentioned below tables.

Table I Materials Stable Towards Electrical and Mechanical Action

Materials	$T_c °C$	$\varepsilon^T_{33}/\varepsilon_0$	K_p	d_{33} pC/N	g_{33} mV·m/N	$tg\delta·10^2$ E=5kV/m	$tg\delta·10^2$ E=100kV/m	Q_m	$Q_{m,\sigma}$ σ=13MPa
Materials with High FerroHardness(*Produced by conventional technology)									
PCR-8	325	1400	0.58	130	10.5	0.35	0.70	2000	700
PCR-8*	320	1300	0.57	125	10.8	0.33	0.50	1800	-
PCR-77	345	1350	0.62	140	11.7	0.30	0.40	1200	-
PCR-78	350	1250	0.60	130	11.7	0.30	0.60	1000	-
PCR-23	325	900	0.58	85	10.7	0.85	-	1500	1000
Analogs									
PZT-8	300	1000	0.50	93	10.5	0.40	0.70	1000	-
ZTSS-3	260	1470	0.54	125	9.6	0.60	1.10	630	-
Materials with Moderate FerroHardness									
PCR-12	285	2000	0.66	185	10.4	0.35	0.85	1300	-
PCR-22	240	2100	0.57	155	8.4	0.35	0.90	1800	750
PCR-86	235	2300	0.60	170	8.4	0.35	0.60	1250	-
PCR-6	230	2300	0.64	195	9.6	0.40	1.00	1100	300
Analogs									
PZT-4	328	1300	0.58	123	10.7	0.40	1.00	500	110
ZTBS3	180	2300	0.45	158	7.8	1.20	2.00	350	-

Group 2. Materials with high dielectric permittivity (Table II)

This group comprises the ferrosoft materials PCR-66, PCR-7, PCR-7M, PCR-73. The corresponding analogs are PZT-5H, PCM-33A (Matsushita, Japan) and ZTSNB-1(Russia). In addition to the high value of $\varepsilon^T_{33}/\varepsilon_0$, these materials are characterized by high K_{ij} and d_{ij}. In particular, PCR-7M and PCR-73 are far superior to the existing commercial analogs.

Their high dielectric permittivity predetermines the principal application of these materials, namely, in low-frequency receiving devices such as hydrophones, microphones or seismoreceivers. In order to fully evaluate the effectiveness of an ultrasonic receiver, its specific sensitivity has to be considered, which takes into account the internal resistance of the receiver and is proportional to $|d_{33}|/\sqrt{(\varepsilon^T_{33}/\varepsilon_0)}$. The PCR-7 materials, particularly, PCR-7M and

Dielectric Ceramic Materials

PCR-73, exhibit higher values of that parameter compared to the analogs. PCR-7M is at present effectively used in seismoreceivers.

Furthermore, the above materials may be utilized in medical diagnostics instruments working at a load with low ohm input resistance which provides matching of the high-frequency transducer with that load.

The materials in question may also be utilized in transducers that make use of the inverse piezoeffect, such as vibration exciters of motion. These are linear and low-power step-motors, transducers for adjusting mirrors in optical communication systems and in astronomical systems. The PCR-7M material is presently being tested in some of the above devices.

Table II. Parameters of Piezoceramic Materials with High Dielectric Permittivity.

Materials	$T_c{}^\circ C$	$\varepsilon^T{}_{33}/\varepsilon_0$	K_p	K_{15}	d_{31} pC/N	d_{33} pC/N	$d_{33}/\sqrt{(\varepsilon^T{}_{33}/\varepsilon_0)}$ pC/N	$tg\delta\cdot10^2$ E=5kV/m	E=100kV/m
PCR-66	280	2800	0.68	0.72	245	535	10.2	1.20	70
PCR-7	220	3500	0.68	0.76	280	610	10.3	1.50	80
PCR-7M	175	5000	0.71	0.78	350	760	10.8	2.0	60
PCR-73	155	6000	0.70	0.77	380	860	11.1	2.9	35
Analogs									
PZT-5H	183	3400	0.65	0.75	274	593	10.2	2.0	65
PCM-33A	-	3200	0.66	0.71	262	572	10.1	1.7	80
ZTSHB-1	240	2200	0.54	0.69	205	445	9.5	1.9	70

Group 3. Highly sensitive materials

The materials of this group(Table III) have high sensitivity with respect to mechanical stresses which is described by the piezoelectric coefficient g_{ij}. This quantity is proportional to the sensitivity of the idling ultrasonic receiver. Two materials PCR-1 and PCR-37 represent this group. The former possesses a greater value of g_{ij}, which exceeds considerably the same parameter in the industrial analogs PZT-5A, "Piezolan S" ("Keramische Werke") and ZTS-19(Russia).

This group includes the subgroup of high-sensitivity materials characterized by great anisotropy of their piezoelectric parameters: PCR-40, PCR-69, PCR-72. Of particular interest is PCR-72 which, in virtue of its very high anisotropy $(K_t/K_p \sim \infty, d_{33}/d_{31} \sim \infty)$ and low mechanical figure of merit, makes it possible to reduce sharply the number and level of non-genuine vibrations.

The above materials may effectively be applied in accelerometers, ultrasonic defectoscopes, devices for non-fracture materials control by the acoustic emission method and in instruments for ultrasonic medical diagnostics. At present PCR-1 occupies among other PCR-type materials one of the first places as to the scope and versatility of their practical application. For example, it is used in devices of ultrasonic control of equipment and products, in particular, in atomic power stations. Furthermore, it may be used for generating high voltage in disposable ignition systems. Its high K_{15} and relatively low $\varepsilon^T{}_{33}/\varepsilon_0$ have made it a desirable material for ultrasonic delay lines in color TV-sets.

Table III. Parameters of High Sensitive Piezoceramic Materials.

Materials	$T_c°C$	$\varepsilon^T_{33}/\varepsilon_0$	K_p	K_{15} (K_t)	d_{31} pC/N	d_{33} pC/N	g_{31} mV·m/N	g_{33} mV*m/N	$tg\delta\cdot10^2$ E=5kV/m	Q_m
PCR-1	355	650	0.62	0.70	95	220	16.5	38.0	2.0	90
PCR-37	345	1400	0.68	-	170	375	13.7	30.3	1.6	105
Analogs (* - Sintered by Hot-Pressing method)										
Piezolan-S	350	800	0.48	0.60	92	206	13.0	29.4	2.5	80
PZT-5A	365	1700	0.60	0.68	171	374	11.4	24.8	2.0	75
ZTS-19*	290	1600	0.60	-	150	340	10.6	24.0	2.5	50
Materials with great anisotropy of piezoelectric parameters (*measuring by the quasi-static regime)										
PCR-40	440	180	0.07	(0.44)	5	52	3.3	33	1.0	2000
PCR-69	350	170	0.04	(0.57)	3.5	90*	2.3	60*	2.2	50
PCR-72	277	150	0	(0.63)	0	100*	0	75*	1.9	6

Group 4. Materials with high stability to the resonance frequency

Materials of this group (Table IV) are intended to be used in filtering devices.

Table IV. Parameters of Piezoceramic Materials with High Stability of Resonant Frequency.

Materials	$T_c°C$	$\varepsilon^T_{33}/\varepsilon_0$	K_p	K_{15}	d_{31} pC/N	$\delta f_0/f_t, \%$ (- 60 + 85°C)	C^D_{31}/C^E_{44}	$tg\delta\cdot10^2$ E=5kV/m	Q_m
PCR-15	325	950	0.45	-	75	0.25	-	0.6	2500
PCR-31	305	940	0.45	-	81	0.25	-	1.5	1100
PCR-13	335	800	0.40	0.41	65	0.2-0.25	-	0.7	2800
PCR-63	235	1170	0.37	0.38	60	< 0.15	-	3.0	1100
PCR-80	310	800	0.36	-	51	0.25	-	0.6	4000
PCR-30	305	780	0.32	-	47	< 0.15	-	1.3	2000
PCR-62	300	660	0.32	0.35	42	< 0.15	-	1.4	2700
PCR-28	325	600	0.28	-	36	0.25	-	0.5	2000
PCR-83	360	400	0.28	-	28	0.10	-	0.6	4000
PCR-84	360	180	0.10	0.25	6	0.15	-	0.2	12000
PCR-21	310	1400	0.53	0.45	100	0.25	-	0.7	800
PCR-74	365	850	0.52	0.47	90	0.20*	6	1.6	400
Analogs (*measuring at the thickness vibration mode)									
PZT-6A	335	1050	0.42	0.39	80	< 0.20	-	2.0	450
PZT-6B	350	460	0.25	0.30	27	< 0.20	-	0.9	1300
ZTS-35	300	1000	0.46	0.49	95	< 0.5	-	2.0	700
ZTS-35Y*	290	800	0.50	0.50	90	0.5*	6	2.5	650

They are PCR-13, PCR-15, PCR-21, PCR-30, PCR-31, PCR-74 (the last three materials are prepared by conventional method). Their industrial analogs are the materials PZT-6A, PZT-6B, ZTS-35 and ZTS-35Y. Majority of these PCR materials (with $K_p>0,4$) are used in broad-band filters. Many of them (except the last two in the table) have high mechanical quality Q_m. Their comparison with PZT-6A has shown that, while their temperature stability is close ($\delta f_o/f_t = 0.2$-0,25% in the -60 to +85°C range), the PCR materials (except PCR-21 and PCR-74) have a much greater figure of merit. In addition PCR-13 possesses a better time stability. Some materials are superior to PZT-6A in the value of K_p. The ZTS-35(Russia) is inferior to the above PCR materials in all principal parameters.

Among the materials for narrow-band filters, PCR-30 may be noted which in its temperature stability ($\delta f_o/f_t \leq 0,15\%$ in the -60 to +85°C range) and in Q_m is superior to PZT-6B.

The PCR-74 material characterized by anisotropy of elastic constants (C^D_{31}/C^E_{44}) may be used for the mode of thickness vibrations with energy trapping. Its industrial analog ZTS-35Y is much inferior as to the temperature stability of the resonance frequency at this mode.

Group 5. Materials with low dielectric permittivity

These materials are presented in Table V and designed to be applied in high-frequency acousto-electric transducers and in pyroreceivers. Industrial analogs of these materials are PCM-75, PZT-2, PCD-233/18.

Table V. Parameters of Piezoelectric Materials with Low Dielectric Permittivity.

Materials	T_c°C	$\varepsilon^T_{33}/\varepsilon_0$	K_p	K_{15} (Ks)	d_{31} pC/N	g_{31} mV·m/N	$tg\delta \cdot 10^2$ E=5kV/m	Q_m	V km/s
PCR-53	240	260	0.20	- (0.16)	16	7.0	0.3	4500	4.20
PCR-11	280	290	0.31	0.45 (0.14)	27	10.5	0.3	4000	4.05
PCR-3	280	350	0.38	-	37	12.0	0.5	2000	3.95
PCR-10	320	380	0.47	0.65	49	14.6	0.3	2500	3.80
PCR-24	320	480	0.53	0.68	64	15.1	1.0	200	3.60
PCR-20	315	510	0.54	0.72	70	15.5	0.8	700	3.60
Analogs									
PCM-75	359	390	0.23	-	23.5	6.8	0.25	4520	-
PZT-2	370	450	0.47	0.70	60	15.1	0.5	680	-
Subgroup with high speed of sound									
PCR-35	370	120	0.22	0.31	12	10.5	1.6	1000	5.90
PCR-34	420	460	0.42	~0.60	45	10.0	2.5	150	5.40
There are no commercial analogs									

When these materials are employed in ultrasonic delay lines using space waves, their high K_{15} and relatively low $\varepsilon^T_{33}/\varepsilon_0$ and Q_m, play an important role. The PCR-24 and PCR-20 materials possess precisely this combination of parameters in which they are not inferior to the PZT-analog and are much better than PCM-75. Note the role of hot pressing in this case which produces very thin high frequency transducers.

For devices using surface acoustic waves, the most important parameters are the electromechanical coupling coefficient of the surface waves (K_S) and the temperature delay coefficient. Also important are low values of $\varepsilon^T_{33}/\varepsilon_0$ and high values of Q_m. The PCR-53 material has the best combination of mentioned above parameters, its temperature delay coefficient is $\approx 40*10^{-6}$ 1/K.

The above-mentioned materials with high Q_m, may also be effectively used in high-efficiency devices. Thus, for the piezoelectric motor with a piezoelement made from PCR-10 a record-breaking efficiency coefficient of 85% has been achieved.

PCR materials of this group have high pyrocoefficient due to that they can be referred to pyroelectric materials. The use of PCR-3 has made it possible to double sensitivity of pyroreceivers..

The PCR-34 and PCR-35 materials are characterized in addition to low dielectric permittivity by a high speed of sound (V). This property simplifies the manufacture of elements since they may be made thicker even in the case when they are intended to work at high frequencies. Moreover, this property ensures good matching of the elements with the external circuit. Low specific weight of these materials (4 500 kg/m^3) permits their utilization in devices where the weight characteristics are decisive.

Group 6. High-temperature materials

The materials of this group (Table VI) PCR-26, PCR-40, PCR-45, PCR-50, PCR-61 are notable for high values of the Curie point and operating temperatures. They may successfully be applied in high-temperature transducers in such areas as atomic energetics or space technology. The highest operating temperature can be achieved using PCR-50 and, particularly, PCR-61. The latter material has a very low dielectric permittivity, which makes it suitable for high frequency devices. Known industrial analogs are, in the parameters mentioned, inferior to these PCR materials.

Table VI. Parameters of High Temperature Piezoceramic Materials.

Materials	$T_c°C$	$T_w°C$	$\varepsilon^T_{33}/\varepsilon_0$	$tg\delta \cdot 10^2$ E=5kV/m	K_p	K_t	d_{31} pC/N	d_{33} pC/N	g_{33} mV·m/N	Q_m
PCR-26	400	300	455	1.0	0.32	-	35	90	22.5	200
PCR-45	420	350	380	0.1	0.26	0.46	30	100	30.0	3000
PCR-40	440	350	180	1.0	0.07	0.44	5	52	33.0	2000
PCR-50	670	500	150	0.4	0.04	0.32	3	25	19.0	4000
PCR-61	>1200	950	48	1.9	0.015	0.29	0.51	12	28.5	<100
Analogs										
ZTS-21	400	300	420	1.8	0.27	-	30	75	20.0	100
TNaB-1	670	500	130	1.0	-	-	-	14	13.0	-

PRESENT TASKS.

At present, the Institute of Physics is developing the directions in the field of ferro-electricity as follows.

1. Designing of FPCM with limiting characteristics: ultrahigh operating temperatures (>1000°C), high piezoelectric modulus (>1000pC/N), infinite anisotropy of properties (K_t/Kp, $d_{33}/d_{31} \rightarrow \infty$), ultra low mechanical quality factor ($Q_m < 10$) for different piezotechnical applications.

2. Creation of novel composite materials, in particular, of shear type and the composites of the type "single crystal - ceramics" on the basis of a wide spectrum of the materials developed.

3. Designing of the materials with large values of the piezoelectric modulus $|d_{31}|$. Development of the devices on their base ensuring large transfers including the range of ultra low temperatures. In particular, the Institute of Physics together with the North-West University (USA) participated in the creation of a tunnel microscope on the basis of the developed piezoelectric materials.

4. Creation, together with "Technology Compilations, Inc", of the database on ferroelectric materials designed at the Institute of Physics.

5 Development of X-ray and electron techniques to investigate the ceramic materials, including the composition and the electric state of their surface. Development of the method using the spectra of low-energetic electrons excited by X-ray is now performed.

6. Replacement of the existing Pb-containing FPCM by the ecologically pure ones. Among these materials of most importance are FPCM based on the complex compositions including alkali and alkaline-earth metals niobates, Bi niobate-titanate, Bi titanates, (Bi, Na), (Bi, Ca) and their modifications. The fact that these materials do not contain toxic Pb is also of importance from the point of view of "ecological comfort" in the process of manipulation with them.

These materials seem to be very promising for industrial use because they possess

- high Curie temperatures (~ 600 - 1800°C) and operating temperatures (~400 - >1000°C);
- a stability of the main characteristics in the case of simultaneous action of high temperatures (600-800°C) and pressures (to 300 MPa);
- the practical absence of time aging, the sufficiently pronounced piezoelectric effect (K_t=0.3 - 0.45; d_{33}=12-100 pC/N);
- the increased piezoelectric sensitivity (g_{33}= $2*10^{-2}$ - $3*10^{-2}$ mV·m/N), at the expense of the relatively high d_{33}, and the low dielectric permittivity $\varepsilon^T_{33}/\varepsilon_0$ (<50 - 380);
- the good matching of the elements with an external circuit in the electrical resistance (at the expense of low $\varepsilon^T_{33}/\varepsilon_0$ and the high sound speed $V_R > 6$ km/s of individual compositions);
- a possibility of the maneuver by the thickness of a piezoelectric element, (because of high V_R) that is of importance for work at high and ultrahigh frequencies (allowed by the high V_R and low $\varepsilon^T_{33}/\varepsilon_0$);
- the infinite (in some cases) anisotropy of properties ($K_t/Kp \rightarrow \infty$, $d_{33}/d_{31} \rightarrow \infty$) in combination with the extremely low mechanical quality factor (Q_m<10) ensuring the high resolution and the narrow diagram of direction of transducers;
- low dielectric losses ($tg\delta < 2\%$) and the high coercive force determining their increased electrical strength;
- the non hygroscopicity and stability the action of most of acids;
- a simplicity and mass character of the technology (ceramic with solid-phase synthesis).

Their high metrological characteristics in combination with the low specific weight (p<4 500 kg/m^3) enabling one "to facilitate" and to micro miniaturize the sensors on their basis , promote the capacity for work of these materials as active elements of measuring transducers, operating in a wide range of temperatures, frequencies, pressure, including the simultaneous action of several external factors. The Institute of Physics has developed several materials from the above group used in practice for diagnostics of the work of internal combustion engines.

References

1. G. Shirane and K. Suzuki, "Crystal Structure of Pb(Zr,Ti)O$_3$," Journal of The Physical Society of Japan, 7 [3], 333 (1952).
2. E. Sawaguchi, "Ferroelectricity Versus Antiferroelectricity in the Solid Solutions of PbZrO$_3$ and PbTiO$_3$," Journal of The Physical Society of Japan, 8, 615-629 (1953).
3. G. Jaffe and D. A. Berlincourt, "Piezoelectric Materials for Transducers," Proceeding of The Institute of Electrical and Electronics Engineers, 53 [10], 1552-1567 (1965)
4. B. Jaffe, W. Cook, G. Jaffe, "Piezoelectric Ceramics," Mir, Moscow, Russia, 1974
5. H. Ouchi, K. Nagano, S. Hayakawa, "Piezoelectric Properties of Pb(Mg$_{1/3}$Nb$_{2/3}$)O$_3$ - PbTiO$_3$ - PbZrO$_3$ Solid Solution Ceramics," Journal of The American Ceramic Society, 48 [12] 630-635 (1965).
6. Ferroelectric Components Catalog, Matsushita Electric Co. LTD, Kadoma, Osaka, Japan, 624 (1975).
7. E. G. Fesenko, A. Y. Dantsiger, O. N. Razymovskaya, Novel Piezoelectric Materials, Rostov State University, Rostov-on-Don, Russia, 1983
8. A. Y. Dantsiger, O. N. Razymovskaya, L. A. Reznichenko, etc., High-efficiency Piezoelectric Ceramic Materials. Reference Book, Kniga, Rostov-on-Don, 1994
9. A. Y. Dantsiger, O. N. Razymovskaya, L. A. Reznichenko, S. I. Dudkina, High-efficiency Piezoelectric Ceramic Materials. Optimization of the Search, Pake, Rostov-on-Don, 1995

GRAIN GROWTH OF BaTiO₃ DOPED WITH ALIOVALENT CATIONS

M. N. Rahaman
University of Missouri-Rolla, Department of Ceramic Engineering, Rolla, Missouri 65401

ABSTRACT

The effect of three aliovalent cations, Nb^{5+}, La^{3+} and Co^{2+} on the grain growth kinetics of nearly fully dense $BaTiO_3$ (Ba/Ti atomic ratio = 1.001) was measured in O_2 at 1300 °C. For the donor cation Nb^{5+}, the boundary mobility initially increased with cation concentration but then decreased markedly above a doping threshold of 0.3-0.5 atomic %. The boundary mobility of the $BaTiO_3$ doped with the acceptor cation Co^{2+} decreased monotonically with dopant concentration. At a cation concentration of 0.75 at%, the boundary mobility was reduced by a factor of approximately 25, 10 and 50 times by Nb^{5+}, La^{3+} and Co^{2+}, respectively. A major role of the dopants is seen to be their ability to influence the boundary mobility. The effects of the dopants on the boundary mobility are discussed in terms of the defect chemistry and the space-charge concept.

INTRODUCTION

Following the work of Coble[1] on MgO-doped Al_2O_3, the use of dopants has been shown to be a very effective approach for the production, by conventional sintering, of ceramics with high density and fine grain size such as are normally required for most advanced technological applications.[2] However, the role of the dopant has been the subject of considerable debate. While progress has been made in a few recognized examples,[3-9] e.g., Al_2O_3, $BaTiO_3$, ZrO_2 and CeO_2, a considerable gap exists in the understanding of the dopant role. A major reason for this gap has been discussed in terms of the multiplicity of functions that a dopant can display.[2-4] However, recent studies indicate that the most significant effect of dopants on sintering is their ability to influence the grain boundary mobility. In the case of MgO-doped Al_2O_3, the work of Harmer and coworkers[3,11] indicates that while the MgO affects the lattice and surface diffusion coefficients, the single most important effect is its ability to reduce the boundary mobility significantly. For CeO_2 doped with divalent and trivalent cations,[7-9] the importance of the additive concentration, cation size and cation charge for controlling the grain growth has been reported. Detailed studies[8] indicate that at lower dopant concentration (intrinsic regime), the boundary mobility is controlled by the grain boundary diffusion of the host cations. At higher concentration (extrinsic regime), the mobility is controlled by solute drag through the lattice.

Barium titanate, $BaTiO_3$, is an important ferroelectric material that has been studied extensively for nearly 50 years. Generally,[12,13] optimization of the ferroelectric behavior of $BaTiO_3$ requires a high density and controlled, homogeneous microstructure with an average grain size of ≈ 1 μm. Since the use of dopants forms a very effective approach for microstructural control,[14] considerable work has been performed to characterize the defect structure of $BaTiO_3$ and its effect on sintering and grain growth behavior. Previous research has provided information on the defect structure[15-20] and the grain boundary chemistry[21] of $BaTiO_3$. This background suggests that $BaTiO_3$ would be an excellent model system for investigating the influence of dopants on sintering.

Donor cations, i.e., ions of higher valence than the host cation (e.g., La^{3+} for Ba^{2+} and Nb^{5+} for Ti^{4+}) at a concentration of a few tenths of 1 at% lead to a dramatic change in the behavior of $BaTiO_3$. This phenomenon is sometimes referred to as the "doping anomaly" or "grain size anomaly". Below this doping threshold, donor dopants enhance grain growth while above the threshold, they are effective in inhibiting grain growth. Acceptor dopants (i.e., ions of a valence lower than that of the host cation) produce mixed behavior,[4] i.e., some acceptors (e.g., Co^{2+}) inhibit grain growth while others (e.g., Cu) enhance grain growth. The origins of the mixed behavior are unclear. However, it has been found[4] that if a donor (e.g., Ta^{5+}) and an acceptor that inhibits grain growth (e.g., Co^{2+}) are present in compensating concentrations (i.e., the sum of the effective charges is equal to zero), then grain growth is not inhibited. This indicates that defect charge compensation may have a significant effect on grain growth.

The sintering of $BaTiO_3$ is also dependent on the stoichiometry of the compound. The solubility of TiO_2 and of BaO in $BaTiO_3$ is very limited,[22] i.e., less than ≈ 100 ppm. For $BaTiO_3$ materials with a small excess of TiO_2 (i.e., Ba/Ti atomic ratio less than 0.999), the excess TiO_2 reacts with $BaTiO_3$ to form a second phase, $Ba_6Ti_{17}O_{40}$, which has a eutectic with $BaTiO_3$ at ≈ 1320 °C. The liquid phase which forms above the eutectic temperature is beneficial for densification but can also enhance grain growth. An excess of BaO (i.e., Ba/Ti atomic ratio greater than 1.001) leads to the formation of a second phase, Ba_2TiO_4, and to a reduction in the densification and grain growth rates.[23]

It is clear that dopants have a profound effect on the microstructural evolution of $BaTiO_3$. However, in addition to the variety of functions that a dopant can display, an understanding of the dopant role is further complicated by the presence of impurities, the effect of the Ba/Ti atomic ratio and the formation of a second phase, particularly a liquid phase above the eutectic temperature for Ba/Ti less than 0.999.

In the present work, the effect of three aliovalent cations, Nb^{5+}, La^{3+} and Co^{2+}, on the grain growth kinetics of nearly fully dense $BaTiO_3$, prepared from a high-purity, fine-grained powder (Ba/Ti atomic ratio = 1.001) was measured in O_2 at 1300 °C. An important feature of the work is the ability to obtain samples with nearly full density at a sintering temperature below that of the eutectic (1563 °C) in the $BaTiO_3/Ba_2TiO_4$ system as well as that (1320 °C) in the excess TiO_2 side. In this way, the effects associated with impurities and the presence of a liquid phase are significantly reduced.

EXPERIMENTAL PROCEDURE

The $BaTiO_3$ powder used in the present work was provided by Sakai Chemical Industry Co., Japan. The characteristics of the powder (Lot BT-01) as described by the manufacturer are as follows: average particle size = 0.1 μm, specific surface area = 13.5 m^2/g, Ba/Ti atomic ratio = 1.001, SrO = 0.01 wt%, CaO < 0.001 wt%, Na$_2$O = 0.002 wt%, SiO$_2$ = 0.003 wt%, Al$_2$O$_3$ < 0.001 wt% and Fe$_2$O$_3$ = 0.001 wt%. X-ray analysis revealed that the powder was structurally identical to a standard tetragonal BaTiO$_3$.

Niobium chloride, lanthanum nitrate and cobalt nitrate were used to achieve the required doping at cation concentrations ranging from 0.25 to 1.25 at%. (The niobium and lanthanum salts, purity 99.999% and 99.99%, respectively, were obtained from AESAR/Johnson Matthey, Ward Hill, MA, USA and the cobalt nitrate, purity 99.999% was obtained from Aldrich Chemical Co., Milwaukee, WI, USA.) The procedure described by Kahn[24] was used for doping the BaTiO$_3$ powder with Nb^{5+}. For doping with the other cations, the nitrate was dissolved in anhydrous ethanol and the BaTiO$_3$ powder was added to the solution and mixed thoroughly by stirring. The mixture was dried, under vigorous stirring, to ensure homogeneity of the additive. All processing was carried out using Teflon labware in a dust-free environment. The doped powders were calcined in air for 2 hours at 900 °C in a Pt crucible and then ground in a plastic mortar and pestle. For comparison, the same procedure was used for the undoped BaTiO$_3$ powder but without the addition of the dopant.

Compacts (6 mm in diameter by 6 mm) were formed by uniaxial pressing of the powder in a tungsten carbide die ($\approx$20 MPa), the contact surfaces of which were coated with stearic acid. The green densities of the samples were within a narrow range of $\approx$0.48-0.52 of the theoretical density of BaTiO$_3$ (assumed to be 6.02 g/cm^3).

The powder compacts were sintered in O$_2$ in a dilatometer (1600 C; Theta Industries, Port Washington, NY, USA) that allowed continuous monitoring of the shrinkage kinetics. The sample holder, push rods and protection tube of the dilatometer were made from high purity Al$_2$O$_3$ and the compacts were separated from the contact surfaces of the Al$_2$O$_3$ by Pt foil. The samples were heated at a constant rate of 5 °C/min to 1300 °C and held at this temperature for times ranging from 0 to 24 h after which they were cooled (at ~25 °C/min). The final densities were determined from the initial density of the sample and the measured shrinkage and the values were checked using the Archimedes method.

The microstructure of the sintered materials was observed by scanning electron microscopy, SEM (JEOL-2000FXII), of fractured surfaces or of polished and thermally etched surfaces. Thermal etching was conducted for 2 h at 1200 °C. The average grain size was determined by the linear intercept method[25,26] from $\approx$150 grains for each sample. In addition, for dense samples of undoped and Nb-doped BaTiO$_3$ (relative density greater than 0.98), the average grain size and the distribution in sizes were determined from micrographs of thermally etched, polished surfaces by computerized image analysis (NIH Image software). At least 150 grains were analyzed for each sample.

RESULTS

The density data and the microstructural observations indicated that the compacts reached densities greater than 98% of the theoretical value after 1 to 2 h at 1300 °C for undoped and Nb-doped BaTiO$_3$ and after 3 to 4 h at the same temperature for Co-doped and La-doped BaTiO$_3$. The initial time for grain growth was taken as the time when the density of the sample was greater than 98% of the theoretical, thereby reducing the effects of pores on the grain growth kinetics significantly.

Figure 1 shows SEM micrographs of undoped $BaTiO_3$ held for 1 and 16 at 1300 °C. Some porosity is evident after 1 h but it disappeared at longer heating times. Microstructures of $BaTiO_3$ doped with 0.25 and 0.75 at% Nb and held for 8 h at 1300 °C are shown in Fig. 2. The significant difference in grain size is evident. This pronounced change-over in the grain growth rate with the donor cation concentration (≈ 0.3 to 0.5 at%) is the so-called doping anomaly in $BaTiO_3$. Microstructural observations by SEM as well as an analysis of the grain size distribution obtained from image analysis showed no evidence for the growth of abnormally large grains in the undoped and doped $BaTiO_3$.

The grain boundary mobility, M_b, can be estimated from the grain growth kinetics by the equation[14]:

$$\frac{dG}{dt} = \frac{M_b \gamma_b}{G} \tag{1}$$

where G is the grain size, γ_b is the specific energy of the grain boundary and t is the time. Assuming that M_b and γ_b are independent of time, then integration of Eq. (1) leads to the parabolic grain growth law:

$$G^2 - G_o^2 = 2M_b \gamma_b (t - t_o) \tag{2}$$

where G_o is the grain size at an initial time t_o. In practice, it is common to find grain growth exponents varying from 2 to 4. This variation can be interpreted simply as being due to a decreasing M_b. According to Eq. (2), a plot of $G^2 - G_o^2$ versus $t - t_o$ yields a curve with a slope equal to $2M_b\gamma_b$. In the present work, a straight line passing through the origin was fitted to the data to provide the average value of $2M_b\gamma_b$ for the duration of the grain growth experiment. Furthermore, to avoid any significant influence of porosity on the grain boundary mobility, the time t_o at 1300 °C was chosen so that the sintered density of the samples was at least 98% of the theoretical value.

Fig. 1. SEM of the undoped $BaTiO_3$ after heating for (a) 1h and (b) 16 h at 1300 °C.

Dielectric Ceramic Materials

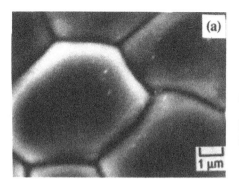

Fig. 2. SEM of BaTiO$_3$ doped with (a) 0.25 at% Nb^{5+} and (b) 0.75 at% Nb^{5+} after heating for 8 h at 1300 °C.

Figure 3 shows that the grain growth data for the undoped BaTiO$_3$ and for the BaTiO$_3$ doped with up to 1.25 at% Nb^{5+}. Assuming that γ_b is approximately constant, the normalized boundary mobility, $M_b{'}$, taken as the boundary mobility of the doped BaTiO$_3$ relative to that for the undoped BaTiO$_3$, is plotted in Fig. 4 as a function of the Nb^{5+} concentration. The value of $M_b{'}$ increases initially with Nb^{5+} concentration, reaching a value of ≈ 4 at 0.25 at% but then decreases dramatically with increasing Nb^{5+} concentration to a value of 0.04 at 0.75 at%, i.e., for a change in Nb^{5+} concentration from 0.25 to 0.75 at%, $M_b{'}$ decreases by a factor of ≈ 100. A further increase in the Nb^{5+} concentration to 1.25 at% produces only a small reduction in $M_b{'}$ compared to the value at 0.75 at%.

A plot of the normalized grain boundary mobility, $M_b^{'}$, as a function of Co^{2+} concentration is shown in Fig. 5. Compared to the Nb-doped BaTiO$_3$, the value of $M_b^{'}$ decreases monotonically with the Co^{2+} concentration, i.e., Co^{2+} inhibits grain growth for all of the concentrations investigated and there is no doping threshold.

Figure 6 shows the grain growth kinetics for undoped BaTiO$_3$ and BaTiO$_3$ doped separately with 0.75 at % of Nb^{5+}, Co^{2+} and La^{3+}. While all three cations are very effective for the inhibition of grain growth, Nb^{5+} and Co^{2+} are more effective than La^{3+}. At this dopant concentration, the boundary mobility of BaTiO$_3$ is reduced by a factor of approximately 50, 25 and 10 for Co^{2+}, Nb^{5+} and La^{3+}, respectively.

DISCUSSION

The results indicate that a major role of the dopants is their ability to influence the boundary mobility. Furthermore, the difference in the dependence of the boundary mobility on the cation concentration (Figs. 4 and 5) indicates that the dopant role for the donor Nb^{5+} is different from that for the acceptor Co^{2+}. Some observations pertinent to the understanding of the dopant role may be made at this stage.

As mentioned earlier, the grain boundary chemistry of BaTiO$_3$ has been investigated

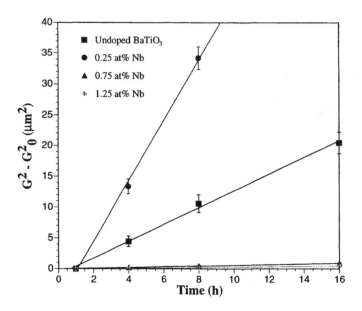

Fig. 3. Grain growth kinetics for undoped $BaTiO_3$ and Nb-doped $BaTiO_3$ at 1300 °C.

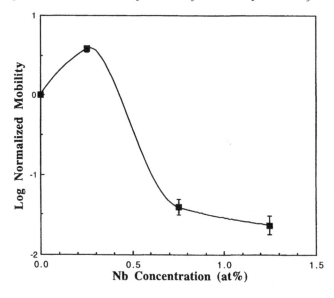

Fig. 4. Grain boundary mobility of Nb-doped $BaTiO_3$, normalized to that of undoped $BaTiO_3$, as a function of Nb^{5+} concentration.

Dielectric Ceramic Materials

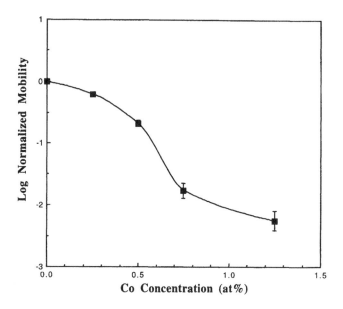

Fig. 5. Grain boundary mobility of Co-doped $BaTiO_3$, normalized to that of undoped $BaTiO_3$, as a function of Co^{2+} concentration.

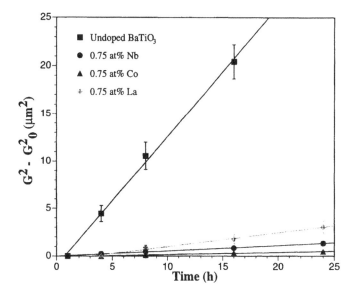

Fig. 6. Grain growth kinetics for undoped $BaTiO_3$ and $BaTiO_3$ doped with 0.75 at% Nb^{5+}, Co^{2+} and La^{3+} at 1300 °C.

in earlier work. An analysis using scanning transmission electron microscopy (STEM) performed by Chiang and Takagi[21] revealed the segregation of acceptor cations but not donor cations at the grain boundaries in $BaTiO_3$ and $SrTiO_3$. According to the grain boundary model proposed by Chiang and Takagi, accumulation of the acceptor solute in the negative space charge is expected in the case of Co-doped $BaTiO_3$. Assuming that the segregated cations control the boundary mobility, then the solute drag mechanism proposed by Cahn[27] may provide an interpretation of the reduction in M_b (Fig. 5) in Co-doped $BaTiO_3$. According to the Cahn model, the boundary mobility is given approximately by an equation of the form:

$$M_b = D/(\Delta C \, k \, T) \qquad (3)$$

where D is the diffusivity of the dopant cation, ΔC is the excess concentration of the dopant cation in the grain boundary, k is the Boltzmann constant and T is the absolute temperature. It may be expected that ΔC will increase with the Co^{2+} concentration, leading to a monotonically decreasing M_b (Fig. 5).

In the case of donor dopants, Fig. 4 shows a dramatic variation in M_b with Nb^{5+} concentration. The occurrence of the doping threshold is an indication that donor cations influence M_b by a solid solution mechanism. Furthermore, the absence of measurable segregation of donor cations[21] at the $BaTiO_3$ grain boundaries can be taken to rule out a solute drag mechanism by the donor cations. Below the doping threshold, a possible defect reaction for the incorporation of Nb^{5+} into $BaTiO_3$ can be written in the Kroger-Vink notation as[19]:

$$4BaO + Nb_2O_5 \rightarrow 4Ba_{Ba}^{x} + 4Nb_{Ti}^{\cdot} + 12O_O^{x} + O_2 + 4e^{'} \qquad (4)$$

Above the doping threshold, assuming that ionic compensation occurs by the formation of Ti vacancies[15], a possible defect reaction is:

$$4BaO + 2Nb_2O_5 + BaTiO_3 \rightarrow 5Ba_{Ba}^{x} + 4Nb_{Ti}^{\cdot} + 15O_O^{x} + TiO_2 + V_{Ti}^{''''} \qquad (5)$$

In the grain boundary model of Chiang and Takagi[21] discussed earlier, electrons and titanium vacancies are expected to accumulate in the negative space charge. Below the doping threshold, the mechanism for the small enhancement of M_b with increasing Nb^{5+} concentration is not clear. Above the threshold, the accumulation of Ti vacancies in the space charge may be associated with a depletion of oxygen vacancies. Because of their relatively large size, the diffusivity of oxygen ions is expected to be slow. The low diffusivity of oxygen ions across the grain boundary provides a possible mechanism for the significant reduction in M_b.

In earlier work,[4] the use of double dopants consisting of a donor (Ta^{5+}) and an acceptor (Co^{2+}) indicated that charge compensation has an important effect on M_b. Similar results[28]

Dielectric Ceramic Materials

were also found for co-doping with Nb^{5+} and Co^{2+}. As long as the Nb^{5+} concentration was approximately twice that of the Co^{2+} concentration, the boundary mobility was approximately equal to that of the undoped the $BaTiO_3$. If, as discussed above, the single dopants Nb^{5+} and Co^{2+} influence M_b by different mechanisms, then an important question is concerned with why charge compensation should have the significant observed effect. A possible explanation for the effect of the double dopants on M_b is that segregation of the acceptor cations and the cation vacancies [given in Eq. (5)] does not occur. This may be caused by a reduction of the grain boundary potential to a value considerably smaller than that for $BaTiO_3$ doped with single dopants. At other concentrations, the excess of donor or acceptor cations (relative to the charge-compensated values) leads to effects similar to those produced by the single dopants at the equivalent concentration.[28]

CONCLUSIONS

In $BaTiO_3$, a major role of donor cations (Nb^{5+} for Ti^{4+} and La^{3+} for Ba^{2+}) and acceptor cations (Co^{2+} for Ti^{4+}) is their ability to influence the boundary mobility. For the acceptor dopant, the boundary mobility decreases monotonically with increasing Co^{2+} concentration. In the case of Nb-doped $BaTiO_3$, the boundary mobility increases slowly with the cation concentration but then decreases significantly above a doping threshold of 0.3-0.5 at%. For the acceptor dopant and the donor dopant above the doping threshold, the reduction in the boundary mobility can be rationalized in terms of the segregation of defects (acceptor solutes or ionic vacancies) at the grain boundaries by a space charge mechanism.

REFERENCES

[1]R. L. Coble, "Sintering of Crystalline Solids: II. Experimental Test of Diffusion Models in Porous Compacts," *J. Appl. Phys.*, **32**(7) 793-99 (1961).

[2]R. J. Brook, "Fabrication Principles for the Production of Ceramics with Superior Mechanical Properties," *Proc. Brit. Ceram. Soc.*, **32** 7-24 (1982).

[3]S. J. Bennison and M. P. Harmer, "A History of the Role of MgO in the Sintering of α-Al_2O_3,"; pp. 13-49 in *Ceramic Transactions*, Vol.7. Eds. C.A. Handwerker, J.E. Blendell, and W.A. Kaysser. American Ceramic Society, Westerville, OH, 1990.

[4]R. J. Brook, W. H. Tuan, and L. A. Xue, "Critical Issues and Future Directions in Sintering Science, pp. 811-23 in *Ceramic Transactions*, Vol.1B. Eds. G.L. Messing, E.R. Fuller, and H. Hausner. American Ceramic Society, Columbus, OH, 1988.

[5]L. A. Xue, Y. Chen, R. J. Brook, "The Effect of Lanthanide Contraction on Grain Growth in Lanthanide-Doped $BaTiO_3$, *J. Mater. Sci. Lett.*, **7** 1163-65 (1988).

[6]S.-L. Hwang and I.-W. Chen, "Grain Size Control of Tetragonal Zirconia Polycrystal using the Space Charge Concept," *J. Am. Ceram. Soc.*, **73**(11) 3269-77 (1990).

[7]P. Lin, I.-W. Chen, J. E. Penner-Hahn, and T. Y. Tien, "X-ray Adsorption Studies of Ceria with Trivalent Dopants," *J. Am. Ceram. Soc.*, **74**(5) 958-67 (1991).

[8]P.-L. Chen and I.-W. Chen, "Role of Defect Interaction in Boundary Mobility and Cation Diffusivity of CeO_2," *J. Am. Ceram. Soc.*, **77**(9) 2289-97 (1994).

[9]M. N. Rahaman and Y. C. Zhou, "Effect of Solid Solution Additives on the Sintering of Ultra-Fine CeO_2 Powders," *J. Europ. Ceram. Soc.*, **15** 939-50 (1995).

[10]R. J. Brook, "Frontiers of Sinterability," pp. 3-12 in *Ceramic Transactions*, Vol.7. Eds. C.A. Handwerker, J.E. Blendell, and W.A. Kaysser. American Ceramic Society, Westerville, OH, 1990.

[11]K. A. Berry and M. P. Harmer, "Effect of MgO Solute on Microstructure Development in Al_2O_3," *J. Am. Ceram. Soc.*, **69**(2) 143-49 (1986).

[12]K. Kinoshita and A. Yamaji, "Grain Size Effects on Dielectric Properties in Barium Titanate Ceramics," *J. Appl. Phys.*, **47**(1) 371-73 (1976).

[13]G. Arlt, D. Hennings, and G. DeWith, "Dielectric Properties of Fine-Grained Barium Titanate," *J. Appl. Phys.*, **58**(4) 1619-25 (1985).

[14]R. J. Brook, "Controlled Grain Growth," pp. 331-64 in *Treatise on Materials Science and Technology*, Vol. 9. Ed. F. F. Y. Wang, Academic Press, New York, 1976.

[15]H. M. Chan, M. P. Harmer, and D. M. Smyth, "Compensating Defects in Highly Donor-Doped $BaTiO_3$," *J. Am. Ceram. Soc.*, **69**(6) 507-10 (1986).

[16]G. H. Jonker and E. E. Havinga, "The Influence of Foreign Ions on the Crystal Lattice of Barium Titanate," *Mater. Res. Bull.*, **17**(3) 345-50 (1982).

[17]A. M. J. H. Seuter, "Defect Chemistry and Electrical Transport Properties of Barium Titanate," *Philips Res. Repts. Suppl.*, **3** 1-84 (1974).

[18]J. Nowotny and M. Rekas, "Defect Chemistry of $BaTiO_3$," *Solid State Ionics*, **49**(12) 135-54 (1991).

[19]N.-H. Chan and D. M. Smyth, "Defect Chemistry of Donor-Doped $BaTiO_3$," *J. Am. Ceram. Soc.*, **67**(4) 285-88 (1984).

[20]N.-H. Chan, R. K. Sharma, and D. M. Smyth, "Non-Stoichiometry in Acceptor-Doped $BaTiO_3$," *J. Am. Ceram. Soc.*, **65**(3) 167-70 (1986).

[21]Y.-M. Chiang and T. Takagi, "Grain Boundary Chemistry of Barium Titanate and Strontium Titanate: I. High Temperature Equilibrium Space Charge," *J. Am. Ceram. Soc.*, **73**(11) 3278-85 (1990).

[22]R. K. Sharma, H. M. Chan, and D. M. Smyth, "Solubility of TiO_2 in $BaTiO_3$," *J. Am. Ceram. Soc.*, **64**(8) 448-51 (1981).

[23]Y. H. Hu, M. P. Harmer, and D. M. Smyth, "Solubility of BaO in $BaTiO_3$," *J. Am. Ceram. Soc.*, **68**(7) 372-76 (1985).

[24]M. Kahn, "Preparation of Small-Grained and Large-Grained Ceramics from Nb-Doped $BaTiO_3$," *J. Am. Ceram. Soc.*, **54**(9) 452-54 (1971).

[25]M. I. Mendelson, "Average Grain Size in Polycrystalline Ceramics," *J. Am. Ceram. Soc.*, **52**(8) 443-46 (1969).

[26]J. C. Wurst and J. A. Nelson, "Lineal Intercept Technique for Measuring Grain Size in Two-Phase Polycrystalline Ceramics," *J. Am. Ceram. Soc.*, **55**(2) 109 (1972).

[27]J. W. Cahn, "The Impurity-Drag Effect in Grain Boundary Motion," *Acta Metall.*, **10**(9) 789-98 (1962).

[28]R. Manalert, "Dopants and the Sintering of Fine-Grained Barium Titanate Powder," *Ph.D. Thesis*, University of Missouri-Rolla, 1996.

EFFECT OF RARE-EARTH OXIDES ON FORMATION OF CORE-SHELL STRUCTURES IN BaTiO$_3$

Hiroshi Kishi, Noriyuki Kohzu, Yoshikazu Okino, Yoshinao Takahashi and Yoshiaki Iguchi, Taiyo Yuden Co., Ltd., Gunma, Japan
Hitoshi Ohsato, Kazutaka Watanabe, Junichi Sugino and Takashi Okuda
Nagoya Institute of Technology, Japan

ABSTRACT

The effect of rare-earths, Dy and Ho, on the formation of core-shell structure in BaTiO$_3$(BT)-MgO-rare-earth oxide based system was studied. It was found that the Dy-doped sample indicated the collapse of the core-shell structure with lower content of rare-earth oxide compared with Ho-doped sample. Also, replacement modes of rare-earths and Mg in BT lattice were investigated. It was found that Dy and Ho ions dissolved in both Ba- and Ti-sites, and the ratio of Ba-site substitution of Dy ions was larger than that of Ho. This suggest that Dy ions behave more likely as donor dopants than Ho ions in the shell phase.

INTRODUCTION

Multilayer ceramic capacitors (MLCs) with Ni electrodes have been widely used in electronic equipments to meet ever-increasing demands for high-volumetric-efficiency, high reliability and reduced manufacturing costs. It is well known that temperature-stable X7R dielectrics have chemically heterogeneous microstructure in a grain, so-called core-shell structures[1-4]. The control of dielectric microstructure has become more important for reducing dielectric layer thickness in MLCs.

Okino et al.[5] previously showed that non reducible dielectrics based on BaTiO$_3$(BT)-MgO-rare-earth oxide(R_2O_3), developed for X7R MLCs, have highly reliable electrical characteristics. The longer lifetime was obtained for the smaller ion (Dy, Ho, Er)-doped dielectrics. By the computer simulation study, Lewis et al.[6] reported that rare-earth ions of intermediate size should occupy both A- and B-sites in the perovskite lattice of generic formula ABO_3. Ohsato et al.[7] confirmed that Dy and Ho dissolved both sites in BT-Ho$_2$O$_3$ and BT-Dy$_2$O$_3$ system by high temtemperature powder X-ray diffraction analysis (XRD).

However, substitution modes of BT containing rare-earth oxide and MgO are not clear yet.

Kishi et al.[8] also reported the influence of Mg and Ho on the formation behavior of core-shell structure in BT. It was confirmed that Mg dissolved B-site and suppressed the diffusion of Ho into the core region. It suggested that solubility modes of dopants in the shell phase affected on the formation mechanism of core-shell structure. But the formation behavior of core-shell structure using other rare-earth oxides has not been reported so far.

The purpose of this study is to clarify the influence of rare-earth oxides, Dy and Ho, on the formation behavior of core-shell structure in BT-MgO based system, and to investigate the solubility modes of Dy-Mg, and Ho-Mg into BT.

EXPERIMENTAL

In order to determine the influence of Dy_2O_3 and Ho_2O_3 on the formation of the core-shell structure, BT-0.006MgO-xR_2O_3-0.01BaSiO$_3$ ($x=0\sim0.02$) samples were prepared by the conventional method as follows. Reagent-grade $BaCO_3$, $MgCO_3$, TiO_2 and SiO_2, 99.9%Dy_2O_3 and Ho_2O_3 were used as raw materials. $BaSiO_3$ was used as a sintering aid. BT was precalcined at 1200°C after $BaCO_3$ and TiO_2 were mixed in equimolar ratios. The obtained particle size of BT was about 0.4μm. $BaSiO_3$ was also precalcined at 1000°C. Additives and BT were weighed, mixed by ball milling and then dried. The mixture added an organic binder was pressed into disks and after the binder was burned out, the disks were fired at various temperatures in a low oxygen atmosphere controlled by H_2, N_2, O_2 and H_2O. The microstructures of the ceramics were observed by scanning electron microscopy (SEM). The dielectric properties of the ceramics were measured by LCR meter. The ceramics were crushed and ground into powder, and then the phase transition of the samples was characterized by differential scanning calorimetry (DSC).

In order to analyze solubility modes of Dy-Mg and Ho-Mg in BT, another samples were also prepared according to following formulae, $(Ba_{1-2x}R_{2x})(Ti_{1-x}Mg_x)O_3$:(A-site: rare-earth, B-site: Mg replacement model) ($x=0\sim0.1$). The raw materials, $BaCO_3$, $MgCO_3$, TiO_2, Dy_2O_3 and Ho_2O_3 were mixed and then calcined at 1250°C. Identification of precipitated phases was performed by powder XRD analysis. The solubility mode was monitored by the behavior of the lattice parameters as a function of the doping amount x. In order to avoid the influence on the lattice parameters by crystal structure change according to compositions, high-temperature powder XRD analysis was also carried out at higher temperature (300°C) than the Curie point. The lattice parameters were determined precisely using the whole-powder-pattern decomposition method (WPPD[9]) program as described in the previous paper[7].

Figure 1. SEM micrographs of BT-MgO-R_2O_3-BaSiO$_3$ based samples sintered at 1350°C. (R=Dy, Ho)

RESULTS AND DISCUSSION

The influence of rare-earths on the formation behavior of core-shell structure

Figure 1 shows SEM micrographs of BT-MgO-R_2O_3-BaSiO$_3$ based samples sintered at 1350°C. Grain growth occurred in Dy-doped samples with increasing Dy content above 1.6 atomic%. On the other hand, no grain growth was observed in Ho-doped samples. The DSC profiles for the samples sintered at 1250°C are shown in Figure 2(a) and 2(b). These DSC peaks were due to the Curie temperature of the core phase. The peak temperature was unchanged practically for the Ho-doped samples. However, in the case of the Dy-doped samples, the peak temperature shifted down to around 100°C at 1.6atomic% Dy. Figure 3 shows the temperature dependence of the dielectric constant of the disk samples sintered at 1350°C. The dielectric properties for the Dy doped samples

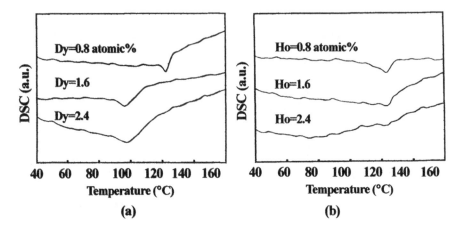

Figure 2. DSC profiles of BT-MgO-R_2O_3-BaSiO$_3$ based samples sintered at 1250°C; (a) R=Dy, (b) R=Ho.

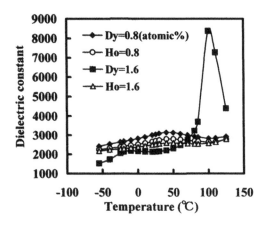

Figure 3. Temperature dependence of the dielectric constant of the disk samples sintered at 1350°C.

were strongly dependent on the amount of rare-earth oxide. However, they were almost independent for the Ho-doped samples.

From these results, it was found that the Dy-doped samples indicated the collapse of the core-shell structure with lower content of rare-earth oxide compared with Ho-doped samples. It seemed that the diffusivity of Dy ions into the core phase is higher than that of Ho ions.

Solubility modes of Dy-Mg and Ho-Mg in BaTiO₃

The electrical properties of BT is strongly influenced by the substitution mode of the dopant, which is dependent on the valence state and ionic radius. The ionic radii of Ba, Ti, Dy, Ho and Mg are summarized as follows; A-site (12 coordinate): Ba^{2+}=1.61Å, Dy^{3+}=1.253Å, Ho^{3+}=1.234Å; B-site (6 coordinate): Ti^{4+}=0.605Å, Dy^{3+}=0.912Å, Ho^{3+}=0.901Å, Mg^{2+}=0.720Å. The ionic radii of rare-earth ions in 12 coordinate are based on the relationship between coordination number and effective ionic radii after Shannon's table[10]. The ionic radii of Dy and Ho locate just intermediate of those of Ba and Ti. Therefore, Dy and Ho are considered to dissolve in both A- and B-sites. On the other hand, the radius of Mg is close to that of Ti ion. Mg is considered to dissolve in the B-site. In the previous study[8], we confirmed that Mg reacts with BT at low temperatures compared with Ho and dissolves into B-site. Therefore, taking the compensation of electrical charge into consideration, we synthesized the samples according to the following compositions: $(Ba_{1-2x}R_{2x})(Ti_{1-x}Mg_x)O_3$.

The precipitated phases of the samples were as follows. The single phase of BT solid solution was obtained up to x=0.005. Pyrochlore $(R_2Ti_2O_7)$ was appeared as a secondary phase from x=0.01 for the both samples. As more phase, R_2O_3 appeared from x=0.05 and then R_2MgTiO_6 appeared from x=0.10. The crystal structure of BT solid solutions changed from tetragonal to cubic at x=0.02 for the Ho-Mg substituted sample and x=0.025 for the Dy-Mg substituted sample.

Figure 4 shows the X-ray diffraction patterns at 300°C of calcined samples including diffraction lines of Pt-heating filament for internal standard. Figure 5 shows the lattice parameters at 300°C of a-axis of Dy-Mg and Ho-Mg substituted samples, as determined by WPPD. Although the ionic radii of Dy and Ho is very close, it appeared that the solubility behavior between the Dy-Mg and Ho-Mg sample was clearly different. In the case of the Dy-Mg substituted sample, the lattice parameter changed on the three stages as follows. The lattice parameter decreased linearly up to x=0.01 in the first stage. In the second stage up to x=0.05, they decreased gradually and then they showed no change in the last stage up to x=0.10. On the other hand, the change of the lattice parameter of the Ho-Mg substituted sample is divided into five stages as follows. The lattice parameter increased up to x=0.002 in the first stage and then decreased up to x=0.01 in the second stage. In the third stage up to x=0.025, they showed no change and increased up to x=0.05 in the fourth stage. In the last zone, they showed little change.

These phenomena can be explain by the different substitution modes of rare-earth elements at cation sites. The increase in lattice parameter is based on B-site replacement by larger cation (Mg, Dy, Ho) than Ti ion. The decrease in lattice parameter is based on A-site replacement by smaller cation (Dy, Ho) than Ba ion. In the case of the Dy-Mg substituted sample, the change of the lattice parameter is considered as follows. In the first stage, Dy mainly dissolved in A-site and then the substitution ratio of Dy into A-site decreased gradually in the second stage. On the other hand, the change of the lattice parameter of the Ho-Mg substituted sample is considered as follows. In the first stage, Mg

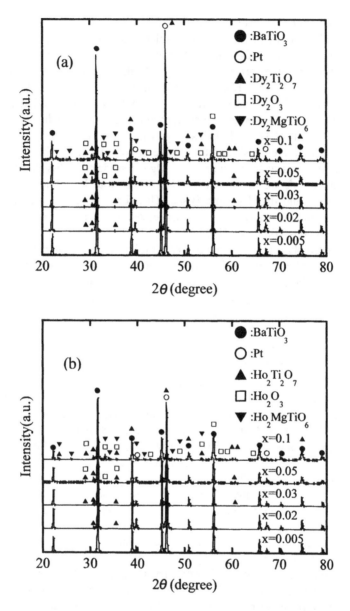

Figure 4. X-ray diffraction patterns at 300°C of the samples and Pt-filament for internal standard.

(a): $(Ba_{1-2x}Dy_{2x})(Ti_{1-x}Mg_x)O_3$, (b): $(Ba_{1-2x}Ho_{2x})(Ti_{1-x}Mg_x)O_3$.

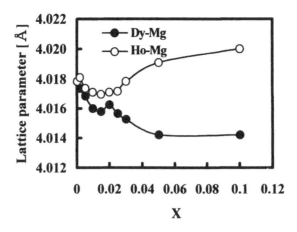

Figure 5. Lattice parameters at 300°C of Dy-Mg and Ho-Mg substituted samples. Compositions: $(Ba_{1-2x}R_{2x})(Ti_{1-x}Mg_x)O_3$ (R=Dy, Ho)

ions mainly substituted for B-site and then Ho ions mainly substituted for A-site in the second stage. The substitution ratio of Ho into A-site decreased steeply in third stage and then Ho mainly dissolved in B-site in the fourth stage. It appeared that the substitution ratio of Dy into A-site was larger than that of Ho. This can be attributed to decrease of radius difference between rare-earth and Ba ion. It seemed that the higher diffusivity of Dy ions into the core phase in the BT-MgO-Dy_2O_3 based system is also related to it's ionic radius. The present results suggest that Dy ions behave more likely as donor dopants than Ho ions. It was found that rare-earth elements play an important role in both the dielectric microstructure and electrical properties. Thus further investigation of the solubility modes of dopants in the shell phase is necessary to improve reliability of MLCs with nickel electrodes.

CONCLUSION

The effects of Dy and Ho addition on formation behavior of core-shell structure in BT-MgO based system were confirmed. It was found that the Dy-doped sample indicated the collapse of the core-shell structure with lower content of rare-earth oxide compared with Ho-doped.

The solubility modes of rare-earths and Mg into BT lattice were also investigated by high-temperature powder XRD. The lattice parameters of Dy-Mg and Ho-Mg substituted samples were determined at higher temperature than the Curie point. It was confirmed that Dy and Ho dissolved both A- and B-sites in BT-MgO-R_2O_3 system. It appeared that the substitution ratio of Dy into A-site in BT lattice was larger than that of Ho. This indicates that Dy ions behave more likely as donor dopants than Ho ions in the shell phase.

REFERENCES
[1]H. Saito, H. Chazono, H. Kishi, and N. Yamaoka, "X7R Multilayer Ceramic Capacitors with Nickel Electrodes," *Japanese Journal of Applied Physics*, **30** [9B] 2307-10 (1991).

[2]B.S. Rawal, M. Kahn, and W.R. Buessem, "Grain Core-Grain Shell Structure in Barium Titanate-Based Dielectrics"; pp.172-88 in *Advances in Ceramics*, Vol. 1, Grain Boundary Phenomena in Electronic Ceramics. Edited by L.M. Levinson . American Ceramic Society, Columbus, OH, 1981.

[3]D. Hennings and G. Rosenstein, "Temperature-Stable Dielectrics Based on Chemically Inhomogeneous $BaTiO_3$," *Journal of the American Ceramic Society*, **67** [4] 249-54 (1984).

[4]H. Chazono and M. Fujimoto, "Sintering Characteristics and Formation Mechanism of "Core-Shell" Structure in $BaTiO_3$-Nb_2O_5-Co_3O_5 Ternary System," *Japanese Journal of Applied Physics*, **34** [9B] 5354-59 (1995).

[5]Y. Okino, H. Shizuno, S. Kusumi, and H. Kishi, "Dielectric Properties of Rare-Earth-Oxide-Doped $BaTiO_3$ Ceramics Fired in Reducing Atmosphere," *Japanese Journal of Applied Physics*, **33** [9B] 5393-96 (1994).

[6]G.V. Lewis and C.R.A. Catlow, "Defect Studies of Doped and Undoped Barium Titanate Using Computer Simulation Techniques," *Journal of Physics and Chemistry of Solids*, **47** [1] 89-97 (1986).

[7]H. Ohsato, M. Imaeda, Y. Okino, H. Kishi, and T. Okuda, "Lattice Parameters of $BaTiO_3$ Solid Solutions Containing Dy and Ho at High Temperature," to be published in *Advances in X-ray Analysis*, **40**

[8]H. Kishi, Y. Okino, M. Honda, Y. Iguchi, M. Imaeda, Y. Takahashi, H. Ohsato, and T. Okuda, "The Effect of MgO and Rare-Earth Oxide on Formation Behavior of Core-Shell Structure in BaTiO3" *Japanese Journal of Applied Physics*, **33** [9B] 5954-57 (1997).

[9]H. Toraya, "Whole-Powder-Pattern Fitting without Reference to A Structual Model: Application to X-ray Powder Diffraction Data," *Journal of Applied Crystallography*, **16** 440 (1986)

[10]R.D. Shannon, "Revised Effective Ionic Radii and Systematic Studies of Interatomic Distance in Halides and Chalcogenides," *Acta Crystallographica A*, **32** 751-67 (1976).

NON-LINEAR MICROWAVE QUALITY FACTOR CHANGE BASED ON THE SITE OCCUPANCY OF CATIONS ON THE TUNGSTENBRONZE-TYPE $Ba_{6-3x}R_{8+2x}Ti_{18}O_{54}$ (R = RARE EARTH) SOLID SOLUTIONS

Hitoshi Ohsato, Motoaki Imaeda, Atsushi Komura and Takashi Okuda
Nagoya Institute of Technology
Showa-ku, Nagoya 466-8555, Japan
Susumu Nishigaki
Daiken Chemical Co., Ltd., Joto-ku, Osaka 536-0011, Japan

ABSTRACT

The $Ba_{6-3x}R_{8+2x}Ti_{18}O_{54}$ (R = rare earth) solid solutions in the BaO-R_2O_3-TiO_2 system belong to a kind of microwave dielectric ceramics with high dielectric constant ε_r. The structure of the solid solutions has the new tungstenbronze-type with 2 by 2 perovskite blocks ($A1$-site) and pentagonal columns ($A2$-site). The fundamental structural formula is expressed by $[A1]_{10}[A2]_4B_{16}X_{54}$. The quality factor $Q \cdot f$ changes non-linearly as a function of composition x, though ε_r and the temperature coefficient τ_f of resonant frequency are linearly. It is cleared that ε_r is proportional to the volume of the unit cell and the $Q \cdot f$ depends on the distribution of the cations Ba and R. When the rare earth ions and the Ba ions occupy separately the perovskite blocks and pentagonal columns, respectively, the $Q \cdot f$ reaches high values such as 10000GHz. Internal stress around Ba ions which occupy the smaller A_1 sites might lead to dielectric loss.

INTRODUCTION

Recently, microwave dielectric ceramics have received much attention due to the rapid progress in microwave telecommunication such as car telephone system, potable telephone and satellite broadcasting.[1-2] Microwave dielectric ceramics contribute to this development by enabling the miniaturization of resonators. The desirable properties in microwave dielectric resonators are a high dielectric constant ε_r, high quality factor $Q \cdot f$ and low temperature coefficient τ_f of the resonant frequency. The ε_r reduces the dimensions of resonators according to the $\lambda = \lambda_0 / \sqrt{\varepsilon_r}$ equation, where λ_0 is the wavelength in vacuum and λ that in ceramics. The Q, which is the inverse of the dielectric loss $\tan\delta$, is required to achieve high frequency selectivity and stability in microwave transmitter components. $Q \cdot f$ is used to evaluate the quality instead of the Q value, because the Q values varies inversely with the frequency f in the microwave region. The τ_f is required to be as close to 0 ppm/°C as possible in the working temperature range required for the application.

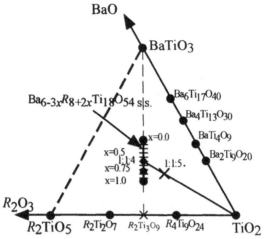

Figure I.   Tungstenbronze-type $Ba_{6-3x}R_{8+2x}Ti_{18}O_{54}$ (R = rare earth) solid solutions in the BaO-R_2O_3-TiO_2 ternary system.

The microwave dielectric ceramics with high quality in the TiO_2-rich region of the BaO-R_2O_3-TiO_2 (R = rare earth) ternary system were reported by Kolar et al.[3] (1978).   It is worth noting that ten years before Kolar's report, Bolton[4] had systematically studied the same region of the ternary system. He proposed two compounds with high dielectric constant and low temperature coefficient. Following Kolar et al.'s work,[3] the microwave dielectric ceramics were investigated by many researchers such as Wakino et al.,[5] Kawashima et al.,[6] Nishigaki et al.[7] and so on. The last three groups in Japanese companies produce the resonators for commercial products.

Kolar et al.[8] also reported in detail the crystal data such as lattice parameters and space group but reported wrong composition formula as $BaO \cdot R_2O_3 \cdot 5TiO_2$ which composition is out of the tungstenbronze-type $Ba_{6-3x}R_{8+2x}Ti_{18}O_{54}$ (R = rare earth) solid solution range. The solid solutions were reported as ternary compounds existing on the tie line between $BaTiO_3$ and $R_2Ti_2O_9$ composition in the BaO-R_2O_3-TiO_2 ternary system as shown in Figure.I.[9-11] The $BaO \cdot R_2O_3 \cdot 4TiO_2$ ternary compound[12-16] known as microwave resonator materials corresponds exactly to $x=0.5$, and $Ba_{3.75}Pr_{9.5}Ti_{18}O_{54}$ reported by Matveeva et al.[17] corresponds to $x=0.75$ in the $Ba_{6-3x}R_{8+2x}Ti_{18}O_{54}$ solid solution formula. These solid solutions are formed   by the substitution of $2R^{3+}$ for $3Ba^{2+}$ maintaining electrostatic stability. The crystal structure will be described in the next section.

The dielectric properties of $Ba_{6-3x}R_{8+2x}Ti_{18}O_{54}$ solid solutions were reported as a function of composition in our previous paper.[18-20] The $Q \cdot f$ changed non-linearly as a function of composition x, though ε_r and τ_f are linearly. In this study, we will discuss the reason why the $Q \cdot f$ changes non-linearly as a function of composition based on the crystal structure.

Dielectric Ceramic Materials

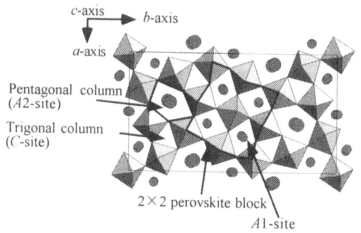

c-axis ▸ b-axis

a-axis

Pentagonal column
($A2$-site)

Trigonal column
(C-site)

2×2 perovskite block

$A1$-site

Figure II. Structure of tungstenbronze-type $Ba_{6-3x}R_{8+2x}Ti_{18}O_{54}$

(R = rare earth) solid solutions

EXPERIMENTAL

The samples with the desired composition were prepared as disks for resonators with 10 mm diameter and 5 mm height by the solid state reaction method at 1450°C as described in the previous paper.[20,21] The dielectric properties of the disks were measured in the range of 4 to 5 GHz by the method of Hakki and Colleman[22] using network scalar analyzer produced by HP company. The disks were inserted among a pair of parallel conducting Ag plates where the TE_{011} mode of microwave was applied.

STRUCTURAL FORMULAE

The $Ba_{6-3x}R_{8+2x}Ti_{18}O_{54}$ solid solutions have a superlattice with a c-axis two times larger than the fundamental lattice: $Pbam$ (No.55) or $Pba2$ (No.32), orthorhombic, $a=12.17$, $b=22.34$, $c=3.84$ Å .[8,23] The crystal structure of the solid solutions is a new tungstenbronze-type with 2 by 2 perovskite blocks, pentagonal columns and trigonal columns which are called as $A1$-site, $A2$-site and C-site, respectively, as shown in Figure II. The solid solutions are located on a tie line with Ti:O=1:3 ratio including perovskite $BaTiO_3$, because three-dimensional $[TiO_3]^{2-}$ framework of octahedra (B-site) connected with each other at all apices such as in the perovskite structure.[10] There are three different cations with different diameter in the crystal structure, which occupy different sites. The largest Ba ions mainly occupy pentagonal $A2$-sites. If the composition is Ba rich, small amount of Ba ions occupy also $A1$-sites. The middle-sized R ions mainly occupy the $A1$-sites and the smallest Ti ions alone occupy octahedra B-sites. The trigonal C-sites are now empty. The basic structural formula is $[A1]_{10}[A2]_4B_{18}X_{54}$ in a fundamental unit cell including 10 sites for $A1$ and four sites for $A2$. Here, B is a cation in an octahedron and X is an anion, occupied by Ti and O ions, respectively. The final composition $Ba_6R_8Ti_{18}O_{54}$ of the solid solutions with $x=0$ is represented as the structural formula $[R_8Ba_2][Ba_4]Ti_{18}O_{54}$ in which the two largest Ba ions are included in $A1$-sites accompanied with eight middle-sized R ions. The $A2$-sites are only occupied by four Ba ions. The structural formula of the solid solutions is derived as $[R_{8+2x}Ba_{2-3x}V_x][Ba_4]Ti_{18}O_{54}$ in the range of $0<x<2/3$ from the terminal composition according to the substitution mode $3Ba^{2+} \rightarrow 2R^{3+} + V_{A1}$. When Ba ions in $A1$-sites are all substituted by R ions, x is $2/3$. Moreover, beyond the $x=2/3$, Ba ions in $A2$-sites are substituted by R ions entered in $A1$-sites and vacancies might be created in $A2$-sites. The structural formula is $[R_{9+1/3+2(x-2/3)}V_{2/3-2(x-2/3)}][Ba_{4-3(x-2/3)}V_{3(x-2/3)}]Ti_{18}O_{54}$ in the range of $2/3<x<1$. According to this substitution of $2R$ for $3Ba$, the cell volume of the solid solution decreases as shown in Figure III.

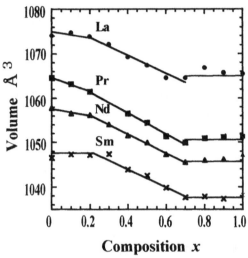

Figure III. Lattice volumes of the $Ba_{6-3x}R_{8+2x}Ti_{18}O_{54}$ (R = Sm, Nd, Pr and La) solid solutions as a function of composition x.

RESULT AND DISCUSSION

The dielectric constants ε_r, the quality factors $Q \cdot f$ and the temperature coefficients τ_f of resonant frequency as a function of composition x for the $Ba_{6-3x}R_{8+2x}Ti_{18}O_{54}$ solid solutions are shown in our previous paper.[20] Here, the ε_r and τ_f for the solid solution regions with $0.2<x<0.7$ for the Sm system, $0<x<0.7$ for the Nd, Pr and La systems are replotted as a function of cell volume in Figure IV. These values are almost proportional to the cell volumes which depend on the composition x as shown in Figure III. The cell volumes of Pr-system are sited from Fukuda's paper.[19] The ε_r values on the Sm system are the smallest. They increase according to the cell volume on the Nd, Pr and La systems. In each system, the values are valid linearly as a function of cell volume. The τ_f are also replotted as a function of cell volume in Figure IVb. Though similar tendency with ε_r is observed, the reason has not been clarified yet. The τ_f values of the Sm system are usually negative but close to zero. As they increase according to the substitution by elements with large ionic radii, they can be reduced to zero by adding a small amount of other rare earth elements with larger diameter as reported in the previous paper.[26]

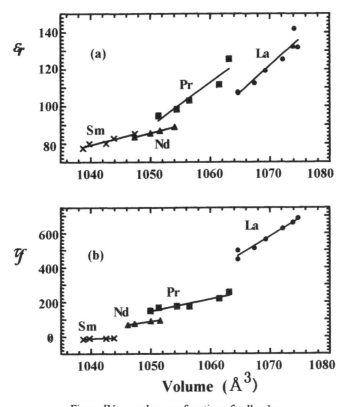

Figure IV. ε_r and τ_f as a function of cell volume.

It is a characteristic phenomenon that the $Q \cdot f$ values vary non-linearly as shown in Figure V, though ε_r and τ_f increase linearly as a function of composition or cell volume. The ε_r and τ_f are correlated with the lattice parameters as mentioned in the above section. On the other hand, the $Q \cdot f$ might have a correlation with the distribution of cations in the perovskite blocks ($A1$-sites) and pentagonal columns ($A2$-sites). The structural formula is $[R_{8+2x}Ba_{2-3x}V_x][Ba_4]Ti_{18}O_{54}$ in the range of $0<x<2/3$ as mentioned before. In the region of around $x=0.2$ where the system contains a lot of Ba ions and the Ba ions with large ionic radius occupy the $A1$-sites, the $Q \cdot f$ values are low. On the other hand, the $Q \cdot f$ values steeply increase in the vicinity of $x=0.5$ with decreasing amount of Ba ions in the $A1$-sites. In the case of $x=0.6$ the $Q \cdot f$ has its highest value near 10000GHz as shown in Figure V. The reason that the $Q \cdot f$ value become high might be explained by reducing the internal

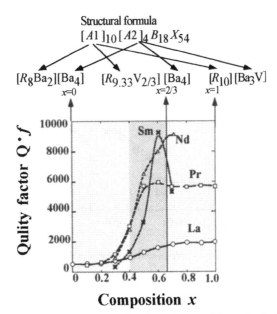

Structural formula

$$[A1]_{10}[A2]_4 B_{18} X_{54}$$

$[R_8 Ba_2][Ba_4]$
$x=0$

$[R_{9.33}V_{2/3}][Ba_4]$
$x=2/3$

$[R_{10}][Ba_3V]$
$x=1$

Figure V. $Q \cdot f$ varied non-linearly as a function of composition and relationship with site occupancy of Ba.

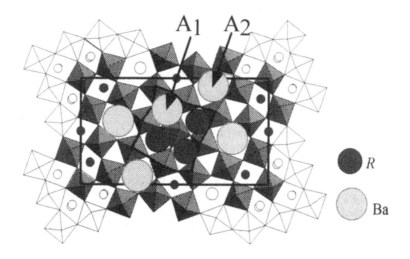

Figure VI. Ba ions located in perovskite blocks, which produce internal strain leading less $Q \cdot f$ values. Ba and R ions with large circle are represented by effective ionic radii according to Shanonn.[27]

Dielectric Ceramic Materials

stress due to the large cation distribution. To see the generation mechanism of internal stress, R and Ba ions in Figure VI are drawn just by ionic radius. The Ba ion entered in the cuboctahedron (A1-site) in the perovskite blocks, which expands this site. Usually, most of the cuboctahedra are occupied by R ions with smaller sizes. As the cuboctahedron containing a Ba ion becomes large in volume in comparison with other cuboctahedra with a R ion, internal stress might be generated around the Ba ion in perovskite blocks. The internal stress might bring lower $Q \cdot f$ values as seen in Figure V. Finally, at x=2/3, the amount of Ba ions in A1-sites become zero, that is when Ba and R ions occupy separately A1 and A2 sites, respectively. The large cation distribution occupied with same size ions might cause no internal strain and shows high $Q \cdot f$ values, which are the same as in the case of ordering of cations in the complex perovskite. Moreover, the vacancies generated by substituting $2R$ for 3Ba might be the second reason for the high $Q \cdot f$ values in the region for x<2/3. This structure is maintained stable with different sizes of large ions such as Ba and R, in order to reduce internal strain. Differences of the ionic radius between Ba and R ions also lead to less internal strain. The compounds containing Sm ions with the largest difference in radius against Ba have high $Q \cdot f$ values compared with La ions with the smallest difference.

CONCLUSION

The microwave dielectric properties on the tungstenbronze-type $Ba_{6-3x}R_{8+2x}Ti_{18}O_{54}$ (R = rare earth) solid solutions with 2 by 2 perovskite blocks are studied as a function of composition. The dielectric constant ε_r and the temperature factor τ_f of resonant frequency changed linearly as a function of cell volume. But the quality factor $Q \cdot f$ changed non-linearly such as steep increasing in the vicinity of composition x=0.5. The $Q \cdot f$ depends on the distribution of the cations with large ionic radius. When the rare earth ions with middle radius and Ba ions with large size are sitting separately in the perovskite blocks and in the pentagonal columns, respectively, the $Q \cdot f$ reaches a high value such as 10000GHz. The vacancies generated by the substitution of $2R$ for 3Ba also contribute to such a high value. The occupation of Ba ions in the perovskite blocks might generate the internal stress which will increase the dielectric loss and the vacancies generated might reduce the internal stress.

ACKNOWLEDGEMENTS

The authors thank President Akio Harada in Daiken Chemical Co., Ltd. and Dr. H. Tamura in Murata Manufacturing Co., Ltd. for their assistance with the measurement of the dielectric properties and Mr. R. D. Saxena and Dr. W. Wunderlich for discussion of technical expressions. The work was supported in part by the Japan Sheet Glass Foundation for Materials Science and Engineering.

REFERENCES

1) S. Nishigaki, "Microwave Dielectric," FC Rep. 5, 413-422 (1987). Translation, FC Annual

Report for Overseas Readers 32-41(1988).

2) S. Nishigaki, "My Research and Development in Microwave Dielectrics". New Ceramics **No.9**, 25-36 (1996).

3) D. Kolar, Z. Stadler, S. Gaberscek and D. Suvorov, "Ceramic and Dielectric Properties of Selected Compositions in the $BaO\text{-}TiO_2\text{-}Nd_2O_3$ system," *Ber. Dt. Keram. Ges.* **55**, 346-347 (1978).

4) R. L. Bolton, "Temperature Compensating Ceramic Capacitors in the System Baria-Rare-Earth-Oxide Titania," 1968, Ph. D. Thesis, Ceramic Engineering, University of Illinois, Urbana, Illinois, (University Microfilms International, A Bell & Howell Information Company).

5) K. Wakino, K. Minai and H. Tamura, "Microwave Characteristics of $(ZaSn)TiO_4$ and BaO-PbO-Nd_2O_3-TiO_2 Dielectric Resonators," *J. Am. Ceram. Soc.* **67**, 278-281 (1984).

6) S. Kawashima, M. Nishida, I. Ueda and H. Ouchi, "Dielectric Properties at Microwave Frequencies of the Ceramics in $BaOSm_2O_3TiO_2$ System," presented *at the 87th Annual Meeting, American Ceramic Society*, Cincinnati, OH, May 6, 1985 (Electronics Division Paper No.15-E-85).

7) S. Nishigaki, H. Kato, S. Yano and R. Kamimura, "Microwave Dielectric Properties of $(Ba,Sr)O\text{-}Sm_2O_3\text{-}TiO_2$ Ceramics," *Am. Ceram. Soc. Bull.* **66**, 1405-10 (1987).

8) D. Kolar, S. Gaberscek and B. Volavsek, "Synthesis and Crystal Chemistry of $BaNdTi_3O_{10}$, $BaNd_2Ti_5O_{14}$, and $Nd_4Ti_5O_{24}$," *J. Solid State Chem.* **38**,158-164 (1981).

9) M. B. Varfolomeev, A. S. Mironov, V. S. Kostomarov, L. A. Golubtova, and T. A. Zolotova, "The Synthesis and Homogeneity Ranges of the Phases $Ba_{6-x}Ln_{8+2x/3}Ti_{18}O_{54}$," *Zh. Neorg. Khim.* **33**, 1070-1071 (1988), *Translation, Russ. J. Inorg. Chem.* **33**, 607-609 (1988).

10) H. Ohsato, T. Ohhashi, S. Nishigaki, T. Okuda, K. Sumiya and S. Suzuki, "Formation of Solid Solution of New Tungsten Bronze-Type Microwave Dielectric Compounds Ba_{6-3x} $R_{8+2x}Ti_{18}O_{54}(R=Nd$ and Sm, $0<x<1)$," *Jpn. J. Appl. Phys.* **32** 4323-4326 (1993).

11) H. Ohsato, T. Ohhashi, K. Sumiya, S. Suzuki and T. Okuda, "Lattice Parameters of Bronze-Type $3BaO$ $2R_2O_39TiO_2(R=Sm$ and Nd) Solid Solutions for Microwave Dielectric Ceramics," *Advances in X-ray Analysis*, **37**, 79-85 (1994).

12) E. S. Razgon, A. M. Gens, M. B. Varfolomeev, S. S. Korovin and V. S. Kostomarov, "The Complex Barium and Lanthanum Titanates," *Zh. Neorg. Khim.* **25**, 1701-1703 (1980).

Translation, Russ. J. Inorg. Chem. **25**, 945-947 (1980).

13) E.S. Razgon, A. M. Gens, M. B. Varfolomeev, S. S. Korovin and V. S. Kostomarov, "Some Barium Lanthanide," *Zh. Neorg. Khim.* **25**, 2298-2300 (1980). *Translation, Russ. J. Inorg. Chem.* **25**, 1274-1275 (1980).

14) A. M. Gens, , M. B. Vafolomeev, V. S. Kostomarov and S. S. Korovin, "Crystal-chemical and Electrophysical Properties of Complex Titanates of Barium and the lanthanides," *Zh. Neorg. Khim.* **26**, 896-898 (1981). *Translation, Russ. J. Inorg. Chem.* **26**, 482-484 (1981).

15) J. Takahashi, T. Ikegami and K. Kageyama, "Occurrence of Dielectric 1:1:4 Compound in the Ternary System $BaO-Ln_2O_3-TiO_2$ (Ln=La, Nd and Sm): **I.** An Improved Coprecipitation Method for Preparing a Single-Phase Powder of Ternary Compound in the $BaO-La_2O_3-TiO_2$ System," *J. Am. Ceram. Soc.* **74**, 1868-1872 (1991).

16) J. Takahashi, T. Ikegami and K. Kageyama, "Occurrence of Dielectric 1:1:4 Compound in the Ternary System $BaO-LnO_3-TiO_2$ (Ln=La, Nd and Sm): II. Reexamination of Formation of Isostructural Ternary Compounds in Identical systems," *J. Am. Ceram. Soc.* **74**, 1873-1879 (1991).

17) R.G. Matveeva, M. B. Varforomeev and L. S. Il'yuschenko, "Refinement of the Composition and Crystal Structure of $Ba_{3.75}Pr_{9.5}Ti_{18}O_{54}$," *Zh. Neorg. Khim.* **29**, 31-34 (1984). *Translation, Russ. J. Inorg. Chem.* **29**, 17-19 (1984).

18) H. Ohsato, T. Ohhashi, H. Kato, S. Nishigaki and T. Okuda, "Microwave Dielectric Properties and Structure of the $Ba_{6-3x}Sm_{8+2x}Ti_{18}O_{54}$ Solid Solutions," *Jpn. J. Appl. Phys.* **34**, 187-191 (1995).

19) K. Fukuda and R. Kitoh, "Microwave Characteristics of $BaPr_2Ti_4O_{12}$ and $BaPr_2Ti_5O_{14}$ Ceramics," *J. Mater. Res.* **10**, 312(1995).

20) H. Ohsato, M. Mizuta, T. Ikoma, Z. Onogi, S.Nishigaki and T. Okuda, "Microwave Dielectric Properties of Tungsten Bronze-type $Ba_{6-3x}R_{8+2x}Ti_{18}O_{54}$ (R = La, Pr, Nd and Sm) Solid Solutions," *J. Ceram. Soc. Japan*, **106(2)**, 178-182 (1998).

21) M. Imaeda, M. Mizuta, K. Ito, H. Ohsato, S. Nishigaki and T. Okuda, "Microwave Dielectric Properties of $Ba_{6-3x}Sm_{8+2x}Ti_{18}O_{54}$ Solid Solutions Substituted Sr for Ba," *Jpn. J. Appl. Phys.*, **36, 9B**, 6012-6015 (1997).

22) B.W. Hakki and P. D. Coleman, "A Dielectric Resonator Method of Measuring Inductive in the Millimeter Range," *IRE Trans. Microwave Theory & Tech.* **MTT-8** 402-410 (1960).

23) H. Ohsato, S. Nishigaki and T. Okuda, "Superlattice and Dielectric Properties of Dielectric Compounds," *Jpn. J. Appl. Phys.* **31**, 3136-3138 (1992).

24) H. Ohsato, T. Ohhashi and T. Okuda, "Structure of $Ba_{6-3x}Sm_{8+2x}Ti_{18}O_{54}$ (0<x<1)," *Ext. Abstr. AsCA '92 Conf., Singapore, November*, **14U-50** (1992).

25) H. Ohsato, H. Kato, M. Mizuta and T. Okuda, "Superstructure of $Ba_{6-3x}Sm_{8+2x}Ti_{18}O_{54}$ (x=0.71)," *Ext. Abstr. AsCA '95 Conf., Thailand, November*, **1P40** (1995).

26) H. Ohsato, H. Kato, M. Mizuta, S. Nishigaki and T. Okuda, "Microwave Dielectric Properties of the $Ba_{6-3x}(Sm_{1-y}R_y)_{8+2x}Ti_{18}O_{54}$ (R=Nd and La) Solid Solutions with Zero Temperature Coefficient of the Resonant Frequency," *Jpn. J. Appl. Phys.* **34**, 5413-17 (1995).

27) R. D. Shanonn, "Revised Effective Ionic Radii and Systematic Studies of Interatomic Distances in Halides and Chalcogenides," *Acta Cryst.*, **A32**, 751-67 (1976).

DIELECTRIC PROPERTIES OF BaTiO$_3$-BASED CERAMICS WITH GRADIENT COMPOSITIONS

Toshitaka Ota, Masaki Tani, Yasuo Hikichi, Hidero Unuma and Minoru Takahashi
Nagoya Institute of Technology, CRL
Tajimi 507, Japan
and
Hisao Suzuki
Shizuoka University
Hamamatsu 432, Japan

ABSTRACT

Gradient Ba$_{1-x}$Sr$_x$TiO$_3$ (BST) ceramics, which had a continuously varying composition from one surface to the other in a single body, were examined with respect to dielectric property. They were prepared by sintering piles of layered green compacts of BST solid solutions with different values of x. The electric permittivity (ε) of single-phase BST ceramics showed a peak at each Tc, which shifted to lower temperatures with increasing x. On the other hand, a profile of ε vs temperature (T) of gradient BST ceramics gave a linear characteristic.

INTRODUCTION

BaTiO$_3$ (BT) is characterized by a high electric permittivity which assumes maximum at a Curie temperature (Tc) of 120° C. However, a sharp peak in the profile of ε vs T at Tc is undesirable for application as a capacitor material. Practically, as shown in Fig. 1 (a) and (b), the peak is shifted to lower temperatures by additives such as SrTiO$_3$ (ST) or BaZrO$_3$, and depressed by additives such as MgTiO$_3$ or CaTiO$_3$. Besides, the peak is also depressed and broadened by mixing two phases with different Tc. Figure 1 (c) shows the profiles of ε vs T of single-phase materials and a two-phase mixture calculated by the so-called logarithmic mixture rule,

$$\log \varepsilon = v_1 \log \varepsilon_1 + v_2 \log \varepsilon_2 \tag{1},$$

where v_1 and v_2 are the volume fraction of each phase. The two-phase mixture gives a lower variation of permittivity with temperature than single-phase materials. This is due to the paraelectric/ferroelectric phase transition occuring at different temperatures for different regions of the sample. Consequently, a distribution of different transition temperatures in the polyphase mixture should yield an overall broadened peak in the profile of ε vs T.

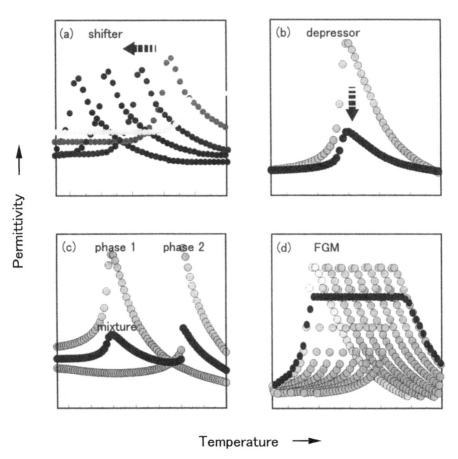

Fig. 1   Control of temperature characteristics for capacitor materials

Dielectric Ceramic Materials

Mixtures of ideal dielectrics can be most simply considered on the basis of layer materials with the layers either horizontal or vertical to the applied field (Fig. 2).[1] When layers are horizontal to the electrodes, the structure corresponds to capacitive elements in series, and the inverse capacitances are additive:

$$1 / \varepsilon = v_1 / \varepsilon_1 + v_2 / \varepsilon_2 \qquad (2),$$

where v_1 and v_2 equal to the relative thicknesses of each layer. In contrast, when layers are arranged vertical to the electrodes, the structure corresponds to capacitive elements in parallel, and the capacitances are additive:

$$\varepsilon = v_1 \varepsilon_1 + v_2 \varepsilon_2 \qquad (3).$$

As shown in Fig. 2, the logarithmic mixture gives a value intermediate between the extremes illustrated in Eq. (2) and (3). The highest permittivity is given by the parallel arrangement.

In this study, the polyphase mixtures with the layered structure were prepared by laminating BST ceramics with different compositions. That is to say, gradient BST ceramics, which had a continuously varying composition from one surface to the other in a single body, were prepared. It is expected that the gradient BST ceramics would exhibit a linear characteristics of permittivity with temperature as shown in Fig. 1 (d). However, functionally gradient materials (FGM) have hardly been investigated with respect to dielectric property.[2-6]

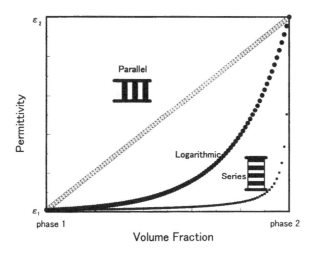

Fig. 2 Permittivity plotted as a function of composition for series and parallel mixing

EXPERIMENTAL PROCEDURE

Commercial $BaTiO_3$ and $SrTiO_3$ powders (KYORIX BT-HP3 and ST-HP1) were used as starting materials. They were mixed with appropriate ratios in ethanol. Further, polyvinyl-alcohol binder solution of 4 wt% was added to the starting powder mixtures. Single-phase BST samples were formed as disks of 12 mm in diameter and 1 to 2 mm thick. Gradient BST samples were formed into cylindrical shape of 12 mm in diameter and 5 to 15 mm thick by uniaxially compressing piles of green compacts containing 2 to 26 layers at 30 MPa for 5 min. Each layer was pre-compressed at 15 MPa for 3 min. The minimal thickness of each layer was about 0.5 mm. After the binder was burned out (heat treatment at $450°C$ for 1 h), the cylindrical samples were hydrostatically pressed at 100 MPa for 5 min and then sintered at $1250°$ to $1350°C$ for 2 h in O_2 gas. After sintering, the cylindrical samples were cut into rectangular shapes, $8 \times 8 \times 5$-15 mm, polished and electroded by painting silver paste.

The electric permittivity (ε) was measured at 1 kHz with LCR meter (HP-4248A) from room temperature to $180°C$. XRD analysis was used to determine the phases present. Microstructures were observed and analyzed by SEM-EDS.

RESULTS AND DISCUSSION
(1) Preparation of single-phase BST ceramics and their dielectric properties

Single-phase BST ceramics ($Ba_xSr_{1-x}TiO_3$) for x in the range 0 to 0.5 were prepared by sintering at $1250°C$ to $1350°C$. Every single-phase BST ceramic obtained was not a mixture of ST and BT but a single phase BST solid solution. Relative density of about 95% was achieved by sintering both at $1300°C$ and $1350°C$ for 2 h. It slightly decreased with increasing x. Higher sintering temperatures resulted in grain growth. Further, the grains became exaggeratedly larger as the BT content was increased. For example, single-phase BST ceramics for x in the range of 0.2 to 0.5, sintered at $1300°C$, were made up of only fine grains of 1 to 2 μ m, while samples for x in the range of 0 to 0.1 had a duplex structure of fine grains of 1 to 2 μ m and large grains of 10 to 30 μ m.

Figure 3 shows ε vs T measured for single-phase BST ceramics with different compositions sintered at $1300°C$ and $1350°C$. The profiles of ε vs T exhibited peaks around Tc. The peak shifted to lower temperatures with increasing x. Since ST exhibited much weaker changes of permittivity with temperature than BT, the width of peaks increased and the height of peaks decreased with increasing x. The maximum values of ε at Tc increased from 4000-5000 to 6000-8000 with increasing sintering temperature. This may be attributed to the effect of grain size.

Dielectric Ceramic Materials

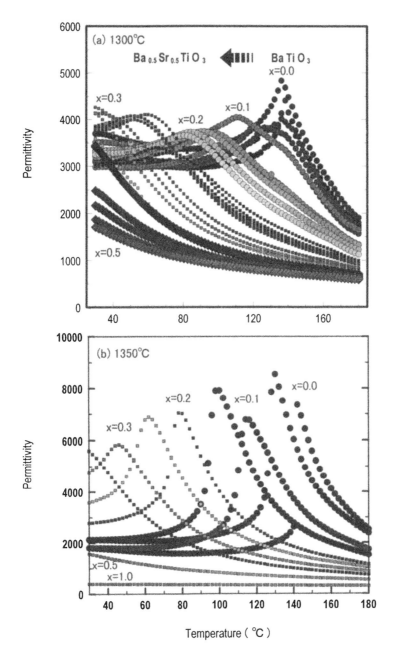

Fig. 3 ε vs T measured for single-phase BST ceramics

(2) Preparation of gradient BST ceramics and their dielectric property

Figure 4 illustrates ε vs T estimated for gradient BST ceramics, for x in the range 0 to 0.5, calculated by equations (2) and (3) from ε values in Fig. 3. The profile of ε vs T calculated by eq. (2) for the series arrangement was flatter but lower than that calculated by eq. (3) for the parallel arrangement. For the parallel arrangement, it is obvious that the profile of ε vs T levelled out with increasing numbers of layers. It is expected that the gradient BST ceramics would exhibit a linear profile when a pile of ten layers or more was sintered at 1300° C. On the other hand, gradient BST ceramics sintered at 1350° C would exhibit an irregular profile even at twenty layers or more.

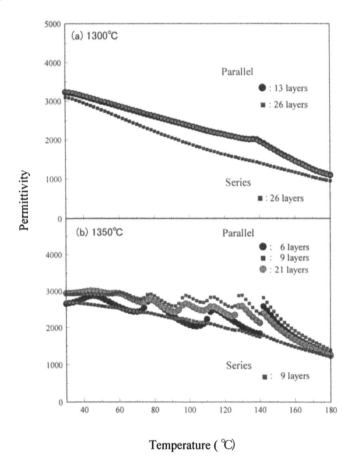

Temperature (℃)

Fig. 4 ε vs T predicted for gradient BST ceramics

Dielectric Ceramic Materials

In this experimental study, the maximum number of layers which it was possible to compress was 26 layers, each about 0.5 mm thick. Figure 5 (a) shows a gradient BST ceramic prepared by sintering a pile of 26 layers at $1300°$ C. The layers bonded tightly together. There was no split at the interfaces between layers. Figure 5 (b) shows SEM-EDS X-ray microanalysis of the gradient BST ceramic. The distribution of Sr increased from one surface (x=0) to the other (x=0.5) with a slope. Figure 6 compares the XRD patterns of cross-sections shaved from the surface of the gradient BST ceramic. The XRD peaks shifted from $BaTiO_3$ to $Ba_{0.5}Sr_{0.5}TiO_3$ with the distance from surfaces. They demonstrate that each cross-section was not a mixture of BT and ST, but a single phase of BST solid solutions, and that the composition continuously varied with position from one surface (x=0) to the other (x=0.5). In addition, Fig. 7 shows SEM photographs of cross-sections of the gradient BST ceramic. Similarly, the microstructures in single-phase BST ceramics, large grains were observed in the Ba-rich layers.

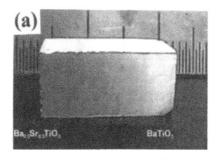

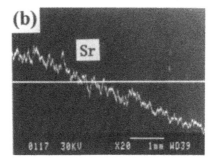

Fig. 5 (a) Vertical section of a gradient BST ceramic and
(b) the distribution of Sr determined by X-ray microanalysis

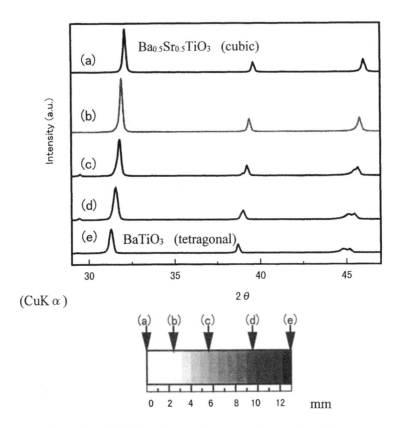

Fig. 6 Variation of XRD patterns of cross section with position in a gradient BST ceramic

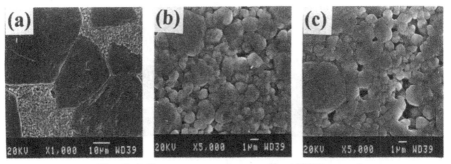

Fig. 7 Microstructures in a gradient BST ceramic:
(a) BT side; (b) center; (c) Sr-rich side

Figure 8 shows ε vs T measured for gradient BST ceramics for x in the range 0 to 0.5 sintered at 1300° C and 1350° C. The variation of ε vs T is in good agreement with the predicted profiles presented in Fig. 4. Figure 8 (b) demonstrates that the profile for the series arrangement was flatter but lower than that for the parallel arrangement, and that the profile for the parallel arrangement became more linear with increasing the number of layers. Figure 8 (a) demonstrates that the gradient BST ceramic prepared by sintering 26 layers at 1300° C had a complete linear characteristic of permittivity with temperature, and that the profile perfectly fitted in with the calculated profile in Fig. 4. However, the profile sloped, because the profiles for the single-phase BST ceramics corresponding to the individual layers had different shapes. It may be effectively modified by an appropriate change of thickness of the layers, or by control of grain size.

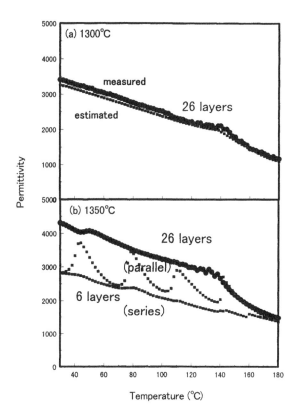

Fig. 8   ε vs T measured for gradient BST ceramics

CONCLUSION

Gradient BST ceramics, which had a continuously varying composition from one surface to the other in a single body, were prepared by sintering piles of layered green compacts of BST solid solutions with different compositions. The gradient BST ceramics obtained gave a linear characteristic in profile of ε vs T in agreement with the profile derived from the calculations for the polyphase mixtures with the layered structure. It is concluded that this materials would exhibit substantially lower variations of permittivity with temperature, despite higher values of ε, than capacitor materials prepared by conventional methods.

REFERENCES

[1] W. D. Kingery, H. K. Bowen and D. R. Uhlmann, "Dielectric Properties"; pp. 913-974 in *Introduction to Ceramics*, 2nd ed. John Wiley & Sons, New York, 1976.

[2] M. Sasaki and T. Hirai, "Fabrication and Properties of Functional Gradient Materials", *Journal of the Ceramic Society of Japan*, **99** [10] 1002-13 (1991).

[3] I. Yamai and T. Ota, "Thermal Expansion of Homogeneous and Gradient Solid Solutions in the System $KZr_2(PO_4)_3$-$KTi_2(PO_4)_3$", *Journal of the American Ceramic Society*, **75** [8] 2276-82 (1992).

[4] T. Kawai and S. Miyazaki, "Development of a Piezo-Ceramic Actuator with Functionally Gradient Material", *Journal of the Ceramic Society of Japan*, **98** [8] 900-04 (1990).

[5] C. C. M. Wu, M. Kahn and W. Moy, "Piezoelectric Ceramics with Functional Gradients: A New Application in Material Design", *Journal of the American Ceramic Society*, **79** [3] 809-12 (1996).

[6] T. Ota, T. Fukaya, N. Ikeda, Y. Hikichi, H. Unuma, M. Takahashi and H. Suzuki, "Preparation and Electrical Properties of PZT Ceramics with Gradient Compositions", *Journal of the Ceramic Society of Japan*, **106** [1] 119-23 (1998).

THE CHARACTERISTICS OF ZINC OXIDE VARISTOR WITH DIFFERENT ADDITIVES

Hyung Sik Kim*, Se won Han, In Sung Kim and Han Goo Cho
Korea Electrotechnology Research Institute
28-1 Sungju Dong, Changwon, Kyungnam, Korea 641-120

ABSTRACT

ZnO based varistor is generally used to protect electrical apparatus from harmful surge attack. Besides Bi_2O_3 many kinds of additives like Sb_2O_3, CoO, NiO, MnO_2 and Cr_2O_3 are mixed together with ZnO to produce highly reliable and surge-absorbing varistor. First of all the influence of Al-addition on the characteristics of ZnO varistor was examined. The average size of ZnO grains was decreased from about 7 μm to 2.4 μm with increase of Al-contents, and leakage current and non-linearity of samples changed. Also with the variation of Si-contents the electrical stability of ZnO varistor was investigated.

INTRODUCTION

Since the discovery of ZnO varistor in the end of 1960s many kinds of additives, Bi_2O_3, MnO, CoO, Sb_2O_5, Cr_2O_3, and so on, are introduced to improve the characteristics of ZnO varistor[1]. ZnO has normally the electrical property as a n-type semiconductor with a cation excess. The electrical barriers which are located between, or near, the grain boundaries of ZnO grains are known as double Schottky barriers. Until now many researchers have reported that the predominant point defects, like ionized zinc interstitial atoms, are responsible for the bulk electrical behaviour of zinc oxide varistor[2,3]. Besides those donor ions such as Al-, Ga- and In- could indeed delay the onset of upturn-voltage to higher current density. Other studies on the influence of oxygen have proved that the nonlinearity of ZnO varistor was reduced to some extent. Also the different types of oxygen compounds and the heat-treatment atmosphere would change the electrical behaviours of the varistor. The effects of different additives are not unique but complex. In order to withstand a high current surge and to have a long life time various oxide additives are comprised in zinc oxide varitstor. But until now the characteristics of zinc oxide varistor are not studied satisfactory, although the basic of experimental principal to produce varistor is generally acceptable and this type of product is sold commercially all over the world.

In this study, first of all, the influence of Al addition to ZnO varistor was investigated among many kinds of oxides. Moreover some addition of silica to zinc oxide varistor was studied in terms of the change of induced current-voltage characteristics under different conditions.

EXPERIMENTAL PROCEDURE

The ZnO based varistor specimens were fabricated through conventional mixed oxide ceramic processing. Raw materials were prepared with different additives, Bi_2O_3, Sb_2O_5, and so on, in reagent grade. The batch of raw materials was mixed in ethyl alcohol through wet ball milling. Because of small amount of solution the mixture was dried carefully at 380 K instead of spray drying. Then the so prepared powders were pressed into small disc with a diameter of 20 mm. The green samples of cylinder form were sintered at the temperature of 1473 K in a electrical muffle furnace. Both of the surfaces of sintered sample were polished well enough to be decked with silver paste as a electrode. The I-V curves were measured most of all by using a voltage-current meter(Keithly 237). And a pulse power generator(8 x 20 μs) was used to test at higher current(over 10 mA). The microstructure and crystal structure of sintered specimens were investigated using a scanning electron microscope(SEM, Hitachi) and a X-ray diffractormeter (Philips).

RESULTS AND DISCUSSION

- Microstructure

The microstructure of sintered ZnO samples was composed of ZnO grains under 10 μm in size typically surrounded by thin intergranular Bi_2O_3 layers and other secondary phases of ZnO grains. With addition of Al- or Si-contents the grain size in samples changed. The grain growth was hindered first of all by forming of spinel- or pyrochlore-phase with addition of Al-compound. The more Al-content was added to a sample, the smaller grain size was in a sintered specimen(fig. 1). But such kind of tendency was not found remarkably in this study with addition of Si-contents(fig. 3). The higher the amount of Si-addition was, the finer the grain size of the secondary phases were found among normal

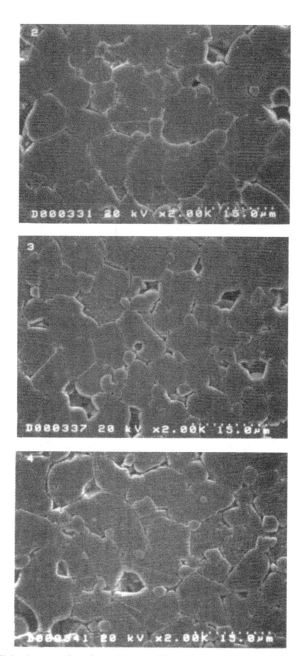

Fig. 1. SEM micrographs of specimens with various Al-contents:
2(0.005 %), 3(0.02 %), 4(0.1 %)

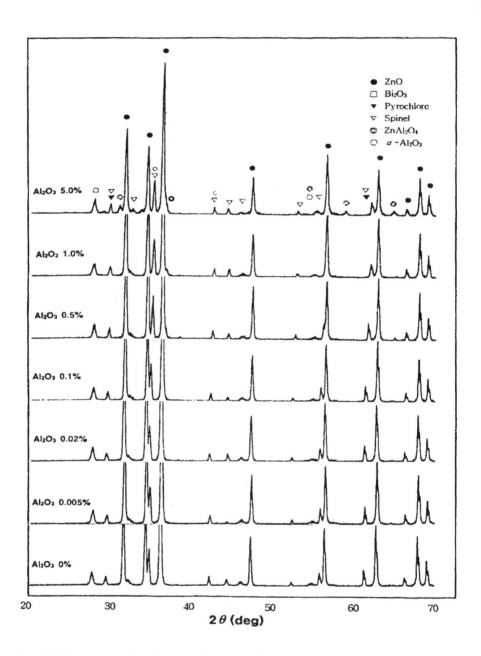

Fig. 2. XRD patterns of specimens with various Al-contents

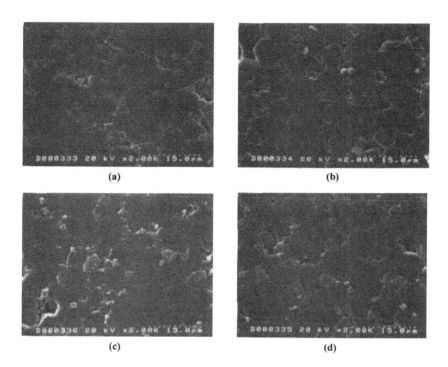

Fig. 3. SEM micrographs of ZnO samples with various Si-contents: a(G3) b(GS1) c(GS3) d(GS5)

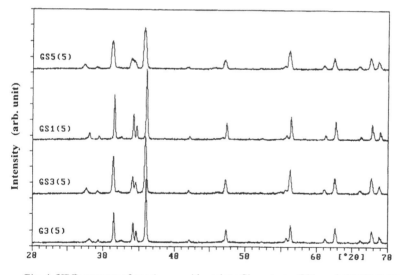

Fig. 4. XRD patterns of specimens with various Si-contents: G3(none) GS1(0.1) GS3(0.3) GS5(0.5)

ZnO grains. In general these grains were composed of zinc oxide based spinel, pyrochlore, glass phase and so on.

The crystal structure of sintered samples examined by X-ray diffraction method showed some differences from variation of chemical compositions through small quantity of minor additives. The intensity of main peaks changed to some extent in both cases of Al- and Si-addition(fig.2, 4). By the forming of secondary phases like Bi-rich glassy phase the result showed a little increased base line under $30(=2\theta)$. Apparently a little shift and overlapping of peaks were registered. Also the intensity of peaks changed according to a variation of composed crystal structure. Besides main peaks of ZnO the peaks of $ZnAl_2O_4$ phase was detected clearly in the case of Al-addition higher than 0.5 wt %.

Table I. Electrical test results on the ZnO samples with various Al-contents

Al_2O_3, wt %	0	0.005	0.013	0.02	0.1	0.5	1.0	5.0
Leakage current, $\mu A/cm$	30	20	67	80	210	300	240	1040
Nonlinear coefficient	52	55	48	46	30	30	35	17
Breakdown voltage, V/cm	2907	2701	3076	3044	2811	2890	3536	4035

- I-V characteristics

The results in table I showed that the electrical properties, leakage current, nonlinear coefficient and breakdown voltage, were dependent on Al-contents in ZnO varistor. Nonlinear coefficient was decreased proportional to the Al-contents over 0.005 %, while the leakage current increased proportional to that. The breakdown voltage changed not so regularly under 0.1 %. But it was increased proportionally over 0.1 %. These phenomena were influenced not only by the effect of grain size but also by the doner- or acceptor-effect of Al-ion.

With addition of Si-contents the leakage current of varistor was decreased to some extent and the nonlinear coefficient increased as indicated(fig. 5, 6, 7, 8). The effect of Si-addition on I-V curves was not so impressive recorded at room temperature, for the diameter of ZnO grains showed little change with Si-addition under 0.5 mol %. But the

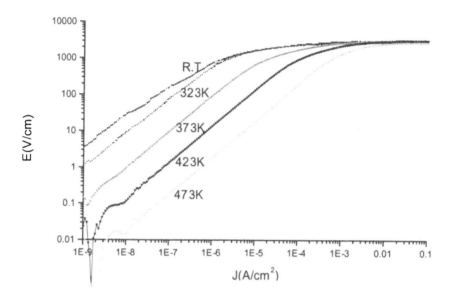

Fig. 5. I-V curves of specimen G3 at different test temperatures

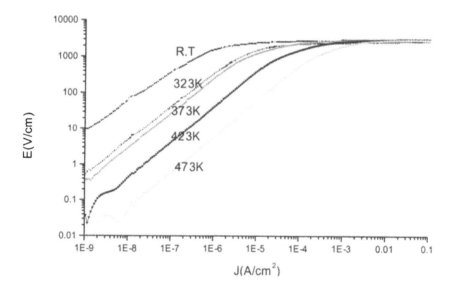

Fig. 6. I-V curves of specimen GS1 at different test temperatures

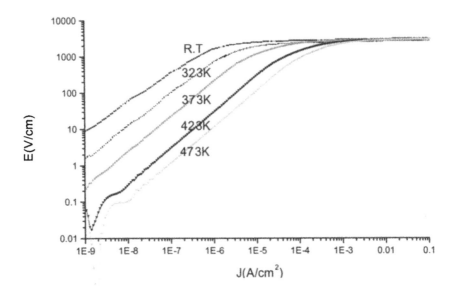

Fig. 7. I-V curves of specimen GS3 at different test temperatures

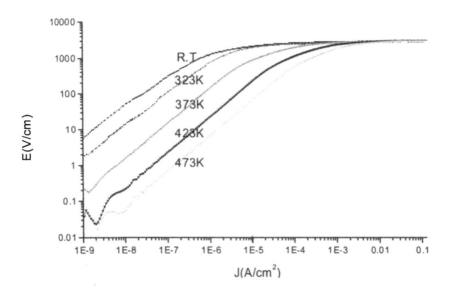

Fig. 8. I-V curves of specimen GS5 at different test temperatures

Dielectric Ceramic Materials

result of I-V characteristics showed that the current density was increased a little differently with various addition of Si-contents at higher testing temperatures. Also the time dependent resistive current were recored absolutely different with variation of additives.

The leakage current of all samples was increased proportional to the test temperature in the prebreakdown range. But in the range of current density over about 0.5 mA/cm2 the I-V curves indicated little difference with various test temperatures. As a result of this experiment the stability of ZnO varistor under electrical field was not proportional to the Si-content in samples. In the beginning of accelerating test at 388 K the resistive current of all samples increased rapidly(fig. 9). About after 100 hrs it stayed almost constant.

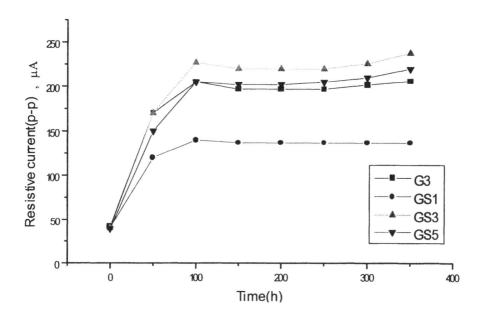

Fig. 9. Change in leakage current of various samples with time at 388K

But leakage currents were recorded differently with various Si-contents. The sample GS1(0.1 mol % Si-content) indicated the lowest value. This effect was strongly dependent on the microstructure of samples originated from the addition of Si.

The breakdown voltage was dependent on mainly ZnO grain size, because the overall varistor structure could be described as a network of semiconducting ZnO grains and insulating intergranular barriers. These barriers were thought to be responsible for the nonlinear behavior of a varistor.

CONCLUSIONS

With Al-addition the leakage current of a varistor was increased because of doping effect of Al-content, while the nonlinear coefficient decreased.

The leakage current of varistor was reduced to some extent by Si-addition to samples, but the thermal stability was mainly dependent on the microstructure of varistors. By forming of the secondary phases in small size the electrical property of varistor was influenced negatively.

More work is planned in the future:
- effect of additives on the high power surge absorbility
- relationship between microstructure and thermal stability.

REFERENCES

[1]K. Eda, "Zinc Oxide Varistors", IEEE Electrical Insulation Mag., 5(6) 28 (1989)

[2]T.K. Gupta, "Application of Zinc Oxide Varistors", J. Am. Ceram. Soc., 73(7) 1817-40(1990)

[3]A.Laglange, "Present and Future of Zinc Oxide Varistors", in Electronic Ceramics edited by B. C. H. Steele, Elsevier Science Publisher (1991)

LOW-LOSS DIELECTRIC CERAMICS WITH TUNGSTEN-BRONZE STRUCTURE IN BaO-Nd$_2$O$_3$-TiO$_2$-Ta$_2$O$_5$ SYSTEM

X.M. Chen*, J.S. Yang and J. Wang
Department of Materials Science and Engineering, Zhejiang University, Hangzhou 310027, CHINA

ABSTRACT

In order to search new microwave dielectric ceramic system, ceramics with tungsten-bronze structure were prepared and characterized in BaO-Nd$_2$O$_3$-TiO$_2$-Ta$_2$O$_5$ system. There were five possible tungsten-bronze compounds in the present system, BaNd$_5$Ti$_7$Ta$_3$O$_{30}$, Ba$_2$Nd$_4$Ti$_6$Ta$_4$O$_{30}$, Ba$_3$Nd$_3$Ti$_5$Ta$_5$O$_{30}$, Ba$_4$Nd$_2$Ti$_4$Ta$_6$O$_{30}$ and Ba$_5$NdTi$_3$Ta$_7$O$_{30}$. When the densification was performed well, almost frequency-independent high dielectric constant (70-160) and low dielectric loss (on the order of 10^{-4} at 1MHz) were determined in all compositions mentioned above. The dielectric constant increased with increasing BaO and Ta$_2$O$_5$ concentration, and Ba$_5$NdTi$_3$Ta$_7$O$_{30}$ indicated the highest dielectric constant (>160). Modification of temperature coefficient should be the primary issue when the microwave applications of the present ceramics were considered.

INTRODUCTION

With the rapid development of wireless communication, high-ε microwave dielectric ceramics are attracting more and more scientific and commercial interests. However, the dielectric constant is usually less than 100 among the microwave dielectric ceramics developed[1-6], and only a few ceramics with dielectric constant greater than 100, e.g. ceramics of (PbCa)(MeNb)O$_3$[6], have been investigated. Therefore, it is a challenge issue to search microwave dielectric ceramics with dielectric constant greater than 110 combined with low dielectric loss and near-zero temperature coefficient of resonant frequency.

The microwave dielectric ceramics developed generally belong to perovskite, complex perovskite or perovskite-like structures, and they must indicate the paraelectric nature. On the other hand, although most of the tungsten-bronze compounds are usually known as ferroelectric, pyroelectric, piezoelectric and electrooptic materials[8], there are also a number of tungsten-bronze compounds with high-ε and paraelectric nature[9]. So, the question is, whether some high-ε microwave dielectric ceramic candidates can be developed from the tungsten-bronze family.

In the present work, a new ceramic system, BaO-Nd$_2$O$_3$-TiO$_2$-Ta$_2$O$_5$, has been proposed as the promising candidates for high-ε microwave dielectric ceramics, where the tungsten-bronze structure is generally formed. The preparation and dielectric characterization are performed for the compositions of BaNd$_5$Ti$_7$Ta$_3$O$_{30}$, Ba$_2$Nd$_4$Ti$_6$Ta$_4$O$_{30}$, Ba$_3$Nd$_3$Ti$_5$Ta$_5$O$_{30}$, Ba$_4$Nd$_2$Ti$_4$Ta$_6$O$_{30}$ and Ba$_5$NdTi$_3$Ta$_7$O$_{30}$, especially for the latter three compositions.

EXPERIMENT PROCEDURE

High-purity powders of $BaCO_3$ (>99.95%), Nd_2O_3 (>99.95%), TiO_2 (>99.7%), and Ta_2O_5 (>99.99%) were adopted as the starting materials, and the ceramic powders with five typical compositions mentioned above were synthesized respectively by calcination under the conditions given in Table I, after mixing by ball-milling with zirconia media in ethanol for 24h. The disc compacts with dimensions of 12mm in diameter and 2 to 4mm in height were created by pressing under 98MPa from such powders, and they were sintered at 1310 to 1450°C in air for 3h for preparing dense ceramics.

Table I. Calcination conditions for ceramics with various compositions

Composition	Calcination conditions
$BaNd_5Ti_7Ta_3O_{30}$	1300°Cx3h in air
$Ba_2Nd_4Ti_6Ta_4O_{30}$	1300°Cx3h in air
$Ba_3Nd_3Ti_5Ta_5O_{30}$	1280°Cx3h in air
$Ba_4Nd_2Ti_4Ta_6O_{30}$	1300°Cx3h in air
$Ba_5NdTi_3Ta_7O_{30}$	1300°Cx3h in air

X-ray diffraction (XRD) analysis using Ni-filtered Cuα radiation was carried out for structure characterization. The dielectric properties at room temperature were determined by an LCR meter (HP4284A) at 1kHz, 10kHz, 100kHz, 500kHz and 1MHz, respectively, and the temperature dependence of dielectric constant was evaluated at 10kHz by another LCR meter (WK4210) equipped with a thermalstat.

RESULTS AND DISCUSSION

In the system of BaO-Nd_2O_3-TiO_2-Ta_2O_5, there are five possible tungsten-bronze structure compounds, $BaNd_5Ti_7Ta_3O_{30}$, $Ba_2Nd_4Ti_6Ta_4O_{30}$, $Ba_3Nd_3Ti_5Ta_5O_{30}$, $Ba_4Nd_2Ti_4Ta_6O_{30}$ and $Ba_5NdTi_3Ta_7O_{30}$, where ten oxygen-octahedra containing Ti and Ta ions share corner atoms, and larger ions of Ba and Nd fill six cages-four 15-coordinated and two 12-coordinated sites, similar to that described previously by Ikeda etc[9]. Fig. 1 shows the XRD pattern of $Ba_5NdTi_3Ta_7O_{30}$ powders, where the typical tungsten-bronze structure is indicated.

For five compositions mentioned above, the dense ceramics can be created with easy, although the suitable sintering temperature varies in the range of 1310 to 1450°C with composition. In Table II, the bulk density, dielectric constant and dielectric loss for the well densified ceramics with various composition are illustrated, where the dielectric constant varies from 70 to 160 combined with the low dielectric loss even less than 10^{-4}. There is a clear tendency that the dielectric constant increases with increasing concentration of Ba and Ta and decreasing concentration of Nd and Ti.

Dielectric Ceramic Materials

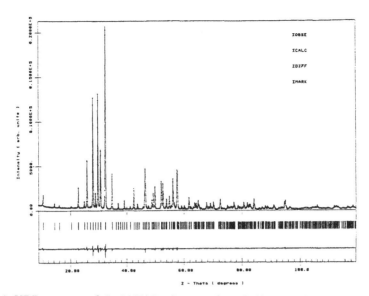

Fig. 1 XRD pattern of $Ba_5NdTi_3Ta_7O_{30}$ powders, indicating the tungsten-bronze structure.

Table II. Bulk Density and Dielectric Characteristics of Ceramics in $BaO-Nd_2O_3-TiO_2-Ta_2O_5$ System (at 1MHz)

Composition	Sintering Conditions	Density (Kg/m^3)	ε	tan δ
$BaNd_5Ti_7Ta_3O_{30}$	1400°Cx3h in air	6,080	70	0.0150
$Ba_2Nd_4Ti_6Ta_4O_{30}$	1400°Cx3h in air	6,375	140	0.0019
$Ba_3Nd_3Ti_5Ta_5O_{30}$	1400°Cx3h in air	6,680	110	0.0012
$Ba_4Nd_2Ti_4Ta_6O_{30}$	1310°Cx3h in air	6,700	140	0.0007
$Ba_5NdTi_3Ta_7O_{30}$	1450°Cx3h in air	7,100	160	0.000018

Fig. 2-4 give the frequency dependence of dielectric characteristics of dielectric ceramics of $Ba_3Nd_3Ti_5Ta_5O_{30}$, $Ba_4Nd_2Ti_4Ta_6O_{30}$ and $Ba_5NdTi_3Ta_7O_{30}$, respectively, and the dielectric constant is almost frequency independent for all compositions, and this is also the situation for other two compositions. This suggests the possibility of microwave application of such ceramics.

Fig. 5 and 6 give the temperature dependency of dielectric constant of $Ba_3Nd_3Ti_5Ta_5O_{30}$ and $Ba_5NdTi_3Ta_7O_{30}$ ceramics. The relatively larger temperature coefficient of dielectric constant ($\tau_\varepsilon < -1500$ppm/°C) is indicated in these two compositions, and the temperature coefficient of resonant frequency ($\tau_f = -\tau_\varepsilon/2-\alpha$, α is the thermal expansion coefficient, ~10ppm/°C) is estimated greater than 740ppm/°C. Moreover, it seems that the temperature coefficient increases with increasing dielectric constant in the present system, and this relatively larger temperature coefficient is the

Dielectric Ceramic Materials   73

primary problem of the present ceramics for microwave application. Even though, because of the large degree of freedom to adjust composition, phase constitution and microstructures, the temperature coefficient is expected to be controlled into an acceptable level through the   preferred modifications.

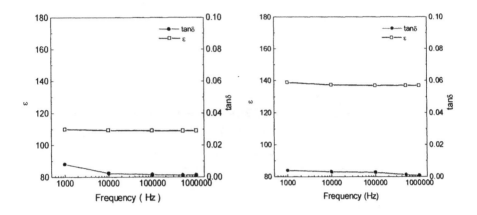

Fig.2   Frequency dependency of dielectric character of $Ba_3Nd_3Ti_5Ta_5O_{30}$ ceramics sintered at 1400℃ in air for 3h.

Fig.3 Frequency dependence of dielectric character of $Ba_4Nd_2Ti_4Ta_6O_{30}$ ceramics sintered at 1310℃ in air for 3h.

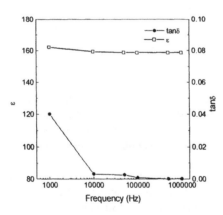

Fig.4 Frequency dependency of dielectric character of $Ba_5NdTi_3Ta_7O_{30}$ ceramics sintered at 1450℃ in air for 3hr.

Dielectric Ceramic Materials

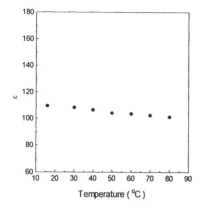

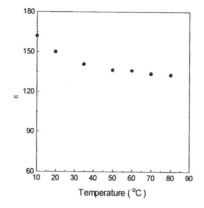

Fig.5 Temperature dependence of dielectric constant of $Ba_3Nd_3Ti_5Ta_5O_{30}$ ceramics sintered at 1400℃ in air for 3h.

Fig.6 Temperature dependence of dielectric constant of $Ba_5NdTi_3Ta_7O_{30}$ ceramics sintered at 1450℃ in air for 3h.

CONCLUSIONS

The dielectric ceramics with tungsten bronze structures in $BaO-Nd_2O_3-TiO_2-Ta_2O_5$ system indicate almost frequency-independent high dielectric constant of 70 to 160, combined with low dielectric loss even less than 10^{-4}. For this promising new microwave dielectric ceramic system, the control of temperature coefficient should be the primary issue in the future work.

ACKNOWLEDGEMENT

This work was financially supported by Chinese National Hi-tech Project under contract number 863-715-06-0072, and the assistance of Professor G.L. Lu, Central Laboratory, Hangzhou University for structural analysis was appreciated.

REFERENCES

1.J.K. Plourde, D.F. Linn, H.M. O'Bryan, Jr. and J. Thomson, Jr., "$Ba_2Ti_9O_{20}$ as a Microwave Dielectric Resonator", J. Am. Ceram. Soc., **58** [9-10] 418-420 (1975).

2. P.K. Davies, J. Tong and T. Negas, "The Effect of Ordering Induced Domain Boundaries on Low Loss $Ba(Zn_{1/3}Ta_{2/3})O_3$-BaZrO3 Perovskite Microwave Dielectrics", J. Am. Ceram. Soc., **80** [7] 1724-40 (1997).

3. X.M. Chen, Y. Suzuki & N. Sato, "Sinterability Improvement of $Ba(Mg_{1/3}Ta_{2/3})O_3$ Dielectric Ceramics", J. Mater. Sci.: Mater. in Electronics, **5**, 244-247 (1994).

4.K. Wakino, K. Minai and H. Tamura, "Microwave Characteristics of $(Zr,Sn)TiO_4$ and $BaO-PbO-Nd_2O_3-TiO_2$ Dielectric Resonators", J. Am. Ceram. Soc., **67** [4] 278-281 (1984).

5.X.M. Chen, Y. Suzuki & N. Sato, "Microstructure and Microwave Dielectric Characteristic of Ceramics with Composition of BaO.Nd$_2$O$_3$.5TiO$_2$", J. Mater. Sci.: Mater. in Electronics, **6,** 10-16 (1995).

6.M. Valant, D. Suvorov and D. Kolar, "Role of Bi$_2$O$_3$ in Optimizing the Dielectric Properties of Ba$_{4.5}$Nd$_9$Ti$_{18}$O$_{54}$ Based Microwave Ceramics", J. Mater. Res., **11** [4] 928-931 (1996).

7.J. Kato, H. Kagata K. Nishimoto, "Dielectric Properties of (PbCa)(MeNb)O$_3$ at Microwave Frequencies", Jpn. Appl. Phys., Part 1, **31** [9B] 3144-3147 (1992).

8.R.R. Neurgaonkar, W.F. Hall, J.R. Oliver, W.W. Ho & W.K. Cory, "Tungsten Bronze Sr$_{1-x}$Ba$_x$Nb$_2$O$_6$: A Case History of Versatility", Ferroelectrics, **87** , 167-179 (1988).

9.T. Ikeda, T. Haraguchi, Y. Onodera and T. Saito, "Some Compounds of Tungsten-Bronze Type A$_6$B$_{10}$O$_{30}$", Japanese Journal of Applied Physics, **10** [8] 987-994 (1971).

10.X.M. Chen, "Dielectric Characteristics of Composite Ceramics in Ba(Mg$_{1/3}$Ta$_{2/3}$)O$_3$ - BaO·Nd$_2$O$_3$·5TiO$_2$ System", J. Mater. Sci., **31,** 4853-4857 (1996).

NOVEL FERROELECTRICITY IN PIEZOELECTRIC ZnO BY Li-SUBSTITUTION

A. Onodera, N. Tamaki, H. Satoh*, H. Yamashita+ and A. Sakai#
*Department of Physics, Faculty of Science, Hokkaido University,
Sapporo 060, Japan,*
Hakodate National College of Technology, Hakodate 042, Japan
+*Department of Electronics and Information Engineering, Faculty of Engineering,
Hokkai-Gakuen University, Sapporo 064, Japan*
#*Department of Electrical and Electronic Engineering, Muroran Institute of
Technology, Muroran 050, Japan*

ABSTRACT
A new ferroelectric phase transition was found in Li-substituted II-VI semi-conductor ZnO. The dielectric constant shows a broad anomaly around 330 K in $Zn_{0.9}Li_{0.1}O$ ceramics. The observation of a ferroelectric hysteresis loop was successful. However, the obtained P_s varies from 0.05 to 0.59 [$\cdot 10^{-2}$ C/m^2], which depends on samples. Taking into account the preferred orientation in ceramic samples, spontaneous polarization was obtained for $Zn_{0.9}Li_{0.1}O$ as 0.9 [$\cdot 10^{-2}$ C/m^2] at room temperature. Structural changes associated with this phase transition are very small (0.003 Å) compared with those in typical ferroelectric phase transition, which is induced by the introduction of substitutional Li instead of the host Zn in wurtzite-type ZnO. This novel ferroelectricity may be due to changes in d-p hybridization in the Zn-O bond. The mechanism of this novel ferroelectricity is discussed phenomenologically.

INTRODUCTION
Zinc oxide, ZnO, is an n-type piezoelectric II-VI semiconductor [1]. Because of its high piezoelectricity and electromechanical coupling properties, ZnO has been applied for many practical devices such as ultrasonic transducers, SAW (surface-acoustic-wave) devices and oxygen sensors [1-3]. The crystal structure of ZnO is hexagonal (C_{6v}^4 - P6$_3$mc): each Zn atom is tetrahedrally coordinated to four O atoms and the Zn d-electrons hybridize with the O p-electrons [4-6]. As ZnO usually contains excess zinc atoms, its physical properties such as electrical conductivity, piezoelectricity and defect structure depend sensitively on the

excess zinc atoms [7, 8]. The resistivity (ρ) of ZnO is about 300 Ω·cm. It increases drastically to 10^{10} Ω·cm with Li^+ doping [7, 8] whereas it is 10^{-4} Ω·cm with Al^{3+}, Ga^{3+} and In^{3+}[9]. The change in ρ by substitutional doping exceeds over 14 orders of magnitude. Many first-principles studies of ZnO have been published recently, focusing on structural, electronic and piezoelectric properties [6, 10-13]. No phase transition has been found in the pure crystal at atmospheric pressure. Pure ZnO is a polar crystal, while it is not possible to reverse its moment by an external electric field until its melting point (2248 K).

Recently a ferroelectric phase transition was discovered in Li-substituted ZnO ceramics [14-17]. Because of the easy fabrication of good-quality thin films, ZnO may be one promising material for integrated devices. The replacement of host Zn atoms by substitutional Li atoms plays an important role for the appearance of ferroelectricity, although the driving mechanism of this phase transition has not been well understood. The observation of ferroelectric D-E hysteresis loops was successful, but the observed spontaneous polarization (P_s) shows sample dependence. As the value of the P_s is essential for understanding the mechanism of phase transition, we studied the P_s in $Zn_{0.9}Li_{0.1}O$, taking account of the preferred orientation of ceramic samples. The mechanism of this novel ferroelectric phase transition is discussed phenomenologically.

EXPERIMENTAL

Ceramic samples of $Zn_{1-x}Li_xO$ with $x = 0.1$ were synthesized by solid reaction of ZnO and Li_2O at 1073 K for 72 hours. The purity of ZnO (Kanto Chemical Co.) was 99.95%. When Zn atoms (ionic radius 0.74 Å) are replaced partially by Li atoms (ionic radius 0.60 Å), Li atoms may locate at off-centered positions in $Zn_{1-x}Li_xO$. The resistivity in ZnO is improved drastically by Li doping as reported by Kolb and Laudise [7, 8]. The resistivities of present ceramic samples were more than 10^{10} Ω·cm. The homogeneity and preferred orientation of ceramic samples were checked by X-ray diffraction (CuKα, 50 kV 260 mA). The diffraction patterns showed that all samples were single phase. Ferroelectric D-E hysteresis loops were measured by using the Sawyer-Tower circuit with a triangular signal generator (10 Hz) and a digital oscilloscope. Dielectric constants were measured by an LCR meter (Hewlett Packard HP-4284A).

DIELECTRIC PROPERTIES IN $Zn_{1-x}Li_xO$

Dielectric anomaly was found at about 330 K for $Zn_{0.9}Li_{0.1}O$, whereas pure ZnO did not show any anomaly as reported by Kobiakov [18]. Dielectric behavior of $Zn_{0.9}Li_{0.1}O$ is shown in Figure 1, in which the anomaly is very small and broad in contrast to those of typical ferroelectrics.

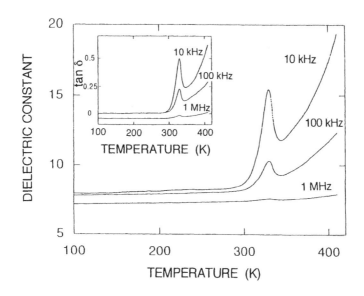

Figure 1. Temperature dependence of dielectric constant of $Zn_{0.9}Li_{0.1}O$.

FERROELECTRIC D-E LOOPS

Ferroelectric D-E hysteresis loops were measured for ceramic samples with $x = 0.1$ as shown in Figure 2. The spontaneous polarization is $0.52 \cdot 10^{-2}$ C/m^2 at room temperature, which is about ten times as large as the previously reported $0.0044 \cdot 10^{-2}$ C/m^2 [14]. The transition temperature of $Zn_{0.9}Li_{0.1}O$ determined by dielectric measurements is 330 K (T_c), which is the same for both samples. The observed P_s was found to be 0.044, 0.0056, 0.52 and 0.59 [$\cdot 10^{-2}$ C/m^2] for four ceramic samples. The variation from sample to sample is probably due to a preferred orientation, as ZnO has a crystallographic nature to orient preferably in ceramic samples. In the following section, we determined the preferred orientation of ceramic samples based on reflection intensity by X-ray diffraction and estimated the exact P_s of $Zn_{0.9}Li_{0.1}O$.

PREFERRED ORIENTATION AND P_s

The X-ray diffraction pattern of powdered $Zn_{0.9}Li_{0.1}O$ is almost the same as that of pure ZnO. The systematic check of absence of possible reflections shows no changes in the space group (wurtzite, P6$_3$mc) at r. t., although the lowering of symmetry is accompanied generally by a ferroelectric phase transition. X-ray diffraction patterns of powdered and pellet samples were measured. Compared

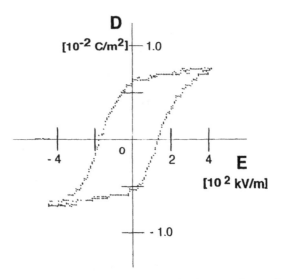

Figure 2. Ferroelectric D-E hysteresis loop for $Zn_{1-x}Li_xO$ ($x = 0.1$) ceramics. The measurement was performed at 10 Hz and r.t.

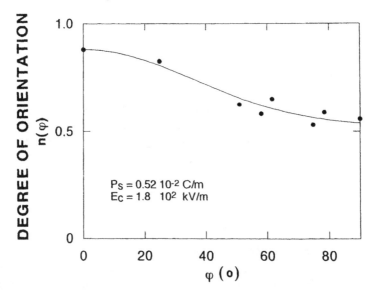

Figure 3. Distribution of the degree of orientation of the hexagonal c-axis of $Zn_{0.9}Li_{0.1}O$ estimated from X-Ray intensities. The inclined angle of the c-axis from the perpendicular plane of the sample is denoted as φ.

Dielectric Ceramic Materials

with the powder sample, several strong lines such as the (0, 0, 6), (0, 1, 5), (1, 0, 10) and (0, 0, 15) reflections were observed in the pellet sample; this evidence means that this ceramic sample is oriented preferably along the c-axis. The relative degree of this orientation of the c-axis was calculated from the ratio I_{exp}/I_{cal}, where I_{exp} and I_{cal} are the observed X-ray Bragg intensity of the ceramic sample and the calculated intensity of an ideal powder sample, respectively. In the calculation, the following atomic parameters were adopted; (1/3, 2/3, 0) and B=0.63 for Zn(Li) and (1/3, 2/3, u) u=0.3825 and B= 0.68 for O [4]. The distribution of the degree of preferred orientation, n(φ), is shown in Figure 3, where φ is the inclined angle of the c-axis from the perpendicular direction of the sample plate. The tetra-hedron ZnO_4 has a permanent dipole moment P_0 along the polar c-axis in the wurtzite structure. The direction of the induced dipole moment of the ZnO_4 tetra-hedron, μ, is restricted to the polar c-axis by the $P6_3mc$ symmetry. Here we assumed that +c- and - c-oriented ZnO_4 tetrahedra have the same induced dipole moment m as $(P_0 - \mu)$ and $(P_0 + \mu)$, respectively. The host and induced dipole moments align along the c-axis as schematically described in Figure 4. The observed P_s is given from the following relation,

$$(P_s)_{obs} = \frac{Z\,2\pi\,\mu}{N\,V_0} \int_0^{\pi/2} n(\varphi)cos\varphi\; sin\varphi\; d\varphi ,$$

$$N = 2\pi \int_0^{\pi/2} n(\varphi)\; sin\; \varphi\; d\varphi ,$$

where N, Z, and V_0 are the total number of μ, the number of μ in the unit cell and the unit cell volume. The P_s for the single crystal, $Z\mu/V_0$, is about $0.9 \cdot 10^{-2}$ C/m^2 for four samples as shown in Table I. Corso et al. estimated the value of P_s of pure ZnO based on a first-principles study as $5 \cdot 10^{-2}$ C/m^2 [6], which is five times as large as the present observed value in $Zn_{0.9}Li_{0.1}O$. The temperature dependence of the corrected spontaneous polarization is shown in Figure 5.

Table I. Observed and corrected P_s of $Zn_{0.9}Li_{0.1}O$ ceramics

Sample	#1	#2	#3	#4
$(P_s)_{obs.}$ $[\cdot 10^{-2} C/m^2]$	0.044	0.0056	0.52	0.59
$(P_s)_{cor.}$ $[\cdot 10^{-2} C/m^2]$	0.784	0.853	0.984	0.904

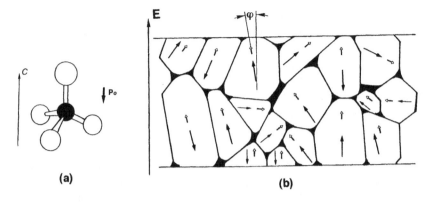

Figure 4. Model of dipole moment alignment in ZnO ceramics. (a) Configuration of ZnO$_4$ tetrahedron and dipole moment, (b) alignment of dipole moments in grains in a ceramic sample. Thin and thick arrows are original (P_o) and induced (μ) dipole moments.

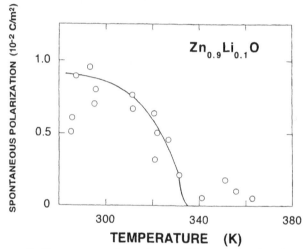

Figure 5. Temperature dependence of the corrected spontaneous polarization P_s in Li-doped ZnO ceramics.

DISTORTION OF ZnO$_4$ TETRAHEDRON

The lattice constants are $a = 3.244$ (1) and $c = 5.199$ (1) Å for Zn$_{0.9}$Li$_{0.1}$O, and

a=3.253(1) and c=5.212(1) Å for pure ZnO at room temperature. These values are in good agreement with those reported in the literature, a= 3.249858 and c= 5.206619 Å for pure ZnO [5]. The ratio a/c of $Zn_{0.9}Li_{0.1}O$ is 0.6239_7, which is a bit smaller than pure ZnO (0.6241_8 in pure ZnO) [26]. The a/c is related to the Zn-O bond length in ideal wurtzite as $\sqrt{\frac{a^2}{3} + \left(\frac{1}{2} - u\right)^2 c^2}$ and cu. Thus, the average Zn-O bond lengths are 1.974_5 Å for $Zn_{0.9}Li_{0.1}O$ and 1.977_8 Å for pure ZnO. The Zn-O bond length is only 0.003 Å shorter than that in $Zn_{0.9}Li_{0.1}O$. Based on the assumption of a simple point charge model, the displacement of O atoms (δ) is estimated as 0.002_6 Å from the relation $\delta=P_s V_0/Zec$, where e is the electron charge and P_s= $0.9 \cdot 10^{-2}$ C/m². This δ coincides well with the change of the lattice constant c at T_c (0.002 Å) along the polar direction in the wurtzite structure as shown in Figure 6. Small structural changes on the order of 0.003 Å were found at this novel ferroelectric phase transition.

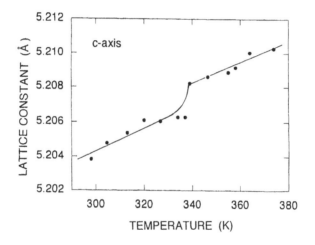

Figure 6. Temperature dependence of the lattice parameter c in $Zn_{0.9}Li_{0.1}O$.

DISCUSSION

Polar semiconductors with off-centered ions were found to exhibit interesting ferroelectric properties [19]. A narrow-gap IV-VI semiconductor, $Pb_{1-x}Ge_xTe$, is one representative compound. This phase transition was explained by a vibronic model which takes an electron-phonon interaction into account, on the assumption of a local instability of dilute Ge ions on host Pb sites. In this model, the strong electron-phonon coupling, soft TO mode and small band gap energy are important for the occurrence of ferroelectric instability. Similar ferroelectric phase transitions have been discovered even in II-VI semiconductors with a wide band gap, $Cd_{1-x}Zn_xTe$ [20, 21] and $Zn_{1-x}Li_xO$. In the II-VI compounds, the covalent character is much weaker than those in the IV-VI compounds.[22] Dielectric properties are summarized in Table II.

Table II. Dielectric properties of AB-type ferroelectric semiconductors.

	$Pb_{1-x}Ge_xTe$	$Cd_{1-x}Zn_xTe$	$Zn_{1-x}Li_xO$
Crystal Structure			
Paraelectric Phase	Rock salt	Zinc-blend	Wurtzite
Ferroelectric Phase	Rhombohedral	Rhombohedral	Wurtzite
T_c (K) (x=0.1)	200	390	330
E_g (eV)	0.3	1.53	3.2
ρ (Ω cm)	10	10^3	10^{10}
P_s ($\cdot 10^{-2}$ C/m^2)	-	5	0.9
Soft Mode	yes	no	?
ε_{peak}	10^3	50	16

No soft mode has been observed in $Cd_{1-x}Zn_xTe$: the anomalous behavior of the Debye-Waller temperature factor of Zn atoms [23] and optical absorption analysis suggested an order-disorder nature [24]. Thus it is clear that a new mechanism of phase transition different from the vibronic model is necessary for these II-VI ferroelectric semiconductors.

The ferroelectricity results from small structural distortion on the order 10^{-3} Å induced along the polar c-axis by substitutional Li ions, instead of the host Zn in the wurtzite structure. Because of the large ionic size mismatch between Zn ions and Li ions or the change in covalency caused by Li-doping, it is expected that Li ions occupy off-centered positions, forming additional electric dipoles. According to the recent first-principles study [6], hybridization of the O 2p-electron and Zn 3d-electron plays an important role in ZnO. As the Li atom has no d-electron, the replacement of Zn by Li should increase ionic character and affect the Zn-O bond length along the c-axis. This novel ferroelectricity may be related primarily to changes in d-p hybridization by Li substitution, which causes slight structural changes through an electron-phonon coupling interaction. According to the phase transition theory for a polar crystal [25], the local additional dipole moment p is in the double-well potential of the host system. The potential energy is given in terms of the positional parameter z along the polar c-axis and p as

Dielectric Ceramic Materials

$$G = G_o - az^2 + bz^4 - \beta p^2 ,$$

where $a > 0$, $b > 0$, $\beta > 0$ and $p = N\mu = Nq$ $(z - z_o)$; N and q are the density of dipoles and the effective charge, respectively. In this model, p locates at the off-centered position (z) in the double-well potential of the host system $(z_o$ is the center position $(z_o = \sqrt{a/2b})$). Rewriting G using $z = z_o + p/Nq$, we have

$$G = (G_o - \frac{a^2}{4b^2}) + (\frac{2a}{N^2 q^2} - \beta)p^2 + \frac{2\sqrt{2a\,b}}{N^3 q^3}p^3 + \frac{b}{N^4 q^4}p^4 ,$$

Minimizing G on p, we have paraelectric and ferroelectric solutions when $0 < a/N^2q^2 < \beta/2$. In this model, ferroelectric instability can not result from the host system but is induced with the help of local p.

ZnO undergoes a new type of phase transition from the polar high-temperature phase to the low-temperature ferroelectric phase via the existence of dilute local μ without lowering of the symmetry. Recently ferroelectric thin films have been used for random access memory (RAM) and other integrated devices. The advantage of $Zn_{1-x}Li_xO$ as a functional ferroelectric material is that it is easy to control its resistivity by Li doping. The direction of P_s is unique in wurtzite while eight kinds of domains are possible in rhombohedral $Cd_{1-x}Zn_xTe$ and $Pb_{1-x}Ge_xTe$. This is also of great advantage to fabricate good-quality thin films. Because of its simple crystal structure and the high quality of its thin films, ZnO is a promising material not only for piezoelectric films but also for ferroelectric thin films for integrated ferroelectric devices.

ACKNOWLEDGMENT
This work was supported by a Grand-in-Aid for Scientific Research from the Ministry of Education, Science and Culture, Japan.

REFERENCES
[1]G. Heiland, E. Mollwo and F. Stockmann, "Electronic Processes in Zinc Oxide", Solid State Phys. **8**, 193-326 (1959).
[2]C. Campbell, "Surface Acoustic Wave Devices and Their Signal Processing Applications", Academic Press, San Diego, 1989.
[3]W. Hirshwald, "Zinc Oxide", Current Topics in Materials Science **7**, 143-303 (1981).
[4]S.C. Abrahams and J.L. Bernstein, "Refinement of the Structure of Hexagonal ZnO", *Acta Crystallogr.* **B25**, 1233-1236 (1969).
[5]J. Albertsson, S.C. Abrahams and A. Kvick, "Atomic Displacement, Anharmonic Thermal Vibration, Expansivity and Pyroelectric Coefficient Thermal Dependencies in ZnO", *Acta. Crystallogr.* **B45**, 34-40 (1989).

[6]A.D. Corso, M. Posternak, R. Resta and A. Baldereschi, "Ab *initio* Study of Piezoelectricity and Spontaneous Polarization in ZnO", *Phys. Rev.* **B50**,10715-10721 (1994).

[7]R.A. Laudise, E.D. Kolb and A.J. Caporaso, "Hydrothermal Growth of Large Sound Crystals of Zinc Oxide", *J. Am. Ceram. Soc.* **47**, 9 (1964).

[8]E.D. Kolb and R.A. Laudise, "Hydrothermally Grown ZnO Crystals of Low and Intermediate Resistivity", *J. Am. Ceram. Soc.* **49**, 302 (1966).

[9]T. Minami, H. Nanto and S. Tanaka, "Highly Conductive and Transparent Aluminium Doped Zinc Oxide Thin Films Prepared by RF Magnetron Sputtering", *Jpn. J. Appl. Phys.* **23**, L280-L282 (1984).

[10]S. Massidda, R. Resta, M. Posternak and A. Baldereschi, "Polarization and Dynamical Charge of ZnO within Different One-Particle Schemes", *Phys. Rev.* **B52**, R16977-R16980 (1995).

[11]J. E. Jaffe and A.C. Hess, "Hatree-Fock Study of Phase Changes in ZnO at High Pressure", *Phys. Rev.* **B48**, 7903-7909 (1993).

[12]P. Schroer, P. Kruger and J. Pollmann, "First-principle Calculation of the Electronic Structure of the Wurtzite Semiconductor ZnO and ZnS", *Phys. Rev.* **B47**, 6971-6980 (1993).

[13]O. Zakharov, A. Rubio, X. Blase, M.L. Cohen and S.G. Louie, "Quasiparticle Band Structure of Six II-VI Compounds: ZnS, ZnSe, ZnTe, CdS, CdSe and CdTe", *Phys. Rev.* **B50**, 10780-10787 (1994).

[14]A. Onodera, N. Tamaki, Y. Kawamura, T. Sawada and H. Yamashita, "Dielectric Activity and Ferroelectricity in Piezoelectric Semiconductor Li-doped ZnO", *Jpn. J. Appl. Phys.* **35**, 5160-5162 (1996).

[15]N. Tamaki, A. Onodera, T. Sawada and H. Yamashita, "Measurements of D-E Hysteresis Loop and Ferroelectric Activity in Piezoelectric Li-doped ZnO", *J. Kor. Phys.* (Proc. Suppl.) **29**, S668-S671 (1996).

[16]A. Onodera, N. Tamaki, K. Jin and H. Yamashita, "Ferroelectric Properties in Piezoelectric Semiconductor $Zn_{1-x}M_xO$ (M=Li, Mg)", *Jpn. J. Appl. Phys.* **36**, 6008-6010 (1997).

[17]A. Onodera, N. Tamaki, Y. Kawamura, T. Sawada, N. Sakagami, K. Jin, H. Satoh and H. Yamashita, "Ferroelectric Phase Transition in Piezoelectric Semiconductor ZnO", *J. Kor. Phys.* (Proc. Suppl.) **32**, S11-S14 (1998).

[18]I.B. Kobiakov, "Elastic, Piezoelectric and Dielectric Properties of ZnO and CdS Single Crystals in a Wide Range of Temperatures", *Solid State Commun.* **35**, 305-310 (1980).

[19]H. Bilz, A. Bussmann-Holder, W. Jantsch and P. Vogl, "Dynamical Properties of IV-VI Compounds", Springer-Verlag, Berlin, 1983.

[20]R. Weil, R. Nkum, E. Muranevich and L. Benguigui, "Ferroelectricity in Zinc Cadmium Telluride", *Phys. Rev. Lett.* **62**, 2744-2746 (1989) .

[21]L. Benguigui, R. Weil, E. Muranevich A. Chack and E. Fredj, "Ferroelectric Properties of $Cd_{1-x}Zn_xTe$ Solid Solutions", *J. Appl. Phys.* **74**, 513-520 (1993).

[22]J.C. Phillips, "Bonds and Bands in Semiconductors", Academic Press, New York,1973, Ch.2.

[23]H. Terauchi, Y. Yoneda, H. Kasatani K. Sakaue, T. Koshiba, S. Murakami, Y. Kuroiwa, Y. Noda, S. Sugai, S. Nakashima and H. Maeda, "Ferroelectric Behaviors in Semiconductive $Cd_{1-x}Zn_xTe$ Crystals, *Jpn. J. Appl. Phys.* **32**, 728-730 (1993).

[24]D.N. Talwar, Z.C. Feng and P. Becla, "Impurity-induced Phonon Disordering in $Cd_{1-x}Zn_xTe$ Ternary Alloys", *Phys. Rev.* **B48**, 17064-17071 (1993).

[25]S. Sawada, "Fundamental Prospect in Ferroelectric Research", *Bussei* **6**, 211-221 (1965) (in Japanese).

NOVEL PROCESSES FOR SYNTHESIS OF DIELECTRIC CERAMICS USING METALLO-ORGANIC PRECURSORS

E. H. Walker, G.D. Georgieva, E.M. Holt, L. E. Reinhardt, and A.W. Apblett*,
Oklahoma State University, Stillwater, OK. 74078.

ABSTRACT

The synthesis of dielectric ceramics using metallo-organic precursors has had a long and fruitful history. We have investigated several types of metal carboxylate systems that are suitable for the preparation of these ceramics by metallo-organic deposition. These include metal salts that are liquids at room temperature as well as metal carboxylates that are unusually hydrolytically-stable. Additionally, water-stable single source precursors for a variety of ferroelectric materials were synthesized. The preparation of ferroelectric thin films using water-based or solventless processes has significant environmental and economic advantages and is a significant application of these precursors.

INTRODUCTION

Metallo-organic deposition, MOD [1, 2], is a non-vacuum, solution-based method of depositing thin films. In the MOD process, a suitable metallo-organic precursor dissolved in an appropriate solvent is coated on a substrate by spin-coating, screen printing, or spray- or dip-coating. The soft metallo-organic film is then pyrolyzed in air, oxygen, nitrogen or other suitable atmosphere to convert the precursors to their constituent elements, oxides, or other compounds. Shrinkage generally occurs only in the vertical dimension so conformal coverage of a substrate may be realized. Metal carboxylates with long slightly-branched alkyl chains (e.g. 2-ethylhexanoate or neodecanoate) are often used as precursors for ceramic oxides since they are usually air-stable, soluble in organic solvents, and decompose readily to the metal oxides. MOD processes for the generation of many oxide-based materials have already been developed: e.g. indium tin oxide [3], SnO_x [4], $YBa_2Cu_3O_7$ [5] and ZrO_2 [6]. Barium titanate has been synthesized previously using a variety of metallo-organic precursors [7, 8]. These include routes based solely on metal alkoxides [9] or metal carboxylates (e.g. the Pechini (or citrate) process [10] and mixed carboxylate/alkoxide precursors [11].

The increasing demand for environment-friendly processes places stringent requirements on precursors for ceramic materials. In particular, the avoidance of organic solvents necessitates the development of preceramic compounds that are either water-soluble or which are amenable to solventless processing. We report herein several metal carboxylate precursors for a variety of ferroelectric ceramics that are either liquids or which are water-soluble and are therefore suitable for such processing techniques.

EXPERIMENTAL

All metal salts and carboxylic acids were commercial products and were used as supplied. Water was deionized and distilled before use. All reagents were commercial products and were used without further purification with the exception of 2-[2-(2-methoxyethoxy)ethoxy] acetic acid, MEEAH which was dried over activated 3A molecular sieves for ca. 24 hours. Water was distilled and deionized in a Modulab UF/UV Polishing apparatus before use. Thermogravimetric studies were performed using 20-30 mg samples under a 100 ml/minute flow of dry air in a Seiko TG/DTA 220 instrument or a TA Instruments Hi-Res TGA 2950 Thermogravimetric Analyzer. The temperatures were ramped from 25 °C to 1025 °C at a rate of 2 °C per minute or from 25 °C to 650 °C at a rate of 5 °C/min. Bulk pyrolyses at various temperatures were performed in ambient air in a temperature-programmable muffle furnace using 1-2 g samples, a temperature ramp of 5 °C/minute and a hold time of 6-12 hours. X-ray powder diffraction patterns were obtained using copper K_α radiation on a Scintag XDS 2000 diffractometer equipped with an automated sample changer and a high resolution solid state detector. Jade, a search/match software package, was used in the identification of XRD spectra.

RESULTS AND DISCUSSION

BaTiO(Ox)$_2$ (Ox=oxalate, $C_2O_4^{2-}$) is an example of a very successful precursor for barium titanate that is used widely by industry is [12]. If prepared carefully, this material comes very close to the stoichiometry of 1 barium: 1 titanium required for BaTiO$_3$. Unfortunately, BaTiO(Ox)$_2$ is quite insoluble rendering it useless for preparing BaTiO$_3$ thin-films. Previously we developed a method for the preparation of liquid precursors from BaTiO(Ox)$_2$ and SrTiO(Ox)$_2$ by reaction with refluxing methoxyacetic acid [13, 14]. At the elevated temperatures used (ca. 200°C), the oxalate ions decompose so that the metal salts are converted to methoxyacetates (Eq. 1). Thus, the insoluble oxalate salts gradually dissolve to yield pale yellow solutions of the methoxyacetates.

$$MTiO(C_2O_4)_2 + 4\ CH_3OCH_2CO_2H \rightarrow MTiO(O_2CCH_2OCH_3)_4 + 2\ CO_2 \\ + 2\ CO + 2\ H_2O \qquad (1)$$

Removal of the methoxyacetic acid in vacuo yields MTiO(O$_2$CR)$_2$ (HO$_2$R)$_2$, {M= Sr or Ba, R= CH$_2$OMe} as viscous yellow liquids that have high solubility in water and a variety of organic solvents such as ethanol, acetonitrile, and chlorocarbons. The extra equivalents of coordinated acid play an important role in stabilizing the compound and attempts to remove them result in decomposition to an insoluble material. Using these materials, it is possible to prepare BaTiO$_3$ and SrTiO$_3$ by metallo-organic deposition using the pure precursors and Ba$_x$Sr$_{1-x}$TiO$_3$ (BST) from their mixtures.

The presence of the excess acid in the precursors and its corrosive nature led us to seek alternative carboxylic acids which would be liquid without the extra equivalents of acid. It was apparent that greater coordination of the metal was a necessary requirement so carboxylates with additional ether chains such as 2-[2-(2-methoxy)ethoxy]ethoxyacetate, MEEA were investigated. This lead to the discovery of a family of metal carboxylates are liquids at room temperature and are therefore amenable to the preparation of dielectric ceramics via liquid phase processing [14-20].

$$H_3C \diagdown_O \diagdown \substack{CH_2 \\ CH_2} \diagdown_O \diagdown \substack{CH_2 \\ CH_2} \diagdown \substack{CH_2 \\ O} \diagdown \substack{CH_2 \\ CH_2} \diagdown_O \diagdown \substack{C \\ CH_2} \diagup O^-$$

<u>MEEA</u>

Furthermore, these liquids are excellent solvents for other metal salts so that precursor solutions for multi-metallic ceramic materials are readily prepared. The main challenge for the preparation of dielectric titanate phases via a carboxylate route is the preparation of a hydrolytically-stable titanium carboxylate. Fortunately the reaction of $Ti(O^iPr)_4$ with four equivalents of 2-[2-(2-methoxy)ethoxy] ethoxyacetic acid in anhydrous ethanol readily yields $Ti(MEEA)_4$ as viscous, water-stable pale yellow liquid [14]. The desired protolysis reaction (Eq. 2) is accompanied by an esterification reaction that yields the ethyl ester of the carboxylic acid and a titanium hydroxide moiety (Eq. 3). The ester side-product is readily removed from the liquid titanium salt by extraction with diethyl ether but this step can be neglected since the ester does not adversely affect ceramic synthesis. The formula $Ti(MEEA)_4$ will be used to represent the titanium carboxylate despite the fact that esterification has altered the actual formula by replacing a number of the MEEA groups with hydroxyls. The purified product was found to be monomeric by freezing-point depression measurements of benzene solutions.

$$Ti(OR)_4 + 4\ HMEEA \rightarrow\ Ti(MEEA)_4 + 4\ ROH \qquad (2)$$

$$Ti\text{-}OEt + 4\ HMEEA \rightarrow\ Ti\text{-}OH + EtMEEA \qquad (3)$$

A liquid barium carboxylate, $Ba(MEEA)_2$ is also readily prepared by neutralization of barium hydroxide with two equivalents of HMEEA. This material is a colorless air-stable liquid but its aqueous solutions are prone to a reaction with carbon dioxide leading to deposition of insoluble barium carbonate. Thus, aqueous solutions should be freshly prepared before use. A liquid precursor for barium titanate can be simply prepared by mixing appropriate amounts of the $Ba(MEEA)_2$ and $Ti(MEEA)_4$ liquids. The weighing out of stoichiometric amounts is facilitated by first preparing 10 weight percent aqueous solutions in order to overcome the high viscosity of the MEEA complexes. Subsequent removal of the water by rotary evaporated yields a viscous liquid that is readily spin or dip-coated on substrates such as silicon. Pyrolysis and sintering of the precursor yields crystalline barium titanate at 700°C as demonstrated by X-ray powder diffraction (Figure 1). Since the MEEA salts are very good solvents, a more economical precursor for barium titanate has been synthesized using a liquid metal carboxylate precursor that was prepared by dissolving barium acetate in $Ti(MEEA)_4$ [14]. This material also yielded $BaTiO_3$ at 700°C. Similarly, liquid strontium titanate and barium strontium titanate precursors are readily prepared by dissolving strontium acetate or a mixture of barium and strontium acetates in $Ti(MEEA)_4$.

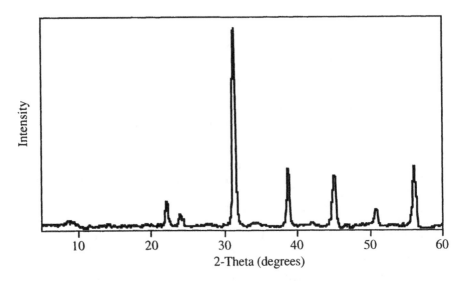

Figure 1. XRD Pattern of BaTi(MEEA)$_4$(CH$_3$CO$_2$)$_2$ Heated to 700°C

Gluconic acid, is a very inexpensive carboxylic acid that is also capable of conferring significant water solubility on metal carboxylate salts.

D-Gluconic acid

A water-stable titanium gluconate can be prepared by stirring a mixture of four molar equivalents of gluconic acid and TiCl$_3$ in air [14]. This results in gradual oxidation of the purple Ti^{3+} ions to Ti^{4+} and eventual formation of a light yellow solution. Removal of the volatiles in vacuo without heating yields a grayish-yellow, extremely hygroscopic solid. This material is very heat sensitive and, above 60°C, converts to a dark brown insoluble material. The success of this preparative route for titanium gluconate is likely attributable to chelation of the Ti^{3+} ions by gluconate before oxidation. A precursor for barium titanate was prepared by mixing aqueous solutions of titanium gluconate and barium gluconate and then removing the water by rotary evaporation to yield an amorphous gray solid. Heating this solid to 800°C yielded crystalline barium titanate, as demonstrated by XRD. Unfortunately, the presence of chloride in the titanium gluconate resulted in the formation of traces of barium chloride that were observed as a crystalline phase at 600°C but were not detectable in the final product. It is likely that the chloride ions will be detrimental to the application of this material as a MOD precursor and their presence is likely the cause of the higher temperature required for preparation of phase-pure $BaTiO_3$. Therefore, improved synthetic methods are necessary for the preparation of titanium gluconate if a successful MOD route to ferroelectric titanates based on gluconate salts is to be developed.

It is also possible to utilize metal gluconates to prepare water-soluble precursors for bimetallic molybdates such as the lanthanide molybdate ferroelectric phases, $Ln_2(MoO_4)_3$ [14]. These have typically been prepared in the past by elevated temperature (700 °C) reactions of the metal oxides[15], a process that is not amenable to the synthesis of thin films. We have recently demonstrated that gluconates are capable of forming stable complexes with MoO_2^{2+} ions. An example of such a complex is shown in Figure 2 [21].

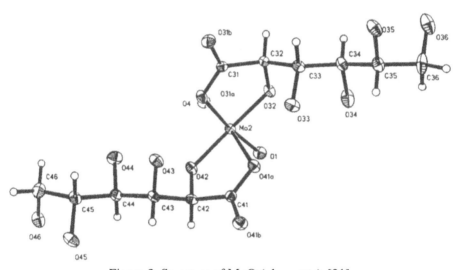

Figure 2. Structure of $MoO_2(gluconate)_2$ [21].

The coordinated hydroxides that are adjacent to the carboxylate groups are acidic so that it is possible to deprotonate this complex to generate metal salts of the dianionic complexes. In actual fact, bimetallic complexes can readily be prepared directly from a metal gluconate and MoO_3 rather than by reaction of $MoO_2(gluconate)_2$ with bases. For example, a bimetallic single-source precursor for neodymium molybdate was readily prepared by a single pot reaction [14] between neodymium gluconate and molybdenum trioxide. Using the same approach, gadolinium gluconate was synthesized by reaction of gadolinium carbonate with gluconic acid in refluxing water. After the gadolinium carbonate had completely reacted, molybdenum trioxide was added to the refluxing reaction mixture. Under these conditions, MoO_3 quickly dissolved as a soluble complex. The thermal gravimetric analysis trace for this precursor in air (Figure 3) shows two main weight loss steps that correspond to dehydration (25 to 170 °C) and pyrolysis of the gluconate anion above 175°C. A gradual loss of weight is observed up to 550 °C where it becomes quite rapid as the organic materials quickly oxidize. The ceramic yield at the end of this step (630 °C) is 34.8 %. The XRD pattern of the bulk precursor heated to 640 °C shown in Figure 4 indicates that the material is phase-pure $Gd_2(MoO_4)_3$ (JCPDS Powder Diffraction File #24-0428) a promising ferroelectric material [15]. The high water-solubility and low-temperature conversion of the gadolinium molybdenum gluconate to gadolinium molybdate make it an excellent precursor for water-based preparation of $Gd_2(MoO_4)_3$ thin films. This methodology has been extended to the preparation of $Sm_2(MoO_4)_3$ and $Ce(MoO_4)_2$ and appears to be suitable for the preparation of most lanthanide molybdate phases.

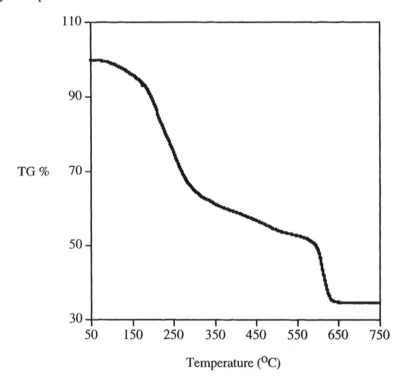

Figure 3. TGA Trace for $Gd_2(gluconate)_6(MoO_3)_3$ in Air

Dielectric Ceramic Materials

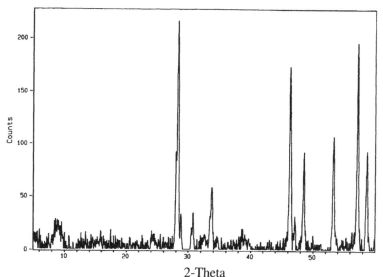

2-Theta

Figure 4. XRD Pattern of $Gd_2(MoO_4)_3$ Heated to 640°C

CONCLUSION

Complexes of metals with ether-containing carboxylates such as methoxyacetate and 2-[2-(2-methoxyethoxy)ethoxy]acetate can be utilized as liquid precursors that have significant potential for the synthesis of dielectric ceramic thin films. Alternatively, metal gluconates also show promise for preparation of barium titanate and lanthanide molybdates powders and films.

ACKNOWLEDGEMENTS

We thank the Louisiana Board of Regents for support of this research through Grant # LEQSF(1996-99)-RD-B-08.

REFERENCES

[1] J. V. Mantese, A. L. Micheli, A. H. Hamdi, and R. W. Vest, "Metal Organic Deposition" *M.R.S. Bull., (XIV)*, , 1173 (1989).

[2] R. W. Vest, "Electronic Thin Films from Metallo-Organic Precursors"; pp. 303-347 in *Ceramics Films and Coatings,*Edited by J. B. Wachtman and R. A. Haber, Noyes Publications, Park Ridge, N.J., 1993.

[3] J. J. Xu, A. S. Shaikh, and R. W. Vest, "Indium Tin Oxide Films from Metallo-Organic Precursors"*Thin Solid Films,* **161**, 273-280 (1988).

[4] T. Maruyama and K. Kitamura, "Fluorine-Doped Tin Oxide Thin Films Prepared by Thermal Decomposition of Metallic Complex Salts" *Jpn. J. Appl. Phys.,* **28**, L312-L313 (1989).

[5] A.H. Hamdi, J.V. Mantese, A.L. Micheli, R.C.O. Laugal, D. F. Dungan, Z. H Zhang, and K.R. Padmanabhan, "Formation of Thin-Film High T$_c$ Super-conductors by Metalorganic Deposition" *Appl. Phys. Lett.,* **51**, 2152-54 (1987).

[6] V. Hebert, C. His, J. Guille, S. Vilminot, and T. L. Wen, "Preparation and Characterization of Precursor of Y_2O_3 Stabilized ZrO_2 by Metal Organic Compounds" *J. Mater. Sci.,* **26**, 5184 (1991).

[7] J. J. Xu, A. S. Shaikh, and R. W. Vest, "High K $BaTiO_3$ Films from Precursors" *IEEE Trans UFFC,* **36**, 307-312 (1989).

[8] P. P. Phule and S. H. Risbud, "Low Temperature Synthesis and Processing of Barium Titanates" *J. Mater. Sci.,* **25**, 1169-1183 (1990).

[9] Z. Xu, H. K. Chae, M. H. Frey, and D. A. Payne, "Chemical Processing and Properties of Nanocrystalline $BaTiO_3$" *Mat. Res. Soc. Symp. Proc,* **271**, 339-344 (1992).

[10] M. P. Pechini, "Ceramic Dielectric Materials", U.S. Patent **3,330,697**, 1967.

[11] C. D. Chandler, M. J. Hampden-Smith, and C. J. Brinker, "Single-Component Routes To Perovskite Phase Mixed Metal Oxides" *Mat. Res. Soc. Symp. Proc.,* **271**, 89-100 (1992).

[12] Y. Enomoto and A. Yamaji, "Preparation of Uniformly Small-Grained $BaTiO_3$," *Amer. Ceram. Soc. Bull.,* **60**, 566 (1981).

[13] A. W. Apblett, G. D. Georgieva, and M. I. Raygoza-Maceda, "Rational Design of Molecular Precursors for Barium Titanate" *Ceram. Trans.,* **43**, 73-74 (1994).

[14] A. W. Apblett, G. D. Georgieva, L. E. Reinhardt, and E. H. Walker, "Precursors for Aqueous and Liquid-Based Processing of Ferroelectric Thin Films"; pp. 95-105 in *Synthesis and Characterization of Advanced Materials,* , Edited by M. Serio, D. M. Gruen, and M. Ripudaman, American Chemical Society, Washington, D.C, 1998.

[15] M. Takashige, S. I. Hamazaki, N. Fukurayi, F. Shimuzu, and S. Kojima, "Surface-Morphology of Ferroelectric Gd2(MoO4)3 Observed By Atomic Force Microscope" *Ferroelectrics,* **203**, 221-225 (1997).

[16] A. W. Apblett, J. C. Long, E. H. Walker, M. D. Johnston, K. J. Schmidt, and L. N. Yarwood, "Metal Organic Precursors for Yttria" *Phosph. Sulf. Silicon Rel. Elements,* **93-94**, 481-481 (1994).

[17] A. W. Apblett, S. M. Cannon, G. D. Georgieva, J. C. Long, M. I. Raygoza-Maceda, and L. E. Reinhardt, "Polymeric Precursors for Yttria." *Mat. Res. Soc. Symp. Proc,* **346**, 679-683 (1995).

[18] A. W. Apblett, L. E. Reinhardt, and E. H. Walker , "Novel Liquid Precursors for Spinel and Alumina Ceramic Coatings" *Proceedings of Unified International Technical Conference on Refractories,* **Vol III**, 1503-1507 (1997).

[19] A. W. Apblett, L. E. Reinhardt, and E. H. Walker, "Liquid Metal Carboxylate Precursors for Yttrium Aluminum Garnet" *Proceedings of Unified International Technical Conference on Refractories,* 1525-1529 (1997).

[20] A. W. Apblett, M. L. Breen, and E. H. Walker, "The Application of Liquid Metal Carboxylates to the Preparation of Aluminum-Containing Ceramics" *Comments on Inorganic Chemistry,* in press (1998).

[21] A. W. Apblett, E. M. Holt, and L. E. Reinhardt, "Single-Component Routes To Molybdate Materials" *Chem. Mater.* Manuscript in preparation.

Hydrothermal Synthesis of Pure and Dy:BaTiO$_3$ Powders at 90°C and Their Sintering Behavior

Ersin E. Ören and A. Cüneyt Taş
Department of Metallurgical and Materials Engineering, Middle East Technical University, Ankara 06531, Turkey.

ABSTRACT

Sub-micron (150-200 nm), monodisperse and spherical powders of pure and Dy-doped (0.8 at%) BaTiO$_3$ have been prepared by "hydrothermal synthesis" at 90°C in an air atmosphere. The powder preparation technique adopted in this work did not necessitate the utilization of strict and expensive measures which were commonly required for the removal of free CO$_2$ present in the atmosphere. The synthesized powders were crystalline, pure, and did not contain BaCO$_3$ as an impurity phase. Pure and Dy-doped BaTiO$_3$ powders synthesized at 90°C had the cubic (space group: Pm-$3m$) crystal structure. Grain growth characteristics of pure and Dy-doped BaTiO$_3$ pellets were investigated during sintering in air and compared with one another over the temperature range of 1200 to 1500°C. Sample characterization was achieved by XRD, SEM, EDXS, and Rietveld Analysis.

INTRODUCTION

The ceramic compositions (pure and doped) of BaTiO$_3$ have been one of the main constituents of the "type II" dielectric materials and of multilayer ceramic capacitors. It has been a widely acclaimed conception since the early 1960's that the dielectric properties of these ceramics are closely related to the grain size in their microstructures. Yamaji, et al. [1-3] have demonstrated the strong effect of Dy in controlling and reducing the sintered grain size of BaTiO$_3$, however, they have used conventional solid-state reactive firing routes in their synthesis experiments. On the other hand, pure BaTiO$_3$, depending on synthesis procedure used and temperature, may exhibit four different polymorphic forms [4]. Among these four polymorphs of BaTiO$_3$, the cubic form (space group: Pm-$3m$) is "paraelectric," and the other three (tetragonal: P$4mm$, orthorhombic: P$mm2$, and rhombohedral: R$3m$) are "ferroelectric." BaTiO$_3$ is of the tetragonal symmetry from room temperature up to its Curie temperature (T$_c$: ~ 128°C), and above T$_c$ it adopts the cubic symmetry. When the ambient temperature is below T$_c$, BaTiO$_3$ is ferroelectric, and when the temperature is above T$_c$ it becomes paraelectric [5]. The change observed in the crystal structure of barium titanate at its T_c could also be observed by the significant change (from about 1000 to 11000) which simultaneously occurs in its dielectric constant. It has also been shown that the average grain size in the microstructure of $BaTiO_3$ turns out to be quite influential on its dielectric constant [1-3, 6]. It has now been a well-established fact that the decrease to be achieved in the average grain size (from 50 μm to ~ 1 μm) of the sintered ceramic microstructure of pure $BaTiO_3$ would show itself up in the form of an increase in its dielectric constant (RT) from 1000 to about 5000 [7], whereas for Dy-doped $BaTiO_3$ samples it would increase to about 10000 [1]. For this reason, it has been an important concern in the synthesis of $BaTiO_3$ ceramics that any precautions which would be taken towards the precise control to be gained over the particle (pre-sintering) and grain (post-sintering) sizes would directly influence the electronic properties of the final product.

The preparation of monosize and crystalline $BaTiO_3$ powders by using hydrothermal synthesis has been known for a long time. This process commonly uses the starting materials of water-soluble inorganic $Ba(OH)_2.8H_2O$ salts and insoluble TiO_2 powders to be mixed in an aqueous solution kept at a temperature near to its boiling point for prolonged times. Hydrothermal synthesis involves the formation of crystalline materials from the starting materials in such aqueous media under strongly alkaline conditions. Hydrothermal processing of $BaTiO_3$ powders has always been prescribed to utilize a certain quantity of excess barium hydroxide in the starting mixture to speed up the hydrothermal reactions [8-9]. Lencka and Riman [10] showed that successful preparation of $BaTiO_3$ by hydrothermal processing required a pH>12 in the aqueous solutions kept near the boiling point. They also underlined the importance of eliminating CO_2 from the reaction vessel to avoid the formation of $BaCO_3$.

The present study focuses on the quest for finding an economical and alternative solution to the problem of "$BaCO_3$-contamination" in hydrothermally synthesized (in air atmosphere) $BaTiO_3$ powders. The nominal addition of small amounts of a RE dopant (i.e., 0.8 at% Dy) has been achieved by incorporating prescribed amounts of Dy-nitrate solutions into the processing route of hydrothermal synthesis, in contrast to the addition of Dy_2O_3 into $BaTiO_3$ by mixing and milling [1-3] in the conventional schemes of synthesis.

EXPERIMENTAL PROCEDURE

Sub-micron, pure and Dy-doped $BaTiO_3$ powders were synthesized from the mixtures of proprietary amounts of $Ba(OH)_2.8H_2O$ (+99.9%, Riedel-de Haën AG, Germany) and TiO_2 (+99.9%, Riedel-de Haën AG). The preparation conditions and parameters [11] of pure barium titanate powders were given in the flowchart of Fig. 1(a). Dy_2O_3 (+99.9%, Merck, Germany) powder was dissolved by reacting it with a stoichiometric amount of HNO_3 (99%, Merck, Germany) to form 0.1 M stock solutions of $Dy(NO_3)_3$. Similarly, the preparation conditions of Dy-doped (0.8 at%) barium titanate powders were given in the flowchart of Fig. 1(B). The constant temperature of 90°C, required for prolonged times (48 to 72 h) of aging (in a Teflon beaker placed in an ordinary closed glass jar), was maintained in a microprocessor-controlled (± 1°C) laboratory oven.

The pellets of pure and Dy-doped $BaTiO_3$ were heated in platinum crucibles for 6 hours at temperatures in the range of 1200 to 1500°C, in an air atmosphere. The pellets were heated to the peak sintering temperatures at the rate of 5°C/min, and cooled back to RT at the same rate. The effect of Dy-doping in $BaTiO_3$ powders on the sintered grain sizes and morphology was studied by scanning electron microscopy (SEM) (JEOL, JSM6400, Tokyo, Japan) micrographs taken directly from the surfaces of 0.5 cm-diameter pellets which were uniaxially pressed in hardened steel dies at a pressure of 200 MPa. The samples for SEM studies were, first, sputter coated with an approximately 25 nm-thick layer of gold-palladium alloy. Energy-dispersive X-ray spectroscopy (EDXS) (Kevex, Noran, CA, USA) analysis were performed on our samples to determine the elemental distribution in the powders.

Powder X-ray diffraction spectra were obtained from the 90°C-dried samples for phase characterization purposes. An X-ray powder diffractometer (Rigaku, D-Max/B, Tokyo, Japan) was used with FeK_α radiation at the step size of 0.02° and a preset time of 1 second to check the purity of $BaTiO_3$ powders. The possible presence of other polymorphs of $BaTiO_3$ in our hydrothermally synthesized powders were examined by Rietveld Analysis [12].

RESULTS AND DISCUSSION

The precipitates of $BaTiO_3$ aged at 90°C for 72 hours were already crystalline and had the cubic crystal structure. The small amounts of $BaCO_3$ present in these precipitates were easily removed by the dilute HCl-washing step included in the flowchart of Fig. 1(a). Figure 2 shows the comparative XRD charts of the "as-is" (trace-**A**, $BaCO_3$ present) and "HCl-washed" (trace-**B**, pure $BaTiO_3$) powders of pure $BaTiO_3$. XRD analysis showed that the cubic unit cell of pure $BaTiO_3$ powders had the lattice parameter, a = 4.0186 Å, with a cell volume of 64.89 Å^3. The experimental XRD pattern of Table I, generated from our HCl-washed samples displayed a better crystallographic quality [13] than the already present ICDD PDF (i.e., 31-174) for this phase.

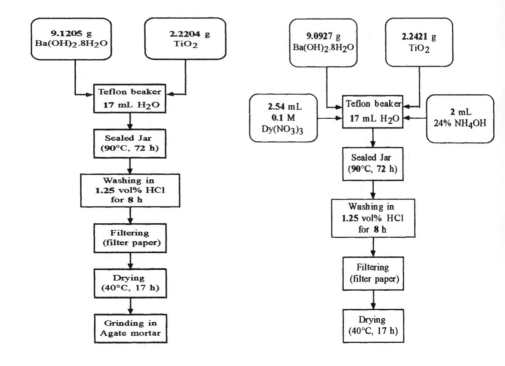

Fig. 1(a) Synthesis flowchart of pure BaTiO₃ Powders

Fig. 1(b) Synthesis flowchart of Dy:BaTiO₃ Powders

Table I. Tentative XRD Pattern of BaTiO₃ Powders

hkl	d_{calc}	d_{obs}	I/I_0
100	4.0185	4.0200	21
110	2.8727	2.8420	100
111	2.3201	2.3205	22
200	2.0093	2.0099	28
210	1.7972	1.7973	7
211	1.6406	1.6406	24
220	1.4208	1.4209	12
300	1.3395	1.3395	4
310	1.2708	1.2708	9
311	1.2116	1.2116	5
222	1.1601	1.1600	5
320	1.1145	1.1145	2
321	1.0740	1.0740	11

Dielectric Ceramic Materials

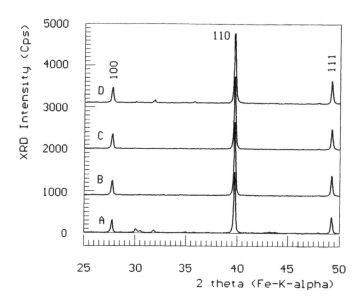

Fig. 2 XRD traces of pure and Dy:BaTiO₃ Powders

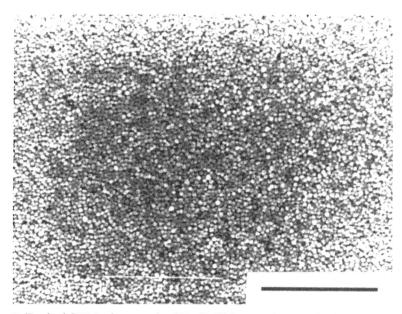

Fig. 3 Typical SEM micrograph of Dy:BaTiO₃ powders synthesized at 90°C
(Bar = 10 μm)

The powder samples of cubic BaTiO₃ synthesized (and then dried) at 90°C were checked for phase purity (and for the possible existence of its tetragonal polymorph in the powders) by Rietveld analysis. XRD data to be used in these analysis were collected by using a step size of 0.02° and a count time of 5 seconds. A Pseudo-Voigt profile function was used, and the lattice parameters, as well as the preferred-orientation, asymmetry, temperature, scale, mixing, half-width and background parameters were refined to a final R_{wp} of 5.51%. These analysis showed that the prepared BaTiO₃ powders were pure and did not contain the tetragonal phase. The FWHM values (for peaks 110 and 111) obtained from the Rietveld runs were used to determine the average crystallite size of the synthesized BaTiO₃ powders by using the Scherrer formula [14], and it was found to be around 28 nm.

Dy-doped (0.8 at%) BaTiO₃ powders were synthesized according to the flowchart given in Fig. 1(b). We have observed that the addition of the small volume of 0.1 M $Dy(NO_3)_3$ caused a decrease in the pH value (from 12.5 to 11.9) of the Ba-hydroxide and TiO_2 mixture heated to 90°C. This drop in the pH value of the precipitation suspensions also caused the poisoning of the resultant $Dy:BaTiO_3$ powders by a second phase of TiO_2. It was found that the addition of a 2 mL aliquot of 24% NH_4OH solution (after the addition of Dy-nitrate solution to the barium hydroxide and titania suspension in water) into the Teflon reaction beaker removed that second phase by increasing the pH to above 12.5. The influence of the initial NH_4OH addition on the phase purity of $Dy:BaTiO_3$ powders is depicted in the XRD spectra of Fig. 2 (trace-**D**: w/o NH_4OH (TiO_2 present), trace-**C**: w/ NH_4OH addition). The particle morphology of as-recovered Dy-doped BaTiO₃ powders (dispersed in isopropanol by an ultrasonic disrupter (Misonix, XL2015, USA) and then evaporated to dryness on a small piece of Al-foil) is shown in the SEM micrograph of Fig. 3. Dy-doped BaTiO₃ powders had monodisperse, spherical particles of about 200 nm average particle diameter.

The SEM micrographs given in Figs. 4(a) through 4(f) provide a chance of visual comparison to the strong role of dysprosium (Dy) added at the nominal level of 0.8 at% into the hydrothermally synthesized BaTiO₃ powders. Pure BaTiO₃ powders produced according to the processing flowchart of Fig. 1(a), which were heated at the temperatures of 1200, 1300, and 1400°C in an air atmosphere for 6 hours, displayed anomalous grain growth as shown in Figs. 4(a), 4c, and 4(e). However, in contrast to this behavior, the $Dy:BaTiO_3$ powders heated at the same temperatures, under exactly similar conditions, did not display exaggerated grain growth as given in the micrographs of Figs. 4(b), 4(d), and 4(f), respectively. On the other hand, $Dy:BaTiO_3$ pellets heated at 1500°C, for 6 h, had the typical microstructure given in Fig. 4(g).

Dielectric Ceramic Materials

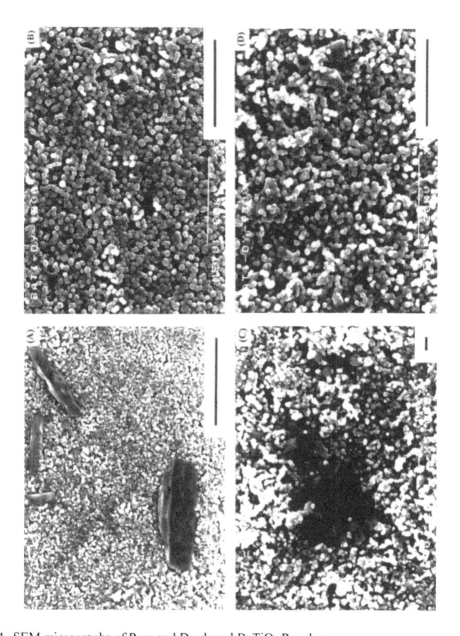

Fig. 4   SEM micrographs of Pure and Dy-doped BaTiO$_3$ Powders
(A) Pure BaTiO$_3$, 1200°C, Bar = 10 µm, (B) Dy:BaTiO$_3$, 1200°C, Bar = 10 µm,
(C) Pure BaTiO$_3$, 1300°C, Bar = 1 µm, (D) Dy:BaTiO$_3$, 1300°C, Bar = 10 µm

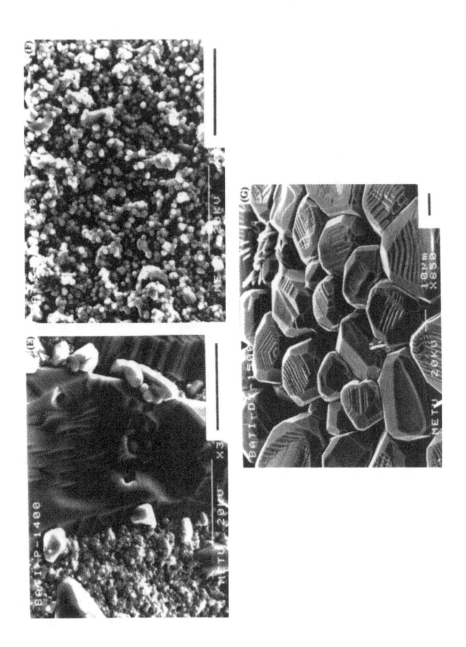

Fig. 4 SEM micrographs of Pure and Dy-doped BaTiO₃ Powders
(E) Pure BaTiO₃, 1400°C, Bar = 10 μm, (F) Dy:BaTiO₃, 1400°C, Bar = 10 μm,
(G) Dy:BaTiO₃, 1500°C, Bar = 10 μm

Dielectric Ceramic Materials

It was known [15] that even the slightest excess of TiO_2 in the initial, powder processing suspensions of barium hydroxide and titania to cause exaggerated grain growth (with occasional grains of sizes in the range of 20 to 40 μm) in the final, sintered microstructures. The presence of such an excess of TiO_2 was believed to induce the formation of a $BaTiO_3$-$Ba_6Ti_{17}O_{40}$ eutectic at high temperatures. The slightest presence of such a eutectic liquid at the sintering temperatures might then promote grain growth, especially at slow heating rates. Keeping the hydrothermal synthesis solutions in strongly alkaline conditions may also be helpful in the reduction of the amount of excess TiO_2 in the final powder bodies.

CONCLUSIONS

The experimental conditions and parameters of the hydrothermal synthesis of pure and 0.8 at% Dy-doped $BaTiO_3$ at 90°C, in an air atmosphere, were investigated. The dilute HCl-washing step included in the hydrothermal synthesis of pure and doped $BaTiO_3$ powders seemed to eliminate the need for carrying out the aging and washing stages of the process in controlled (i.e., free of CO_2) atmospheres. Sub-micron, monodisperse, spherical particles (with an average particle size of 150 nm) of $BaTiO_3$ were obtained.

The recovered precipitates were already crystalline and cubic with the space group of *Pm-3m*. A tentative XRD pattern (instead of the already present PDF 31-174 of the ICDD database) for the cubic form of pure $BaTiO_3$ was suggested in this study. 0.8 at% Dy-doping and the hydrothermal synthesis conditions of these powders were found to be quite effective in inhibiting the grain growth in $BaTiO_3$ samples heated over the temperature range of 1200 to 1400°C.

REFERENCES

1) A. Yamaji, Y. Enomoto, K. Kinoshita, and T. Murakami, "Preparation, Characterization, and Properties of Dy-Doped Small-Grained BaTiO₃ Ceramics," *J. Am. Ceram. Soc.*, **60**, 97-101 (1977).
2) T. Murakami and A. Yamaji, "Dy-Doped BaTiO₃ Ceramics for High Voltage Capacitor Use," *Am. Ceram. Soc. Bull.*, **55**, 572-575 (1976).
3) Y. Enomoto and A. Yamaji, "Preparation of Uniformly Small-Grained BaTiO₃," *Am. Ceram. Soc. Bull.*, **60**, 566-570 (1981).
4) N. W. Thomas, "Crystal Structure-Physical Property Relationships in Perovskites," *Acta Crystall.*, **B45**, 337-344 (1989).
5) R. C. Buchanan, Ceramic Materials for Electronics, Marcel Dekker, Inc.,

New York, 1986.

6) K. Kinoshita and Y. Yamaji, "Grain-size Effects on Dielectric Properties in Barium Titanate Ceramics," *J. Appl. Phys.*, **47**, 371-373 (1976).

7) S. Malbe, J. C. Mutin, and J. C. Niepce, "Distribution of Crystalline Cell Parameters in Powder Samples of $BaTiO_3$," *J. Chem. Phys.*, **89**, 825-843 (1992).

8) J. H. Peterson, "Process for Producing Insoluble Titanates," U.S. Patent, No: 2,216,655 (Oct. 22, 1940).

9) E. B. Slamovich and I.A. Aksay, "Structure Evolution in Hydrothermally Processed (<100°C) $BaTiO_3$ Films," *J. Am. Ceram. Soc.*, **79**, 239-247 (1996).

10) M. M. Lencka and R. E. Riman, "Thermodynamic Modeling of Hydrothermal Synthesis of Ceramic Powders," *Chem. Mater.*, **5**, 61-70 (1993).

11) A. C. Taş, "Hydrothermal Synthesis of $BaCO_3$-free $BaTiO_3$ Powders at 90°C," *Patent Pending*, Turkish Patent Institute, Turkey, Appl. No: 96-00539.

12) A. Sakthivel and R. A. Young, "Rietveld Analysis of X-Ray Powder Diffraction Patterns: *Program DBWS-9411 PC*," Version: March 1995. School of Physics, Georgia Institute of Technology, Atlanta, GA 30332, USA.

13) D. E. Appleman and H. T. Evans, "Least-squares and Indexing Software for XRD Data," U. S. Geological Survey, *Computer Contribution No. 20*, U.S. National Technical Information Service, Document PB-216188 (1973).

14) H. Hsiang and F. Yen, "Effect of Crystallite Size on the Ferroelectric Domain Growth of Ultrafine $BaTiO_3$ Powders," *J. Am. Ceram. Soc.*, **79**, 1053-60 (1996).

15) A. C. Caballero, C. Moure, P. Duran, and J. F. Fernandez, "High Density Zn-Doped $BaTiO_3$ Ceramics," Ceramic Transactions, Vol. 32, pp. 201-210, *Dielectric Ceramics: Processing, Properties, and Applications*, (Eds.) K. M. Nair, J. P. Guha and A. Okamoto, The American Ceramic Society, OH, USA, 1993.

Preparation of Lead Zirconate Titanate $(Pb(Zr_{0.52}Ti_{0.48})O_3)$ by Homogeneous Precipitation and Calcination

E. Emre Ören and A. Cüneyt Taş

Department of Metallurgical and Materials Engineering, Middle East Technical University, Ankara 06531, Turkey

ABSTRACT

Aqueous solutions of lead chloride $(PbCl_2)$, zirconium oxychloride $(ZrOCl_2.8H_2O)$ and titanium tetrachloride $(TiCl_4)$, in appropriate volumetric amounts, were used as the starting materials in the synthesis of phase-pure $Pb(Zr_{0.52}Ti_{0.48})O_3$ powders. Preparation of the phase-pure PZT powders were achieved, in the presence of urea (CH_4N_2O), by the chemical powder synthesis route of homogeneous precipitation.

Calcination and phase evolution behaviors of PZT precursor powders were studied as a function of temperature by powder XRD (X-ray diffraction) in an air atmosphere, over the temperature range of 90 to 750°C. Morphological properties of the precipitated $Pb(Zr_{0.52}Ti_{0.48})O_3$ powders were investigated by the SEM (scanning electron microscopy) studies. Semi-quantitative chemical analysis of the samples were performed by EDXS (energy-dispersive X-ray spectroscopy).

INTRODUCTION

Lead zirconate titanate, $PbZr_xTi_{1-x}O_3$ (PZT), ceramics are of great technological interest due to their excellent piezoelectric and ferroelectric properties [1, 2]. PZT ceramics are extensively used as electromechanical transducer materials. The electromechanical response of these ceramics is maximum when x corresponds to the composition of the morphotrophic phase boundary (MPB) which separates the tetragonal (T) and rhombohedral (R) phases towards Ti-rich and Zr-rich sides, respectively. The precise determination of the MPB composition range, which is believed to be quite narrow, has, therefore, attracted immense interest.

Conventionally, the PZT phase is prepared by solid state reactive firing of the constituent oxides (PbO, ZrO_2 and TiO_2). However, due to intermediate reactions which lead to the formation of $PbTiO_3$ (PT) and $PbZrO_3$ (PZ), the PZT formed by this method is chemically heterogeneous, including compositional fluctuations, which modify several electrical properties [3, 4].

The variations in composition may lead to a diffuse MPB between the tetragonal and rhombohedral PZT phases. Another negative feature of such a practice is the production of agglomerated powders with low sinterability. Moreover, the completion of the reactions by long-range diffusion also requires temperatures in excess of 1000°C. A lower calcination temperature, on the other hand, is always desirable since it yields fine powders of improved surface reactivity.

PZT's synthesized by chemical methods have resulted in powders with low compositional fluctuations, narrow MPB and of high reactivity [5-7]. Among these chemical methods were the sol-gel technique [6], and the partial chemical methods as described by Kakegawa [5] and Yamamoto [8]. In the method proposed by Yamamoto [8], lead-oxalate was precipitated on ZT particles processed by hydrothermal reaction, or on ZrO_2 and TiO_2 particles. However, the use of zirconia and titania particles also resulted in intermediate reaction products and a consequent deviation in the PZT stoichiometry. It was also reported by Singh, et al. [9] that single-phase $PbZr_xTi_{1-x}O_3$ powders were synthesized at about 600°C by a semi-wet procedure.

In the present study, the experimental details of the synthesis of lead zirconate titanate phase from water-soluble salts of Pb-, Zr- and Ti-chlorides by homogeneous precipitation, via urea decomposition, were presented. The decomposition of urea in aqueous solutions was accompanied by the slow and controlled supply of ammonia and carbon dioxide into the solution [10]. The smooth pH increase obtained by the decomposition of urea, in unison with the steady supply of OH^- and CO_3^{2-} ions, typically leads to the precipitation of metal hydroxycarbonates of controlled particle morphology [11-13]. Homogeneous precipitation from aqueous solutions, in the presence of urea, has previously been used to produce dispersed spherical particles of basic lanthanide carbonates [14], CeO_2 [15], YAG ($Y_3Al_5O_{12}$) [16], $LaAlO_3$ [17], YIG ($Y_3Fe_5O_{12}$) [18], $BaTiO_3$ [19], and $PbZrO_3$ [20]. In this study homogeneous precipitation techniques, similar to those described in the literature [11-20], were employed and shown to be successful for the preparation of phase-pure PZT after calcination in air at 500°C.

EXPERIMENTAL PROCEDURE

$PbCl_2$ (99.5%, Merck, Darmstadt, Germany), $ZrOCl_2.8H_2O$ (99%, Merck, Darmstadt, Germany), and $TiCl_4$ (Merck, Darmstadt, Germany) were used as the starting chemicals. 0.2 M $ZrCl_4$ and 0.02 M $PbCl_2$ aqueous stock solutions were first prepared in de-ionized water. The as-received $TiCl_4$ (9.1191M) was diluted to 2 M $TiCl_4$ by de-ionized water. Reagent-grade urea (CH_4N_2O, 99.5% Riedel-De Haen AG, Germany) was used as the precipitation agent.

(1) Homogeneous Precipitation

A total of 160 mL of cation stock solutions (i.e., lead, zirconium and titanium chlorides) were thoroughly mixed in a glass beaker. 27 g of urea dissolved in de-ionized water (350 mL) was then mixed with the above solution. The resultant clear solution was slowly heated in 1 h to 90°C in a water bath. Precipitation started by the end of the next 70 min with a slight turbidity in the clear solution. Precipitates were aged for 2 hours at 90°C. pH-values were continuously recorded as a function of time.

Precipitation experiments were also repeated for varying urea/cation ratios (i.e., 75, 60, 45, 30, 15, and 10) and for different aging times (120, 60, 30, and 10 min). Following aging, the precursors were quenched to room temperature. Precipitates were then separated from the mother liquors by filtering. Recovered precursors were washed four times with deionized-water. The washed precipitates were oven-dried overnight at 90°C in an air atmosphere.

(2) Calcination

Dried precursors were slightly ground in an agate mortar and then calcined in an air atmosphere, in alumina boats or crucibles, over the temperature range of 100 to 750°C for calcination times of up to 6 hours at the peak temperatures.

(3) Powder Sample Characterization

Powder XRD spectra were obtained from the dried precipitates, as well as from powders heated at 100 to 750°C, in open alumina crucibles in a dry air atmosphere. An X-ray powder diffractometer (Rigaku, D-Max/B, Tokyo, Japan) was used in this study with CuK_α radiation at the step size of 0.02° and a preset time of 1 second to perform phase characterization, and to check the phase purity of the synthesized PZT powders.

Particle size and morphological characteristics of the powders were monitored through scanning electron microscopy (SEM) photomicrographs (Model JSM6400, JEOL, Tokyo, Japan). An ultrasonic disrupter (Misonix, Inc., Model: XL 2015, NY, USA) was used to form suspensions of PZT powders. 1 g of powder was ultrasonicated in 15 mL of isopropyl alcohol for 10 minutes, prior to evaporation to dryness on tiny glass slides. Energy dispersive X-ray spectroscopy (Kevex, Foster City, CA) analysis were performed on green PZT pellets to acquire semi-quantitative elemental-distribution information. The EDXS analysis were believed to be accurate to about ± 3 wt%.

RESULTS AND DISCUSSION

The high temperature behavior of PZT precursors produced by homogeneous precipitation experiments were monitored by the XRD charts of the isothermally heated powder samples. Fig. 1 shows the phase distribution and crystallization paths of these powders as a function of increasing calcination temperature.

It was observed that the precursor powders calcined (in an air atmosphere) at 100 and 200°C were already crystalline, and composed of the phases of Cerussite ($PbCO_3$: ICDD PDF 5-417), Hydrocerussite ($Pb_3(CO_3)_2(OH)_2$: ICDD 13-131) and Laurionite (PbClOH : ICDD 31-680).

Following calcination at 300°C for 6 h, the PZT precursors were composed of the phases of litharge (PbO : ICDD 5-561), zirconia (ZrO_2 : ICDD 37-31), and titanium oxide (Ti_7O_{13} : ICDD 18-1403). This observation showed that over the temperature range of 200 to 300°C, lead hydroxycarbonate phases (i.e., Cerussite and Hydrocerussite) have totally been converted to their oxides (i.e., Litharge). It was also interesting to note that at this temperature the free phases of zirconium and titanium oxides did crystallize from the precursor powders. Upon calcination at 400°C, the phases of litharge and zirconia completely disappeared, and the formation of $PbZrO_3$ (ICDD 35-739 and 20-608) was observed, and a small amount of PZT ($Pb(Zr_{0.52}Ti_{0.48})O_3$: ICDD 33-784) evolved. This temperature may be regarded as the starting temperature for the formation / crystallization of the PZT phase.

Calcination of our precursor powders at temperatures higher than 400°C (in air atmosphere), as seen in Fig.1 (urea/cation ratio = 45), all yielded single-phase PZT. A pyrochlore phase was not observed. Therefore, it can be stated that "500°C" is the temperature of formation for PZT powders prepared by homogeneous precipitation in aqueous solutions in the presence of urea. The EDXS analysis performed on the 700°C-calcined samples gave the mean atomic percentages of 20.04 : 10.29 : 9.67 (for Pb : Zr : Ti).

The XRD traces given in Fig. 2 show the phase formation behavior of precursor powders (all being heated at 700°C) as a function of changing "Urea/Cation" ratios in aqueous solutions. It was hereby shown that there must be a minimum level for this ratio for the formation of PZT phase. At the ratio of 10, the PZT phase can not be formed, and the resulting phases are a mixture of Cerussite and Hydrocerussite.

Figure 3 shows the solution "pH" versus "time" curves recorded in the precipitation solutions on a real-time basis. Various Urea/Cation ratios in the solutions have been studied and the final pH values required for the

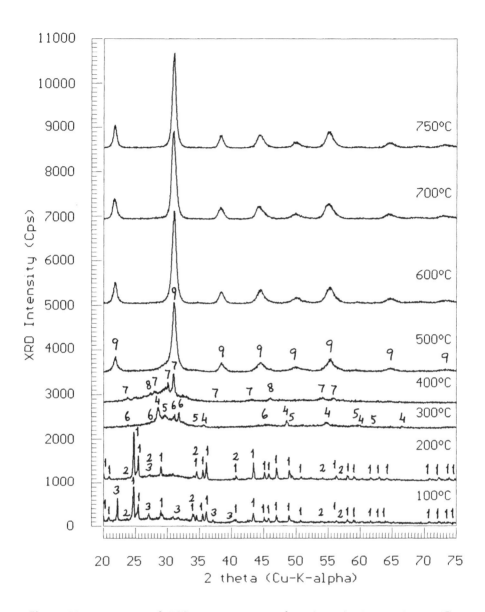

Fig. 1 XRD spectra of PZT precursor powders (*urea/cation ratio = 45*) heated at various temperatures for 6 h in an air atmosphere (ICDD PDF **1**: 5-417, **2**: 13-131, **3**: 31-680, **4**: 5-561, **5**: 37-31, **6**: 18-1403, **7**: 35-739, **8**: 20-608, **9**: 33-784)

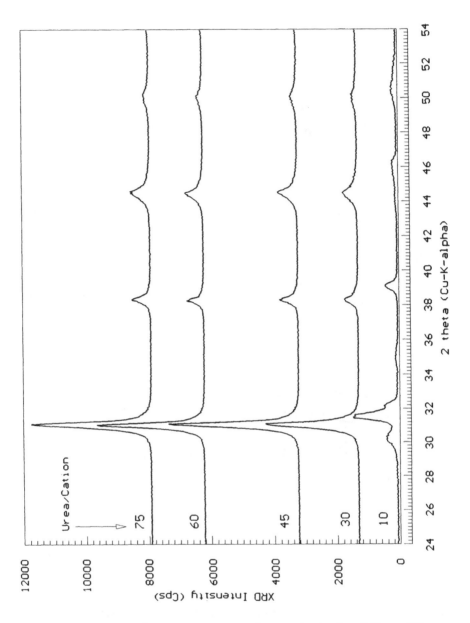

Fig. 2 XRD spectra of PZT precursor powders synthesized at different "Urea / Cation" ratios (calcination temperature : 700°C, 6 h)

Dielectric Ceramic Materials

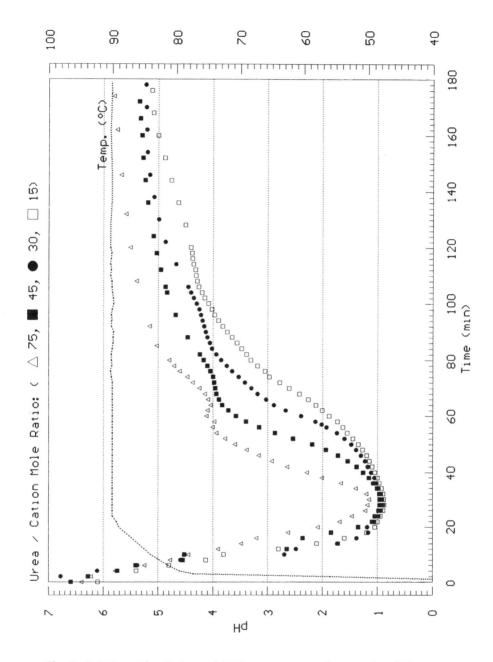

Fig. 3 "pH" vs. "time" plots of PZT precursor powders aged at different "Urea / Cation" ratios

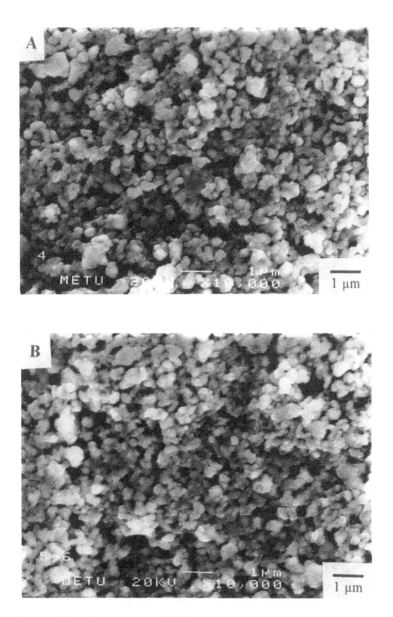

Fig. 4 SEM micrographs of PZT powders calcined at 700°C for 6 h
Urea/Cation ratio: (A) 45, (B) 30

precipitation reactions to proceed were determined. It was apparent from Fig. 3 that as the amount of urea in the solutions increased, the final pH values of the solutions were also increased. The "shoulders" displayed in these charts indicated the precipitation-start points, as a function of pH and aging time.

By using the data obtained in Figs. 2 and 3, it can be stated that the critical "urea/cation ratio" in the mother liquors must be located somewhere between 10 and 15, for the homogeneous precipitation of PZT powders in aqueous solutions. The morphology of the produced powders were examined by the SEM micrographs given in Fig. 4. The PZT powders had sub-micron particles, although they were highly agglomerated.

CONCLUSIONS

The lead zirconate titanate $(Pb(Zr_{0.52}Ti_{0.48})O_3)$ phase has been synthesized from aqueous solutions, following calcination at 500°C in an air atmosphere. This study showed that it is possible to obtain chemically homogeneous sub-micron powders of $Pb(Zr_{0.52}Ti_{0.48})O_3$ by homogeneous precipitation from the chlorides of the respective cations via the decomposition of urea. A considerable decrease in the processing temperature may thus be obtained in comparison to the processes which involve powder mixing and reactive firing in the solid-state. This decrease in synthesis temperature is also believed to help in eliminating the need for the strict use of atmosphere control to compensate for the problems of lead volatilization in large-scale manufacturing processes.

REFERENCES

1) B. Jaffe, W. R. Cook, and H. Jaffe, "*Piezoelectric Ceramics*," pp. 135-83, Academic Press, New York (1971).
2) H. Kanai, O. Furukawa, H. Abe, and Y. Yamashita, "Dielectric Properties of $Pb(X)ZrTiO_3$ (X=Ca,Sr,Ba) Ceramics," *J. Am. Ceram. Soc.*, **77**, 2620-24 (1994).
3) S. S. Chandatreya, R. M. Fulrath, and J. A. Pask, "Reaction Mechanisms in the Formation of PZT Solid Solutions," *J. Am. Ceram. Soc.*, **64**, 422-25 (1981).
4) K. Kakegawa, J. Mohri, T. Takahashi, H. Yamamura, and S. Shirasaki, "A Compositional Fluctuation and Properties of $Pb(Zr,Ti)O_3$," *Solid State Commun.*, **24**, 769-72 (1977).
5) K. Kakegawa, K. Arai, Y. Sasaki, and T. Tomizawa, "Homogeneity and Properties of Lead Zirconate Titanate Prepared by a Combination of Thermal Spray Decomposition Method with Solid-state Reaction," *J. Am. Ceram. Soc.*, **71**, C49-C52 (1988).

Dielectric Ceramic Materials

6) G. Toimandl, A. Stiegelschmitt, and R. Bohner, "Lowering the Sintering Temperature of PZT Ceramics by Sol-gel Processing;" p. 56, in *Science of Ceramic Chemical Processing. Part 1, Sol-Gel Science*. Edited by L. L. Hench and D. R. Ulrich, J. Wiley and Sons, New York (1986).

7) M. A. Zaghete, C. O. Paiva-Santos, J. A. Vareta, E. Longo, and Y. P. Mascarenhas, "Phase Characterization of Lead Zirconate Titanate obtained from Organic Solution of Citrates," *J. Am. Ceram. Soc.*, **75**, 2089-93 (1992).

8) T. Yamamoto, "Optimum Preparation Methods for Piezoelectric Ceramics and Their Evaluation," *Am. Ceram. Soc. Bull.*, **71**, 978-85 (1992).

9) A. P. Singh, S. K. Mishra, D Pandey, C. D. Prasad, and R. Lal, "Low-Temperature Synthesis of Chemically Homogeneous Lead Zirconate Titanate (PZT) Powders by a Semi-wet Method," *J. Mat. Sci.*, **28**, 5050-55 (1993).

10) H. H. Willard and N. K. Tang, "A Study of the Precipitation of Aluminum Basic Sulphate by Urea," *J. Am. Chem. Soc.*, **59**, 1190-1192 (1937).

11) B. C. Cornilsen and J. S. Reed, "Homogeneous Precipitation of Basic Aluminum Salts as Precursors for Alumina," *Am. Ceram. Soc. Bull.*, **58**, 1199-1200 (1979).

12) J. E. Blendell, H. K. Bowen, and R. L. Coble, "High-Purity Alumina by Controlled Precipitation from Aluminum Sulphate Solutions," *Am. Ceram. Soc. Bull.*, **63**, 797-802 (1984).

13) J. Sawyer, P. Cairo, and L. Eyring, "Hydroxy-Carbonates of the Lanthanide Elements," *Revue de Chim. Miner.*, **10**, 93-104 (1973).

14) D. J. Sordelet and M. Akinc, "Preparation of Spherical, Monosized Y_2O_3 Precursor Particles," *J. Coll. Int. Sci.*, **122**, 47-59 (1988).

15) P. Chen and I-Wei Chen, "Reactive Cerium (IV) Oxide Powders by the Homogeneous Precipitation Method," *J. Am. Ceram. Soc.*, **76**, 1577-1583 (1993).

16) D. J. Sordelet, M. Akinc, M. L. Panchula, Y. Han, and M. H. Han, "Synthesis of Yttrium Aluminum Garnet Precursor Particles by Homogeneous Precipitation," *J. Eur. Ceram. Soc.*, **14**, 123-127 (1994).

17) E. Taspinar and A. C. Tas, "Low Temperature Chemical Synthesis of Lanthanum Monoaluminate," *J. Am. Ceram. Soc.*, **80**, 133-141 (1997).

18) Y. S. Ahn and M. H. Han, "Synthesis of Yttrium Iron Garnet Precursor Particles by Homogeneous Precipitation," *J. Mat. Sci.*, **31**, 4233-40 (1996).

19) S. Kim, "Preparation of Barium Titanate by Homogeneous Precipitation," *J. Mat. Sci.*, **31**, 3643-45 (1996).

20) E. E. Oren, E. Taspinar, and A. C. Tas, "Preparation of Lead Zirconate by Homogeneous Precipitation and Calcination," *J. Am. Ceram. Soc.*, **80**, 2714-16 (1997)

PREPARATION AND LOW-FREQUENCY ELECTRICAL PROPERTIES OF LaFeO₃ BY CHEMICAL COPRECIPITATION

CHEN-FENG KAO AND KOU-FANG HUANG
Department of Chemical Engineering, National Cheng Kung University
Tainan, 70101, TAIWAN

ABSTRACT
Various mole ratios of $La(NO_3)_3 \cdot 6\ H_2O$ and $Fe(NO_3)_3 \cdot 9\ H_2O$ aqueous solutions with sodium hydroxide solution were used to prepare the precursors with OH^- ligands by using chemical coprecipitation for controlling stoichiometry accurately. After freeze drying, the precursors were calcined in the oven with a heating rate of 10 K/min to 1073 K for 4 h to obtain the corresponding compounds. Then the powders were sintered in the different atmosphere(air or nitrogen) and temperature (1373, 1473 or 1573 K) for 8 h.   XRD, FTIR, SEM and XPS(ESCA) were used to analyze the crystallinity and microstructure of the calcined powders and the sintered bodies.
The DC resistivity decreases with increasing charged voltage. And LaFeO₃ is a semiconductor. The DC resistivity increases with increasing doping amount of ZrO_2. But the resistivity and dielectric constant are smaller in sintering atmosphere of air than that in nitrogen. The dielectric constant increases with increasing temperature.

INTRODUCTION
Kao et al. have studied the preparation and characterization of fine La-Cr-Zr oxides ceramics by chemical coprecipitation at high temperatures[1]. LaBO₃ (B= transition metal element, such as Cr, Fe, Mn, Co etc.) is a perovskite-type oxide, having the specific properties of highly nonstoichiometry and mixed conductivity by both ionic

Dielectric Ceramic Materials

and electronic charge carriers[2,3]. It was mainly used as the functional materials, such as an electronic conductive semiconductor[4], a catalyst[5], a gas sensor[6], an ethanol sensor[7] and an electrode in high temperature fuel cell[8]. Owing to perovskite structure with strong nonstoichiometry, it is easy to make the oxygen deficiency in perovskite-type oxide[9]. Therefore perovskite-type oxide has good electrical conductive properties,, and its properties change with synthetic method, sintering temperature and atmosphere control. Zirconia[10,11] is biocompatible, non-toxic and of high strength ,therefore it is used to improve the toughness of the ceramics.Chemical coprecipitation has many advantages: (1) It can control the chemical stoichiometric ratio of each element accurately; (2) It is mixed with atomic scale; (3)It is very uniform in the powder; and (4) Its calcining temperature is lower than that in solid-state reaction and it can save energy. There is no literatures describing the electrical properties of La-Fe and La-Fe-Zr oxides ceramics until now.   In this study, the preparation and electrical properties of La-Fe and La-Fe-Zr oxides by chemical coprecipitation were investigated.

EXPERIMENTAL
The chemical equation of La-Fe-Zr oxides in the chemical coprecipitation can be written as follows:
For $La^{3+}/Fe^{3+} > 1$
$La(NO_3)_3 \cdot 6 H_2O + (1-x) Fe(NO_3)_3 \cdot 9 H_2O + x ZrOCl_2 \cdot 8 H_2O +(6-x) NaOH$
$\rightarrow La(OH)_3 +(1-x)Fe(OH)_3 + x ZrO(OH)_2 + (6-3x)NaNO_3 + 2x NaCl + (15-x)H_2O$
For $La^{3+}/Fe^{3+} < 1$
$(1-x)La(NO_3)_3 \cdot 6 H_2O + Fe(NO_3)_3 \cdot 9 H_2O + x ZrOCl_2 \cdot 8 H_2O +(6-x) NaOH$
$\rightarrow (1-x)La(OH)_3 +Fe(OH)_3 + xZrO(OH)_2 +(6-3x)NaNO_3 + 2x NaCl + (15+2x)H_2O$
The chemicals used were all of high purity. $La(NO_3)_3 \cdot 6 H_2O$, $Fe(NO_3)_3 \cdot 9 H_2O$ and $ZrOCl_2 \cdot 8 H_2O$ were used as the starting reagents of La(III), Fe(III) and Zr(IV), respectively.
Figure 1 shows the flow chart of experiment. The sample index and mole ratio of La:Fe:Zr for this study are listed in Table I.

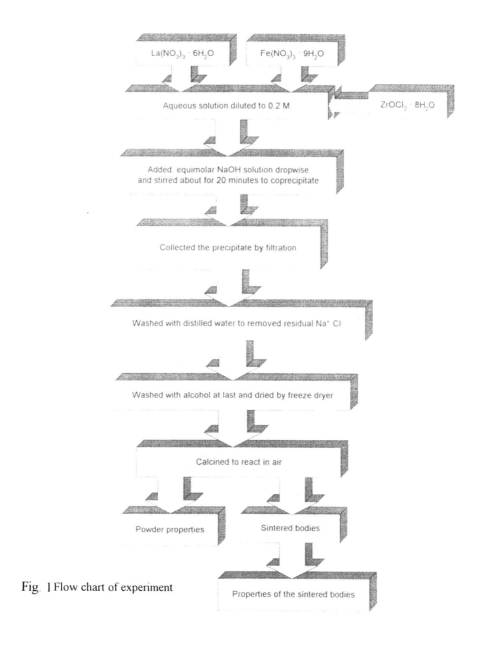

Fig. 1 Flow chart of experiment

La(NO₃)₃ · 6H₂O   Fe(NO₃)₃ · 9H₂O

Aqueous solution diluted to 0.2 M   ZrOCl₂ · 8H₂O

Added equimolar NaOH solution dropwise and stirred about for 20 minutes to coprecipitate

Collected the precipitate by filtration

Washed with distilled water to removed residual Na⁺ Cl⁻

Washed with alcohol at last and dried by freeze dryer

Calcined to react in air

Powder properties   Sintered bodies

Properties of the sintered bodies

Table I The Sample Index and Mole Ratio of the Solution for Coprecipitation

Mole Ratios	La^{3+}	Fe^{3+}	Zr^{4+}	NaOH
101000	1.00	1.00	0.00	6.00
109901	1.00	0.99	0.01	5.99
109802	1.00	0.98	0.02	5.98
109505	1.00	0.95	0.05	5.95
991001	0.99	1.00	0.01	5.99
981002	0.98	1.00	0.02	5.98
951005	0.95	1.00	0.05	5.95

RESULTS AND DISCUSSION

TGA/DTA and XRD analyses of the coprecipitated powder
There is a large weight loss between 303 K and 633 K for the powder of La:Fe=1:1 in TGA/DTA, owing to OH⁻ reacted into water to give heat and the vaporization of water from absorption heat. There is also a weight loss in the vicinity of 633 K(about 12 %) beginning, because of the the vaporization of the fused sodium chloride and sodium nitrate. The weight becomes constant at 1073 K. The total weight loss is about 19.5 % of its original weight. It means that one-fifth of the powder is lost due to melting. Compared with Joint Committee on Powder Diffraction Standard (JCPDS) card of LaFeO$_3$(Card No. 37-1493), there is no reaction occurring for La:Fe=1:1 until 773 K in XRD analysis(Fig.2). There is the phase of LaFeO$_3$ ($2\theta = 32.19^0$) appeared at 773K, but this diffractive peak is no so clear. As the temperature increases, the intensity of the diffractive peaks also increases. Heated to 1073 K, the product of LaFeO$_3$ is complete for the reaction(Fig.2). Therefore 1073 K was chosen as the calcining temperature.

FTIR analysis
Fig. 3 showed that the FTIR spectra of the precursor LaFeO$_3$ and calcined powder

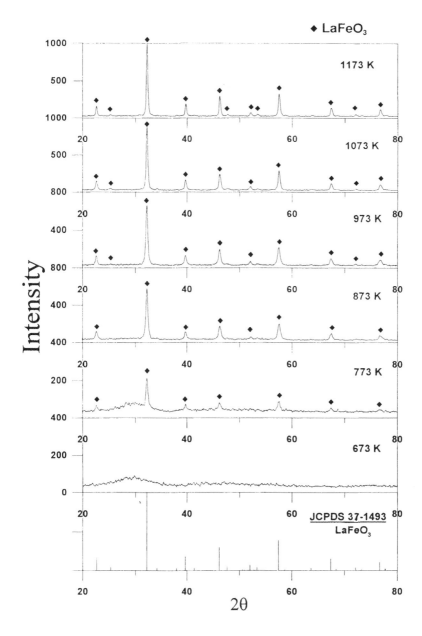

Fig. 2 XRD patterns of the coprecipitated sample La:Fe=1:1 calcined at 673, 773, 873, 973,1073 and 1173 K for 4 h in air

La:Fe=1:1 at 673,773,873,973,1073 and 1173 K for 4 h in air. The absorptive bands of OH⁻ at 1200-1450 cm⁻¹ and 3200-3700 cm⁻¹(Fig.3) change clearly with temperature. There are absorptive peaks, 400 cm⁻¹ for Fe-O-Fe bending and 595 cm⁻¹ for Fe-O stretching, appeared but not clear at 773 K, indicating that the crystal phase starts to form. Up to 973 K OH⁻ reaction is near to be complete, and NO₃⁻ & organics vaporize completely seen from the two absorptive positions at 850 cm⁻¹ and 1480 cm⁻¹. And at 1073 K the absorption of the perovskite La-Fe oxide is very great. The above results are in very good agreement with those of TGA

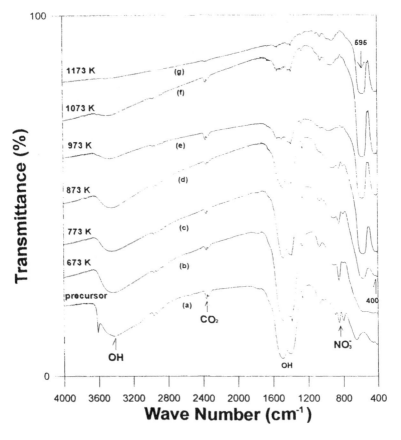

Fig.3 FTIR spectra of the precursor LaFeO₃ and the powder calcined at various temperatures:(a) precursor (b)673K (c)773K (d)873K (e)973K (f)1073K (g)1173K

The SEM and TEM images of La:Fe=1:1 at various temperatures for 4 h in air are shown in Figs. 4 and 5, respectively. It is obvious from Fig. 4 that at T≤973 K the particle size is small and there is a clear phenomenon of agglomeration. This agglomeration is attributed to Van der Waals forces between molecules, the absorption of vapor, and the defect for the surface of the grain or having no balance in electricity to produce aggregation and tendency of adsorptive outer substances [12-13]. Above the reaction temperature(T>973 K), the grain size increases quickly and the agglomeration decreases. The particle size increases with increasing temperature. Therefore, the calcining temperature was set at 1073 K for 4 h in air.

Phase analysis of the calcined powder

Added different ratios of zirconia to La-Fe oxide at 1073 K for 4 h in air(Fig. 6). From Fig. 6, there are only two phases $LaFeO_3$ and ZrO_2(JCPDS 37-1484) existed, and no new compound $La(Fe_{1-x}Zr_x)O_3$ or $(La_{1-x}Zr_x)FeO_3$ forms. And the intensity of the diffractive peak of ZrO_2 increases with increasing doped zirconia. This phenomenon is attributed to insufficient calcining temperature and time to diffuse zirconium ion to react with other ions.

SEM, mapping and particle size analyses

Fig. 7 showed that the SEM images of the sample with various zirconium mole percents calcined at 1073 K for 4 h in air. There is no so big difference between those in Fig. 7 and that at 1073 K in Fig. 4. There is also some agglomeration occurred in Fig.7. The mapping images of the powder La:Fe=1:1 calcined at 1073 for 4 h in air were shown in Fig.8. It is obvious from Fig. 8 that the powder is very uniform in chemical coprecipitation.

Fig. 9 showed that the probability number density graph for the coprecipitated powder of La:Fe=1:1 calcined at 1073 K for 4 h in air. The particle size range is at 10-20 μm, median diameter is 11.8 μm and modal diameter is 17.5 μm.

Determination of the property of the sintered body

XRD analysis

The XRD patterns of the sintered bodies with various zirconium mole percents at 1573 K for 8 h in air and in nitrogen are shown in Figs. 10 and 11, respectively.

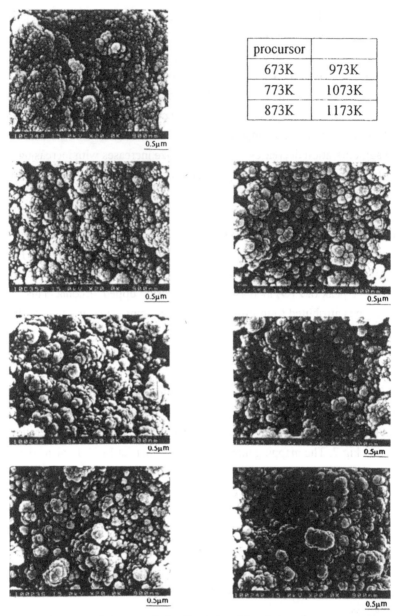

procursor	
673K	973K
773K	1073K
873K	1173K

Fig. 4 SEM images of the sample La:Fe=1:1 calcined at various temperatures for 4 h in air

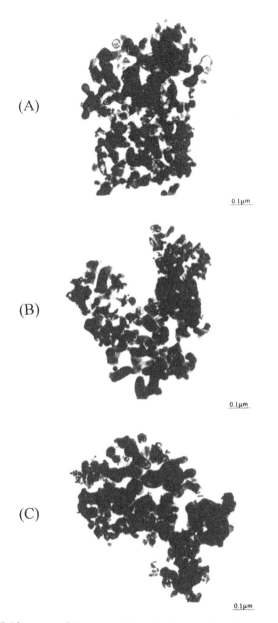

(A)

0.1μm

(B)

0.1μm

(C)

0.1μm

Fig. 5 TEM images of the sample La:Fe=1:1 calcined at various temperatures
    for 4 h in air (A)973 K   (B)1073 K   (C)1173 K

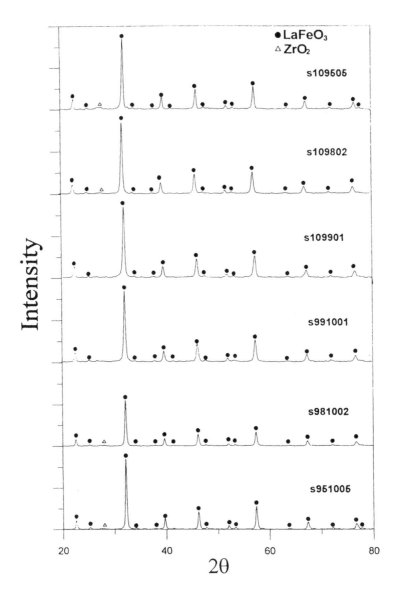

Fig. 6 XRD patterns of the powder with various zirconium mole percents calcined at 1073 K for 4 h in air

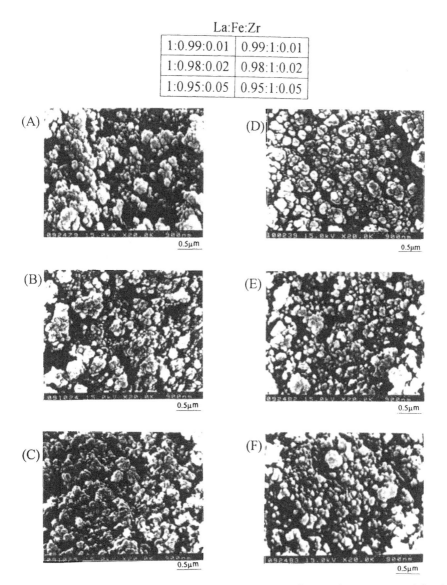

La:Fe:Zr	
1:0.99:0.01	0.99:1:0.01
1:0.98:0.02	0.98:1:0.02
1:0.95:0.05	0.95:1:0.05

Fig. 7 SEM images of the sample with various zirconium mole percents calcined at 1073 k for 4 h in air.

La   Fe

Fig. 8 Mapping images of the powder La:Fe=1:1 calcined at 1073 K for 4 h in air

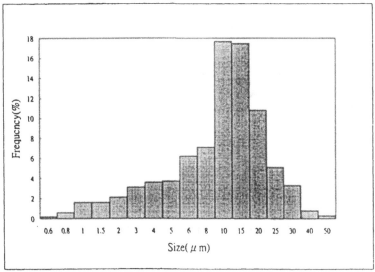

Fig. 9 Probability number density graph for the coprecipitated powder of La:Fe=1:1
calcined at 1073 for 4 h in air

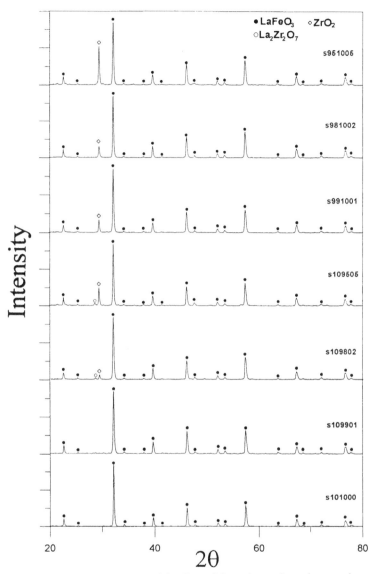

Fig. 10 XRD patterns of the sintered bodies with various zirconium mole percents at 1573 K for 8 h in air

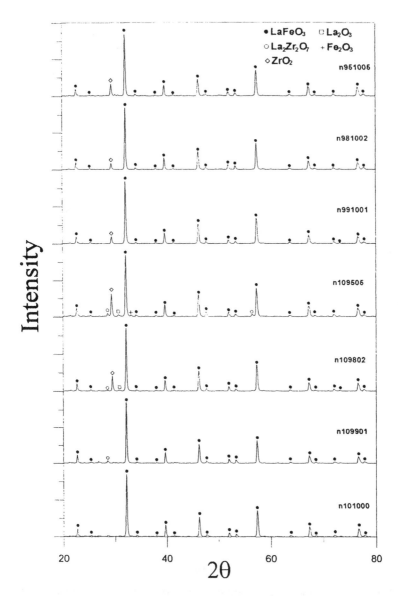

Fig. 11 XRD patterns of the sintered bodies with various zirconium mole percents at 1573 K for 8 h in nitrogen

Dielectric Ceramic Materials

It is obvious from Fig. 10 that in air at $La^{3+}/Fe^{3+}<1$, its major compound is $LaFeO_3$ and there is a small amount of ZrO_2 present. The diffractive peak of ZrO_2 increases with increasing doping ZrO_2 content, indicating that there is no substitution for La^{3+} or Fe^{3+} to form a new compound and only ZrO_2 mixed with $LaFeO_3$. This phenomenon is very clear at 1573 K in sintering. At $La^{3+}/Fe^{3+}>1$ there is still a small amount of $La_2Zr_2O_7$ formed (JCPDS 17-450) due to small amount of Zr^{4+} with an excess of La^{3+}, specially clear at doping 5 % mole of zirconium. The XRD spectra in nitrogen at $La^{3+}/Fe^{3+}<1$ (Fig.11) are similar to those in air. But that in nitrogen at $La^{3+}/Fe^{3+}>1$ there are La_2O_3(JCPDS 22-641) and Fe_2O_3(JCPDS 33-664) phases existed. Because the sintering in air the diffusion of O^{2-} in sample can be made up from oxygen in air to attain uniform sintering easily. That the sintering in nitrogen the diffusion of O^{2-} is from the sample itself completely to make other phases or deficiencies present.

XPS(ESCA) analysis
X-ray photoelectron spectroscopy(XPS) is also called electron spectroscopy for chemical analysis(ESCA). It is used widely in chemical surface analysis, film detection, the nuclei of transion metal oxides and electron-transfer mechanism between atomic valence orbitals.   The $Fe_{(2p3)}$XPS of the sintered body with various zirconium mole percents at 1573 K for 8 h in air were shown in Fig. 12. Although there are 2+ and 3+ valences in iron, compared with the spectra of Fe and Fe_2O_3[14] and iron oxides[15],there is no 2+ valence in iron existed from the $Fe_{(2p3)}$XPS for $LaFeO_3$ in Fig. 12. For $La^{3+}/Fe^{3+}>1$ system, the main peak of $Fe_{(2p3)}$ shifts to lower energy as the zirconium content increases. And for $La^{3+}/Fe^{3+}<1$ system, the peak of $Fe_{(2p3)}$ shifts to lower energy at doping 1% zirconium and then to higher energy as the increasing zirconium. From the above, the conclusion is that the ratio of Zr^{4+} in La^{3+}/Fe^{3+} affects the bonding energy of $Fe_{(2p3)}$. From the literature[14] there are three different types of $O_{(1s)}$ bonding energies, lattice oxygen at 529.37-530.05 eV, chemisorbed oxygen at 531.18-531.80 eV and physical adsorbed oxygen at 532.46-533.69 eV. There are lattice and chemisorbed oxygens from the $O_{(1s)}$XPS analysis in the $LaFeO_3$. And there is no physical adsorbed oxygen in the vacuum system.

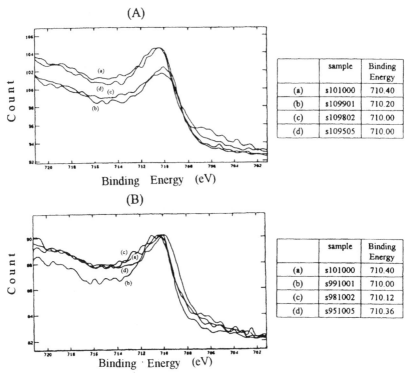

(A)

	sample	Binding Energy
(a)	s101000	710.40
(b)	s109901	710.20
(c)	s109802	710.00
(d)	s109505	710.00

Binding Energy (eV)

(B)

	sample	Binding Energy
(a)	s101000	710.40
(b)	s991001	710.00
(c)	s981002	710.12
(d)	s951005	710.36

Binding Energy (eV)

Fig. 12 $Fe_{(2p3)}$ of the sintered bodies with various zirconium mole percents at 1573 K for 8 h in air (A)La^{3+}/Fe^{3+}>1 system (B)La^{3+}/Fe^{3+}<1 system

<u>Measurement of resistivity for sintered body at room temperature</u>
The DC electrical resistivity vs. charged voltage for sintered bodies with various zirconium mole percents sintering at 1373 K for 8 h in air and in nitrogen were shown in Fig. 13 (A) & (B), respectively.   It is obvious from Fig. 13 that the DC resistivity decreases with increasing charged voltage. And the resistivity of $LaFeO_3$ is 10^7 to 10^8 ohm-cm, belonged to a semiconductor. The resistivity of $La_2Zr_2O_7$ at La^{3+}/Fe^{3+}>1 is larger than that of ZrO_2 at La^{3+}/Fe^{3+}<1. When the content of doping zirconium increases, the second phase of ZrO_2 or $La_2Zr_2O_7$ formed increases and the resistivity also increases. The resistivity for the sintered body in air is smaller than that in nitrogen, for the body sintered in air atmosphere can be supplied with oxygen to make the diffusion rate of the inner O^{2-} ion increase. Therefore the uniformity after sintering in air is better than that in nitrogen, and there is higher conductivity after sintering in air than that in nitrogen.

Dielectric Ceramic Materials

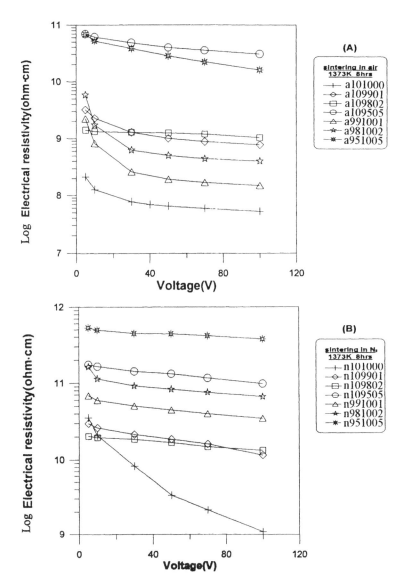

Fig. 13 DC electrical resistivity vs. charged voltage for sintered bodies with various zirconium mole percents sintering at 1373 K for 8 h   (A)in air   (B)in nitrogen

The AC resistivity of the sintered bodies with various zirconium mole percents sintering at 1373,1473 and 1573 K for 8 h in air and in nitrogen were listed in Tables II and III, respectively. It is indicated from Tables II and III that AC

Dielectric Ceramic Materials   131

resistivity decreases as increasing temperature and with increasing zirconium content. The AC resistivity is smaller than that DC resistivity by several orders. The AC resistivity of LaFeO$_3$ in sintering atmosphere of air is smaller than that in nitrogen.

Table II AC Resistivity of the Sintered Bodies With Various Zirconium Mole Percents Sintering at 1373,1473 and 1573 K for 8 h in Air
(Measured Frequency : 1MHz,   Unit: ohm-m)

Sample\Temperature	1373K	1473K	1573K
s101000	68.91	66.13	61.24
s109901	241.44	202.57	167.69
s109802	236.90	157.76	124.48
s109505	180.93	125.83	105.86
s991001	277.66	188.02	166.29
s981002	215.11	184.28	140.96
s951005	197.14	178.02	135.93

Table III AC Resistivity of the Sintered Bodies With Various Zirconium Mole Percents Sintering at 1373,1473 and 1573 K for 8 h in Nitrogen
(Measured Frequency : 1MHz,   Unit: ohm-m)

Sample\Temperature	1373K	1473K	1573K
n101010	98.88	97.23	70.22
n109901	56.95	38.08	25.46
n109802	46.12	22.95	14.03
n109505	25.62	20.37	11.70
n991001	45.54	33.64	19.93
n981002	41.41	25.25	10.08
n951005	35.15	24.34	9.55

Measurement of dielectric constant for the sintered body

Dielectric constant can be obtained by $\varepsilon_r = (t_a c_p)/(\pi r^2 \varepsilon_o)$, where t_a is the thickness of the sample, r is the measured radius of the holder, 2.5×10^{-3} m, ε_o is the the dielectric constant in air, 8.854×10^{-12} F/m and c_p is the measured capacitance. The dielectric constant vs. frequency for sintered bodies with various zirconium mole percents sintering at 1373 K for 8 h in air and in nitrogen were shown in Fig. 14 (A) and (B), respectively.

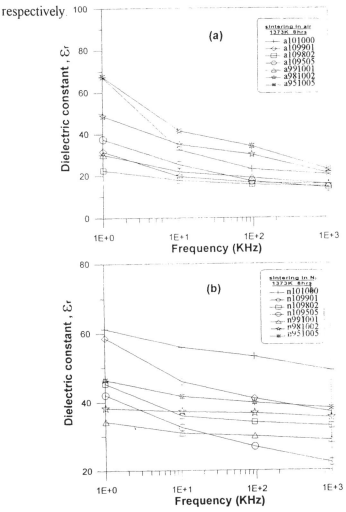

Fig. 14 Dielectric constant vs. frequency for sintered bodies with various zirconium mole percents sintering at 1373 K for 8 h (A)in air (B)in nitrogen

It is obvious from Fig. 14 that the dielectric constant decreases with increasing frequency, and the dielectric constant sintering in nitrogen is higher than that sintering in air. The dielectric constant increases with increasing temperature.

CONCLUSION

After freeze drying, the precursors were calcined in the oven to 1073 K for 4 h to obtain the corresponding compounds. There are lattice and chemisorbed oxygens from the $O_{(1s)}$XPS analysis in the $LaFeO_3$. There is no physical adsorbed oxygen in the vacuum system. The DC resistivity decreases with increasing charged voltage. And the resistivity of $LaFeO_3$ is at 10^7-10^8 ohm-cm, belonging to a semiconductor. The resistivity of $La_2Zr_2O_7$ at $La^{3+}/Fe^{3+}>1$ is larger than that of ZrO_2 at $La^{3+}/Fe^{3+}<1$. The AC resistivity is smaller than that DC resistivity by 10^4 to 10^6 orders. The DC and AC resistivities of $LaFeO_3$ in sintering atmosphere of air are smaller than those in nitrogen. The dielectric constant decreases with increasing frequency, and the dielectric constant sintering in nitrogen is higher than that sintering in air. The dielectric constant increases with increasing temperature.

REFERENCES

[1] Chen-Feng Kao and Jyh-Shen Jou, " Preparation and Characterization of Fine La-Cr-Zr Oxides Ceramics by Chemical Coprecipitation at High Temperatures", Ceramic Transactions, 71,521-538, in "Mass and Charge Transport in Ceramics", Edited by Kunihito Koumoto, Laurel Sheppard and Hideaki Matsubara, The American Ceramic Society, USA, 1996.

[2] Y.Sadaoka, K.Watanabe, Y.Sakai and M.Sakamoto, " Preparation of Perovskite-type Oxides by Thermal Decomposition of Heteronuclear Complexes,{Ln[Fe(CN)$_6$]·n H$_2$O}$_x$, (Ln=La-Ho) ", J. Alloys and Compounds, 224, 194-198(1995).

[3] S.Nakayama, M.Sakamoto, K.Matsuki, Y.Okimura, R.Ohsumi, Y.Nakayama an Y.adaoka, " Preparation of Perovskite-type LaFeO$_3$ by Thermal Decomposition of Heteronuclear Complexes, {La[Fe(CN)$_6$]·5 H$_2$O}$_x$ ", Chem. Lett., 2145-2148(1992).

[4] J.L.G.Fierro and L.G.Tejuca, " Non-stoichiometric Surface Behaviour of LaMO$_3$ Oxides as Evidenced by XPS", Appl. Surf. Sci., 27,453-457(1987).

[5] R.J.H.Voorboeve, D.W.Johnson Jr., J.P.Remeika and P.K.Gallagher, "Perovskite Oxides: Materials Science in Catalysis", Science,195,827-833 (1977).

[6] T.Arakawa, H.Kurachi and J.Shiokawa, "Physicochemical Properties of Rare Earth Perovskite Oxides Used as Gas Sensor Material", J. Mat. Sci., 20, 1207-1210(1985).

[7] H.Obayashi, Y.Sakurai and T.Gejo, "Perovskite-type Oxides as Ethanol Sensor", J. Solid State Chem., 177, 299-303(1976).

[8] J.Mizusaki, W.R.Cannon and H.K.Bowen, "Electrochemical Degradation of Ceramic Electrodes", J. Am. Ceram. Soc., 63[7-8], 391-397(1980).

[9] J.A.M.Van Roosmalen and E.H.P.Cordfunke, "A New Defect Model to Describe the Oxygen Deficiency in Perovskite-type Oxides", J. Solid State Chem., 93, 212-219(1991).

[10] C.Morterra, V.Boils, B.Fubini, L.Orio and T.B.Williams, "A FTIR and HEAR Study of Some Morphological and Absorptive Properties of Monoclinic ZrO_2 Microcrystal", Surface Sci., 251/252,540-545(1991).

[11] H.Nishizawa, T,Tani and K.Matsuoka, "Crystallization Process of Cubic ZrO_2 in Calcium Acetate Solution", J. Mat. Sci., 19,2921-2926(1984).

[12] R.G.Darrie, W.P.Doyle and I.Kirkpatrick, "Spectra and Thermal Decomposition of Chromate(VI) of Magnesium, Lanthanum, Neodymium and Samarium", J. Inorg. Nucl. Chem., 29, 979-992(1967).

[13] G.V.Subba Rao and C.N.R.Rao and J.R.Ferraro, "Infrared and Electronic Spectra of Rare Earth Perovskite: Ortho-Chromites, -Manganites and -Ferrites ", Applied Spectroscopy, 24,436(1970).

[14] X.Li, H.Zhang, X.Liu, S.Li and M.Zhao, "XPS Study on $O_{(1s)}$ and $Fe_{(2p)}$ for Nanocrystalline Composite Oxide $LaFeO_3$ with the Perovskite Structure", Mater. Chem. and Phys., 38, 355-362(1994).

[15] N.S.McIntyre and D.G.Zetaruk, "X-ray Photoelectron Spectroscopic Studies of Iron Oxides", Anal. Chem., 49[11], 1521-1529(1977).

Structure and optical property of translucent BaTiO₃ gels consisted of crystalline fine particles

Hirofumi Matsuda, Takeshi Kobayashi, Makoto Kuwabara、and Hirokazu Shimooka*
Department of Materials Science, University of Tokyo, Tokyo 113-8656, Japan
Department of Applied Chemistry, Kyushu Institute of Technology, Fukuoka 804-8550, Japan*

ABSTRACT

Pure, translucent and highly crystallized BaTiO₃ monolithic gels were synthesized by sol-gel technique at down to 90 °C by using high concentration solution of Ba- and Ti-alkoxides. The gels were consisted of fine particles having an average diameter of ∼10 nm and showed an X-ray powder diffraction pattern of pseudo-cubic BaTiO₃ system at room temperature, although tetragonal structure was more stable for bulk crystals. At the temperature, Raman scattering lines were observed but whether the lines are assigned to the excitations of tetragonal lattice vibrations or cubic lattice vibrations by higher order processes is under investigation. Rayleigh scattering of visible light at interfaces of fine particles were drastically reduced and optical transmission spectra of the polycrystalline gels were similar to those for BaTiO₃ single crystal. The estimation of optical gap energies for the gels revealed ∼0.1 eV larger value than those for single crystal may due to a size effect. Surface area and porosity analyses by nitrogen adsorption / desorpotion measurements showed that distribution of pore centered at larger diameter and the gels lost translucency at some extent with increasing firing temperature due to growth of aggregates and roughening of surfaces. The micropores with the diameter below 1 nm were not observed.

INTRODUCTION

Traditional sol-gel method has been applied mostly to synthesize monodispersive powdery oxides to sinter ceramics at lower temperatures or to form thin films and fibers of high purity. To obtain denser, more homogenous ceramics, much finer powders are need. However, it is difficult to prevent fine powders from aggregation. Recently, we reported improved sol-gel process by increasing the concentration of precursor solutions about an order than before (typically 0.1 mol/L) to obtain highly crystallized and dense monolithic gels[2,3] instead of powdery xerogels. Directly firing the monolithic gels at considerably low temperatures of 1000 to 1100 °C, we succeeded to obtain dense BaTiO₃ ceramics exceeding their relative densities of 95 %. This powder-less process will give a route to prepare dense and homogenous ceramic bodies with little aggregated structure because average diameter of primary particles is 10∼20 nm.

We then slowed down, homogenized hydrolysis reaction, and succeeded to synthesize polycrystalline translucent BaTiO$_3$ monolithic gels. Our process may have possibilities for polycrystalline electro-optic device applications because sol-gel process has advantages of high purity preparations and ease control of constituents and compositions.

In this article, we report details of synthesizing polycrystalline, translucent BaTiO$_3$ monolithic gels, the results of optical absorption measurements for BaTiO$_3$ fine particles, aggregated structure measured by nitrogen adsorption/desorption studies, and discuss mainly on the structural effect on the optical gap energies and translucency.

EXPERIMENTAL

Equimolar, Ba-ethoxide (99.9 % of purity) and Ti-iso-propoxide (99.999 % of purity) were dissolved at concentration of up to 1.2 mol/L in mixed solvent of 60 vol. % ethanol and 40 vol. % 2-methoxyethanol in N$_2$ dry box. Hydrolysis of the well-stirred precursor solutions were performed at 0 °C by adding water vapor carefully for over 10 hours. Hydrolyzed sol solutions were held at 30 °C then at 50 °C for several days to proceed gelation and consolidation reaction (aging process). After aging process wet gels were dried in N$_2$ gas at 90 °C for several days and disk-shaped crystalline BaTiO$_3$ dried monolithic gels were obtained. Dried monolithic gels were fired at temperature up to 500 °C in O$_2$ flow, or the gels turned to be opaque after firing over the temperature. X-ray diffraction (XRD) patterns for dried and fired gels were studied for confirming single phase and crystallinity. Raman scattering light intensity of Stokes component excited by Ar laser (514 nm emission) by back scattering geometry was also measured to study crystal structure of fine particle. Details of crystallization and sintering behaviors of BaTiO$_3$ monolithic gels were reported elsewhere.[2,3] To study the effect of residual organic ligands on optical absorption and microstructure, samples with shorter aging time or smaller amount of hydrolysis water were also prepared. Optical transmission spectra for the gels dried at 90 °C and fired at 500 °C were measured at room temperature in the excitation wavelength range of 200~900 nm by using VT-550 UV/Visible Light Spectrometer (Nihon Bunko, Japan). The disk-shaped sample with typical thickness of 0.3 mm was immersed into liquid paraffin to reduce scattering of light at the surfaces. The transmission spectra for BaTiO$_3$ single crystal were also measured at room temperature. Relative surface areas and porosity of gels were analyzed from isothermal adsorption / desorption curves for nitrogen using Autosorb 1 (Quanta-Chrome, USA). Gels were evacuated down to 10^{-7} Torr at 80 °C for 8 hours before measurements. Porosity of gels was analyzed by using Dollimore-Heal (DH) method[4] which is a model suitable for analyzing mesopores.

RESULTS AND DISCUSSION
Sample characterizations

Figure 1 shows a top view of translucent BaTiO$_3$ monolithic gel only after drying at 90 °C and Fig. 2 shows the XRD patterns for the gels at room temperature. Quasi-stable, pseudo-cubic BaTiO$_3$ single phase was observed. By TEM study, the average diameter of primary crystalline particles was ∼10 nm. It has been pointed out theoretically that the instability of tetragonal phase arises from delicate cancellation of favorable long-range electrostatic

Fig. 1 Pure, translucent and highly crystallized BaTiO$_3$ monolithic gel, by hydrolyzing high concentration precursor solution (1.0 mol/L) of Ba- and Ti-alkoxide, aging the wet gel at 30 °C then 50 °C, and drying at 90 °C.

force from ionicity of Ti and O ions by short-range overlap force from covalency of Ti-O bond.[5] In fine particles, long range electrostatic interaction may be suppressed and cubic phase may become stable. After firing the gels at 500 °C, pseudo-cubic reflections of sharp peaks may be resulted from growth of crystallites were observed. Contrary to XRD results, Raman scattering lines were observed as shown in Fig. 3, though lattice vibrations in cubic perovskite were inactive for first order Ramam process.[6] These lines were commonly observed in chemically grown BaTiO$_3$ fine crystallites,[7] on the other hand, even in cubic perovskite structure higher order Raman active scattering was observed.[8] Temperature dependence of Raman scattering spectra for gel fired at 500 °C which showed Raman inactive cubic perovskite structure in XRD, apparent change in peak shape at ∼520 cm^{-1} was not observed. In contrast, this peak for gel fired at 1000 °C flattened at ∼130 °C may be suggesting tatragonal-to-cubic transition. Thus this transition above room temperature may be absent in up to 20 nm-crystallites.

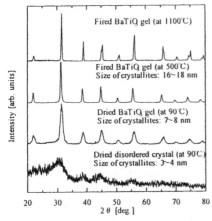

Fig. 2 X-ray diffraction patterns for dried and fired BaTiO₃ monolithic gels. Only the reflections of pseudo-cubic BaTiO₃ are observed fired below 900 °C. The gel prepared for short period of aging shows broad peaks of BaTiO₃ structure indicating imperfect consolidation and disordered crystal structure.

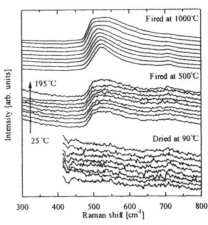

Fig. 3 Temperature dependence of Raman scattering spectra for gels fired at various temperatures. Though, strong scattering was observed for gel fired at 500 °C which showed Raman inactive cubic perovskite structure in XRD, apparent change in peak shape at ∼520 cm⁻¹ was not observed. In contrast, this peak for gel fired at 1000 °C flattened at ∼130 °C may be suggesting tatragonal-to-cubic transition. Thus this transition above room temperature may be absent in up to ∼20 nm-crystallites.

Optical absorption

Figure 4 shows transmission spectra for dried and fired BaTiO₃ monolithic gels and BaTiO₃ single crystal.[9] Due to scattering of incident light at sample surfaces and grain interfaces, transmission of incident light with longest wavelength (900 nm) through the gels decreased, though, the overall structures similarly, but the positions of absorption edges appeared at shorter wavelength with decreasing process temperature (decreasing size of crystallites).

It was predicted theoretically by electronic structure calculations for bulk BaTiO₃ that lowest excitation was indirect transition from valence band edge to conduction band edge,[10,11] Fig. 5 shows excitation energy dependence of square root of absorption coefficient α to evaluate optical gap. The α for indirect transition is proportional to density of states (DOS) at top of valence band (initial state) and DOS at bottom of conduction band (final state). Assuming parabolic dispersion of bands,

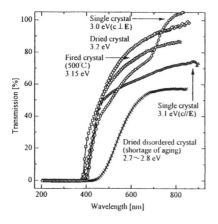

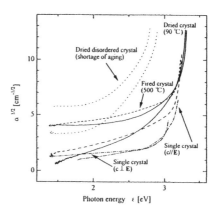

Fig. 4 Optical transmission spectra for dried and fired BaTiO₃ monolithic gels and single crystals at room temperature. E//c and E ⊥ c denote the directions of polarization vector parallel and perpendicular to the c-axis, respectively.

Fig. 5 Excitation energy dependence of absorption coefficient for dried and fired BaTiO₃ monolithic gels and single crystals. In crystalline samples, the position of absorption edge energy hardly depends on light scattering in lower energy region.

$$\alpha\,(\epsilon) \propto\, <n(\epsilon_p)>(\epsilon+\epsilon_p-\epsilon_0)^2+<n()>\,(\epsilon-\epsilon_p-\epsilon_0)^2 \qquad (1)$$

where ϵ is excitation energy, ϵ_0 is gap energy which is energy difference between top of valence band and bottom of conduction band, ϵ_p is energy of phonon which is absorbed or emitted at transition, and $<n(\epsilon_p)>$ is average phonon density. When $\alpha^{1/2}$ is plotted as a function of ϵ, extrapolation of tangential line from high excitation energy region intersects $\alpha=0$ at $\epsilon=\epsilon_0+\epsilon_p$.

As in Fig. 5, excitation energy dependence of α for disordered BaTiO₃ crystal showed large tail absorption compared to steeple absorption for crystallized gels. Structural disorder or dangling bond states observed in XRD introduce localized states between the gap and this tail region absorption process may be associated with the localized states. The optical gap energies for disordered crystal ranged 2.7 ~2.8 eV. Considering that samples with shorter aging time in preparation showed larger TG loss than for highly crystallized samples,[2,3] larger number of residual organic ligands may exist. Optical absorption might be associated with these organic legand states between the gap and absorption edge energies might decrease

Table I shows the estimated optical gap energies $\epsilon_0+\epsilon_p$ for polycrystalline BaTiO₃ gels and BaTiO₃ single crystal. The optical gap energies excited by light

Dielectric Ceramic Materials

141

with polarization parallel to c-axis (E ∥ c) and that by light with polarization perpendicular to c-axis (E⊥c) are 3.1 and 3.0, respectively, and the difference of these energies (anisotropy of optical gap energy) ∼0.1 eV. Since these values are consistent with reported values for BaTiO$_3$ single crystals reported before,[12,13] we made similar estimation for polycrystalline BaTiO$_3$ gels (Table I).

According to these literatures,[12,13] the optical gap energy for cubic phase is 0.05∼0.1 eV smaller than that for E⊥c. The observed optical gap energies for dried gels, however, are apparently 0.15∼0.2 eV larger than that for E⊥c in BaTiO$_3$ single crystal. Though for fine crystallites, this energy shift may not be introduced by the confinement of center-of-mass motion of excitons as explained for nanocrystalline silicon.[14,15] The ratio of the size of crystallites to effective Bohr diameter $2a_B$* of electron-hole pair created by light excitation could be estimated down to ∼5. This value is still large to confine weakly bound electron-hole pair with huge reduced mass (same order of magnitude to electron effective mass of 6.5m$_0$) in BaTiO$_3$.[16]

As an alternative interpretation, since valence electrons lose the gain of kinetic energy by reducing the size of crystal and the valence band width may become narrower, the difference between valence band and conduction band may become larger. In addition, the optical gap energies for fired gels are smaller than those for dried gels, while the absorption edge structure for fired gels is steeper than that for dried gels. These facts may also indicate size effect induced blue shift because by heating at higher temperature the crystalline primary particles may grow larger and may be more ordered.

We have to draw attention that the optical absorption edge for fine particles (Fig. 4) are not directly comparable to theoretical results of bulk electronic structures. In addition, localized states of dangling bonds at surface of crystallites may exist and makes it difficult to define band gap. Thus more precise experiments using much thinner samples are needed to investigate excitation energy dependence of absorption coefficient for much higher value.

Table I Estimated optical gap energies for BaTiO$_3$ gels (in eV)

Sample	Dried 90℃	Fired 500℃	Disordered Crystal 90℃	Single crystal (E // c)	Single crystal (E ⊥ c)
$\varepsilon_0 + \varepsilon_p$	3.20	3.15	2.7∼2.8	3.10	3.00

Microstructures

Figure 5 shows distributions of pore volume analyzed from isothermal adsorption curves by DH method[4]. Since, by firing, the gels lose translucency at some extent, the shift of distribution of pore volume to larger diameter indicates light scattering at the pores become more frequent due to growth of aggregates by necking.[1,17] However, assuming that the Rayleigh scattering (by which intensity ratio of transmitted light to incident light $I/I_0 \propto exp(-d^6/\lambda^4)$ where d is diameter of body to scatter light and λ is wavelength of incident light) is dominant, pore diameter distributes 10~30 nm even for the gel fired in air and is too small to scatter visible light. As shown in Fig. 6, the peak of pore diameter distribution is somewhat asymmetric, thus more precise adsorption/desorption curves in the highest pressure region are needed to decompose the asymmetric peaks and evaluate the effect of upper tail region of pore diameter and the structural features of aggregates on light scattering. In Fig. 6, in addition, the gel fired in O_2 flow showed sharper distribution with smaller peak pore diameter than that for the gel fired in air. This fact could be interpreted as that growth of aggregates may take place via oxygen vacancies and firing in O_2 flow may slow the growth of grains, thus scattering of light at rough surfaces and interfaces of grains may be suppressed.

CONCLUSION

Optical absorption measurements for fine-grained, polycrystalline $BaTiO_3$ monolithic gels were reported. The structure of absorption spectra for polycrystalline gels synthesized at only 90 °C are similar to those for $BaTiO_3$ single crystals grown at >1000 °C due to the small size of particles and aggregates prevent from scattering light at these interfaces. The optical gap energies for the gels become larger influenced by a size effect of crystalline primary particles. By firing, the aggregates of crystalline primary particles in the gels grow and more frequently scatter the visible light to lose translucency.

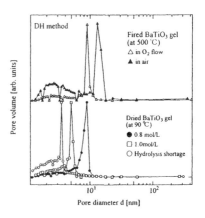

Fig. 6 Pore volume distribution for $BaTiO_3$ gels analyzed from isothermal adsorption curves by DH method.

ACKNOWLEDGMENTS

This research is supported by Grant-in-Aid for Science Research (Grant No. 09305043) from the Ministry of Education, Science, and Culture of Japan, and Special Coordination Funds for Promoting the Science and Technology, "Research on Fundamental Science of Frontier Ceramics" from the Japanese Science and Technology Agency.

REFERENCES

[1] J. R-Ortega and L. Equivias, "Structural Models of Dense Gels", J. Sol-Gel Sci. Tech., **8**, 117 (1997).

[2] H. Shimooka and M. Kuwabara, "Preparation of Dense $BaTiO_3$ Ceramics from Sol-Gel-Derived Monolithic Gels", J. Am. Ceram. Soc., **78**, 2849 (1995).

[3] H. Shimooka and M. Kuwabara, "Crystallinity and Stoichiometry of Nano-Structured Sol-Gel-Derived $BaTiO_3$ Monolithic Gels", J. Am. Ceram. Soc., **79**, 2983 (1996).

[4] D. Dollimore and G. R. Heal, J. Appl. Chem., **12**, 109 (1964).

[5] R. E. Cohen and H. Krakauer, "Lattice dynamics and origin of ferroelectricity in $BaTiO_3$; Linearized-augmented-plane-wave total-energy calculations", Phys. Rev. **B42**, 6416 (1990).

[6] M. DiDomenico, Jr., S. H. Wemple, and P. S. Porto, "Raman Spectrum of Single-Domain $BaTiO_3$", Phys. Rev. **174**, 522 (1968).

[7] M. H. Frey and D. A. Payne, "Grain-size effect on structure and phase transformations for barium titanate", Phys. Rev. **B54**, 3158 (1996).

[8] W. G. Nilsen and J. G. Skinner, "Raman Spectrum of Strontium Titanate", J. Chem. Phys. **48**, 2240 (1968).

[9] H. Matsuda, M. Kuwabara, K. Yamada, H. Shimooka, and S. Takahashi, "Optical absorption in sol-gel derived crystalline $BaTiO_3$ fine particles", submitted to J. Am. Ceram. Soc.

[10] Y.-N. Xu, W. Y. Ching, and R. H. French, "Selfconsistent band structures and optical calculations in cubic ferroelectric perovskites", Ferroelectrics **111**, 23 (1990).

[11] Y.-N. Xu, H. Jiang, X-F. Zhong, W. Y. Ching, "Comparison of electronic structure and optical properties of $BaTiO_3$ in the cubic and tetragonal phases", Ferroelectrics **153**, 19 (1994).

[12] S. H. Wemple, Phys. "Polarization Fluctuations and the Optical-Absorption Edge in $BaTiO_3$", Rev. **B2**, 2679 (1970).

[13] L. Hafid, G. Godefroy, A. E. Idrissi, and F. M.-Calendini, "Absorption spectrum in the near U.V. and electronic structure of pure barium titanate", Solid State Commun. **66**, 841 (1988).

[14] S. Nomura, X. Zhao, Y. Aoyagi, and T. Sugano, "Electronic structure of a model nanocrystalline/amorphous mixed-phase silicon", Phys. Rev. **B54**, 13974 (1996).

[15] G. Allan, C. Delerue, and M. Lannoo, "Electronic Structure of Amorphous Silicon Nanoclusters", Phys. Rev. Lett. **78**, 3161 (1997).

[16] Y. Kayanuma, "Quantum size effect of interacting electrons and holes in semiconductor microcrystals with spherical shape", Phys. Rev. **B38**, 9797 (1989).

[17]E. R. Leite, M. A. L. Nobre, M. Cerqueira, and E. Longo, "Particle Growth during Calcination of Polycation Oxides Sythesized by the Polymeric Precursors Method", J. Am. Ceram. Soc., **80**, 2649 (1997).

LARGE-SIGNAL CHARACTERIZATION OF PRESTRESSED PMN-PT-BA AND PMN-PT-SR AT VARIOUS TEMPERATURES

Elizabeth A. McLaughlin
Naval Undersea Warfare Center
1176 Howell Street
Newport, RI 02841

Robert S. Janus
Naval Undersea Warfare Center
1176 Howell Street
Newport, RI 02841

Kim D. Gittings
Naval Undersea Warfare Center
1176 Howell Street
Newport, RI 02841

Lynn Ewart
Naval Undersea Warfare Center
1176 Howell Street
Newport, RI 02841

ABSTRACT

Quasi-static (10 Hz) electromechanical measurements were made of two lead-magnesium-niobate/lead-titanate relaxor ferroelectrics, one doped with 3% barium (PMN-PT-Ba (90/10/3)) and the other with 2% strontium (PMN-PT-Sr (88/12/2)) under high electric fields (0-1.5 MV/m) and high mechanical prestresses (0-41.4 MPa) over a temperature range (-5 to 40 degrees Celsius). Both materials were designed to operate at 5 degrees Celsius. The measured and derived material properties - the piezoelectric constant, d_{33}, the permittivity, ε_{33}^{T}, the Young's modulus, Y_{33}^{E}, the electromechanical coupling factor, k_{33}, and the energy density, U - are found for both small and large ac drive signal conditions. The material properties for these conditions are of interest for Navy high-power sonar transducer designers and material designers. Of the measured quantities, d_{33}, ε_{33}^{T} and U are shown to decrease in value for increasing prestress and temperature. The PMN-PT-Sr is shown to have better properties for sonar transducers than the PMN-PT-Ba. Compared to Navy Type-III (PZT-8) piezoelectric ceramic, the energy density of PMN-PT-Sr at 5 degrees Celsius and 41.4 MPa is nearly 11 dB higher but the coupling is 23 % lower.

INTRODUCTION

The Navy is investigating doped PMN-PT electrostrictive ceramic for use in sonar transducers because of its high energy density. When operated around their transition temperatures, these electrostrictive materials typically show strains an order of magnitude higher and energy densities

10 dB greater than that of the standard Navy Type-III piezoelectric ceramic material (PZT-8). The higher energy density enables a sonar projector to be built with an order of magnitude increase in power or, conversely, provide an order of magnitude smaller projector (with the same source level) than one built with PZT-8.

Being electrostrictive, PMN-PT requires a dc bias (in addition to the normal ac drive field used in sonar projectors) to prevent frequency doubling of the output strain. Also PMN-PT is strongly temperature dependent requiring that material properties measurements be made over the entire expected operating temperature range.

In addition, PMN-PT has been shown to be dependent on mechanical prestress. Active materials in sonar projectors are normally prestressed in compression to enable the material to be driven to high electric fields without suffering tensile stresses.

Previously, similar large signal, high prestress measurements have been made on other PMN-PT formulations using the measurement methods described here [1,2,3].

The polarization vs. field and the strain vs. field curves are measured at a given temperature and prestress, and the stress vs. strain behavior is measured for a given temperature and electrical bias. From these curves, the following measured and derived material properties are found: the piezoelectric constant, d_{33}, the relative permittivity, $\varepsilon_{33}^T / \varepsilon_0$, the short circuit Young's modulus, Y_{33}^E, the electromechanical coupling factor, k_{33}, and the energy density, U, where, the subscript, 33, denotes extensional mode boundary conditions and the superscripts T and E denote a constant stress and constant electric field boundary condition, respectively [4].

For the small signal case, d_{33} is the derivative of the strain vs. electric field curve at a given field. In this analysis the small signal values are taken at the field (bias) where d_{33} is maximum. Similarly, the small signal permittivity is the slope of the polarization vs. field curve, and the Young's modulus is the slope of the stress vs. strain curve, at the same bias field. The coupling factor for small signal reduces to

$$k_{33} = d_{33}^2 Y_{33}^E / \varepsilon_{33}^T . \qquad (1)$$

It is derived from the ratio of the available mechanical energy over the total of the mechanical energy and the stored electrical energy [5]. It is a measure of how well a material converts electrical energy to mechanical energy. The electric-field-limited energy density per unit volume is a figure of merit for comparing high power active sonar materials, and may be expressed as

$$U = \tfrac{1}{2} Y_{33}^E (S/2)^2, \qquad (2)$$

where S is the peak-to-peak strain [6].

Even when biased, the strain vs. field characteristic of an electrostrictor is not linear over its useful range of drive field (it has a square law dependence at low field and saturates at high field). This means that the large signal properties cannot be obtained from small signal measurements but must be measured directly.

For the large signal cases, the same bias field is used as before, but the ac drive is 0.394 MV/m (10 V/mil) rms. This large ac drive is a typical value used in Navy sonar transducers.

To obtain the desired quantities, the strain and permittivity vs. field curves are fit simultaneously using a phenomenological model of an ideal electrostrictor developed by Piquette and Forsythe [7]. It uses an inverse square root to model saturation, has four fit parameters, includes remanent polarization, and the independent parameters are stress and electric displacement.

EXPERIMENTAL PROCEDURE

The samples consisted of two formulations of lead magnesium niobate, $Pb(Mg_{1/3}Nb_{2/3})O_3$, one was in solid solution with 10 mol% lead titanate, $PbTiO_3$, doped with 3 wt% barium (PMN-PT-Ba (90/10/1)) and the other with 12 mol% lead titanate doped with 3 wt% strontium (PMN-PT-Sr (88/12/2)). They were designed to have optimal properties at 5 degrees Celsius.

The sample dimensions were 2x2x10 mm and 3x3x9 mm for PMN-PT-Ba and PMN-PT-Sr, respectively. The length was kept short to be able to reach high fields with a 20 kV amplifier. The 2x2 and 3x3 mm faces were electroded. The prestress fixture applied the load parallel to the electric field and clamped the small ends. The sample dimensions were chosen such that the aspect ratio was at least 3:1, which allowed the 33-boundary conditions to be met. According to St. Venant's principle[8], this aspect ratio allows the middle of the sample to be free of any lateral stresses resulting from the fixturing. Strain gages were mounted in this region on one pair of opposite sides and wired in series to measure longitudinal strain (eliminating any contribution from bending).

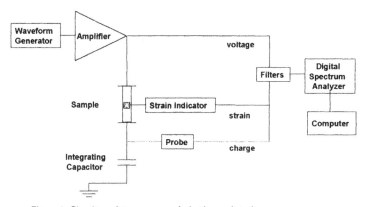

Figure 1. Circuit used to measure polarization and strain curves

Figure 1 shows the measurement circuit designed to measure the strain and polarization induced in the sample by the field. The excitation waveform was generated by a programmable waveform synthesizer (Analogic / Data Precision, Model 2020). It produced a biased 10 Hz sine wave that oscillated from 0 to 10 Vpk. The output of the high voltage amplifier (Trek, Model 20/20) supplied up to 20 kV across the series combination of the sample and the large integrating capacitor (mica, 101.49 nF). A high voltage probe (Tektronix, Model P6015) measured the accumulated charge on the integrating capacitor [9]. The strain was found using a strain indicator (Vishay, Model P-3500). The three time waveforms were filtered with a low-pass cutoff frequency of 100 Hz (Wavetek, Model 852) to reduce noise, then captured with a dynamic signal analyzer (Hewlett Packard, Model HP3562A). Five time-based averages were made of the data for each measurement condition to further reduce noise.

The strain signal was measured twice, once with the gages energized and once with the excitation turned off. The second measurement was subtracted from the first. This technique removes the undesired capacitive crosstalk signal arising because of the proximity of the strain gage to the high field in the sample.

The measurement fixture was placed in an environmental chamber (Thermotron, Model S1.2). The prestress load was applied to the sample using a pneumatic cylinder (Illinois Pneumatics, Inc., model 4LBX-3-BC) and was monitored with a load cell (Omega LCG-500). Four temperatures were investigated, -5, 5, Tmax, and 40 degrees Celsius. Tmax, the temperature at which the capacitance peaks when measured at zero field and no prestress, was found to be 22 degrees Celsius for PMN-PT-Sr and 30 degrees Celsius for PMN-PT-Ba.

The short-circuit Young's modulus data was taken at seven equally spaced dc voltages from 0 to 12 kV at each temperature. The load was brought up to 41.4 MPa and back down while the strain was measured.

RESULTS AND DISCUSSION

Figures (2) and (3) show the typical trend of the polarization vs. field and strain vs. field data, respectively, with increasing prestress at a single temperature. For each successively higher prestress the curves and their slopes are depressed further. This trend is true for both PMN-PT materials.

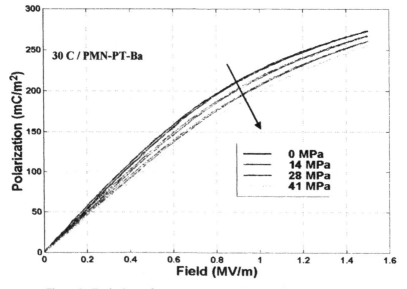

Figure 2. Typical set of constant temperature polarization vs. field curves.

Figure (4) shows the response of decreasing polarization with increasing temperature typical for both materials. This trend is also evident in the strain vs. field curves. The expected lessening of hysteresis with increasing temperature as the material moves from one with slight piezoelectric properties to one with increasingly more electrostrictive qualities can also be seen.

Dielectric Ceramic Materials

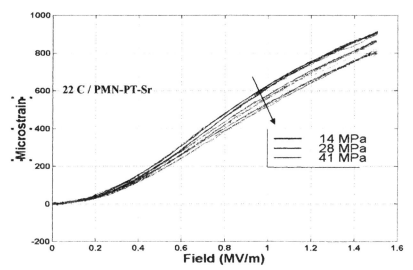

Figure 3. Typical set of constant temperature strain vs. field curves.

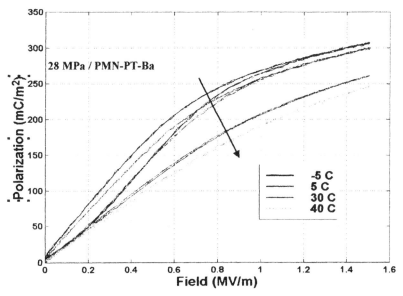

Figure 4. Typical set of constant prestress polarization vs. field curves.

Resulting trends in material properties include lower values for d_{33}, $\varepsilon_{33}{}^T/\varepsilon_0$, and U with increasing prestress and temperature for both the small and large signal cases.

Tables 1 and 2 list the measurement conditions and material properties for PMN-PT-Sr and PMN-PT-Ba, respectively.

For these drive conditions the coupling factor decreases with increasing temperature and in general decreases for increasing prestress. The coupling factor values found are low compared to PZT-8 (0.64).

To reach the maximum d_{33} bias point, the required bias field increases as the prestress and temperature increase. However, the hysteresis is less, and when a higher prestress is used the material can be driven harder to obtain a higher output.

Table I. Material properties for PMN-PT-Sr. (dB // PZT8 = 580 J/m^3)

temp.	prestress	bias	large signal						small signal			
			d_{33}	$\varepsilon_{33}/\varepsilon_0$	Y_{33}	k_{33}	U	U	d_{33}	$\varepsilon_{33}/\varepsilon_0$	Y_{33}	k_{33}
(Celsius)	(kPa)	(MV/m)	(pm/V)		(Gpa)		(J/m^3)	(dB)	(pm/V)		(Gpa)	
-5	14	0.46	1037	28 600	75.9	0.51	10578	13	1427	30 570	76	0.66
	28	0.56	909	25 550	73.7	0.47	9433	12	1232	26 040	74	0.63
	41	0.62	865	22 680	72.7	0.49	8445	12	1104	23 280	73	0.61
5	14	0.54	817	26 010	71.8	0.43	7185	11	1017	26 270	72	0.58
	28	0.65	809	22 640	69.7	0.46	7072	11	1015	22 750	70	0.60
	41	0.76	805	19 970	68.7	0.49	6918	11	959	20 150	69	0.58
22	14	0.70	690	19 880	91.6	0.46	6768	11	830	19 760	92	0.55
(Tmax)	28	0.76	662	18 090	89.9	0.46	6121	10	773	18 100	90	0.54
	41	0.81	632	16 830	88.8	0.46	5491	10	728	16 910	89	0.53
40	14	0.94	521	14 280	112.0	0.46	4716	9	573	14 200	112	0.51
	28	1.00	498	13 390	110.8	0.45	4255	9	543	13 350	111	0.50
	41	1.00	471	12 810	114.4	0.44	3802	8	513	12 810	110	0.48

At the design temperature (5 degrees Celsius), the response of the strontium-doped material is better than that of the barium-doped. As evidenced in the tables, the coupling factor and the energy density are higher for all prestress levels. Also, figure 5 shows polarization vs. field for 14 MPa prestress where PMN-PT-Sr displays less hysteresis (lower losses) and lower values for polarization. The corresponding strain curve also shows lower hysteresis and higher strain values for PMN-PT-Sr.

Dielectric Ceramic Materials

Table II. Material properties for PMN-PT-Ba. (dB // PZT8 = 580 J/m³)

temp. (Celsius)	prestress (kPa)	bias (MV/m)	large signal						small signal			
			d_{33} (pm/V)	$\varepsilon_{33}/\varepsilon_0$	Y_{33} (Gpa)	k_{33}	U (J/m³)	U (dB)	d_{33} (pm/V)	$\varepsilon_{33}/\varepsilon_0$	Y_{33} (Gpa)	k_{33}
-5	0	0.40	850	32 930	85	0.47	7051	11	1159	36 490	85	0.57
	14	0.47	799	31 010	82	0.44	6870	11	1085	32 760	82	0.57
	28	0.55	729	28 770	79	0.41	6434	10	989	29 060	79	0.55
	41	0.61	724	26 640	77	0.43	6274	10	940	26 970	77	0.55
5	0	0.46	745	30 780	88	0.43	6262	10	1005	32 510	88	0.54
	14	0.52	703	29 080	86	0.40	6095	10	948	29 680	86	0.53
	28	0 59	674	26 700	83	0 40	5843	10	877	26 790	83	0 52
	41	0.69	674	23 670	81	0.43	5702	10	825	23 820	81	0.52
30 (Tmax)	0	0.68	533	22 020	108	0.38	4744	9	640	21 790	108	0.46
	14	0.75	524	20 300	106	0.39	4483	9	609	20 170	106	0.46
	28	0.83	510	18 740	104	0.40	4177	9	577	18 620	104	0.45
	41	0.91	494	17 580	102	0.40	3846	8	550	17 560	102	0.45
40	0	0.78	474	19 010	114	0.38	3981	8	542	18 840	114	0.44
	14	0.86	458	17 530	112	0.38	3644	8	512	17 420	112	0.43
	28	0.93	451	16 600	110	0.39	3484	8	498	16 540	110	0.43
	41	1.02	446	15 320	108	0.40	3332	8	485	15 290	108	0.43

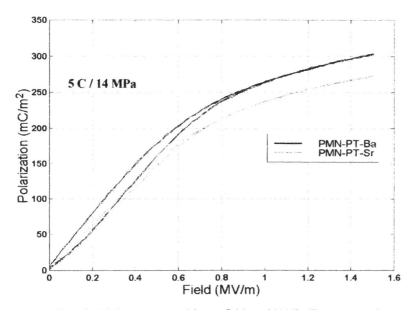

Figure 5. Polarization curves at 5 degrees Celsius and 14 MPa. The upper curve is PMN-PT-Ba and the lower is PMN-PT-Sr.

CONCLUSIONS

Of the two materials investigated, the measured material properties of PMN-PT-Sr were superior to those of PMN-PT-Ba for use in high power sonar transducers. Trends in material behavior included decreasing values of the piezoelectric constant, the permittivity, and the energy density with increasing prestress and temperature. The coupling factor was found not to vary considerably. Compared to Navy Type-III (PZT-8) piezoelectric ceramic, the energy density of PMN-PT-Sr at degrees Celsius and 41.4 MPa is 11 dB higher and the coupling factor 23 % lower.

ACKNOWLEDGMENTS

This work was sponsored by the Office of Naval Research. The authors wish to thank Harold C. Robinson, also of NUWC, for discussions on modeling.

REFERENCES

[1] E.A. McLaughlin, J.M. Powers, M.B. Moffett, H.C. Robinson, and R.S. Janus "Characterization of PMN-PT-La (90/10/1) for use in sonar transducers", presented to ONR Transducer Materials and Transducers Workshop, March 1996. (J. Acous. Soc. Am. article in preparation.)

[2] F.A. Tito, "PMN measurements meeting report", Naval Undersea Warfare Center Division Newport, Newport, RI, meeting held in Stonington CT, 17-18 December 1996.

[3] R.S. Janus, M.B. Moffett, and J.M. Powers, "Large-signal characterization of PMN-PT-Ba (90/10/3)" NUWC-NPT Reprint Report 10,860, 20 October 1997.

[4] W. P. Mason (ed.), Physical acoustics, Vol. 1, Part A, Chap. 3, Academic Press NY, NY (1964).

[5] O.E. Mattiat (ed.) Ultrasonic Transducer Materials, Plenum Press, New York, NY, 66-71, (1971).

[6] R.S. Woollett, "Power limitations of sonic transducers," IEEE Trans. Sonics Ultrason. SU-15, 218-229 (1968).

[7] J.C. Piquette and S. E. Forsythe, "A nonlinear material model of lead magnesium niobate (PMN)," J. Acoust. Soc. Am. 101, 289-296, (1997).

[8] E. P. Popov, Mechanics of Materials, 2nd ed., Prentice-Hall, Inc., Englewood Cliffs, New Jersey, p 49, (1978).

[9] C. B. Sawyer and C. H. Tower, "Rochelle salt as a dielectric," Phys. Rev. 35, 269-273, 1930.

GROWTH AND CHARACTERIZATION OF LaAlO₃ SINGLE CRYSTALS BY THE TRAVELING SOLVENT FLOATING ZONE METHOD

Il Hyoung Jung, Chang Sung Lim and Keun Ho Auh

Department of Ceramic Engineering, College of Engineering, Ceramic Processing Research Center, Hanyang University, Seoul, 133-791, Korea

ABSTRACT

LaAlO₃ single crystal, used as a substrate for thin film depositions of a high temperature oxide superconductor YBa₂Cu₃O₇ and applied to microwave frequencies, were grown by the traveling solvent floating zone(TSFZ) method and characterized. The grown LaAlO₃ crystals was 4 ~ 5 mm in diameter, 30 mm in length and dark brown. The growth rate was 2 ~ 3 mm/h and the rotation speeds were 10 rpm for an upper rotation and 40 rpm for a lower rotation. LaAlO₃ single crystals were grown by the traveling solvent floating zone method were composition in growth axis direction had a homogeneous distributions and superior to the other aluminate group.

INTRODUCTION

LaAlO₃ single crystal has widely applications as a substrate for high Tc oxide superconductors because lattice constants and thermal expansion coefficient of the substrates should be closely matched to those of high Tc oxide superconductor films. Because of the low dielectric constants and low dielectric losses, it is also suitable for many microwave frequency[1]. Bondar and Vinogradova[2] determined the phase diagram for the La₂O₃ - Al₂O₃ system and reported a stoichio - metric 1 : 1 compound melting at 2100 ℃. Lanthanum aluminate crystals, which exhibit twin structures resulted from phase transitions, were in difficulties for deposition of superconductor thin films. It was reported that lanthanum aluminate transforms from rhombohedral to cubic symmetry above 500 ℃ [3].

LaAlO₃ single crystals have generally two problems for use the high Tc temperature superconductor substrates. The first, LaAlO₃ single crystals should be colorless ; actually they have brown colors in air. The second, domain structures have also as twin on the (100) plain is complicated. From now LaAlO₃ crystals have been grown by Vertical Gradient Freeze, Czochralski and Inductively Coupled Plasma method[4-7]. However single crystals grown by these methods are not suitable for a substrate for high Tc oxide superconductor thin films because of a nonstoichiometric compositions.

In this study, we investigated the growth characteristics using the traveling solvent floating zone method in order to obtain the stoichiometric compositions. In more detail, we studied physical and chemical properties of grown crystals.

EXPERIMENTAL

Starting materials for preparation of feed rods were La_2O_3(Aldrich fine chemical company) and α -Al_2O_3(Shinyo pure chemical company) with a purity of 99.99%. Stoichiometric compound of the raw materials for $La_2O_3 : Al_2O_3 = 1 : 1$ were ball-milled for 20 h, dried with a drying oven and calcined at 1100 ℃ for 3 h. The feed rod with a circular and columnar for 15 mm in diameter and 140 mm in length was prepared in a silicon rubber mold using a cold isostatic pressure of about 200 MPa. The feed rod was sintered at various temperature in air and then the physical and dielectric properties of grown crystals were investigated by the powder x-ray diffractometer and the LCR meter(Impedance analyzer).

A floating zone system utilizing a thermal imaging system was used and classified into rf-induction, laser, electron beam, infrared light, halogen and Xe arc according to the heat source[8]. For the LaAlO₃ single crystal growth,   we were used in self-designed FZHY1 with 5.4 kW of Xe arc lamp. The apparatus used is schematically shown in Fig. 1. The main advantage of the floating zone method are the absence of the crucible and the possibility to use a various growth atmospheres. The high temperature gradient is a critical parameter for some oxides because it can give rise to crack formations. The crystal diameter is limited to a few millimeters which reduces the usefulness of this technique for the growth of single crystals[9].

Dielectric Ceramic Materials

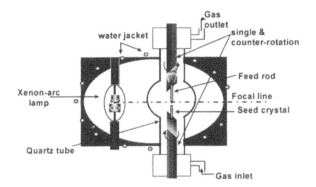

Fig. 1. Configuration of Xe arc type floating zone system

Also, the disadvantage of normal floating zone method is a different composition between crystal and molten zone, so that the control of the solid composition is very difficult[10-11]. So in present studies, $LaAlO_3$ single crystal growth was used by traveling solvent floating zone method in order to prevent the fluctuation of composition between molten zone and crystal part.

RESULTS AND DISCUSSIONS

As shown in Fig. 2 and Fig.3, XRD results needed to investigated the optimum time and temperature conditions for calcination and sintering. In Fig. 2, where $LaAlO_3$ was calcined for 3 h from 1000 ℃ to 1200 ℃ at 100 ℃ intervals to remove organic matter and to synthesized powders

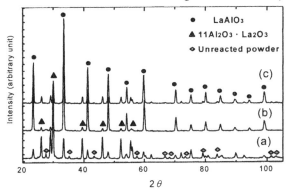

Fig. 2. X-ray diffraction patterns of $LaAlO_3$ powder calcined at (a) 1000 ℃, (b) 1100 ℃ and (c) 1200 ℃ for 3 h, respectively.

Dielectric Ceramic Materials

through the solid state reactions, it was found that LaAlO$_3$ began forming at 1000 °C with numerous peaks showing unreacted powders and these powders dissipating at 1100 °C while a secondary phase, 11Al$_2$O$_3$ · La$_2$O$_3$ was detected. Also, in increase agglomeration and the quantity of secondary phase, 11Al$_2$O$_3$ · La$_2$O$_3$ was found to increase with time. The optimum conditions for the total elimination of CO$_2$ gases adsorbed in air by La$_2$O$_3$ and the ideal calcination condition for powder synthesis was found to be 3 h at 1100 °C.

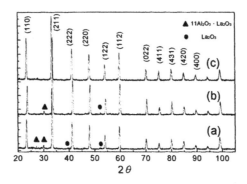

Fig. 3. X-ray diffraction patterns of LaAlO$_3$ feed rod sintered at (a) 1300 °C, (b) 1350 °C and (c) 1400 °C for 4h, respectively.

Fig. 3 shows the XRD pattern of the sintered feed rod. The experiment was carried out at 50 °C intervals from 1300 °C to 1400 °C for 4 h. XRD results showed synthesis of pure LaAlO$_3$ with no other components at 1400 °C, and this condition was chose as the optimum condition in making the feed rod needed for the crystal growth.

Table I. Sintering densities of the LaAlO$_3$ feed rod measured by the Archemedes method

Sintering temperature (°C)	Mass (g)	Volume Wsat	Wsus	Density (g/cm^3)
1300	11.06	3.269	1.459	6.12
1350	10.60	2.984	1.304	6.31
1400	11.29	3.105	1.345	6.42

Dielectric Ceramic Materials

Since the sintering of LaAlO₃ is solid state reaction accompanying thermal energy, the change in density using the Archimedes method at various temperatures with time fixed at 4 h was analyzed with the results shown in table I. Wsat represents the total weight of the liquid and the specimen placed in it and Wsus represents the weight of liquid alone(i.e. they represent the values of the difference in buoyancy converted into a volumetric value). In conclusion, results show a density of 6.12 g/cm³ at 1300 ℃, 6.31 g/cm³ at 1350 ℃ and 6.42 g/cm³ at the optimum temperature, 1400 ℃.

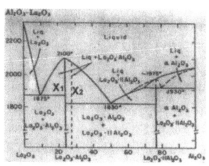

Fig. 4. Phase diagram of the La₂O₃-Al₂O₃ system[2].

When one observes Fig. 4, an La₂O₃-Al₂O₃ phase diagram constructed by Bondar and Vinogradova [2]of Russia, one can see that the binary system melts congruently at 2100 ℃ in a 1:1 mole ratio. This shows that even a little variance in composition will make it difficult to grow an LaAlO₃ single crystal having a uniform composition. On the right hand side of the diagram(i. e. the Al₂O₃ rich region), phases like corundum and 11Al₂O₃ · La₂O₃ as well as LaAlO₃ are known to form, while on the left hand side(i. e. the La₂O₃ rich region), unreacted La₂O₃ is still remnant along with LaAlO₃. In the existing floating zone method, a change in the region between the molten zone and feed rod. To prevent this change in composition of the molten zone and thus grow a single crystal having the uniform LaAlO₃ compound was grown. The feed rod as seen in Fig. 5 was prepared accordingly. First, an Al₂O₃ rich LaAlO₃(X₂) and an LaAlO₃(X₁) powder having a 1:1 mole ratio was packed in a rubber tube. The Al₂O₃ rich LaAlO₃ was initially melted, forming the molten zone, but an effort grow a single crystal having the X₁ compound failed when cracks resulted on the boundaries where the composition differed, thus causing the mentioned region to fall off during the experiment. In the second trial, two feed rods having a composition

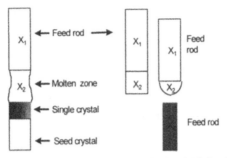

Fig. 5. Principle of the TSFZ method and fabrication of the LaAlO₃ feed rods.

X_1 and X_2 was prepared. The rod having an X_2 composition was melted and joined with the rod of X_1 composition. This was again melted to form a molten zone of an X_2 compound and consequently a single crystal of compound X_1 was grown. Since no seed crystals were available, numerous necking was performed and a single crystal of a dark brown color having dimensions of 4 ~ 5 mm in radius and a length of 30 mm was grown. The deficiency in oxygen is seemed to be the reason for the crystal's brownish color since it was grown in air. The optimum condition for the crystal growth in air was found to be 2 ~ 3 mm/h for the growth rate, 10 rpm and 40 rpm for the speed of the upper and lower shaft respectively in a counter-rotation.

Fig. 6. Laue pattern of the grown LaAlO₃ single crystal.

We must become aware of preferred growth orientation because the crystal was growing without seed crystal. So when the power was 20kV, 25mA and exposed time was 10minute with

Dielectric Ceramic Materials

back reflection Laue camera, Fig. 6 was shown in Laue   pattern of grown LaAlO₃ single crystal.
As a result from analysis of Grenninger chart, we can see that preferred growth orientation was
identified to be in [111] direction of rhombohedral structure. In practice, the pattern was slightly
missed direction in [111] direction. It is considered that sample cutting was not exactly direction.

Specially, in such a case of LaAlO₃ was used with high frequency dielectric material, dielectric
constant and dielectric loss(tan δ) are very important . Dielectric constant and dielectric losses
have a great effect miniaturization of devices and high frequency characteristics, respectively.

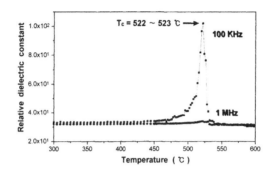

Fig. 7. Dielectric constants and transformation temperatures of
the grown LaAlO₃ single crystals.

Dielectric constants of grown crystal were shown in Fig. 7 using the LCR meter(HP4192A) in
temperature range between 300 ℃ and 600 ℃. Dielectric constants were measured to be 30 ~ 33
between 100 kHz and 1 MHz in the 300 ℃ to 450 ℃ temperature range and 102 in a range of
100 kHz at the phase transformation temperature of 522 ℃. This is due to the phase
transformation of LaAlO₃ crystal at 522 ℃. When the crystal was transformed rhombohedral to
cubic phase, phenomena of transformation from the rhombohedral to the cubic phase occurs
gradually and the crystal has been become isotropic at around curie temperature. Also, dielectric
losses(tan δ) were calculated to be 1.8 $\times$ 10⁻⁴ at the room temperature and 5.7 $\times$ 10⁻³ at the
transformation temperature. As a above result , in comparison with dielectric properties of NdAlO₃
and other aluminate family group, it can give rise to profitable point on miniaturization of devices
because dielectric constant of LaAlO₃ crystal was approximately 30 ~ 40 % higher than that of

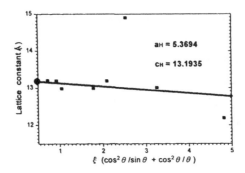

Fig.8. Extrapolation of lattice constant determinations of the grown LaAlO₃ single crystals.

aluminate family group and we were confirmed that LaAlO₃ is suitable for high frequency devices because dielectric losses were measured to be 1.8×10^{-4} at the room temperature and 5.7×10^{-3} at the transformation temperature in the high frequency range is higher than 10MHz.

Fig. 8 gives the lattice constant for LaAlO₃ single crystal using the powder x-ray diffractometer. Because the LaAlO₃ has a rhombohedral structure at room temperature, extrapolation function was selected with $\cos^2 \theta / \sin \theta + \cos^2 \theta / \theta$ and then lattice constants were determined to be $a_R = 5.3806$ Å and $\alpha = 60.043^\circ$ by the least square method. However, as a result of lattice constant data, composition of grown crystal by the TSFZ method in growth axis direction had a homogeneous distributions. LaAlO₃ single crystals can be used as a substrate for high Tc oxide superconductor thin films, as for structure and lattice constant matching are concerned. But it is difficult to grow single crystals by the floating zone method in air. However, as far as crystal quality was only concerned, LaAlO₃ single crystals were grown by the traveling solvent floating zone method are superior to the other aluminate group.

CONCLUSIONS

For the crystal growth, the optimum feed rods which calcined at 1100 ℃ for 3 h and sintered at 1400 ℃ for 4 h after La₂O₃ and Al₂O₃ were mixed in 1:1 mole ratio were fabricated. As for the single crystal growth by the traveling solvent floating zone method , obtained conclusions were as follows.

Calcination and sintering temperature for fabrication of $LaAlO_3$ feed rods were 1100 ℃ and 1400 ℃, respectively. $LaAlO_3$ single crystals having a homogeneous composition were grown by the traveling solvent floating zone method and grown crystals had a dark brown color. It is considered that these result was due to the deficiency of oxygen contents. Optimum crystal growth condition in air was found to be 2 ∼ 3 mm/h for the growth rate and the rotation speeds were 10 rpm for an upper rotation and 40 rpm for a lower rotation. The preferred orientation of grown $LaAlO_3$ single crystals was identified to be [111] direction and dielectric constants were measured to be 30 ∼ 33 between 100 kHz and 1 MHz in the 300 ℃ to 450 ℃ temperature range and 102 in a range of 100 kHz at the phase transformation temperature of 522 ℃. Also dielectric losses were measured to be 1.8×10^{-4} at the room temperature and 5.7×10^{-3} at the transformation temperature. Lattice constants of the grown crystal were determined to be $a_R = 5.3806$ Å and $\alpha = 60.043^O$. As a result from measured lattice constants, we can see that composition of grown crystal by the traveling solvent floating zone method in growth axis direction had a homogeneous distributions.

REFERENCES

[1] T.A.Vanderah, C.K.Lowe-Ma and D.R.Gagnon, "Synthesis and Dielectric Properties of substituted Lanthanum Aluminate," *Journal of the American Ceramic Society*, **77**[12] 3125-30 (1994).

[2] I.A.Bondar and N.V.Vinogradova, "in : Phase Diagrams for Ceramists," *American Ceramic Society*, Vol.2 Fig. 2340 (1969).

[3] M.G.Norton and J. Bentley, "Reflection Electron Microscopy Observations of Twinning in $LaAlO_3$," *Journal of the Material Science Letters*, **15** 1851-53 (1996).

[4] G.W.Berkstresser, A.J.Valentino and C.D.Brandle, "Growth of Single Crystals of Lanthanum Aluminate," *Journal of the Crystal Growth*, **128** 684-88 (1993).

[5] R.E.Fahey, A.J.Strauss and A.C.Anderson, "Vertical Gradient-Freeze Growth of Aluminate Crystals," *Journal of the Crystal Growth*, **128** 672-79 (1993).

[6] I.Tanaka, M.Kobashi and Hironao Kojima, "Single Crystal Growth of Pure and Y-substituted $NdAlO_3$ by the Floating Zone (FZ) Method," *Journal of the Crystal Growth*, **144** 59-64 (1994).

[7] Y.C.Chang, D.S.Hou, Y.D.Yu, S.S.Xie and T.Zhou, "Color Center and Domain Structure in Single Crystals of $LaAlO_3$," *Journal of the Crystal Growth*, **129** 362-64 (1993).

[8]K.Kitazawa, K.Nagashima, T.Mizutani, K.Fueki and T.Mukaibo, "A New thermal Imaging System Utilizing a Xe Arc Lamp and an Ellipsoidal Mirror for Crystallization of Refractory Oxides," *Journal of the Crystal Growth*, **39** 211-15 (1977).

[9]M.G.Norton, J.Bently and R.R.Biggers, "Proceedings of Microscopy and Microanalysis," Edited by G.W.Bailey, M.E.Ellisman, R.A.Hennigar and N.J.Zaluzec. Jones and Begell, New York, 1995.

[10]Z.K.Kun, W.E.Kramer and G.W.Roland, "The Growth of Homogeneous NbO_2 Single Crystals by the float zone method," *Journal of the Crystal Growth*, **58** 122-26 (1982).

[11]C.W.Lan and Sindo Kou, "A Simple Method for Improving the Stability of Float Zones under Normal Gravity," *Journal of the Crystal Growth*, **118** 151-59 (1992).

SPRAY PYROLYSIS DEPOSITION OF ORIENTED, TRANSPARENT AND CONDUCTIVE TIN(IV) OXIDE THIN FILMS

S. Kaneko, K. Nakajima, T. Kosugi, and K. Murakami
Shizuoka University, Hamamatsu 432, Japan

ABSTRACT

Spray pyrolysis deposition using a simple apparatus with good productivity is one of the well-known chemical techniques which facilitate the design of materials on a molecular level in thin film formation. We have succeeded in the preparation of highly [100] oriented SnO_2 thin film on glass substrate by spray pyrolysis of di-n-butyltin(IV) diacetate, but its electrical resistivity was too high to be applied to transparent electrodes. This paper focuses on the lowering of the resistivity by means of the deposition of the Sb- or F-doped SnO_2 film on the [100] oriented film, which results in maintaining the preferred orientation and transparency.

I. INTRODUCTION

Tin(IV) oxide is an oxygen-defect type of semiconductor with a wide band gap and a large mobility. The compound is transparent in the visible region and reflective in the infrared region. Tin oxide thin films have an extremely high chemical stability and their adhesion to glass is very high.[1] Due to these advantages, tin oxide thin films have been used as transparent electrodes in sophisticated electronic devices.[2]

Tin oxide thin films have been prepared by various techniques such as evaporation,[3] sputtering,[4] chemical vapor deposition,[5,6], pyrosol technique,[7] sol-gel dip-coating,[8] and spray pyrolysis deposition.[9-15] Spray pyrolysis deposition, which is one of the well-known chemical techniques, is applied to form a variety of thin films such as noble metals, metal oxides and chalcogenide compounds. Furthermore, this spray pyrolysis deposition can be realized by simple apparatus and with good productivity.

Recently, a high gas sensitivity has been reported for (110)-oriented SnO_2 thin films,[16] and a high electrical conductivity has been obtained in (200)-oriented SnO_2 thin films.[13), 17)] We have succeeded in the preparation of highly [100] oriented SnO_2 thin film on glass substrate through spray pyrolysis deposition of di-n-butyltin(IV) diacetate [DBTDA].[15)] Undoped-SnO_2 (TO) film is transparent in the visible region, but its electrical resistivity is too high to use as transparent electrodes. This paper focuses on an improvement of that the conductivity of tin oxide thin films without degrading the preferred orientation and transparency. For that purpose, we insert the [100] oriented TO film between doped SnO_2 films and glass substrates. The structural, electrical, and optical properties of the doped-SnO_2 films on the oriented TO films (doped-TO/TO films) are also investigated.

II. EXPERIMENTAL PROCEDURE

(1) Film Formation

Tin oxide thin films were deposited using a spray solution of di-n-butyltin(IV) diacetate $((C_4H_9)_2Sn(OCOCH_3))$ in ethanol. Triphenyl antimony $(Sb(C_6H_5)_3)$ and ammonium fluoride (NH_4F) were added to this ethanol solution for antimony and fluorine doping, respectively. The spray apparatus was illustrated schematically in Fig. 1 (a). The spray solution was atomized by compressed air at 0.2 MPa in pressure. The atomized solution was transported onto a heated Corning 7059 glass substrate (25mm × 25mm × 1mm). The solution was atomized not consecutively but intermittently to avoid excessive cooling of the heated substrate. Each spraying period was 0.5 s. The solution feed rate was 1.25 ml/s. The nozzle-substrate distance was fixed to be 300 mm. The substrate was mounted on a cordierite ceramic holder (Fig. 1(b)), and a thermocouple was mounted beside the substrate. The substrate temperature, which was corrected by the deviation in temperature between the center of the substrate surface and the side of the substrate, was kept at 480℃.

(2) Characterization of Films

The crystal structure of the films was determined by X-ray diffraction (XRD) using $CuK\alpha$ radiation (SHIMADZU XD-610 diffractometer). The preferred orientation of the SnO_2 films was evaluated by the texture coefficient (TC), calculated from the equation,

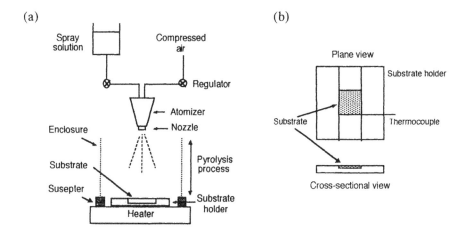

Fig.1.   (a) Spray pyrolysis apparatus   (b) Substrate and holder

$$TC(hkl) = \frac{I_{hkl} / I_{0hkl}}{1/N \; \sum (I_{hkl} / I_{0hkl})} \qquad (1)$$

where TC(hkl) is the texture coefficient of the (hkl) plane, I the measured or normalized intensity, I_0 the corresponding standard intensity of SnO_2 in cassiterite, and N the reflection number of 31. The higher the preferred orientation of the film is, the further the texture coefficient deviates from unity.

Sheet resistance of the films were measured by MITSUBISHI KAGAKU MCP-T410 four probe type resistance meter.   Carrier density and mobility were measured using van der Pauw method.   Film thickness was measured by a stylocontact method using a Slone Dectak IIA after etching the films.[18]   The transmittance in the 200-2600 nm region was measured with a SHIMADZU UV-3100 spectrophotometer.

III.   RESULTS AND DISCUSSION

(1) Oriented growth of tin oxide thin films

Fig. 2 (a) shows a typical XRD profile of undoped-SnO_2 (TO) thin film (650 nm thick) prepared on a glass substrate from 2.0 wt% DBTDA solution.   This film

shows a sharp increase in the peak intensity of the (200) plane. Indeed, the
TC(200) of the film shows a very high value of 27.[15] This result indicates that the
TO film grows preferentially along the |100| direction normal to the glass substrate.

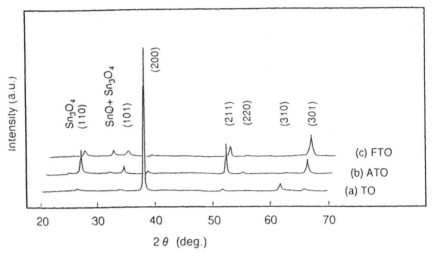

Fig. 2.   XRD profiles of TO and doped-TO films.

Fig. 2 (b) shows the XRD profile of the antimony(Sb)-doped-SnO_2 (ATO)
thin film prepared from 20 at% Sb concentration in the spray solution. Fig. 2 (c)
shows the XRD profile of the fluorine(F)-doped-SnO_2 (FTO) thin film from 160
at% F concentration in the spray solution. In the doped-SnO_2 films, the peak
intensity of the (200) plane decreases drastically, and the peaks of the (110), (101),
(211), and (301) planes are observed. The additional peaks at $2\theta = 24.5°$ and
$2\theta = 31.5°$ could be assigned to Sn_3O_4[19], and to Sn_3O_4 and/or SnO,[17,19]
respectively, which were not observed in the XRD profile of the highly oriented TO
film. The TC(200) of the ATO and FTO films are lower values of 14 (with 330nm
film thickness) and 5.3 (with 380nm film thickness), respectively than the (200)-
oriented TO film. Doped-TO (ATO, FTO) films on the glass substrates do not
exhibit |100| preferred orientation but non-orientation.

In order to maintain the (200) preferred orientation, doped-TO films were
deposited on the oriented TO films. Fig. 3 shows XRD profiles of TO and doped-
TO/TO films. As expected, the peak of the (200) plane increases remarkably in the

doped-TO/TO films. TCs(200) of the ATO/TO and FTO/TO films show very high values of 24 (with 320nm film thickness) and 23 (with 370nm film thickness), respectively. It is concluded from the results that highly (200) preferred orientation of doped-TO film is achieved only by insertion of oriented TO film between doped-TO film and glass substrate.

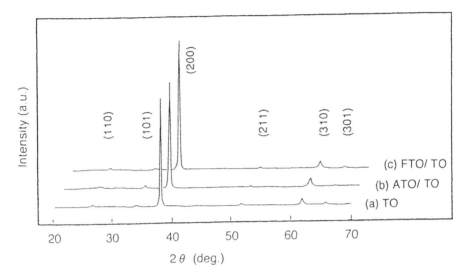

Fig. 3. XRD profiles of TO and doped-TO/TO films.

(2) Electrical conductivity

The electrical conductivity of SnO_2 films is generally caused by lattice imperfections and oxygen deficiencies. The resistivity of the oriented TO film with 300nm film thickness is 8.8×10^{-1} Ωcm. This electrical resistivity is too high for the practical use of oriented TO films as the transparent electrodes.

Fig. 4 shows the variation of the electrical properties with the antimony concentration. Initially, the resistivity decreases rapidly up to 1.7×10^{-3} Ωcm at 20 at% Sb concentration and tends to increase at higher Sb concentrations. As the Sb concentration increases, the carrier concentration has a tendency to increase and saturate. However, the mobility increases rapidly and reaches to 11 $cm^2V^{-1}S^{-1}$ at 9 at%, and it decreases gradually beyond 9 at%.

Fig. 5 shows the variation of the electrical properties with fluorine concentration in the spray solution. The resistivity decreases up to 160 at% F concentration and becomes constant of $7.2 \times 10^{-4}\Omega$cm for higher F concentrations. In the FTO films, the actual fluorine concentration in the film is supposed to be much lower than that in the solution, because of the escape of the thermal decomposition products of NH_4F into the atmosphere.[23)-25)] The carrier concentration and the mobility increase with F concentration up to 160 at% and saturate.

On the other hand, the carrier mobility in the ATO films decreases in the region of higher Sb concentrations. In lower Sb concentrations, Sb^{5+} would predominant and act as an extra charge carrier in the ATO films. As Sb concentration increases, however, Sb^{3+} as well as Sb^{5+} increases in the film and act as an electron trap, which results in the decrease of the carrier mobility.

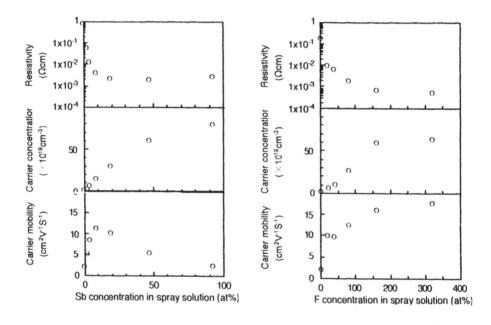

Fig.4. Variation of the electrical properties with Sb concentration.

Fig.5. Variation of the electrical properties with F concentration.

Dielectric Ceramic Materials

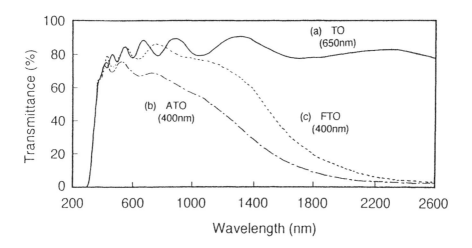

Fig. 6. Transmittance spectra of doped-TO films on glass substrates.

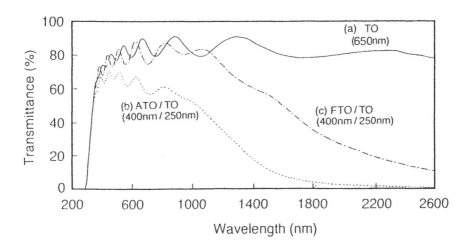

Fig. 7. Transmittance spectra of doped-TO/TO films.

Dielectric Ceramic Materials   171

Finally, Table 1 summarizes the electrical properties for various tin oxide thin films with the lowest resistivity. The oriented doped-TO/TO films have the low resistivities (ATO/TO : 9.9×10^{-4} Ωcm, FTO/TO : 6.3×10^{-4} Ωcm) like the doped-TO films on the glass substrates.

(3) Optical properties

Fig. 6 shows the transmittance spectra of TO, ATO, and FTO films. The transmittance of the oriented TO film is very high ($>$80%) in the wavelength from 500 nm to 2600 nm. The average transmittance (T_{av}) in the visible region from 400 nm to 800 nm is 82 %. In ATO films, the transmittance in the visible region decreases gradually with an amount of Sb included in the film. For the ATO film of the lowest resistivity at 20 at% Sb, T_{av} is 70%. On the contrast, the lowing of the transmittance in the visible region is not observer for the FTO films.[26] T_{av} still keep 80% for the FTO film with 160 at% F. In both ATO and FTO films, the transmittance in the infrared region decreases remarkably due to the plasma resonance, which is caused by a high concentration of free carriers in the films.[27]

In doped-TO/TO films, the similar results are obtained in the transmittance spectra as shown in Fig.7. The number of the interference in both ATO/TO and FTO/TO films increases as a result of the increase of the film thickness.

Table 1. TC(200), electrical and optical properties of various tin oxide films.

Sample	Thickness (nm)	TC(200)	Resistivity (Ωcm)	Carrier concn. ($\times 10^{20}$ cm^{-3})	Mobility (cm^{2}V^{-1}S^{-1})	T_{av} (%)
TO	300	19	8.8×10^{-1}	0.092	0.78	82
ATO	330	14	1.7×10^{-3}	4.7	8.1	72
ATO/TO	320	24	9.9×10^{-4}	5.9	10	70
FTO	380	5.3	7.2×10^{-4}	5.8	16	82
FTO/TO	370	23	6.3×10^{-4}	5.7	18	80

Dielectric Ceramic Materials

IV. CONCLUSIONS

Highly oriented undoped SnO_2 films were grown on glass substrates by spray pyrolysis technique using $(C_4H_9)_2Sn(OOCCH_3)_2$. Sb- or F- doped SnO_2 films were formed to enhance electrical conductivity. However, the doped-TO films showed the degrading of the preferential oriented growth. Therefore, the doped SnO_2 films were prepared on highly oriented SnO_2 films to maintain the (200) preferred orientation like undoped SnO_2 films as well as both the high conductivity and the optical transparency. We have succeeded in the formation of oriented, transparent, and conductive SnO_2 thin films for the practical uses as crystalline thin film substrates or transparent electrodes.

REFERENCES

1. K. L. Chopra, S. Major, and D. K. Pandya, "Transparent Conductors – A Status Review," *Thin Solid Films*, **102**, 1-46 (1983)
2. Z. M. Jarzebski and J. P. Marton, "Physical Properties of SnO_2 Materials," *J. Electrochem. Soc.*, **123**, 199C-205C, 299C-310C, 333C-346C (1976)
3. H. S. Randhawa, M. D. Matthews, and R. F. Bunshah, "SnO_2 Films Prepared by Activated Reactive Evaporation," *Thin Solid Films*, **83**, 267-71 (1981)
4. T. Maruyama and H. Akagi, "Fluorine-doped Tin Dioxide Thin Films Prepared by Radio-Frequency Magnetron Sputtering," *J. Electrochem. Soc.*, **143**, 283-87 (1996)
5. K. H. Kim and S. W. Lee, "Effect of Antimony Addition on Electrical and Optical Properties of Tin Oxide Film," *J. Am. Ceram. Soc.*, **77**, 915-21 (1994)
6. J. Proscia and R. G. Gordon, " Properties of Fluorine-doped Tin Oxide Films Produced by Atmospheric Pressure Chemical Vapor Deposition from Tetramethyltin, Bromotrifluoromethane and Oxygen," *Thin Solid Films*, **214**, 175-87 (1992)
7. J. M. Laurent, A. Smith, D. S. Smith, J. P. Bonnet, and R. R. Clemente, "Morphology and Physical Properties of SnO_2-based Thin Films Deposited by the Pyrosol Process From Dibutyltindiacetate," *Thin Solid Films*, **292**, 145-49 (1997)
8. C. Terrier, J. P. Chatelon, R. Berjoan and J. A. Roger, "Electrical and Optical Properties of Sb:SnO_2 Thin Films Obtained by the Sol-gel Method," *Thin Solid Films*, **295**, 95-100 (1997)
9. I. Yagi, Y. Hagiwara, K. Murakami, and S. Kaneko, "Growth of Highly Oriented SnO2 Thin Films Grown from Tri-n-butyltin Acetate by Spray Pyrolysis Technique," *J. Mater. Res.*, **8**, 1481-83 (1993)
10. I. Yagi, K. Kakizawa, K. Murakami, and S. Kaneko, "Preferred Orientation of SnO_2 Thin Films Grown from Tri-n-butyltin Acetate by the Spray Pyrolysis Technique," *J. Ceram. Soc. Jpn.*, **102**, 296-98 (1994)
11. I. Yagi, and S. Kaneko, "Growth of Oriented Tin Oxide Films from an Organotin Compound by Spray Pyrolysis," *Mater. Res. Soc. Symp. Proc.*, **271**, 407-11 (1992)
12. J. Bruneaux, H. Cachet, M. Froment and A. Messad, "Correlation between Structural and

Electrical Properties of Sprayed Tin Oxide Films with and without Fluorine Doping," *Thin Solid Films*, **197**, 129-42 (1991)

13. G. Gordillo, L. C. Moreno, W. de la Cruz and P. Teheran, "Preparation and Characterization of SnO$_2$ Thin Films Deposited by Spray Pyrolysis from SnCl$_2$ and SnCl$_4$ Precursors," *Thin Solid Films*, **252**, 61-66 (1994)

14. B. Correa-Lozano, Ch. Comninellis and A. de Battisti, "Preparation of SnO$_2$-Sb$_2$O$_5$ Films by the Spray Pyrolysis Technique," *J. Appl. Electrochem.*, **26**, 83-89 (1996)

15. K. Murakami, I. Yagi, and S. Kaneko, "Oriented Growth of Tin Oxide Thin Films on Glass Substrates by Spray Pyrolysis of Organotin Compounds," *J. Am. Ceram. Soc.*, **79**, 2557-62 (1996)

16. J.-S. Ryu, Y. Watanabe, and M. Takata, "Correlation between Gas Sensing Properties of SnO$_2$:F Films Deposited by Spray Pyrolysis," *J. Ceram. Soc. Jpn.*, **100**, 1165-68 (1992)

17. D. Bélanger, and J. P. Dodelet, "Thickness Dependence of Transport Properties of Doped Polycrystalline Tin Oxide Films," *J. Electrochem. Soc..*, **132**, 1398-405 (1985)

18. G. Bradshaw and A. J. Hughes, "Etching Methods for Indium Oxide/Tin Oxide Films," *Thin Solid Films*, **33**, L5-L7 (1976)

19. Y. S. Chung, A. Hubenko, L. Meyering, M. Schade, and J. Zimmer, "Morphology and Phase of Tin Oxide Thin Films during their Growth from the Metallic Tin," *J. Vac. Sci. Technol. A*, **15**, 1108-12 (1997)

20. M. Kojima, H. Kato, and M. Gatto, "Blackening of Tin Oxide Thin Films Heavily Doped with Antimony," *Philos. Mag. B*, **68**, 215-22 (1993)

21. H. Viirola and L. Niinistö, "Controlled Growth of Antimony-doped Tin Dioxide Thin Films by Atomic Layer Epitaxy," *Thin Solid Films*, **251**, 127-35 (1994)

22. C. Terrier, J. P. Chatelon, R. Berjoan and J. A. Roger, "Sb-doped SnO$_2$ Transparent Conducting Oxide from the Sol-gel Dip-coating Technique," *Thin Solid Films*, **263**, 37-41 (1995)

23. J. C. Manifacier, "Thin Metallic Oxides as Transparent Conductors," *Thin Solid Films*, **90**, 297-308 (1982)

24. E. Shanthi, A. Banerjee, and K. L. Chopra, "Dopant Effects in Sprayed Tin Oxide Films," *Thin Solid Films*, **88**, 93-100 (1982)

25. M. Fantini and I. Torriani, "The Compositional and Structural Properties of Sprayed SnO$_2$:F Thin Films," *Thin Solid Films*, **138**, 255-65 (1986)

26. E. Shanthi, A. Banerjee, and K. L. Chopra, "Electrical and Optical Properties of Tin Oxide Films Doped with F and (Sb+F)," *J. Appl. Phys.*, **53**, 1615-21 (1982)

27. E. Shanthi, V. Dutta, A. Banerjee, and K. L. Chopra, Electrical and Optical Properties of Undoped and Antimony-doped Tin Oxide Films," *J. Appl. Phys.*, **51**, 6243-51 (1980)

SOLUTION DERIVED Ni-ZrO$_2$ CERMET THIN FILMS ON Si SUBSTRATES FOR RESISTIVE SENSOR APPLICATIONS

J.E. Sundeen and R.C. Buchanan,
University of Cincinnati, Department of Materials Science and Engineering,
Cincinnati, OH, 45221-0012

ABSTRACT

Ni-ZrO$_2$ cermet thin films were developed using the metalorganic deposition (MOD) process. Precursor films with 16-78 vol.% Ni, were deposited on (100) Si substrates, and sintered in forming gas (10%H$_2$-90%N$_2$) with 10m-2h soaks at temperatures of 600-900°C. Ni-ZrO$_2$ film phase development was studied using XRD and SEM techniques. To obtain full microstructural development, and Ni phase interconnection, minimum thickness of 0.8µm was required. 55 vol.% Ni films, 0.8-1.2µm thick, produced consistent R$_s$ and TCR behavior. Films exhibited sheet resistance (R$_s$) of 2-120Ω/□/1µm, with linear resistance-temperature response from 20-160°C. Linear temperature coefficients of resistance (TCR) up to 4300ppm/°C were achieved, suitable for resistive temperature sensor application. The TCR behavior was related to R$_s$ for specific processing conditions. For 1h sintering, the film TCR is not strongly dependent on sintering temperature from 750-900°C.

INTRODUCTION

Sol-gel processing was previously used to form functionally gradient cermet fibers for catalysts, mechanical and wear applications.[1,2] Typical fibers combine a particular metal (Cr, Fe, Cu, Co, Ni, Pb, Mo, W) with glass, SiO$_2$, or Al$_2$O$_3$. Development of fine particle bulk cermet nanocomposites has also been explored using the sol-gel process.[3,4] These nanocomposites were mainly produced with sintering temperatures of ≤600°C, developing particles <20nm in diameter. For both the fibers and nanocomposites, reduction heat treatment was used to develop cermets from oxide precursor materials. Coatings and thin films produced by this reduction heat treatment method have not been fully explored. MOD (metalorganic deposition) processing has been used to produce high purity Ni-ZrO$_2$ cermet films through reducing (10% H$_2$-N$_2$) ambient heat treatment of oxide precursors.[5]

Multiple phase formation is generally avoided in cermet film development, to avoid changing cermet conduction behavior. To prevent phase reactions, noble and refractive metals such as Pt, Au, W or Ag have been used with glass, Al_2O_3 or SiO_2 insulator.[6,7,8,9] $Ni-ZrO_2$ is a two-phase composite system, with non-noble Ni metal that is non-wetting, nonreactive, and bonds well to ZrO_2.[10]

Thick and thin films containing Pt and Ni are used in resistive temperature detectors (RTD), thermostats, heaters, flow and level sensors.[11-13] Thin films have been deposited by PVD techniques, usually requiring heat treatment. For these sensors, (temperature coefficient of resistance), TCR>3000ppm/°C, and 10-10kΩ resistance is required.[14] Pt film resistors can be used at 600-700°C, but many sensor applications require <200°C. Ni films are useful, based on higher TCR, (bulk Ni 6800ppm/°C, bulk Pt 3900ppm/°C), higher resistance, and low cost.[14] Pure Ni, however, has characteristic of non-linear resistance-temperature response.

For applications, several resistive properties are defined. Resistivity is defined as:

$$\rho = R \cdot A/l \qquad \text{(units: } \Omega \cdot cm) \qquad (1)$$

Where R is resistance, l is length and A is the area, with A = w·d for planar film section geometry, where w is width and d is thickness of the film. For constant resistor film thickness, the convenient design term sheet resistance is:

$$R_s = R \cdot /(l/w) \qquad \text{(units: } \Omega/\square) \qquad (2)$$

where l/w is the number of squares in resistor surface between contacts. Combining (1) and (2), sheet resistance and resistivity relate as:

$$\rho = R_s \cdot d \qquad \text{(units: } \Omega \cdot cm) \qquad (3)$$

To describe R_s equivalently then, film thickness must be uniform or normalized. Thick films for instance are usually described in terms of $\Omega/\square/25\mu m$.

Resistance-temperature response is important for applications. Temperature coefficient of resistance is defined as:

$$TCR = \{\Delta R / [R_0 \cdot (\Delta T)]\} \times 10^6 \qquad \text{(units: ppm/°C)} \qquad (4)$$

Where ΔT is change in temperature, typically measured between 25 and 125°C. ΔR is the change in sheet resistance between temperatures and R_0 is reference resistance, typically at 25°C. For linear resistance-temperature response

TCR is constant.

Percolative conduction describes cermet resistance behavior. At the percolation composition, as insulator content increases, resistance increases by several orders of magnitude, and TCR changes from positive to negative, coincident with loss of metal phase connectivity.

Films with representative thermal sensor characteristics, containing 55 vol.% Ni in the post percolative range, were investigated. Goals were to combine high TCR and moderate R_s in a range controlled by processing, and to illustrate limits of processing in which films with desired characteristics can be formed. To achieve this, film thickness, sintering temperatures and soak times were varied. Structural development variations reflected these variables.

EXPERIMENTAL

Ni-ZrO_2 cermets were deposited from solutions of carboxylate precursor powders. The reaction sequence to form metalorganic powders was

$$C_4H_9COOH + NH_3OH \rightarrow C_4H_9COONH_2 + H_2O \qquad (1)$$
$$C_4H_9COONH_2 + M(NO_3)_{2(aq)} \rightarrow M(C_4H_9COO)_2 + NH_4NO_{3(aq)} \qquad (2)$$

With M = Ni, ZrO. Ni and Zr carboxylate and zirconyl nitrate powders were dissolved in propionic acid and amylamines to form spinnable solutions. Solutions were produced for sintered film Ni contents of 16-78 vol.% (22-84 vol.% ZrO_2). Films with 55 vol.% Ni were analyzed due to desirable electrical resistance characteristics. Solution viscosities were determined using a cone-plate viscometer. The solutions were spin coated onto Si substrates at 3000 RPM, with pyrolysis at 425°C in air. Layers (3-6) were applied consecutively to increase thickness up to 1.2μm. Each layer was pyrolyzed at 425°C for 2 minutes. Sintering was in flowing 10%H_2-90% N_2 (forming) gas at 600-900°C for 10m-2h.

XRD, including Scherrer technique,[15] and SEM were used to determine the phase and microstructural evolution of the Ni-ZrO_2 films. Electrical resistance measurements were made on planar two point specimens $\approx$1.6 – 2 squares (6mm length, 3-4mm width) mainly using sputtered Au electrodes. Resistance-temperature behavior was characterized at <15mW using a DC power supply and electrometer, between 20°C and 160°C in air, with specimen heating rate of 2.5°C/min.

RESULTS AND DISCUSSION

Figure 1 shows viscosity of 55 vol.% Ni precursor solution. This solution exhibited a viscosity of 62 cpoise, dropping to 58 cpoise as shear rate increased from 5 to 25s^{-1}. Solutions all exhibited pseudoplastic shear thinning behavior, and 55-65cpoise solutions resulted in uniform spin deposition. Figure 2 shows

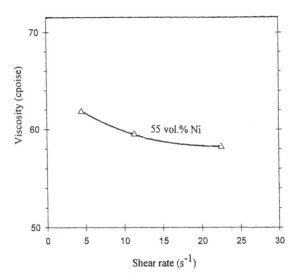

Fig. 1 Viscosity at 25°C for film precursor compositions.

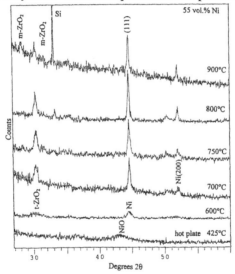

Fig. 2 Phase development of 55 vol.% Ni films sintered 1h (hot plate 10m).

phase development of 1 μm thick 55 vol.% Ni films. Following 10 minutes on hot plate in air, films are mainly amorphous with some nanoscale NiO (particle size ≤ 7-9nm). One hour sintering of the films in forming gas produced Ni and tetragonal ZrO_2 phases. At 600°C, very fine ZrO_2 (≈10nm), and Ni (≈18-20nm) formed. Phase development in the full temperature range shows Ni and ZrO_2

particle growth with sintering temperature, without notable interfacial reaction. Some monoclinic ZrO_2 phase appears following 1h sintering at 900°C.

Figure 3 shows microstructure of the films sintered at 600°C, 700°C, 750°C and 900°C respectively. Following sintering, fine Ni particle agglomerates are dispersed in the structure. These particle agglomerates increase in size into single grains with increased temperature, and are notably faceted with >1μm size following 850-900°C sintering. Localized regions of depleted Ni are observed surrounding larger agglomerates. Film surfaces all appear dense, and structure is well developed at 700°C and above.

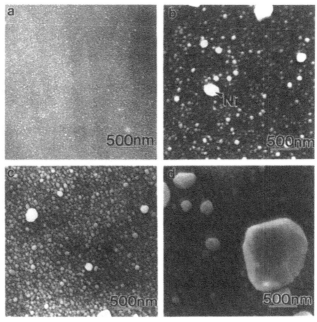

Fig. 3 55 vol.% Ni films sintered 1h at a)600, b)700, c)750 and d)900°C.

Figure 4 shows structural development of a 1μm thick film with 55 vol.% Ni, sintered for 10m-2h at 800°C. The tetragonal ZrO_2 phase is present in significant amounts, and particle size increases with time from ≈30-40nm. The Ni phase is well developed at all sintering times with particle sizes > 40nm, and a sharper Ni (200) peak as sintering time increases.

Figure 5a shows film thickness effect on 55 vol.% Ni film development for 1h sintering at 800°C. Although XRD shows similar Ni phase development for the 0.6-1.0μm thick films, the ZrO_2 particle size decreases as film thickness decreases. Figure 5b and c microstructures show Ni agglomerates, with

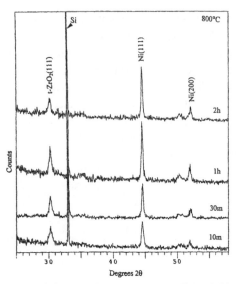

Fig. 4 XRD of 1 μm thick 55 vol.% Ni films with variable sintering time.

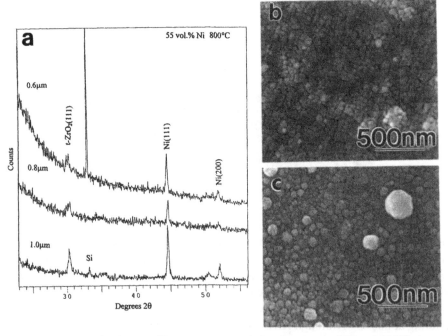

Fig. 5 Film thickness effect on structural development of 55 vol.% Ni at 800°C.

Dielectric Ceramic Materials

fine agglomerate grain development in the 0.6μm thick films (b), compared to the spherical Ni agglomerate structure of the 1μm thick films (c). This structural development can be directly correlated to the resistance properties.

Figure 6 shows the effect on sheet resistance and TCR of sintering temperatures from 750-900°C. For 1 hour sintering, sheet resistance increases with sintering temperature as Ni agglomerate growth causes increased separation of the Ni phase. At sintering temperatures below 750°C, it is more difficult to form conductive cermets with interconnected Ni phase. It is notable that TCR is not significantly decreased with increase in R_s and sintering temperature.

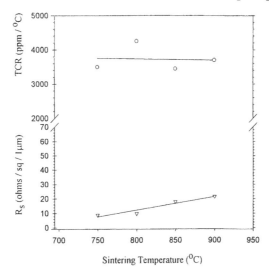

Fig. 6 Sintering temperature effect on TCR and R_s for 1h sintered films.

Figure 7 shows the dependence of TCR on R_s for 1h sintered 55 vol.% Ni films. The value of R_s is linearly related to TCR for the similarly developed cermets. The relationship can be expressed by:

$$\alpha(TCR) = \alpha_o - \beta(R_s) \qquad (1)$$

Where α_o is the TCR at extrapolated zero resistance, and β reflects the decrease of TCR with increased R_s. For the films of Figure 7, $\alpha_o = 4150$ and $\beta = 31.5$. This relationship predicts TCR within ±350 ppm/°C. Values of $\alpha_o = 4010$ and $\beta = 97$ apply for films sintered 10-30m at 750-900°C. The lower α_o and increased β reflect lowered resistance-temperature sensitivity for these films, however, TCR is predicted within ±200ppm/°C, providing a better predictability of TCR from R_s.

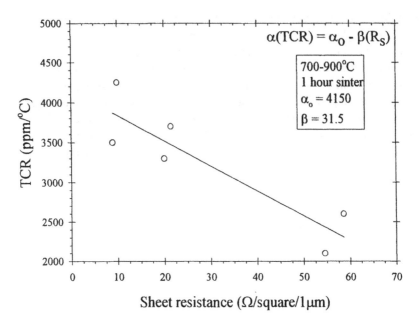

Figure 7 TCR dependence on R_s for films sintered 1h at 700-900°C.

Figure 8 shows response of R_s (normalized for 1 μm thickness), to temperature for the 0.6-1.2μm, 800°C 1h sintered Ni-ZrO_2 films. Response is linear for 0.8-1.2μm films as shown in Table I, with r^2 coefficients > 0.99. The linear resistance-temperature behavior with high TCR makes these films suitable for heaters with simple electrical control. The 0.6μm thick films have a higher resistivity ($R_s \cdot d$), coincident with finer grain structure, as shown in Figure 5. The 1.0 and 1.2μm thick films have variation in resistivity of only 15%, and exhibit larger agglomerate microstructures. Non-linearity of the relationship of R_s (and ρ) to film thickness only becomes significant as film thickness decreases below 0.8μm, so that R_s can be specified for films 0.8-1.2μm thick.

A figure of merit has been defined for temperature and flow sensors based on high dR/dT. The material property figure of merit is expressed as:[11]

$$\rho \cdot TCR \qquad \text{(units: } \Omega \cdot cm/°C) \qquad (2)$$

This value is shown for Ni-ZrO_2 cermet films of different thickness in Table I. A commercial Ni thick film conductor has R_s = 0.40 $\Omega/\square$, and TCR of 5800ppm/°C. Assuming a 25μm thick film, $\rho \cdot TCR$ is 580 ×10^{-8} $\Omega \cdot cm/°C$. The Ni-ZrO_2 thin films of this work compare favorably to Ni conductor films with 1-2 orders of

Dielectric Ceramic Materials

magnitude higher R_s, and TCR of 1500-4250 ppm/°C. This leads to $\rho \cdot$TCR values of 305-1050 $\times 10^{-8}$ $\Omega \cdot$cm/°C. The higher R_s values allow design of sensors on smaller substrate areas, with the added advantage that the films are also compatible with integrated silicon processing.

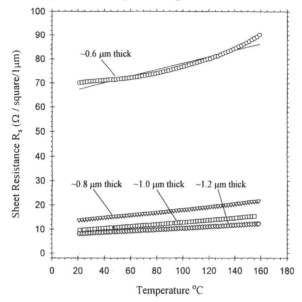

Fig. 8 R_s response to temperature for 0.6-1.2μm thick 55 vol.% Ni films.

Table I Resistive characteristics of 0.6-1.2μm 55 vol.% films (800°C, 1h).

Film thickness (μm±0.1)	Resistivity at 25°C (Ω·cm)	R_s at 25°C (Ω/□)	TCR (ppm/°C) 25-125°C	r^2 coefficient	$\rho \cdot$TCR (Ω·cm/ °C)$\times 10^{-8}$
0.6	70 $\times 10^{-4}$	115	1500	0.932	1050
0.8	14 $\times 10^{-4}$	17	3900	0.997	550
1.0	10 $\times 10^{-4}$	10	4250	0.999	420
1.2	8.4 $\times 10^{-4}$	7.0	3600	0.999	305

CONCLUSIONS

High positive TCR, suitable for resistive sensors, was developed in 55 vol.% Ni (MOD), Ni-ZrO₂ cermets. This composition was chosen for optimal resistance characteristics in post percolative Ni content range. Advantage over Ni films is the linear resistance-temperature response. Resistivity ρ (R_s), and TCR are related to sintering temperature, and to film thickness in this system. Films 0.8-

1µm thick exhibit linear resistance-temperature response, with R_s values that can be specified. For higher resistance range with positive TCR, 0.6µm films are advantageous. Film development during equivalent sintering, specifically the phase evolution, coalescence and Ni phase agglomeration, change with film thickness reflecting resistance changes. TCR can be estimated from film R_s, using the relationship $\alpha(TCR) = \alpha_o - \beta(R_s)$. These Ni-ZrO$_2$ cermet thin films are silicon technology compatible, with sensor properties equivalent to Ni thick film conductor materials used in heaters and thermal sensors. Higher R_s of Ni-ZrO$_2$ also allows decreased device size.

ACKNOWLEDGEMENT

This work was funded by the National Science Foundation under NSF grant ECS-9612122. This support is gratefully acknowledged.

REFERENCES

[1]H.G. Sowman, U.S. Pat. No. 4 797 378, "Internally Modified Ceramic Fiber," Jan. 10, 1989.
[2]H.G. Sowman and D.R. Kaar, U.S. Patent No. 4 713 300, "Graded Refractory Cermet Article," Dec. 15, 1987.
[3]R.A. Roy and R. Roy, "Diphasic Xerogels. I. Ceramic-metal Composites," Mat. Res. Bull.,19, 169-177 (1984).
[4]D. Chakravorty, "Nanocomposites by Sol-gel and Ion-exchange Routes," Ferroelectrics, 102, 33-43 (1990).
[5]J.E. Sundeen and R.C. Buchanan, "Electrical Properties of Nickel-Zirconia Cermet Films for Temperature and Flow Sensor Applications," Sensors and Actuators A, 63 [1] 33-40 (1997).
[6]B. Abeles, "Granular Metal Films," p. 1-117 in Applied Solid State Science, vol. 6, Academic Press, New York, 1976.
[7]N. Gershenfeld, et al., "Percolating Cermet Thin Film Resistor," U.S. Pat. No. 4 906 968, March 6, 1990.
[8]G.R. Witt, "Some Effects of Strain and Temperature on the Resistance of Thin Gold-Glass Cermet Films," Thin Solid Films, 13, 109-115 (1972).
[9]C.A. Neugebauer, "Resistivity of Cermet Films Containing Oxides of Silicon," Thin Solid Films, 6, 443-447 (1970).
[10]D. Sotiropoulous and P. Nikolopoulous, "Work of Adhesion in ZrO$_2$-Liquid Metal Systems," J. Mater. Sci., 28, 356-360 (1993).
[11]V.A. Sandborn, Resistance Temperature Transducers, Metrology Press, Fort Collins, CO, 1972.
[12]P. Baumbach, M.A. Stein, R. Tait and J. Whitmarsh, "Materials, Substrates, and Designs for Manufacturing Heaters Using Thick Film Technology," p. 699-705, ISHM '90 Proceedings.
[13]S.-H. Sheen, A.C. Raptis and M.J. Moscynski, "Automotive Vehicle Sensors," ANL-95/45, Argonne National Laboratory, 1995.
[14]H. Arima, p. 127-150 in Thick Film Sensors, Handbook of Sensors and Actuators 1, Edited by M. Prudenziati, Elsevier Science Publishing Co. Inc., New York, 1994.
[15]B.D. Cullity, Elements of X-ray Diffraction, 2nd Edition, Addison-Wesley, Reading MA, 1978.

FABRICATION OF BISMUTH LAYER STRUCTURED LEAD BISMUTH TITANATE THIN FILMS THROUGH SOL-GEL SPIN COATING

Yong-il PARK and Masaru MIYAYAMA
University of Tokyo
7-3-1 Hongo, Bunkyo-ku, Tokyo 113, Japan

ABSTRACT
Ferroelectric $PbBi_4Ti_4O_{15}$(PBT) thin films were fabricated on $Pt/Ti/SiO_2/Si$ substrates by sol-gel spin-coating and their ferroelectric properties were investigated as a new candidate for ferroelectric nonvolatile memory.

INTRODUCTION
Research and development activities are strongly increasing worldwide in an effort to commercialize devices based upon ferroelectric thin films for nonvolatile memory (NvRAM) applications. However, the ferroelectric thin films such as those of $PbTiO_3$, PZT or PLZT have several problems in their polarization fatigue properties which afford only a limited number of switching cycles to the ferroelectric memory cells.[1-6] Recently, Bi layered perovskite thin films such as $Bi_4Ti_3O_{12}$[7-10], $SrBi_2Ta_2O_9$[11-13] and $SrBi_2Nb_2O_9$[14] which show little or no fatigue have been developed. Among them, $SrBi_2Ta_2O_9$(SBT) films have especially high potential for NvRAM device application. But the SBT films also have some technical problems concerning reliability and processing, *i.e.,* low remanent polarization and high heat-treatment temperature.

The adequate sintering temperature of 700°C and its inherent structure(Fig. 1) and properties of $PbBi_4Ti_4O_{15}$(PBT)[15-17] may avoid some problems in processing of SBT capacitors and some of the degradation problems, such as fatigue and lack of retention which were observed in PZT capacitors. $PbBi_4Ti_4O_{15}$ has a high Curie point of 570°C compared with 335°C of SBT and 385°C of PZT(52/48), allowing poled polycrystals to maintain a remanent polarization state up to at least 300°C with an average temperature coefficient for permittivity of +1200ppm/°C, thereby avoiding the thermal imprint problem. As like other bismuth layer structured ferroelectrics, $PbBi_4Ti_4O_{15}$ shows anisotropy in ferroelectricity, that is, two independently reversible polarization vectors are observed, one with a large dielectric constant and polarization along the a(b)-axis and one with a small dielectric constant along the c-axis. The dielectric properties along the $PbBi_4Ti_4O_{15}$ is very attractive for memory applications from its low dielectric constant avoiding non-switching linear response(εE), faster switching speed, higher retention and higher endurance due to its inherent layered structure. The present paper deals with the sol-gel spin-coating preparation and electrical properties of PBT thin films.

EXPERIMENTAL PROCEDURE
The procedure in this study for preparation of PBT precursor solutions is briefly illustrated

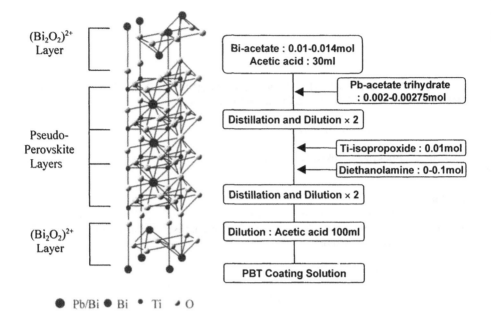

$(Bi_2O_2)^{2+}$ Layer

Pseudo-Perovskite Layers

$(Bi_2O_2)^{2+}$ Layer

● Pb/Bi ● Bi ● Ti ◢ O

Bi-acetate : 0.01-0.014mol
Acetic acid : 30ml

Pb-acetate trihydrate : 0.002-0.00275mol

Distillation and Dilution × 2

Ti-isopropoxide : 0.01mol

Diethanolamine : 0-0.1mol

Distillation and Dilution × 2

Dilution : Acetic acid 100ml

PBT Coating Solution

Fig. 1. One half of $PbBi_4Ti_4O_{15}$ unit cell.

Fig. 2. Flow diagram for the preparation of PBT coating solutions.

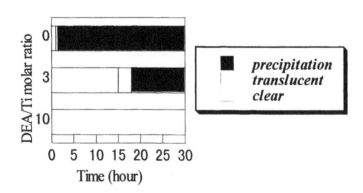

DEA/Ti molar ratio

■ precipitation
| translucent
- clear

Fig. 3. Stability of the PBT coating solutions at different DEA content.

$N_2(C_2H_5OH)_2$ was added as a chelating agent to stabilize hydrolysis and condensation reaction and to enhance the solubility of acetates[18]. The solutions in which DEA was not added were extremely unstable and a precipitate was formed within 1hour. It was found that DEA additions increased the solution stability. The solution which has 10 of DEA/Ti molar ratio showed a life time of over 1month and good adhesion.(Fig. 3) PBT coating solutions

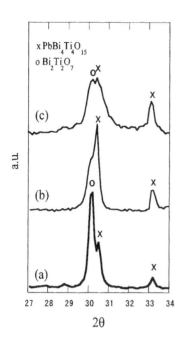

Fig. 4. XRD patterns of $Pb_xBi_{4y}Ti_4O_{15}$ thin films; (a)x=1.0, y=1.0, (b)x=1.0, y= 1.4 and (c) x=1.1, y=1.0.

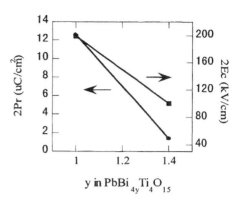

Fig. 5. 2Pr and 2Ec changes of PBT capacitors at different Bi composition.

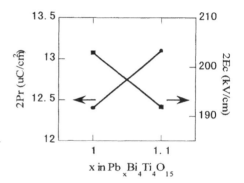

Fig. 6. 2Pr and 2Ec changes of PBT capacitors at different Pb composition.

were prepared through distillation and dilution of these DEA-complexed mixtures. The coating solutions were filtered with 0.2-μm PTFE membrane filter and then deposited onto Pt/Ti/SiO$_2$/Si(100) and Pt/Ti/Si(100) substrates by spin coating at 3000rpm for 30sec., and were dried at 450-500°C for 1min. Coating-drying process was repeated till desired film thickness was obtained. Thickness of each spin-coated layer after final sintering was about 22nm. Final sintering was conducted at 600-800°C for 30min at a heating rate of 35°C/min in an electric furnace flowing O$_2$ gas at a speed of 40ml/min. Pt upper electrode of 300μm diameter was formed by sputtering.

Phase development and film orientation were determined using X-ray diffraction analysis (XRD). Field-emission scanning electron microscope (FE-SEM) was used to observe microstructure. Dielectric measurements were conducted using an YHP4192A Impedance Analyzer. Ferroelectric properties were determined by RT6000HVS(Radiant Technologies, inc.).

RESULTS AND DISCUSSION

Main peaks in the XRD patterns of PBT(Pb/Bi/Ti=x/4y/4 ; x=1.0 and 1.1, y=1.0 and 1.4) films on Pt/Ti/SiO$_2$/Si are shown in Fig. 4. As can be seen in Fig. 4(b), the pyrochlore phase (Bi$_2$Ti$_2$O$_7$) decreases and (0014)(119) and (200) peaks of PBT grows with increasing Bi composition from 4(y=1.0) to 5.6(y=1.4). From this, volatilization of Bi atoms and diffusion of Bi atoms into the substrate can be supposed as reasons of co-existence of pyrochlore phase in stoichiometric composition PBT thin films. These problems were already reported in the study on Bi$_4$Ti$_3$O$_{12}$ thin films by S.Okamura et.al.[7] Notwithstanding large amount of excess Bi content, some extent of remained pyrochlore phase were observed at y=1.4 composition. Remanent polarization(2Pr) and coercive field(2Ec) changes as a function of Bi composition are shown in Fig. 5. In spite of disappearing pyrochlore phase and growing PBT phase with increase of Bi composition, 2Pr and 2Ec extremely decreases as Bi composition increases. This result implies that excess Bi atoms reside at the grain boundaries of the PBT(1/5.6/4) thin film in amorphous form.

Fig. 4(c) shows main peaks in the XRD patterns of the PBT(Pb/Bi/Ti=1.1/4/4) film. With an increase of Pb composition from 1.0 to 1.1, the peak of pyrochlore phase was depressed and the peaks of PBT phase was grown. 2Pr and 2Ec changes as a function of Pb composition are shown in Fig. 6. An increase of 2Pr and a decrease of 2Ec with an increase of Pb composition may be due to the excess Pb effect which was reported in the studies on Pb-based perovskite structured thin films, PbTiO$_3$ or PZT, as the compensation effect of volatilized Pb atom vacancies.[4]

In order to certify, at second hand, the main reason of the phase and ferroelectric property change with Bi composition change, a PBT film with buffer layer was fabricated by a two-step coating process. To prepare the film with a buffer layer, only the first spin layer was deposited with Pb/Bi/Ti=1/5.6/4 composition solution and then Pb/Bi/Ti= 1.1/4/4 composition solution was repeatedly spin-coated onto the substrate up to the desired thickness. (hereafter called PBT film with buffer layer). Fig. 7 shows XRD plots of PBT thin films with and without PBT buffer layer. The pyrochlore peaks observed in the XRD pattern of PBT(1.1/4/4) film were perfectly disappeared in the XRD plots of the PBT film with buffer layer suggesting that the main reason of the phase and ferroelectric property change with Bi composition change is insufficient Bi composition in the first coated layer (about 22nm thick) due to the diffusion of Bi atoms into the substrate and Bi volatilization. FE-SEM images of PBT thin films with and without PBT buffer layer are shown in Fig. 8. Grain size of PBT film without buffer layer is very random compared with homogeneous grain size distribution of the PBT film with buffer layer. To observe the diffusion of bismuth into the substrate through Pt electrode layer, a substrate without SiO$_2$ layer was used. The XRD pattern of the PBT(1/4/4) film with buffer layer deposited onto Pt/Ti/Si substrate is shown in Fig. 9. A large peak of Bi$_2$SiO$_5$ around 2θ of 29° strongly backs up the possibility of Bi diffusion into the substrates. Fig. 10 shows P-E hysteresis loops of PBT thin films with and without buffer layer. 2Pr increased from 12.4 μC/cm^2 to 13.1 μC/cm^2 and 2Ec decreased from 203kV/cm to 192kV/cm in the PBT film with buffer layer compared with PBT capacitor without buffer layer. As seen in SEM and XRD analysis, homogeneous microstructure and depression of second phase seems to cause increase of ferroelectricity in PBT film with buffer layer.

Fig. 11 shows P-E hysteresis loops of the 334nm-thick PBT film with buffer layer at different electric field. Observed non-linear P-E hysteresis loops have confirmed the ferroelectric property of the PBT film with buffer layer. However, the well-saturated hysteresis loops could not be obtained. This may be due to relatively high coercive field of PbBi$_4$Ti$_4$O$_{15}$[19] at room temperature compared with SrBi$_2$Ta$_2$O$_9$[11-13], and depolarizing field and non-switching linear response due to space charge which is originated from somewhat porous

Dielectric Ceramic Materials

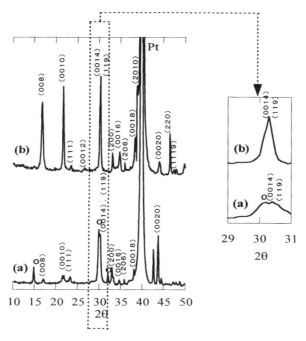

Fig. 7. XRD patterns of PBT(1.1/4/4) thin films (a) without and (b)with buffer layer.

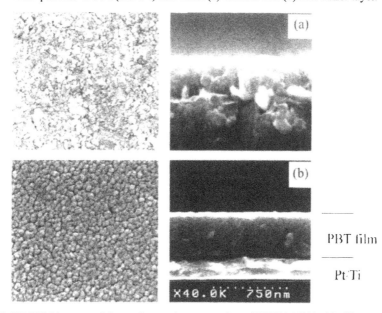

Fig. 8. FE-SEM images of the surface and cross section of PBT(1.1/4/4) thin films (a) without and (b)with buffer layer.

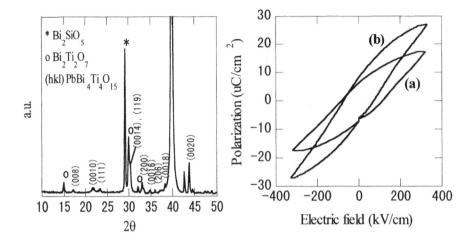

Fig. 9. XRD pattern of the PBT(1/4/4) film deposited on Pt/Ti/Si substrate.

Fig. 10. P-E hysteresis loops of PBT (1.1/4/4) thin films of (a)without and (b) with buffer layer.

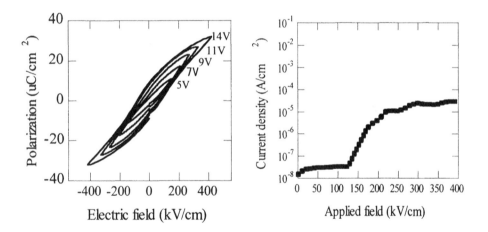

Fig. 11. P-E hysteresis loops of the PBT(1.1/4/4) thin film (thick=334nm) with buffer layer at different applied voltage.

Fig. 12. Leakage current density of the PBT (1.1/4/4) thin film with buffer layer.

Dielectric Ceramic Materials

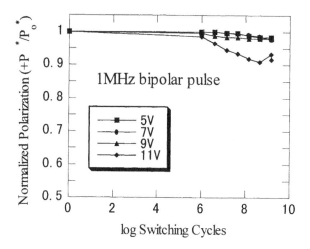

Fig. 13. Fatigue property of the PBT(1.1/4/4) thin film with buffer layer at different applied voltage.

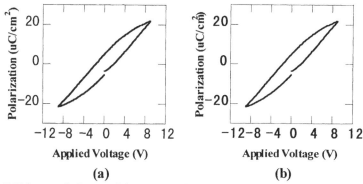

Fig. 14. P-E hysteresis loops of the PBT(1.1/4/4) thin film with buffer layer (a)before and (b)after fatigue test.

surface microstructure due to high organic content of starting solutions. Obtained 2Pr and 2Ec were 15.7μC/cm^2 and 122kV/cm respectively at electric field of 410kV/cm. Fig. 12 shows leakage current density of the PBT film with buffer layer. Leakage current was measured with a voltage step of 0.1V and delay time of 0.1second. The current density of the PBT film with buffer layer heat-treated at 700°C was about 3×10^{-8}A/cm^2 below electric field of 130kV/cm and then increased up to 2×10^{-5}A/cm^2 at electric field of 400kV/cm. Fig. 13 shows fatigue properties of the 334nm-thick PBT film with buffer layer at different applied voltage. Fatigue tests were done using bipolar square waves of 5V~11V at frequency of 1MHz. Almost no fatigue was observed in the fatigue tests conducted with 5V~9V pulses up to 1×10^9 switching cycles. At switching pulses of 11V, the capacitor showed about 9% of depolarization at 1×10^9

switching cycles and then breakdowned at 1×10^{10} switching cycles. The P-E hysteresis loops of PBT capacitor before and after the fatigue test were shown in Fig. 14. The shape of hysteresis loop obtained after fatigue test was practically same with the loop observed before the test. Observed relative permittivity and $\tan\delta$ of the PBT film with buffer layer was 270 and 0.03 respectively at frequency of 1MHz, at room temperature.

CONCLUSIONS

It was found that Bi-rich buffer layer and Pb-rich composition is effective to depress pyrochlore phase growing and to enhance ferroelectric property of the PBT thin film.

The resulting PBT film with buffer layer was of good quality, with low dielectric properties matching bulk ceramic values, and sufficient 2Pr value of $15.7\mu C/cm^2$. Present 2Ec value of 122kV/cm is relatively high for practical memory element switching, however this can be lowered by improved processing and controlling film composition. Especially, modification of PBT through substitution by elements which have different ionic radius and valence with PBT is very hopeful to enhance ferroelectricity at room temperature. A study is now in progress on the modification of PBT.

REFERENCES

[1]R.Takayama, Y.Tomita, K.Iijimaet and I.Ueda, "Pyroelectric Properties and Application to Infrared Sensors of PbTiO$_3$, (Pb,La)TiO$_3$ and Pb(Zr,Ti)O$_3$ Ferroelectric Thin Films," *Ferroelectrics*, **118**, 325-342 (1991).

[2]G.Yi, Z.Wu and M.Sayer, "Preparation of Pb(Zr,Ti)O$_3$ Thin Films by Sol-gel processing : Electrical, Optical, and Electro-optic Properties," *Journal of Applied Physics*, **64**(5), 2717-2724 (1988).

[3]C.D.E.Lakeman, J.F.Campion and D.A.Payne, "Factors Affecting the Sol-gel Processing of PZT Thin Layers" ; pp413-439 *in Ceramic Transactions Volume 25 : Ferroic Materials:Design, Preparation and Characteristics*, Edited by A.S.Bhalla et al., the American Ceramic Society, Westerville, OH, 1992.

[4]B.T.Teowee, J.M.Boulton, D.R.Uhlmann, "Optimization of Sol-gel derived PZT Thin Films by the Incorporation of Excess PbO," *Material Research Society Symposium Proceedings*, **271**, 345 (1992).

[5]C.D.E.Lakeman and D.A.Payne, "Processing Effects in the Sol-gel Preparation of PZT Dried Gels, Powders, and Ferroelectric Thin Layers," *Journal of the American Ceramic Society*, **75**(11), 3091-3096 (1992).

[6]B.A.Tuttle, J.A.Boigt and D.C.Goodnow, "Highly Oriented, Chemically Prepared Pb(Zr,Ti)O$_3$ Thin Films," *Journal of the American Ceramic Society*, **76**(6)1537-1544 (1993).

[7]S.Okamura, Y.Yagi, K.Mori and G.Fujihashi, "Control of the Orientation of Ferroelectric Bi$_4$Ti$_3$O$_{12}$ Thin Films by Multi-step Metal Organic Decomposition Process," *Japanese Journal of Applied Physics*, **36**, 5889-5892 (1997).

[8]R.C.Buchenan and P.Chu, "Processing and Characteristics of Spin-cast Ferroelectric Bi$_4$Ti$_3$O$_{12}$ Thin Films"; pp137-143 *in Ceramic Transactions Volume 55 : Sol-gel Science and Technology*, Edited by E.A.Pope et al., the American Ceramic Society, Westerville, OH, 1995.

[9]P.C.Joshi, A.Manshingh, M.N.Kamalasanan and S.Chandra, "Structural and Optical Properties of Ferroelectric Bi$_4$Ti$_3$O$_{12}$ Thin Films by Sol-gel Technique," *Journal of Applied Physics Letters*, **59**(19), 4 (1991).

[10]M.Azimi and P.K.Ghosh, "Deposition and Dielectric Property of Ion-beam Assisted Bismuth Titanate Film,"; pp15-62 *in Ceramic Transactions Volume 43 : Ferroic Materials:Design, Preparation and Characteristics*, Edited by A.S.Bhalla et al., the American Ceramic Society, Westerville, OH, 1995.

[11]C.A.Paz de Araujo, J.D.Cuchiaro, L.D.McMillan, M.C.Scott and J.F.Scott, "Fatigue-free Ferroelectric Capacitors with Platinum Electrodes," *Nature*, **374**(13), 627-629 (1995).

[12]H.N.Al-Shareef, D.Dimos, W.L.Warren and B.A.Tuttle, "Voltage Offsets and Imprint Mechanism in SrBi$_2$Ta$_2$O$_9$ Thin Films," *Journal of Applied Physics*, **80**(8), 15, 4573-4577 (1996).

[13]Y.Ito, M.Ushikubo, S.Yokoyama, H.Matsunaga, T.Atsuki, T.Yonezawa and K.Ogi, "New

Low Temperature Processing of Sol-gel $SrBi_2Ta_2O_9$ Thin Films," *Japanese Journal of Applied Physics,* **35**, 4925-4929 (1996).

[14]C.A.Paz de Araujo, J.D.Cuchiaro, M.C.Scott and L.D.McMillan : Int. Pat. No. WO93/12542, June. 24, 1993.

[15]G.A.Smollenski, V.A.Isupov and A.I.Agranovskaya, "Ferroelectrics of the Oxygen-Octahedral Type with Layered Structure," *Soviet Physics-Solid State*, **3**(3), 651-655 (1961).

[16]J.A.deverin, "The Dielectric Properties of $PbBi_4Ti_4O_{15}$ Ceramics," *Ferroelectrics*, **19**, 5-7, (1978).

[17]E.C.Subbarao, "Crystal Chemistry of Mixed Bismuth Oxides with Layer-type Structure," *Journal of the American Ceramic Society*, **4**, 166-169 (1962).

[18]L.G.Hubert-Pfalzgraf, M.C.Massiani, J.C.Daran and J.Vaissermann, "Stabilization of Metal Alkoxides (M=Ba, Cu) by Alkanolamines," pp135-140 *in Better Ceramics Through Chemistry V, Edited by M.J.hempden-Smith et al.*, Material Research Society, Pittsburgh, Pennsylvania, 1992.

[19]I.Yi and M.Miyayama, "Anisotropy of Electrical Properties in Single Crystals of Layer-Structured Lead Bismuth Titanate," *Material research Bulletin*, **32**(10), 1349-1357 (1997).

COMPLEX PERMITTIVITY AND PERMEABILITY OF FERRITE CERAMICS AT MICROWAVE FREQUENCIES

RICHARD G. GEYER
National Institute of Standards and Technology
Electromagnetic Fields Division, M.S. 813.08
Boulder, CO 80303

ABSTRACT

Accurate rf dielectric and magnetic characterization of ferrite ceramics requires combined measurement techniques. The chosen techniques depend on the frequency of interest and expected dielectric and magnetic losses of the specimen. Coaxial transmission line waveguide methods are used for magnetic characterization at frequencies less than gyromagnetic resonance, where specimens have high magnetic loss. Specimen complex permittivity is evaluated with either a TM_{0n0} cavity, or by saturating the specimen when operated as a cylindrical H_{011} dielectric resonator. Complex permeability at frequencies above natural gyromagnetic resonance, where magnetic losses are low, are either evaluated with H_{011} dielectric sleeve resonators or by employing the specimen as an H_{011} dielectric resonator in the demagnetized state.

Measurements of various ferrites having different compositions and saturation magnetizations are performed. The tuning range of a ferrite increases above natural gyromagnetic resonance as saturation magnetization increases, but at the expense of increased magnetic loss. Measured real permeabilities compare well with theoretical real permeability predictions of a cylindrical specimen model having 2-domain structure parallel and anti-parallel to the applied rf magnetic, but only at frequencies significantly greater than natural gyromagnetic resonance. Microwave magnetic losses, which depend on composition and ceramic microstructure, must be experimentally evaluated. Cooling the ferrite specimen yields smaller dielectric, but larger magnetic loss.

INTRODUCTION

Ferrite ceramics are used in a variety of rf and microwave applications. At frequencies f less than natural gyromagnetic resonance, $f_M = \gamma_g M_s$, where γ_g is the gyromagnetic ratio (35.19 MHz·m/kA) and M_s is the saturation magnetization (kA/m), ferrites are magnetically lossy. At these frequencies ferrites can be used as absorbant tiles in anechoic chambers or in other shielding applications. For frequencies $f > f_M$, ferrites have low magnetic loss and are often incorporated into microwave tuning devices, such as circulators that control the transmission path of microwave signals, or phase shifters that electronically control the beam pattern of phased antenna arrays.

Accurate dielectric and magnetic property measurements of ferrite ceramics at the intended operational frequency and temperature ranges are needed for optimized development of either shielding materials or tuning devices, as well as to assist in the manufacture of the ferrite itself.

In general, microwave measurement techniques for evaluating complex permittivity and permeability can be classified into various methods:

- Standing wave pattern techniques,

- Transmission and reflection in freespace or waveguide techniques,

- Cavity or dielectric resonator systems.

The chosen measurement method is based on the accuracies required, the amount and shape of the specimen available, the frequency, and expected values for the relative complex permittivity $\epsilon_r^* = \epsilon_r' - j\epsilon_r'' = \epsilon_r'(1 - j\tan\delta_e)$ and complex permeability $\mu_r^* = \mu_r' - j\mu_r'' = \mu_r'(1 - j\tan\delta_m)$, where ϵ_r', μ_r' are the real permittivity and permeability and $\tan\delta_e, \tan\delta_m$ are the dielectric and magnetic loss tangents.

Usually standing wave methods [1] are best suited for liquid property measurements. Freespace techniques can be used for large parallel plate samples; even when this sample geometry is available, these techniques suffer from misalignments of the sample relative to assumed incident angles of electromagnetic waves, as well as from radiation losses.

Waveguide transmission or reflection methods can be used for broadband dielectric and magnetic characterization of relatively small toroidal specimens having moderate to large losses; however they suffer from large uncertainties in measured reflection and transmission coefficients as the specimen permittivity increases. Very small tolerances in the machined dimensions of a waveguide sample are required. If the specimen does not fill the entire cross section of waveguide perfectly, corrections must be made for air gaps present. For high permittivity samples in either rectangular or coaxial transmission lines, the presence of air gaps allow dielectric depolarization effects that can lead to significant underestimation of actual specimen permittivity. These depolarization effects can be mitigated by appropriate metallization of the specimen [2]. In general, transmission line methods cannot accurately measure either dielectric or magnetic loss tangents lower than 10^{-3}. However, coaxial transmission line methods are useful for accurate microwave measurements

Dielectric Ceramic Materials

of complex permeability at frequencies less than natural gyromagnetic resonance, where magnetic losses are relatively high [3].

Cylindrical cavity measurements using TE_{01p} and TM_{0n0} mode structure may be used to evaluate the complex permeability and permittivity of low-loss rod or disk specimens. Real permittivity and permeability evaluations are principally affected by specimen and cavity dimensional uncertainties. For dielectric or magnetic loss determinations, these methods are subject to uncertainties in the characterization of both conductor and coupling losses of the measurement fixture, and any radiative losses arising from sample insertion. Cavity methods, typically fabricated for single-frequency measurements, are also constrained in sensitivity by the sample partial electric and magnetic energy filling factors when specimens are inserted into the fixture.

Dielectric resonator methods are usually most accurate for dielectric and magnetic loss tangent measurements of low-loss materials ($\tan \delta_e < 10^{-3}, \tan \delta_m < 10^{-3}$). One dielectric resonator method is to take a cylindrical specimen of interest and resonate it between two parallel metallic ground planes. Generally, a low-order resonant mode that can be easily identified, such as the H_{011} mode, is used for dielectric measurements. With proper mode usage and identification, dielectric resonator techniques can permit much higher partial electric and magnetic energy sample filling factors because the stored energy exterior to the dielectric resonator specimen is negligible (usually less than 3%). Therefore, they have greater sensitivity to both dielectric and magnetic losses. If the microwave conductive losses of the metallic ground planes are determined, lower measurement uncertainties in dielectric and magnetic material properties can be realized than those attainable in cavity techniques (for comparable specimen dimensional uncertainties). When using the TE_{011} (H_{011}) mode, the effect of any air gap between the sample and the conducting ground planes also becomes negligible (no capacitive coupling between sample and ground planes), since the electric field is tangential to the sample surfaces and since it approaches zero at the ground planes. There is similar lack of capacitive coupling between specimen and conductive endplate or sidewall of a cylindrical cavity for TE_{011} mode structure.

A combination of measurement techniques is frequently necessary. For example, the dielectric and magnetic properties of a ferrite specimen may be required over a broad range in frequency. When ferrite cylindrical rod specimens are available, low-loss dielectric ring resonators may be used to control the H_{011} mode resonant frequency of interest and to evaluate magnetic properties at a large number of frequencies above natural gyromagnetic resonance. Complex permittivities of the same specimen may be measured with TM_{0n0} cavities. If spectral information is needed on ferrite complex permeability both above and below gyromagnetic resonance, coaxial transmission line measurements can be combined with cavity or dielectric resonator data.

THEORETICAL OVERVIEW

Complex Magnetic Permeability and Permittivity Characterization below Gyromagnetic Resonance

Coaxial line techniques may be used to evaluate μ_r^* when $f < f_m$. The use of waveguide transmission and reflection techniques for evaluating complex permittivity and complex permeability has a long history, and the literature describing various techniques is extensive [4-22]. Coaxial lines are broadband in their dominant TEM mode and therefore are attractive for spectral characterization of lossy magnetic materials, despite the problems of measurement uncertainty in complex permittivity determination introduced by potential air gaps between the sample and the coaxial line center conductor. Details of two-port, reference-plane invariant scattering parameter relations that can be used for determining permittivity and permeability are given elsewhere [22]. One set of equations for single sample two-port scattering parameter measurements that can be taken with an automatic network analyzer is

$$S_{11}S_{22} - S_{21}S_{12} = \exp\left[-2\gamma_0(L_{air} - L)\right]\frac{R^2 - T^2}{1 - R^2T^2}, \tag{1}$$

and

$$(S_{12} + S_{21})/2 = \exp\left[-\gamma_0(L_{air} - L)\right]\frac{T(1 - R^2)}{1 - R^2T^2}, \tag{2}$$

where

$$R = \frac{\mu\gamma_0 - \mu_0\gamma}{\mu\gamma_0 + \mu_0\gamma}, \tag{3}$$

$$T = \exp(-\gamma L), \tag{4}$$

$$\gamma_0 = \sqrt{\left(\frac{2\pi}{\lambda_c}\right)^2 - \left(\frac{\omega}{c_{lab}}\right)^2}, \tag{5}$$

$$\gamma = \sqrt{(\frac{2\pi}{\lambda_c})^2 - \frac{\omega^2\mu_r^*\epsilon_r^*}{c_{vac}^2}}, \tag{6}$$

c_{vac} and c_{lab} are the speed of light in vacuum and laboratory, ω is angular frequency, λ_c is cutoff transmission-line wavelength, $\mu = \mu_r^*\mu_0$, L_{air} and L are air-line and sample lengths, and ϵ_0 and μ_0 are permittivity and permeability of freespace (8.854×10^{-11} F/m and $4\pi \times 10^{-7}$ H/m). Equations (1) and (2) may be solved explicitly or implicitly as a system of nonlinear scattering equations at each frequency for μ_r^* and ϵ_r^* or may be solved by using a nonlinear regression model over the entire frequency range. Maximum relative uncertainties in coaxial transmission line measurements for the demagnetized relative real permeability μ_d' and magnetic loss index μ_d'' are $\pm 1\%$ and $\pm 2\%$ up to natural gyromagnetic resonance [22].

Real relative permittivities of most ceramic ferrites typically range from 9 to 20. In order to use waveguide transmission and reflection techniques for accurate evaluation of the real permittivity, specimens must be metallized at all surfaces in contact with the waveguide and coaxial line center conductor. Dielectric losses of most ferrite ceramics are usually low ($\tan \delta_e \ll 0.001$) at all frequencies; hence a cavity or dielectric resonator technique *must* be employed for accurate dielectric loss evaluation.

Dielectric and Magnetic Characterization above Gyromagnetic Resonance

Magnetic losses in ceramic ferrites rapidly decrease at frequencies above f_M, and waveguide transmission line techniques cannot then be used for accurate dielectric or magnetic loss characterization. At these frequencies, two convenient dielectric resonator techniques may be employed for complex permeability and permittivity evaluations. These are:

- H_{011} dielectric ring resonators for complex permeability and TM_{0n0} cavities for complex permittivity evaluation (*single* specimen for multiple measurement frequencies),

- Ferrite specimen as H_{011} dielectric resonator for *both* complex permittivity and permeability evaluation (one specimen for each measurement frequency).

H_{011} *dielectric ring resonators and* TM_{0n0} *cavities*: Dielectric ring resonators may be used for accurate magnetic permeability and loss measurements above natural gyromagnetic resonance [3]. The measurement system is illustrated in Figs. 1 and 2. The relative complex permittivity of the ferrite specimen, ϵ_f^*, as specified by the manufacturer, is first verified with a TM_{0n0} cavity. The complex permittivity is evaluated from measurements of cavity resonant frequencies and Q-factors, with and without sample insertion. Subsequent to complex permittivity evaluation, H_{011} dielectric ring resonators are used for complex permeability characterization.

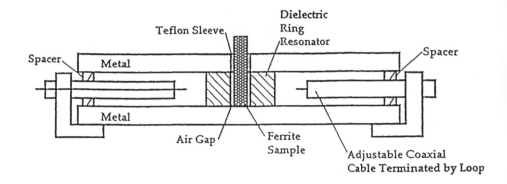

Figure 1: Parallel-plate resonant system used with dielectric ring resonator for demagnetized ferrite ceramic measurements.

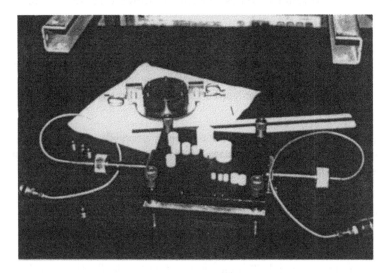

Figure 2: TM_{0n0} cavity and H_{011} dielectric ring resonators used in measuring microwave complex permittivity and complex permeability of ferrite ceramics above natural gyromagnetic resonance.

The low-loss H_{011} dielectric resonators, having nominal relative permittivities $\epsilon'_{r,ring}$, are dimensioned so that the H_{011} mode measurement frequencies (without sample) span frequencies from approximately 2.5 GHz to 25 GHz. Table 1 illustrates the material, nominal relative permittivity, external diameter ϕ_{ext}, height h, and H_{011} resonance f_r of some suitable dielectric ring resonators. The internal diameters of these ring resonators were 2.01 mm.

Table 1: Dielectric ring resonators used for microwave measurements of ferrite ceramics above natural gyromagnetic resonance.

Material	$\epsilon'_{r,ring}$	ϕ_{ext} (mm)	h (mm)	f_r (GHz)
ceramic	37.0	18.98	18.95	2.60
ceramic	37.0	16.89	16.89	2.93
ceramic	37.0	15.89	13.77	3.40
ceramic	30.5	13.40	13.45	4.11
ceramic	30.5	11.97	11.93	4.61
ceramic	30.5	10.65	10.67	5.19
alumina	10.0	11.62	8.40	9.31
alumina	10.0	8.40	8.41	11.28
alumina	10.0	7.60	7.44	12.60
alumina	10.0	5.86	6.04	16.22
quartz	4.42^A	15.61	7.97	12.34
quartz	4.42^A	13.95	6.75	14.31
quartz	4.42^A	12.37	6.41	15.46
quartz	4.42^A	11.28	5.78	17.09
quartz	4.42^A	9.92	5.18	19.18
quartz	4.42^A	8.74	4.52	21.96
quartz	4.42^A	7.95	4.05	24.28

^{A} normal to c-axis

The complex permittivity of each dielectric ring resonator is first found by measuring the resonant frequency and unloaded Q-factor of the (empty) ring resonator operating in the H_{011} mode, given the geometric dimensions of the resonator and taking into account conductive microwave losses of the upper and lower metal ground planes. Values of the dielectric losses of both the ferrite specimens and the ring resonators at other measurement frequencies are calculated assuming a linear increase with frequency. The magnetic permeability, μ'_d, is then evaluated from mea-

Dielectric Ceramic Materials   201

surements of the dielectric ring resonator frequency with completely demagnetized ferrite specimen insertion and solving the H_{011} eigenvalue equation,

$$F(\epsilon'_f, \mu'_d, f) = 0. \tag{7}$$

The imaginary part of the permeability or magnetic loss index, $\mu''_d = \mu'_d \tan \delta_m$, is found from

$$Q_0^{-1} = Q_c^{-1} + p_{\epsilon'_{r,ring}} \tan \delta_{e,ring} + p_{\epsilon'_f} \tan \delta_{e,f} + p_{\mu'_d} \tan \delta_m, \tag{8}$$

where Q_0 is the unloaded Q-factor for the H_{011} mode; Q_c is the Q-factor representing conductor losses in the metal plates for the H_{011} mode; $p_{\epsilon'_r}$ is the electric energy filling factor for the dielectric ring resonator; $p_{\epsilon'_f}, p_{\mu'_d}$ are the ferrite ceramic sample electric and magnetic energy filling factors; $\tan \delta_{e,ring}$ and $\tan \delta_{e,f}$ are the dielectric loss tangents of the ring resonator and ferrite; and $\tan \delta_m$ is the magnetic loss tangent of the ferrite. Discussion of uncertainties in μ^*_d above gyromagnetic resonance using this technique are discussed elsewhere [3,23] and are estimated to be no greater than $\pm 0.8\%$ in μ'_d and $\pm 1 \times 10^{-5}$ in μ''_d. Advantages of using H_{011} dielectric ring resonators for complex permeability evaluations are: (1) a *single* specimen may be tested at a large number of discrete frequencies, and (2) partial magnetic energy filling factors are much greater than those obtainable with cavity measurements for comparably-sized specimens.

Specimen as H_{011} Resonator: Another method may be used for *both* dielectric and magnetic characterization of ferrite ceramic specimens *above* natural gyromagnetic resonance. This approach is one in which the specimen under test constitutes a *single* cylindrical dielectric resonator situated between two conducting metal ground planes as shown in Fig. 3. This commonly used procedure was first introduced by Hakki and Coleman [24], and an uncertainty analysis was later performed by Courtney [25]. Usually, the H_{011} resonance is measured because it is easily identified. For measurement characterization at multiple frequencies, additional specimens must be provided. Some of the details are reviewed here. The general characteristic equation for an open dielectric waveguide is [26],

$$F_1 F_2 - F_3^2 = 0, \tag{9}$$

where

$$F_1 = \frac{\epsilon'_f J'_m(u)}{u J_m(u)} + \frac{K'_m(v)}{v K_m(v)}, \tag{10}$$

$$F_2 = \frac{J'_m(u)}{u J_m(u)} + \frac{K'_m(v)}{v K_m(v)}, \tag{11}$$

Dielectric Ceramic Materials

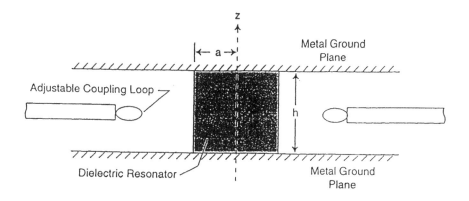

Figure 3: Parallel plate dielectric resonator. Sample relative complex permittivity and permeability are $\epsilon_f^* = \epsilon_f'(1 - j \tan \delta_e)$ and $\mu_d^* = \mu_d'(1 - j \tan \delta_m)$

$$F_3 = \frac{m\beta}{k_0 a}(\frac{1}{u^2} + \frac{1}{v^2}),\tag{12}$$

$u = \sqrt{(k_0 a)^2 \epsilon_f' - \beta^2}$, $v = \sqrt{\beta^2 - (k_0 a)^2}$, $k_0 = \omega\sqrt{\epsilon_0\mu_0}$, a is the radius of the dielectric resonator, and β is the axial propagation constant in the resonator. $J_m(u)$ is the Bessel function of the first kind of order m and $K_m(v)$ is the modified Bessel function of the second kind; the $'$ denotes differentiation with respect to the argument. Equation (9) is generally valid for a dielectric waveguide for azimuthally symmetric $(m = 0)$ or non-symmetric $(m \neq 0)$ modes. When $m = 0$, $F_3 = 0$ and either F_1 or F_2 must vanish. For TM_{0n} modes, $F_1 = 0$ and for TE_{0n} modes, $F_2 = 0$. A resonance condition is established with conducting ground planes placed on both top and bottom of the cylindrical specimen, which forces the transverse electric field to be zero on the surface of each plate; hence the resonance relation

$$\beta h = p\pi,\tag{13}$$

where $p = 1, 2, 3, \dots$ denotes the number of half-wavelength variations along the z-coordinate axis of the sample. The TE_{0np} (H_{0np}) modes are usually chosen for accurate measurements of permittivity since there is no capacitive coupling between

Dielectric Ceramic Materials   203

the dielectric resonator under test and the ground planes for these modes. This permits repeatable measurements with fixture disassembly. Maximum total estimated uncertainty for real relative permittivity is approximately 0.3% with well-machined samples; for loss tangent, it is $\pm 1 \times 10^{-5}$. The resonance condition allows the values of u and v to be written,

$$u = \frac{2\pi a}{\lambda_0}[\epsilon'_f - (\frac{p\lambda_0}{2h})^2]^{1/2},$$ (14)

and

$$v = \frac{2\pi a}{\lambda_0}[(\frac{p\lambda_0}{2h})^2 - 1]^{1/2},$$ (15)

where λ_0 is the free-space wavelength at the measurement resonant frequency. The TE_{011} resonant mode is easily identified as the second lowest-order mode. By measuring the resonant frequency of this mode, knowing the dimensions of the specimen, and solving the equation $F_2 = 0$ for the first eigenvalues $u = u_1$, $v = v_1$, the real relative permittivity is evaluated.

The dielectric loss tangent for any resonant technique may be generally written as,

$$\begin{aligned} \tan \delta_e &= p_e^{-1}(Q_0^{-1} - Q_c^{-1} - Q_r^{-1}) \\ &= p_e^{-1}(Q_0^{-1} - R_s/G), \end{aligned}$$ (16)

where p_e is the partial electric field energy filling factor, Q_0 is the measured unloaded Q-factor, $Q_c^{-1} = R_s/G$ represents conductor losses, Q_r^{-1} denotes radiative losses, $R_s = \sqrt{\pi \mu_0 f/\sigma}$ is the surface resistance of the metal ground planes at the measurement frequency f, and G is the geometric factor of the resonant system. Since there are no radiative losses for TE_{0np} resonant modes, $Q_r^{-1} = 0$. For the TE_{011} mode, the partial electric energy filling factor is given by [24],

$$p_e = \frac{\epsilon'_f}{\epsilon'_f + S(u_1)T(v_1)},$$ (17)

where

$$S(u_1) = \frac{J_1^2(u_1)}{[J_1^2(u_1) - J_0(u_1)J_2(u_1)]},$$ (18)

and

$$T(v_1) = \frac{K_0(v_1)K_2(v_1) - K_1^2(v_1)]}{K_1^2(v_1)}.$$ (19)

The geometric factor for the TE_{011} mode ($p = 1$) is given by,

$$G = \frac{2\pi f^2 \mu_0^2 \epsilon_0 \epsilon'_f h^3}{p_e[1 + S(u_1)T(v_1)]}.$$ (20)

Dielectric Ceramic Materials

The general characteristic equations for a ferrite waveguide in the demagnetized state must be slightly modified. In this case, F_1, F_2, and F_3 are given by,

$$F_1 = \frac{\epsilon'_r J'_m(w)}{w J_m(w)} + \frac{K'_m(v)}{v K_m(v)}, \tag{21}$$

$$F_2 = \frac{\mu'_d J'_m(w)}{w J_m(w)} + \frac{K'_m(v)}{v K_m(v)}, \tag{22}$$

and

$$F_3 = \frac{m\beta}{k_0 a}(\frac{1}{w^2} + \frac{1}{v^2}), \tag{23}$$

where $w = \sqrt{(k_0 a)^2 \epsilon'_f \mu'_d - \beta^2}$. For the TE$_{011}$ mode at measurement frequency f_m, the characteristic equation that involves *both* the real permittivity and real permeability of the *demagnetized* specimen reduces to

$$\mu'_d \frac{J_1(w)}{w J_0(w)} = -\frac{K_1(v)}{v K_0(v)}, \tag{24}$$

with

$$w = \frac{2\pi a f_m}{c}[\epsilon'_f \mu'_d - (\frac{c}{2h f_m})^2]^{1/2}, \tag{25}$$

$$v = \frac{2\pi a f_m}{c}[(\frac{c}{2h f_m})^2 - 1]^{1/2}, \tag{26}$$

where $c = 2.998 \times 10^8$ m/sec. Equation (24) is now solved for its first root, $w = w_1$, given f_m and the specimen dimensions. The magnetic loss tangent is given by,

$$\tan \delta_m = p_m^{-1}(Q_0^{-1} - p_e \tan \delta_e - R_s/G_m), \tag{27}$$

where Q_m is the unloaded measured Q-factor when *both* dielectric and magnetic specimen losses are present. The partial electric and magnetic field energy filling factors in this case are given by

$$p_e = \frac{\epsilon'_f}{\epsilon'_f + S(w_1)T(v_1)}, \tag{28}$$

$$p_m = \frac{p_e c^2}{\epsilon'_f \mu'_d f_m^2}[\frac{1}{4h^2} + \frac{w_1^2}{4\pi a^2}\frac{J_0^2(w_1) + J_1^2(w_1)}{J_1^2(w_1) - J_0(w_1)J_2(w_1)}], \tag{29}$$

and the geometric factor G_m by,

$$G_m = \frac{2\pi f_m^2 (\mu_0 \mu'_d)^2 \epsilon_0 \epsilon'_f h^3}{p_e[1 + (\mu'_d)^2 S(w_1)T(v_1)]}. \tag{30}$$

Equations (11,16,24,27) provide the basis for complex permittivity and permeability evaluation of a *single* cylindrical ferrite specimen if the nature of the complex permeability tensor is considered. The complex permeability tensor is generally written in diagonalized form as,

$$\overline{\overline{\mu}} = \mu_0 \begin{bmatrix} \mu_x^* & 0 & 0 \\ 0 & \mu_y^* & 0 \\ 0 & 0 & \mu_z^* \end{bmatrix}. \tag{31}$$

For the dielectric case, $\mu_x' = \mu_y' = \mu_z' = 1$ and all magnetic losses are 0. For a demagnetized ferrite, $\mu_x^* = \mu_y^* = \mu_z^* = \mu_d^*$. For a partially magnetized ferrite, $\mu_x^* = \mu_y^* = \mu_t^*$ and $\mu_z' \neq 1, \tan \delta_{z,m} \neq 0$, where $\tan \delta_{z,m}$ denotes the magnetic loss tangent in the z-direction. In a saturated ferrite, $\mu_x^* = \mu_y^* = \mu_t^*$ and $\mu_z' = 1, \tan \delta_{z,m} = 0$. The measurement procedure then is to

- Apply a large static magnetic field (beyond ferromagnetic resonance) along the cylindrical z-axis to saturate the ferrite specimen, causing $\mu_z' = 1$ and $\tan \delta_{z,m} = 0$. The ferrite specimen is equivalent to a dielectric resonator, and ϵ_f' and $\tan \delta_e$ may be measured with the TE_{011} (H_{011}) mode resonant frequency f and unloaded Q-factor Q_0.

- Demagnetize the specimen and measure the TE_{011} (H_{011}) mode resonant frequency f_m and unloaded Q-factor Q_m. The real permittivity and dielectric loss tangent are known from the first step, and the real magnetic permeability μ_d' and magnetic loss tangent $\tan \delta_m$ are evaluated from the second.

MEASUREMENT RESULTS AND PREDICTIVE COMPARISONS

Measurements

It is useful to show the results of both dielectric resonator methods for complex permeability evaluation of ferrite ceramic specimens. Example relative permeability and magnetic loss factor data on various aluminum doped, calcium-vanadium, and yttrium garnet ferrites are shown as a function of frequency from 2 GHz to 20 GHz in Figs. 4 and 5. These data, taken at 297 K, were evaluated with dielectric ring resonators. Complex permittivities were determined by TM_{0n0} measurements. All data in Figs. 4 and 5 are taken at frequencies above natural gyromagnetic resonance, where magnetic losses rapidly decrease. As the ferrite saturation magnetization

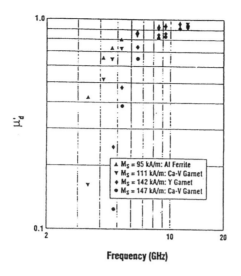

Figure 4: Measured relative permeability μ'_d of several aluminum doped, calcium-vanadium, and yttrium garnets having saturation magnetization M_s between 95 kA/m and 147 kA/m. Dielectric ring resonators were used.

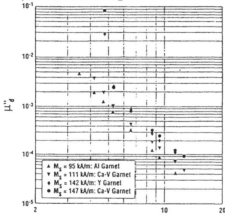

Figure 5: Measured relative magnetic loss factor μ''_d of several aluminum doped, calcium- vanadium, and yttrium garnets having saturation magnetization M_s between 95 kA/m and 147 kA/m. Dielectric ring resonators were used.

Dielectric Ceramic Materials   207

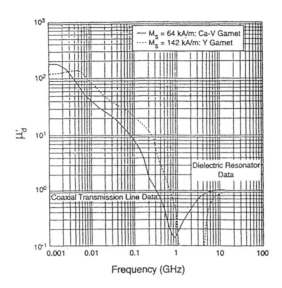

Frequency (GHz)

Figure 6: Combined dielectric ring resonator and coaxial transmission line measurements of relative magnetic permeability. Data shown are for yttrium and calcium-vanadium ceramic garnets at frequencies from 1 MHz to 25 GHz.

increases, μ'_d decreases and the magnetic loss index μ''_d increases for any specified measurement frequency. Magnetic losses decrease nonlinearly with frequency.

Combined coaxial transmission line and dielectric ring resonator measurements for ceramic yttrium and calcium-vanadium garnets are given in Figs. 6 and 7. Specimen natural gyromagnetic resonance is evident as the minimum in the μ'_d data. Both real magnetic permeability and magnetic loss vary nonlinearly with frequency, and the magnetic loss index is observed to change by over 6 orders of magnitude as frequency is increased from 1 MHz to 10 GHz. Maximum relative uncertainties in the coaxial transmission line measurements for μ'_d were $\pm 1\%$ and $\pm 2\%$ up to gyromagnetic resonance. Discussion of uncertainties in μ^*_d above gyromagnetic resonance are discussed in [3,23] and are estimated to be no greater than $\pm 0.8\%$ in μ'_d and $\pm 1 \times 10^{-5}$ in μ''_d.

Microwave dielectric and magnetic properties of various ceramic ferrites were also measured using the second dielectric resonator approach; namely, employing the specimen as an H_{011} dielectric resonator (without the use of ring resonators). The ferrite ceramic specimens, saturation magnetizations at 297 K, and associated gyromagnetic frequencies for these measurements are given in Table 2. Table 3

Dielectric Ceramic Materials

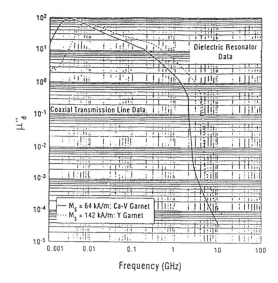

Figure 7: Combined dielectric ring resonator and coaxial transmission line measurements of relative magnetic loss factor. Data shown are for yttrium and calcium-vanadium ceramic garnets at frequencies from 1 MHz to 25 GHz.

summarizes measurement results using the ceramic ferrite specimens as individual H_{011} dielectric resonators in a parallel plate waveguide. The applied static magnetic field for saturating the specimen was approximately 800 kA/m (10kG). In contrast to magnetic losses, dielectric loss tangents linearly increase with frequency. Both dielectric and magnetic properties were measured at X band frequencies on the same polycrystalline magnesium spinel at 297 K and at 77 K. Although dielectric losses decreased by a factor of 1/2 from 297 to 77 K, magnetic losses increased by more than a factor of 10. The real permeability and magnetic loss tangent for these specimens are plotted as a function of normalized frequency f_M/f in Figs. 8. and 9.

Table 2: Ceramic Ferrites Tested as H_{011} Dielectric Resonators, Together with Associated Saturation Magnetizations and Gyromagnetic Frequencies.

Ceramic Under Test	M_s (kA/m)	f_M (GHz)
Al Garnet	95	3.36
Ca-V Garnet 1	111	3.92
Ca-V Garnet 2	147	5.18
Lithium Ferrite 1 (Spinel)	159	5.60
Mg Ferrite (Spinel)	159	5.60
Lithium Ferrite 2 (Spinel)	239	8.40

Table 3: Ceramic Ferrite Measurement Results With Specimen Employed as H_{011} Dielectric Resonator.

Specimen	T (K)	f_M/f	ϵ'_f	$\tan \delta_e$	μ'_d	$\tan \delta_m$
Al Garnet	297	0.42	15.2	7.1×10^{-5}	0.91	8.8×10^{-5}
Ca-V Garnet 1	297	0.48	15.1	4.5×10^{-4}	0.89	2.12×10^{-4}
Ca-V Garnet 2	297	0.63	15.7	9.4×10^{-5}	0.82	3.55×10^{-4}
Lithium Ferrite 1	297	0.62	16.9	9.5×10^{-4}	0.72	2.16×10^{-4}
Mg Ferrite*	297	0.62	13.1	5.1×10^{-4}	0.80	5.56×10^{-4}
Lithium Ferrite 2	297	0.91	12.4	4.4×10^{-4}	0.82	1.19×10^{-3}
Mg Ferrite*	77		12.8	2.4×10^{-4}	0.45	1.65×10^{-3}

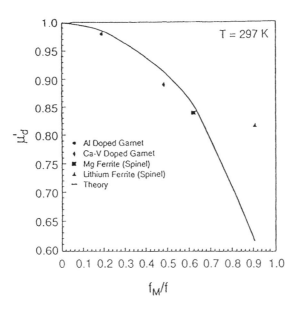

Figure 8: Real relative permeability as function of normalized frequency for several ferrite ceramic specimens. Theoretically predicted values from Schloemann's model shown as solid curve.

Predictive Comparisons

Schloemann [27], using quasi-static theoretical formulations, derived a relation for the permeability of a cylindrical specimen in the demagnetized state with a two-domain model. One domain is parallel and the other antiparallel to the applied rf field. He gives the following result for the demagnetized relative permeability in terms of the saturation magnetization, rf frequency, and gyromagnetic ratio,

$$\mu'_d = \frac{2}{3}\left[1 - \left(\frac{\gamma_g M_s}{f}\right)^2\right]^{1/2} + \frac{1}{3}. \tag{32}$$

Comparative predicted permeability values from eq (32) are shown in Fig. 8. The theoretically predicted μ'_d differs significantly from measured permeability when $f_M/f > 0.75$, which is evident for the lithium ferrite 2 specimen.

SUMMARY

Accurate magnetic characterization of ceramic ferrites over a broad range in fre-

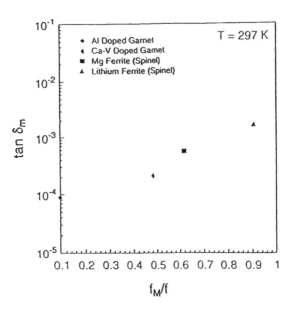

Figure 9: Magnetic loss tangent as function of normalized frequency for several ferrite ceramic specimens.

quency may be accomplished by combining coaxial transmission line with dielectric ring resonator measurements. Dielectric ring resonator measurements operated in the H_{011} mode permit broadband complex permeability evaluations near and above gyromagnetic resonance on a *single specimen*, whereas coaxial transmission line measurements permit evaluation below gyromagnetic resonance. Dielectric ring measurements have greater sensitivity than cavity methods to magnetic loss, since magnetic energy filling factors are greater for comparably-sized samples. Dielectric ring resonator measurements also allow smaller ferrite sample dimensions for measurements at frequencies less than 2 GHz (personal communication system and cellular radio frequencies) than conventional TM_{110} cavities. Accurate dielectric characterization of the same specimen requires additional measurements with a TM_{0n0} cavity.

A cylindrical ceramic ferrite specimen may also be employed as a H_{011} dielectric resonator. Measurements of the H_{011} resonant frequency and Q-factor, when the specimen is alternately saturated and demagnetized, permits accurate microwave evaluation of complex permittivity and complex permeability. With this approach, additional specimens are needed for magnetic and dielectric characterization at multiple frequencies.

Schloemann's theoretical model may be used as a good approximation of the real

Dielectric Ceramic Materials

part of μ_d^*, as long as $f_M/f < 0.75$. However, since both dielectric and magnetic losses depend critically on composition and microstructure, they must be experimentally determined.

Dielectric losses of a ferrite ceramic decrease as the specimen is cooled, albeit with increasing magnetic loss and decreasing real magnetic permeability. Hence, with cooling, the same ferrite ceramic may be used in microwave tuning devices at higher frequencies. The tuning and loss characteristics at different temperatures must also be experimentally evaluated.

REFERENCES

[1]M. Sucher and J. Fox (Editors), *Handbook of Microwave Measurements.* Polytechnic Inst. Brooklyn Series, II, New York: Wiley, 1963.

[2]R.G. Geyer, J. Mantese, and J. Baker-Jarvis, "Effective Medium Theory of Ferrite-Loaded Materials," *Nat. Inst. Stand. Tech. Note 1371*, 1994.

[3]R.G. Geyer, C. Jones, and J. Krupka, "Complex Permeability Measurements of Ferrite Ceramics Used in Wireless Communications," *in* Advances in Dielectric Ceramic Materials, *Am. Cer. Soc. Trans.*, **88** 93-113 (1998).

[4]S. Roberts and A. Von Hippel, "A New Method for Measuring Dielectric Constant and Loss in the Range of Centimeter Waves," *J. Appl. Phys.*, **7** 610-616 (1946).

[5]N. Marcuvitz, *Waveguide Handbook.* Dover Publications, New York, 1951.

[6]G.A. Deschamps, "Determination of Reflection Coefficients and Insertion Loss of a Waveguide Junction," *J. Appl. Phys.*, **2** 1046-1050 (1953).

[7]D.M. Bowie and K.S. Kelleher, "Rapid Measurement of Dielectric Constant and Loss Tangent," *IEE Trans. Microwave Theory Tech.*, **MTT-4** 137-140 (1956).

[8]H.E. Bussey and J.E. Gray, "Measurement and Standardization of Dielectric Samples," *IRE Trans. Instrum.*, **I-11** [3] 162-165 (1962).

[9]G.M. Brydon and D.J. Hepplestone, "Microwave Measurements of Permittivity and tan δ over the Temperature Range 20-700°C," *Proc. Inst. Elec. Eng.*, **112** 421-425 (1965).

[10] G. Franceschetti, "A Complete Analysis of the Reflection and Transmission Methods for Measuring the Complex Permeability and Permittivity of Materials at Microwave Frequencies," *Alta Frequenzia*, **36** 757-764 (1967).

[11] A.M. Nicolson and G.F. Ross, "Measurement of the Intrinsic Properties of Materials by Time Domain Techniques," *IEEE Trans. Instrum. Meas.*, **IM-19** 377-382 (1970).

[12] W.B. Weir, "Automatic Measurement of Complex Dielectric Constant and Permeability at Microwave frequencies," *Proc. IEEE*, **62** 33-36 (1974).

[13] S. Stuchly and M. Matuszewski, "A Combined Total Reflection Transmission Method in Application to Dielectric Spectroscopy," *IEEE Trans. Instrum. Meas.*, **IM-27** 285-288 (1978).

[14] M.S. Freeman, R.N. Nottenburg, and J.B. DuBow, " An Automated Frequency Domain Technique for Dielectric Spectroscopy of Materials," *J. Phys. E: Sci. Instrum.*, **12** 899-903 (1979).

[15] L.P. Ligthardt, "A Fast Computational technique for Accurate Permittivity determination Using Transmission Line Methods," *IEEE Trans. Microwave Theory Tech.*, **MTT-31** 249-254 (1983).

[16] "Measuring Dielectric Constant with the HP 8510 Network Analyzer," *Product Note 8510-3*, Hewlett Packard, 1985.

[17] L. Solymar and D. Walsh, *Lectures on Electrical Properties of Materials*. University Press, New York, 1988.

[18] N. Belhadj-Tahar, A. Fourier-Lamer, and H. de Chanterac, "Broadband Simultaneous Measurement of Complex Permittivity and Permeability Using a Coaxial Discontinuity," *IEEE Trans. Microwave Theory Tech.*, **2** 1-7 (1990).

[19] K.E. Mattar and M.E. Brodwin, "A Variable Frequency Method for Wide-Band Microwave Material Characterization," *IEEE Trans. Instrum. Meas.*, **39** 609-614 (1990).

[20] H.B. Sequeira, "Extracting μ_r and ϵ_r from One-Port Phasor Network Analyzer Measurements," *IEEE Trans. Instrum. Meas.*, **39** 621-627 (1990).

[21] G. Maze, J.L. Bonnefoy, and M. Kamarei, "Microwave Measurement of the Dielectric Constant Using a Sliding Short-Circuited waveguide Method," *Microwave J.*, 77-88 (1990).

[22] J. Baker-Jarvis, M.D. Janezic, J.H. Grosvenor, and R.G. Geyer, "Transmission Reflection and Short-Circuit Line Methods for Measuring permittivity and Permeability," *Nat. inst. Stand. Tech. Note 1355-R*, 1993.

[23] J. Krupka and R.G. Geyer, "Complex Permeability of Demagnetized Ferrites Near and Above Gyromagnetic Resonance," *IEEE Trans. on Magnetics*, **32** [3] 1924-1933 (1996).

[24] B.W. Hakki and P.D. Coleman, "A Dielectric Resonator Method of Measuring Inductive Capacities in the Millimeter Range," *IEEE Trans. Microwave Theory Tech.*, **MTT-8** 402-410 (1960).

[25] W.E. Courtney, "Analysis and Evaluation of a Method of Measuring the Complex Permittivity and Permeability of Microwave Insulators," *IEEE Trans. Microwave Theory Tech.*, **MTT-18** 476-485 (1970).

[26] C.C. Johnson, *Field and Wave Electrodynamics*. New McGraw-Hill, New York, 1965.

[27] E. Schloemann, "Microwave Behavior of Partially Magnetized Ferrites," *J. Appl. Phys.*, **41** 204 (1970).

SINTERING BEHAVIOR OF $SrBi_2Ta_2O_9$ LAYERED FERROELECTRIC CERAMICS

Yi-Chou Chen and Chung-Hsin Lu
Department of Chemical Engineering,
National Taiwan University, Taipei, Taiwan, R.O.C.

ABSTRACT

The sintering behavior of layered ferroelectric $SrBi_2Ta_2O_9$ (abbreviated as SBT) was investigated in this study. Pure SBT powder was found to be hardly sintered at low temperatures. Raising the sintering temperature could increase the density of SBT, but also induced the thermal decomposition of specimens and resulted in the formation of a secondary phase. On the other hand, adding 4 mol% of Bi_2O_3 in SBT was found to markedly improve the densification behavior of specimens at temperature as low as 1000 °C, without causing the decomposition of SBT. The morphology and orientation of SBT were significantly influenced by sintering conditions and sintering additives. The relationship between the processing conditions and ferroelectric properties of SBT was also investigated in this study. The dielectric constant and remanent polarization of densified SBT ceramics were found to be 135 and 5.5 $\mu C/cm^2$, respectively.

INTRODUCTION

Recently ferroelectric materials are being widely investigated in view of their applications in nonvolatile random access memories (NvRAM). The ferroelectric NvRAM possesses low operation voltage, high reading and writing speed as well as nonvolatility. Among the ferroelectric materials, $SrBi_2Ta_2O_9$ (abbreviated as SBT) has attracted worldwide interest for the application in NvRAM due to its extremely high fatigue endurance [1,2]. SBT belongs to the family of layered-type perovskite ferroelectrics, which were first investigated by Aurivillius [3]. The structure of SBT contains a stack of two perovskite-like TaO_6 octahedron units between $(Bi_2O_2)^{2+}$ layers along the c-axis, whereas the strontium cations are located in the space between the TaO_6 octahedrons. The lattice constants of the orthorhombic structure of SBT are 5.306Å, 5.5344Å, and 24.9839Å, and its theoretical density is 8.785 g/cm³ [4].

The SBT films has been successfully prepared using several processes, such as PVD, MOD, and CVD, etc [5-10]. However, the properties of bulk SBT ceramics have not been studied in detail. In order to obtain densified SBT ceramics, SBT powder was synthesized and sintered under various conditions. For improving the sinterability of SBT, excess Bi_2O_3 was doped into SBT powder. The influence of the addition of Bi_2O_3 on the densification behavior and electric properties of SBT was investigated.

EXPERIMENTAL

In this study, SBT powder was synthesized through a new process using $BiTaO_4$ as precursors. The mixtures of equimolar Bi_2O_3 and Ta_2O_5 powder were first ball-milled with ethyl alcohol for 48 hour, using zirconia balls in a poly-ethylene jar. After milling, the slurry was dried in a rotary evaporator under reduced pressure. The dried powder was calcined in an electrical furnace and quenched at 900 °C to obtain pure $BiTaO_4$. Then $SrCO_3$ was mixed with $BiTaO_4$ by ball-milling. The mixed powder was quenched at 900 °C to obtain monophasic SBT powder. On the other hand, 4 mol% of excess Bi_2O_3 was also added into the pure SBT powder. The Bi_2O_3 doped and undoped SBT powder were pressed into disks under a pressure of 98 MPa. These specimens were sintered in air at temperatures ranging from 1000 °C to 1300 °C for 1 h in air. The crystalline phases of the specimens were analyzed via X-ray diffraction (XRD) using CuKα radiation. The microstructures of specimens were investigated by a scanning electron microscope (SEM) operated at 20 kV. The dielectric constants were measured by an impedance analyzer at room temperature. The hysteresis behavior was measured through a Sawyer-Tower circuit and recorded on a digital oscilloscope.

RESULTS AND DISCUSSION

The microstructures of the pure SBT specimens sintered from 1000 °C to 1300 °C are shown in Fig. 1. All specimens contain porous microstructures but exhibit different morphology. At 1000 °C the density of the specimen is only 46% of the theoretical value, which is approximately the pressed density of the specimen. The grain size of this specimen is around 0.5 µm (Fig. 1(a)), similar to that of the specimen before sintering. After sintering at 1100 °C, the relative density of SBT slightly increases to 56%, and its grain size grows to be 0.7 µm (Fig. 1(b)). Once sintering temperature increases to 1200 °C, the grains become significantly coarsened with a size of 0.9 µm, but the microstructure of the specimen remains porous (Fig. 1(c)). With 1300 °C-sintering, the morphology of the specimen markedly turns into an angular shape, implying that the structure of SBT might be varied. The relative density of the specimen increases to 90% at this temperature.

Figure 1. SEM micrograghs of pure SBT sintered at (a) 1000°C, (b)1100°C, (c) 1200°C, and (d) 1300°C.

For investigating the variation in the structure of SBT, the analysis of XRD was performed. Figure 2 shows the XRD patterns of the surface of specimens from 1000 °C to 1300 °C. At 1000 °C only a well-developed crystal structure of SBT is identified, and no secondary phase is found. However, at 1200 °C a secondary phase is found and detected to be $SrTa_2O_6$. This phase belongs to a tetragonal tungsten bronze structure [11]. At 1300 °C only $SrTa_2O_6$ phase is found on the specimen surface. Conclusively, the appearance of $SrTa_2O_6$ was caused by the thermal decomposition of SBT at elevated temperatures. The decomposition process of SBT can be considered as the following equation [12]:

$$SrBi_2Ta_2O_9 (s) \rightarrow SrTa_2O_6 (s) + Bi_2O_3 (l) \tag{1}$$

From the above sintering experiments and XRD results, it is found that SBT was hardly sintered at low temperatures. High-temperature heating could increase the density of specimens, but also induced the decomposition of SBT.

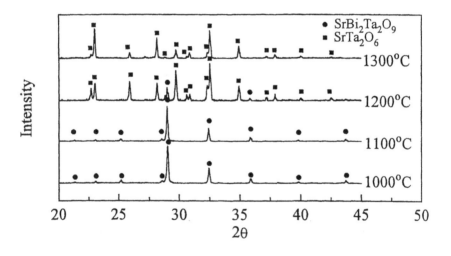

Figure 2. X-ray diffraction patterns of pure SBT sintered for 1 h at various temperatures.

In order to increase the sinterability of SBT, 4 mol% of excess Bi_2O_3 was added into pure SBT powder. The relationship between density and the sintering temperature of Bi_2O_3-doped and undoped SBT is shown in Fig. 3. For pure SBT, the relative densities of SBT are lower than 70% at temperatures below 1150 °C. At 1200 °C and above, the relative densities of SBT are higher than 80%, but the specimens begin to decompose. As for the Bi_2O_3-doped SBT, the sinterability of specimens is significantly improved. After sintering at 1000 °C, the density of SBT is 94.5% of the theoretical value. At 1100 °C, the relative density becomes 96%. From above 1200 °C, the density of specimens inversely decreases slightly.

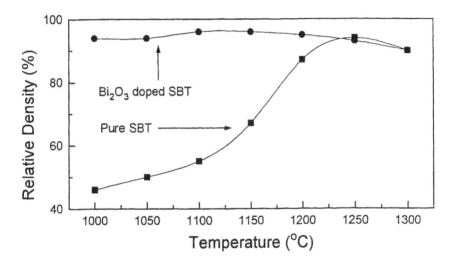

Figure 3. Relative densities of Bi_2O_3 doped and undoped SBT as a function of sintering temperature.

The XRD patterns of the specimen surface of Bi_2O_3-doped SBT are shown in Fig. 4. At 1000 °C and 1100 °C, SBT is the only phase found in the XRD patterns, revealing that addition of Bi_2O_3 does not vary the crystal structure of SBT. At 1200 °C and 1300 °C, the phase of $SrTa_2O_6$ is detected, which is also caused by the decomposition reaction as described previously. Comparing Fig. 2 with Fig. 4 revealing that the *(00l)* peaks of Bi_2O_3-doped SBT is greater than that of pure SBT. In addition increasing sintering temperature further induced an increase in the intensity of the (00*l*) peaks. These results reveals that the addition of Bi_2O_3 in SBT leads to the preferred c-axis orientation in SBT, and the degree of c-axis orientation increases with an increase in sintering temperature.

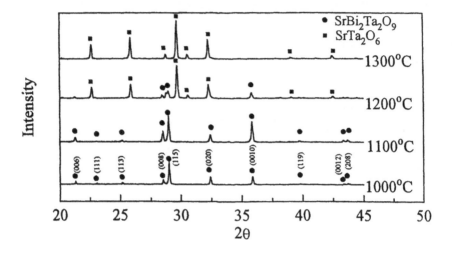

Figure 4. X-ray diffraction patterns of Bi_2O_3 doped SBT sintered for 1 h at various temperatures.

The dielectric constants of the sintered specimens as a function of sintering temperature are shown in Fig. 5. The measuring frequency is 100kHz. After 1000 °C sintering, the dielectric constant of pure SBT is as low as 32. The dielectric constant increases with a rise in sintering temperature, and reaches a maximum value (125) at 1250 °C, decreases at 1300 °C. As for Bi_2O_3-doped SBT, its dielectric constant reaches a maximum of 135 at 1000 °C. The dielectric constant then decreases at higher sintering temperatures. Desu [13] has pointed out that the SBT films with c-orientation have lower spontaneous polarization, and the dielectric constant is a function of the spontaneous polarization [14]. Therefore, the decrease of dielectric constant of SBT might be related to its c-axis orientation.

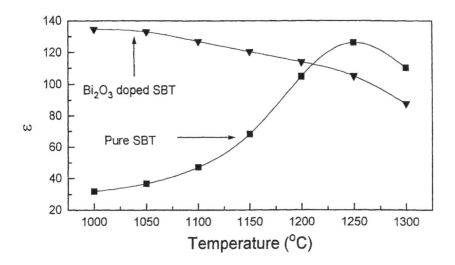

Figure 5. Dielectric constant of Bi_2O_3 doped and undoped SBT measured at 100kHz as a function of sintering temperature.

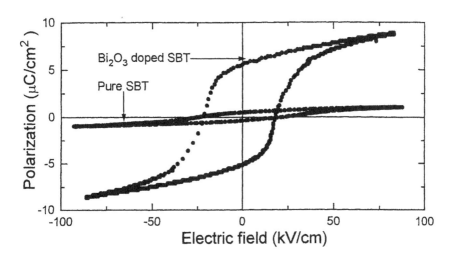

Figure 6. P-E hysteresis loops of Bi_2O_3 doped and undoped SBT

The hysteresis loops (P-E loops) of the specimens sintered at 1000 °C are shown in Fig. 6. The measuring electric-field is about 60 kV/cm. A well saturated P-E loop is obtained for the Bi_2O_3-doped SBT, while a slim P-E loop is found in pure SBT. The remanent polarization of the Bi_2O_3-doped SBT (5.5 $\mu C/cm^2$) is much higher than that of pure SBT (0.5 $\mu C/cm^2$). The addition of Bi_2O_3 in SBT is confirmed not only to significantly improve the sinterability of SBT, but also to result in an increase in the dielectric constant and remanent polarization of SBT ceramics.

CONCLUSIONS

The sintering behavior and dielectric properties of SBT was investigated in this study. Sintering at low temperatures ($\leq$1150 °C) could not obtain densified SBT, and the microstructure contained a large number of pores. Increasing the temperature to 1200 °C resulted in an increase in the density of specimens, but also caused SBT to decompose. When 4 mol% excess Bi_2O_3 was added into SBT, the densification behavior of SBT was significantly improved. After 1000 °C sintering, the relative density of SBT became 94.5%, and crystallization remained unvaried. The dielectric and ferroelectric properties of specimens were found to significantly depend on the density of specimens. Adding Bi_2O_3 into SBT was found to markedly increase the dielectric constant and remanent polarization of SBT to be 135 and 5.5 $\mu C/cm^2$, respectively.

REFERENCES

1. H. Yoshimori, H. Watanabe, C. A. Paz de Araujo, L. D. Mcmillan, J. D. Cuchiaro, and M. C. Scott, "Layered Superlattice Material Applications," U. S. Pat. No. 9310021, June 24, 1993.
2. J. F. Scott, F. M. Ross, C. A. Paz de Araujo, M. C. Scott, and M. Huffman, "Structure and Device Characteristics of $SrBi_2Ta_2O_9$-Based Nonvolatile Random-Access Memories," MRS Bulletin, 21 [7] 33-39 (1996).
3. B. Aurivillius, "Mixed bismuth oxides with layer lattice 1 the structure type of $CaNb_2Bi_2O_9$," Arkiv for Kemi, 1 [54] 463-80 (1949).
4. A. David Rae, J. G. Thompson, and R. L. Withers, "Structure Refinement of Commensurately Modulated Bismuth Strontium Tantalate $Bi_2SrTa_2O_9$," Acta

Crystallographica B, 48 418-28 (1992).

5. S. S.Park, C. H Yang, S. G. Yoon, J. H. Ahn, and H. G. Kim, "Characterizatiion of Ferroelectric $SrBi_2Ta_2O_9$ Thin Films Deposited by a Radio Frequency Magnetron Sputtering Technique," Journal of the Electrochemical Society, 144 [8] (1997) 2855.

6. T. K. Song, J.-K. Lee, and H. J. Jung, "Structural and Ferroelectric Properties of the c-axis Oriented $SrBi_2Ta_2O_9$ Thin Films Deposited by the Radiofrequency Magnetron Sputtering," Applied Physics Letters, 69 [25] 3839-41 (1996).

7. K. Amanuma, T. Hase, and Y. Miyasaka, "Preparation and Ferroelectric Properties of $SrBi_2Ta_2O_9$ Thin Films," Applied Physics Letters, 66 [2] 221-23 (1995).

8. C.A-Paz de Araujo, J. D. Cuchilaro, L.D. McMillan, M. C. Scott, and J. F. Scott, "Fatigue-free Ferroelectric Capacitors with Platinum Electrodes," Nature, 374 [13] 627-29 (1995).

9. T. Li, Y. Zhu, S.B. Desu, C. H. Peng, and M. Nagata, "Metalorganic Chemical Vapor Deposition of Ferroelectric $SrBi_2Ta_2O_9$ Thin Films," Applied Physics Letters, 68 [5] 616-19 (1996).

10. N. J. Seong, S. G. Yoon, and S. S. Lee, "Characterization of $SrBi_2Ta_2O_9$ Ferroelectric Thin Films Deposited at Low Temperatures by Plasma-Enhanced Metalorganic Chemical Vapor Deposition," Applied Physics Letters, 71 [1] 81-83 (1997).

11. F. Galasso, L. Katz, and R. Ward, "Tantalum Analogs of the Tetragonal Tungsten Bronzes," Journal of the American Chemical Society, 81 5898-99 (1959).

12. C. H. Lu, and J. T. Lee, "Strontium Bismuth Tantalate Layered Ferroelectric Ceramics: Reaction Kinetics and Thermal Stability," Ceramics international, in press (1998).

13. S. B. Desu, D. P. Vijay, X. Zhang, and B. P. He, "Oriented growth of SBT ferroelectric thin films," Applied Physics Letter, 69 [12] 1719-21 (1996).

14. L. A. Shuvalov, , "Physical Properties of Crystals"; pp.178-206 in Modern Crystallography, Edited by B. K. Vainshtein. Springer-Verlag, Berlin; New York, 1981.

MICROWAVE DIELECTRIC CERAMICS BASED ON ZINC TITANATES

Hyo Tae Kim and Yoonho Kim
Division of Ceramics
Korea Institute of Science and Technology
Seoul, 131-791, Korea

ABSTRACT

A new microwave dielectric system based on zinc titanates was investigated. The system comprising of $ZnO-TiO_2$, $(Zn, Mg)TiO_3$, and $Zn(Ba, Ca, Sr)TiO_3$ exhibited low sintering temperature, temperature compensating characteristics, and relatively high quality factors compared with those of other low temperature sintering microwave dielectrics. The microwave properties were mainly depended on the phase composition and the stability. The quality factors and temperature coefficients of resonance frequency of pure zinc titanate were improved by the substitution of Zn with Ba, Ca, Sr, and Mg.

I. INTRODUCTION

The development of low temperature cofired ceramics (LTCCs) with low loss and temperature stable characteristics in microwave frequencies draw attention to the ceramic industries in recent days. The primary requirement of LTCCs for microwave applications is low firing temprature which enables to cofire with highly conductive metals such as Ag, Cu, Au or their alloys as an internal electrode of multilayer devices. Another but more important than the primary ones of LTCCs are high frequency characteristics including low loss (high Q), high dielectric constant and stable temperature coefficient of capacitance or resonant frequencies (τ_ε or τ_f). Those features of LTCCs can provide design and functional benefits for the miniaturization of multilayer devices with high electrical performance. The application of LTCCs covers many electronic devices such as multilayer ceramic capacitors (MLCCs), band-pass filters, antenna duplexers, multilayer substrates and packages for multi-chip modules (MCMs).

Most of the commercially known dielectric materials for high frequency applications show high quality factors and dielectric constants, but have high sintering temperatures hence they are not compatible with Ag or Cu electrode. Some of the LTCCs disclosed are (Zr, Sn)TiO$_4$+glass, BaTi$_4$O$_9$+glass, BaO-SiO$_2$-SrO-ZrO$_2$+glass, Bi$_2$O$_3$-Nb$_2$O$_5$ system+additives, CaO-ZrO$_2$+glass, BaO-PbO-Nd$_2$O$_3$-TiO$_2$+glass, (Mg, Ca)TiO$_3$+additives and (Pb, Ca)(Fe, Nb)O$_3$+addditives etc. However, the majority of LTCCs systems with glass and/or additives mixture offer relatively quite lower electrical performance than those of high firing systems.

The objective of this paper is to investigate the LTCCs with high dielectric properties in microwave range. Our basic scheme for a new LTCCs focused on the matrix system with inherently low sintering temperatures as a pure compound. Zinc titanates which have relatively low sintering temperatures with promising dielectric properties were selected as a suitable candidate for the above requirements through the preliminary experiments. This paper covers the structure and microwave properties of zinc titanates and their modifications by A-site substitutions with Ba, Ca, Sr, and Mg which are able to further enhance the sinterability and dielectric properties.

II. EXPERIMENTAL PROCEDURE

The ZnO-TiO$_2$ (Zn/Ti = 0.67~1.75), Zn$_{1-x}$(Ba, Ca, Sr)$_x$TiO$_3$, x = 0~0.09 and Zn$_{1-x}$Mg$_x$TiO$_3$, x = 0~1.0 compositions were prepared by conventional mixed oxide method using reagent grade (>99.9%) oxide form of Zn, Mg, and Ti (as a rutile) and carbonate form of Ba, Ca, and Sr. The starting materials were mixed by ball milling for 12 h and dried at 120°C. The mixed powders were calcined at 1,000°C for 2 h and milled for 24 h. The ground powders were pressed into discs with 10 mm diameter and 4.8 mm thickness. Pellets were sintered at 1,000~1,250°C for 4 h in air.

The phase identification, microstructure and compositional analysis of the sintered specimens were analyzed by XRD (PW1800,Phillips), optical microscopy (BH-2, Olympus), FE-SEM (S-6000, Hitachi) and EPMA-EDS (JXA-8600, Jeol). The microwave dielectric properties were measured using network analyzer (8720C, Hewlett Packard) in s$_{21}$ transmission mode. The unloaded Q factors at microwave frequencies were measured by transmission open cavity method using copper cavity. The dielectric constants were measured by Hakki-Coleman method with silver plate and calculated from the TE$_{01\delta}$ resonance mode value. The τ_f was measured by open cavity method using invar cavity in temperature controlling chamber (Heraeus, VMT/60) at -20~+70°C. The sintered specimens were lapped and polished with aspect ratio of 0.45 for cylindrical type dielectric resonators.

III. RESULTS AND DISCUSSION

III-1. Phase Transformation of $ZnTiO_3$

Three types of compounds are known to exsist in the ZnO-TiO_2 system; Zn_2TiO_4 (face centered cubic), $ZnTiO_3$ (rhombohedral), and $Zn_2Ti_3O_8$ (cubic). However, $ZnTiO_3$ decompose into Zn_2TiO_4 and rutile in the vicinity of $945\,°C$ (Fig. 1)[8]. Recent study revealed that $Zn_2Ti_3O_8$ to be the low temperature form of $ZnTiO_3$. In this work, $ZnTiO_3$ was synthesized using fine particles via high energy milling with YSZ beads ($\varphi = 1.0$ mm) and the phase transformation of the composition was examined by XRD, TG/DSC, and dilatometry. Anatase powders and zinc nitrate, $Zn(NO_3)_2 \cdot 6H_2O$ were used as precursors of TiO_2 and ZnO. The dried mixture of precursor was heat treated at $400\,°C$ for 2 hours for denitration then milled with zirconia beads for 10 minutes. The average diameter of particles thus obtained was about $60 \sim 100$ nm as shown in Fig. 2.

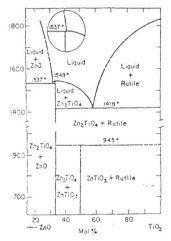

Fig. 1 Phase diagram of ZnO-TiO_2 system.[8]

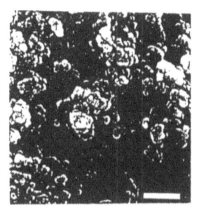

Fig. 2 SEM photographs of $ZnTiO_3$ powders calcined at $700\,°C$-1h. (bar = 500 nm)

The hexagonal $ZnTiO_3$ starts to form at around $600\,°C$ and single phase $ZnTiO_3$ was obtained at near $700\,°C$ (Fig. 3). The stability region of the hexagonal phase exsist up to $925\,°C$ then decompose into cubic and rutile phase at around $950\,°C$, which corresponds to the phase diagram.[8] The $Zn_2Ti_3O_8$ phase was appeared at $850 \sim 925\,°C$ and coexsisted with hexagonal phase, which is somewhat different result from other work.

The phase formation and transition of $ZnTiO_3$ were also investigated by thermal anaysis (Fig. 4). In TG/DSC result, the phase formation and decomposition temperature

Dielectric Ceramic Materials 229

of ZnTiO₃ found to be at around 667℃ and 978℃. respectively. The results are very similar to the XRD data. The evidence of phase formation and decomposition of ZnTiO₃ has been also observed by dilatometry at the temperature up to 1200℃ (Fig. 5). The region (a) designates the phase formation of ZnTiO₃, and (b) designates the small amount of volume change due to the phase decomposition.

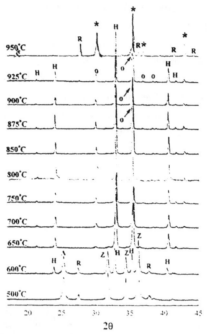

Fig. 3 XRD patterns of 1ZnO-1TiO₂ with various tempratures.
(*: Zn₂TiO₄, A: anatase, R: rutile, H: hexagonal, o: Zn₂Ti₃O₈, z: ZnO)

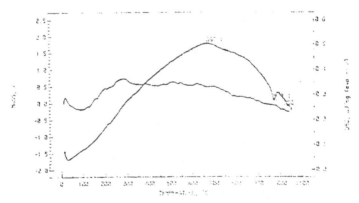

Fig. 4 TG/DSC curves of ZnTiO₃ precursor.

Dielectric Ceramic Materials

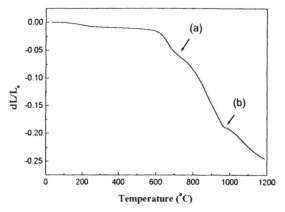

Fig. 5 Dilatometry curves of ZnTiO₃ powder compacts.

III-2. Zn₂TiO₄+TiO₂ System

Figure 6 shows the XRD patterns of the ZnO-TiO₂ (Zn/Ti = 0.67~1.75) specimens prepared by conventional mixed oxide method. The specimens were sintered at 1150°C for 4 h. It was found that the zinc orthotitanate, Zn₂TiO₄ has a rutile solubility up to 0.33 mole. In this solubility region, a single phase of α-Zn₂TiO₄[11] was obtained. With the excess rutile above 0.33 mole, the composite phase with zinc orthotitanate and unreacted rutile was observed. The intensities of the rutile phase were proportional to the Zn/Ti ratio.

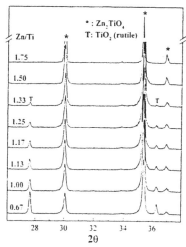

Fig. 6 XRD patterns of ZnO-TiO₂ system sintered at 1150°C for 4h.

The microwave dielectric properties of ZnO-TiO$_2$, (Zn/Ti = 0.67~1.75) specimens sintered at 1150°C for 4h are shown in Fig. 7. The τ_f values changed from positive to negative as the amount of rutile decreased. The zero τ_f value was obtained at around Zn/Ti = 1.15 composition. Pure zinc orthotitanate, Zn$_2$TiO$_4$ has very high loss in the microwave range hence the ε_r, Q*f and τ_f of the specimen was not measurable in this experiment. The best quality factor of the concerned system was obtained at Zn/Ti = 1.75 composition and the value decreased as the amount of rutile increased. The dielectric constants were linearly increased with rutile content.

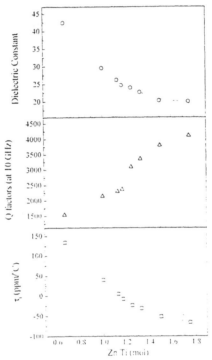

Fig. 7 Microwave dielectric properties of ZnO-TiO$_2$, sintered at 1150°C for 4h.[12]

III-3. ZnTiO$_3$+TiO$_2$ System

Another system in ZnO-TiO$_2$ compounds which is expected to be useful of microwave applications is ZnTiO$_3$+TiO$_2$. ZnTiO$_3$ has rhombohedral symmetry that belongs to C_{3i}^2 ($R\bar{3}$) space group, which is ilmenite structure like MgTiO$_3$, MnTiO$_3$ and NiTiO$_3$. According to our preliminary experiment, the microwave properties of pure ZnTiO$_3$ are ε_r = 21, Q = 3300 (at 10 GHz), and τ_f = -90 ppm/°C, respectively. Considering those properties, it is expected that the zero τ_f composition may exsist in

ZnTiO$_3$+TiO$_2$ system. However, ZnTiO$_3$+TiO$_2$ can exist only at $\leq \cdot 945\,^\circ\text{C}^8$ in contrast to Zn$_2$TiO$_4$+TiO$_2$ that exsist at $\geq 945\,^\circ\text{C}$. Therefore, the synthesis and sintering temperatures of ZnTiO$_3$+TiO$_2$ system are limited at $\leq 945\,^\circ\text{C}$. So, another process is needed to realize the phase composition. As a general approach, two processes such as high energy milling for fine particles via chemical synthesis using nitrates and nano-size starting materials and the addition of sintering additives for the conventional mixed oxide method to reach high density sintered bodies were applied.

The microwave dielectric properties of ZnTiO$_3$+TiO$_2$ system with Zn/Ti mole ratios prepared by the former method are drawn in Fig. 8. The investigated compositions were sintered with dense ceramic compacts in which the ε_r and the τ_f varied in a systematic manner with changing ratios of Zn/Ti. However, the quality factors (Q×f) show maxima where τ_fs were near zero. Examination of this data also shows that the dielectric properties were significantly varied with the sintering temperatures which accompanied by the phase decomposition. The increase in ε_r and the change of τ_f into positive values are attributed to the phase composition change from ZnTiO$_3$+TiO$_2$ to Zn$_2$TiO$_4$+TiO$_2$ such that the relative amount of rutile increased. It was revealed that the quality factors of ZnTiO$_3$+TiO$_2$ system were higher than those of Zn$_2$TiO$_4$+TiO$_2$ system.

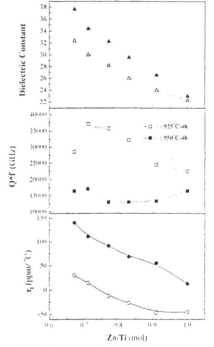

Fig. 8 Microwave dielectric properties of ZnTiO$_3$+TiO$_2$ system.

Boron oxide was chosen as a sintering aid for the powders prepared by CMO methods. Figure 9 shows the dielectric properties of $ZnTiO_3+TiO_2$ system sintered at 825℃ for 4 hours which enables to use Ag or Cu as an internal electrode for multilayer ceramic devices. The addition of boron oxide drastically increased the sinterability hence increased the ε_r and Q factors. The τ_f, however was increased to positive values as the amount of boron oxide increased. Threfore an additional adjustment, decrease in rutile amount, of the composition is needed in order to obtain near zero τ_f when boron oxide is added.

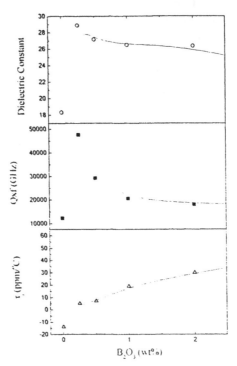

Fig. 9 Microwave dielectric properties of $ZnTiO_3+TiO_2$ with B_2O_3 addition.

III-4. $Zn_{1-x}(Ba, Ca, Sr)_xTiO_3$ System[12]

Typical sintering temperatures of $ZnO-TiO_2$ system were $1100 \sim 1150℃$ in case of conventional mixed oxide methods using industry grade powders ($D_{50} \sim 1 \ \mu m$). In this temperature range, the crystal phases of $ZnTiO_3$ decompose into Zn_2TiO_4 and TiO_2 hence the sign of τ_f changes to positive. One of the ways to adjust the τ_f close to zero is to reduce the rutile phase in the sintered compacts as such small amount (up to 9 mole%) of Ba, Ca or Sr was substituted for Zn to make another titanates with the combination of residue rutile. A comprehensive study on the crystal and microstructure of $Zn_{1-x}(Ba, Ca, Sr)_xTiO_3$ systems will be appeared in elsewhere.[12]

The microwave dielectric properties of the specimens were measured at around 10 GHz and the results are plotted in Fig. 10. The detailed discussion on the structure and microwave properties relationship of the system will be appeared in the other articles.[12]

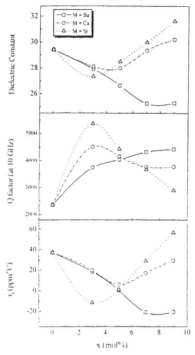

Fig. 10 Microwave dielectric properties of $Zn_{1-x}M_xTiO_3$ system.[12]

According to the previous work[12], the Q factor of stoichiometric Zn_2TiO_4 at the microwave range was too low to measure but that of $Zn_3Ti_2O_7$ form (Zn_2TiO_4 containing $Zn_2Ti_3O_8$ precipitates) was about 3800 at 10 GHz. As a consequence, the Q factors and TCFs of $ZnO \cdot TiO_2$ have been improved by the substitution of Zn with the small amount of Ba, Ca or Sr.

III-5. $(Zn_{1-x}Mg_x)TiO_3$ System

$Zn_{1-x}Mg_xTiO_3$ system was synthesized by CMO[12] and high energy milling methods (HEM). The typical XRD patterns of the specimens obtained by both methods are shown in Fig. 11. The phase compositions in $x = 0 \sim 0.3$ were significantly different from each other according to the temperatures. One of the outstanding features of $Zn_{1-x}Mg_xTiO_3$ system is the increase in decomposition temperature (T_{eo}) of $ZnTiO_3$ to the high temperatures as the amount of Mg increased. The thermal behavior was investigated by TG/DSC analysis. Fig. 12 and Fig. 13.

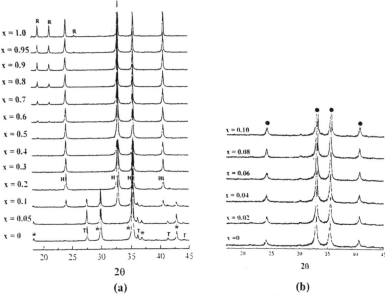

(a)   (b)

Fig. 11 XRD patterns of $(Zn_{1-x}Mg_x)TiO_3$ (a) by CMO^{12} calcined at 1000 ℃ and (b) by HEM calcined at 700 ℃ for 4h (*: α-Zn_2TiO_4, T: rutile, H: hexagonal, and R: rhombohedral phase).

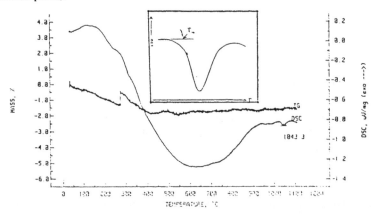

Fig. 12 TG/DSC curves of $(Zn_{0.9}Mg_{0.1})TiO_3$ precursor from HEM method. Inset shows the decomposition temprature determined in this work based on ICTA recommendation.

          Dielectric Ceramic Materials

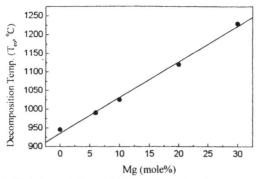

Fig. 13 Variation of T_{eo} with Mg content in $(Zn_{1-x}Mg_x)TiO_3$ system.

Figure 14 shows the microwave dielectric properties of $Zn_{1-x}Mg_xTiO_3$, $x = 0 \sim 1.0$ specimens by CMO method sintered at $1100\,°C$ and $1200°C$ for 4h. The details of the microwave dielectric properties were also discussed in elsewhere.[12] It was found that this system to be the most appropriate candidate for the applications in low temperature sintering microwave devices among other zinc titanate based dielectrics.

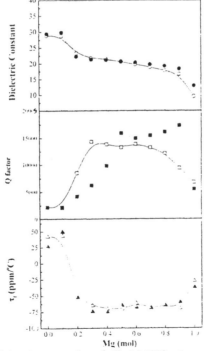

Fig. 14 Microwave dielectric properties of $(Zn,Mg)TiO_3$ (open: $1100\,°C$, solid: $1200\,°C$).

IV. SUMMARY AND CONCLUSION

A consolidated description of zinc titanates based systems which were newly developed has been presented. This work demonstrated that the disadvantage of phase decomposition in zinc metatitanate can be overcomed by the appropriate modification of compositions or powder preparation technique. The compositional modification by the substitution of Ba, Ca, Sr, and Mg for Zn enhanced the sinterability and microwave dielectric properties. Especially, magnesium modified zinc titanates broadened the phase stability region to higher temperatures than zinc metatitanate such that it makes easy to get fully sintered ceramics body with conventionally prepared powders. Sintering of zinc titanates under decomposition temperature using fine particle powders also enables futher improvement of dielectric properties. Zinc titanates system shows the feasibility for low temperature sintering microwave ceramics for multilayer devices with Ag or Cu internal electrodes.

REFERENCES

[1]K. Wakino, T. Nishikawa, Y. Ishikawa and H. Tamura, *Br. Ceram. Trans. J.*, **89**, 39-43 (1990).

[2]H. Creemoolanadhan, M.T. Sebaschan and P. Mohanan, *Mat. Res. Bull.*, **30** [6], pp.653-58 (1995).

[3]S.F. Bartram and R.A. Slepetys, *J. Am. Ceram. Soc.*, **44** [10] 493-99 (1961).

[4]A.T. McCord and H.F. Saunders, U.S. Pat. No.2, 739, 019; *Ceram. Abstr.*, **24** [8] 155 (1945).

[5]L.M. Sheppard, *Am. Ceram. Soc. Bull.*, **70** [9] 1467-77 (1991).

[6]M. Sugiura and K. Ikeda, *J. Jpn. Ceram. Assoc.*, **55** [626] 62-66; *Ceram. Abstr.*, p164e (1950).

[7]K. Haga, T. Ishii, J. Masiyama and T. Ikeda, *Jpn. J. Appl. Phys.*, **31** pp. 3156-59 (1992).

[8]F.H. Dulin and D.E. Rase, *J. Am. Ceram. Soc.*, **43** [3] 125-31 (1960).

[9]M. Tarou, *Electronic Ceramics*, **24** [9], 38-43 (1993).

[10]B.W. Hakki and P.D. Coleman, *IRE Microwave Theor. Tech.*, MTT-8, 402-10 (1960).

[11]JCPDS Card #25-1164 (1980).

[12]Hyo Tae Kim and Yoonho Kim, "Mircostructure and Microwave Dielectric Properties of Modified Zinc Titanates (I) & (II)," *Mat. Res. Bull.*, in press.

BARIUM BISMUTH TANTALATE (BaBi$_2$Ta$_2$O$_9$) AS AN ALTERNATIVE DIELECTRIC FOR DRAM APPLICATIONS

C. R. Foschini, J. A. Varela: Departamento de Engenharia de Materiais, Universidade Federal de São Carlos, SP, Brazil, 13900.
R. Vedula, C. T. A. Suchicital, P. C. Joshi, S. B. Desu: Department of Materials Science and Engineering, Virginia Tech, Blacksburg, VA, 24061.

ABSTRACT

Thin films of BaBi$_2$Ta$_2$O$_9$ (BBT) with up to 20% excess bismuth were deposited over Pt/TiO$_2$/SiO$_2$/Si substrates, followed by pyrolysis and annealing under oxygen atmosphere at 400 to 800°C for 60 minutes. The films were analyzed by XRD and AFM. The crystallization began at 500°C, and at 700°C the films were well crystallized. The BaBi$_2$Ta$_2$O$_9$ phase tended to crystallize as layer-type structure, with an average grain size of 200 nm. The dielectric constant was measured as a function of composition and temperature of annealing. It was found that films with 10 to 20% excess bismuth annealed at 700°C exhibited the highest dielectric constant. The dielectric constant, at 100 kHz, of composition with 10% excess bismuth annealed at 700°C for 60 min. was around 300, tanδ 0.023, and the leakage current less than 1×10^{-9} A/cm^2 at an applied field of 300 kV/cm. These films are useful on RC circuits and DRAM applications.

INTRODUCTION

The rapid increase in DRAM density is placing increasing demands on the materials utilized for the storage capacitor. The parameters of importance for DRAM applications are dielectric constant and leakage current characteristics. The high dielectric constant materials are desirable, as they will lead to simple capacitor design and charge storage density comparable to conventional dielectrics even at much larger thickness. Charge leakage from the capacitor can eventually lead to loss of information, so the capacitor must be periodically refreshed. Since all cells must be refreshed, the time between refresh cycles increases with each new DRAM generation, and the leakage rate must therefore decrease with each generation. In addition, the area available for a memory cells decreases dramatically with each generation, which makes it very difficulty to retain a somewhat constant capacitance. This combination of lower leakage and higher capacitance density with each generation places severe constraints on the materials currently in use[1].

The minimum capacitance necessary for reliable detection in the 256 to 4 Gb DRAM cell is ~ 2.5×10^{-14} Farads (25 fF). Among the more noticeable options to reach the require value of capacitance for the next DRAM generations are: reduction of the dielectric layer thickness, increase of the capacitor area utilizing stacked or trenched capacitor cell structures or introduction of new materials with higher dielectric constant than the current nitride/oxide based systems[2].

SiO_2 with dielectric constant of 3.9 needs a layer thickness of ~ 5 nm with a storage area of ~ 4 μm^2 to achieve a 25 fF capacitance. However with this thickness high leakage currents caused by direct charge tunneling through the layer can occur. In addition, thickness control of very thin layers is somewhat problematic. Reduction of the dielectric thickness is therefore not a long-term solution to compensate for the rapid reduction in surface area of the capacitor.

To increase the storage-area to cell-area ratio several designs have been introduced to capacitor structures, however, here the lithography depth limits the height of the capacitor. With the new structures it is possible to achieve acceptable cell capacitance with (Si_3N_4/SiO_2) to manufacture both 256 Mb and possibly first generation 1 Gb DRAMs.

An alternate approach, namely, replacing the existing 'Si' based dielectrics with new high dielectric constant (ε_r) materials, is under investigation[3]. A variety of materials are under consideration which includes binary oxides such as TiO_2, ZrO_2, Nb_2O_5, Ta_2O_5, Al_2O_3 and complex oxides such as $SrTiO_3$ (ε_r = 230), (Ba, Sr)TiO_3 (ε_r = 400) and Pb(Zr, Ti)O_3 (ε_r = 1300). The advantage of binary oxide dielectrics is that they are simple to process and can be directly deposited on silicon. However they suffer the disadvantage of low dielectric constant compared to complex oxide dielectric materials. The problem with high dielectric constant materials is that much of them can not be directly deposited on poly-silicon and, because of the interaction between the components, the integration requires development of high temperature stable barrier-electrode systems. Among the complex oxides, BST has been identified as a potential candidate for future generation DRAM capacitors, but still has the problems of high leakage current. On the other hand BBT offers the advantage of high dielectric constant compared to binary oxides and much better leakage current characteristics[4] compared to BST. Apart from this there is a possibility that BBT could be directly deposited on to Si substrate.

In this paper we have made an attempt to explore the dielectric properties of $BaBi_2Ta_2O_9$ (BBT) thin films prepared by chemical solution deposition. BBT belongs to layer-perovskite family[5] (Fig. 1) and has a Curie temperature of 110°C, a diffused phase transition and a dielectric constant of 400 has been reported for bulk BBT material[6].

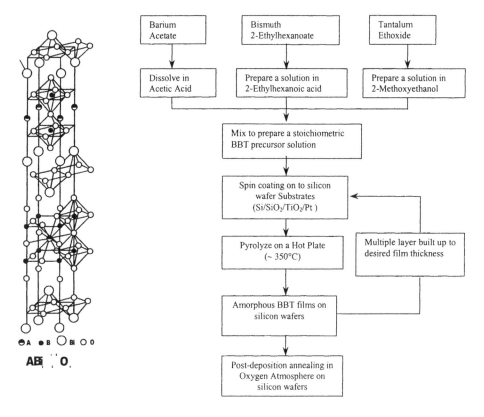

Figure 1: Schematic lattice structure of BaBi$_2$Ta$_2$O$_9$.

Figure 2: Deposition of BBT thin films on Si/SiO$_2$/TiO$_2$/Pt substrates.

EXPERIMENTAL

The BBT films were prepared by chemical solution deposition technique, using an alkoxide-carboxylate precursor solution. Barium acetate, bismuth 2-ehylhexanoate and tantalum ethoxide were selected as the precursors, and acetic acid, 2-ethylhexanoic acid, and 2-methoxyethanol were selected as the solvents. At room temperature, bismuth 2-ethylhexanoate and barium acetate were initially dissolved in 2-ethylhexanoic acid and acetic acid respectively. These solutions were then added to the solution of tantalum ethoxide in 2-methoxyethanol to form a stoichiometric, clear and stable BBT precursor solution. The films were coated on Pt/TiO2/SiO2/Si substrates by spin coating. Then the films were kept on a hot plate (at 350°C) in air for 10 min. This step was repeated after each coating until the desired final film thickness be obtained. The films were annealed in a tube furnace at temperature of 400 to 800°C for 1 h in an oxygen atmosphere to crystallize the films. The process flow chart is given in Figure 2. The dielectric measurements were conducted by a HP 4192A impedance analyzer and the leakage current by a HP 4140B. The microstructure was observed by a D 3000 atomic force microscope (AFM) (Digital Instrument, Inc.), and the crystallographic phases by a Scintag XDS-2000 diffractometer XRD using Cu $K\alpha$ radiation at 40 kV.

RESULTS AND DISCUSSION

The pyrolyzed films at 350°C were found to be amorphous and post deposition annealing was required to develop crystallinity. Figure 3 shows the XRD of BBT thin films as a function of annealing temperature. The films annealed up to 500°C have found to be amorphous. As the annealing temperature was increased to 700°C, a crystalline phase was obtained with peaks attributable to orthorhombic phase. The crystallization of the BBT phase was analyzed based on $BaBi_2Nb_2O_9$ (Powder Diffraction File Card No. 12-0403), which belongs to layered-perovskite family, and crystallizes in the orthorhombic phase [5]. As the annealing temperature was increased, the peak intensity and sharpness were found to increase indicating better crystallinity and also an increasing in grain size with temperature was found. The microstructure of BBT thin films was analyzed by atomic force microscope (AFM) and the Fig. 4 shows the microstructure of the films annealed at 700°C. The film annealed at 650°C showed a smooth surface with very small crystallites while a well defined grain structure was obtained at an annealing temperature of 700°C.

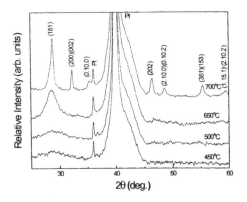

Figure 3: X-ray diffraction of $BaBi_2Ta_2O_9$ thin films annealed at different temperatures for 60 min.

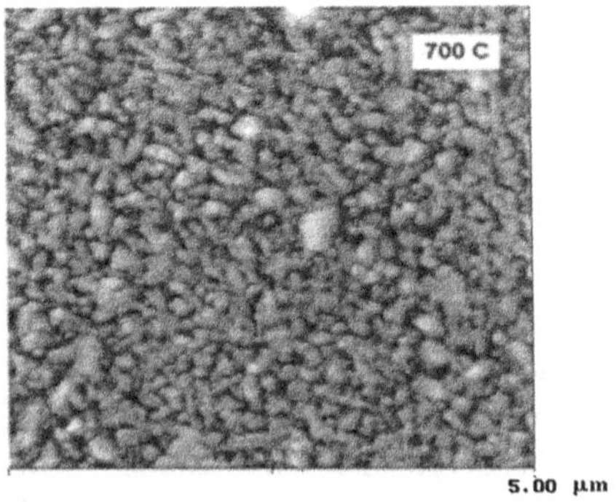

Figure 4: AFM micrograph of $BaBi_2Ta_2O_9$ thin films annealed at 700°C for 60 min.

Dielectric Ceramic Materials

The dielectric properties were found to be strongly dependent on the excess Bi content and the post-deposition annealing temperature. Table I summarizes the effect of excess bismuth and annealing temperature on the dielectric properties of BBT thin films.

Table I. Dielectric constant (ε_r) and dissipation factor (tan δ) as a function of annealing temperature and excess bismuth content.

Excess Bi (%)	650°C		700°C		750°C	
	ε_r	tan δ	ε_r	tan δ	ε_r	tan δ
0	46.5	0.001	62.7	0.009	130.1	0.022
5	54.3	0.002	94.9	0.020	182.5	0.022
10	57.8	0.003	282.6	0.023	151.8	0.023
15	56.2	0.002	211.2	0.021	114.8	0.025
20	67.8	0.003	189.6	0.022	82.4	0.026

The dielectric constant of samples annealed at 700°C was found to increase with the increase in Bi content and a maximum was obtained for samples with 10% excess Bi content. It can be observed from the table that the dielectric constant decreased when the films were annealed above 700°C. This may be attributable to the non-stoichiometry in the film, due to the loss of bismuth. The measured small ac signal dielectric constant and dissipation factor, for films with 10% excess bismuth content and annealed at 700°C for 60 min., were 282 and 0.023, respectively. The dissipation factor did not show any appreciable dependence on excess Bi content and was in the range 0.009 - 0.023 for films annealed at 700°C for 60 min. The dielectric constant, for these films, was found to decrease with further increase in Bi content beyond 10% which may possibly be due to formation of a low dielectric constant bismuth oxide phase, even through no secondary phase was observed in XRD patterns for films with up to 20% excess Bi content. The dielectric constant was found to increase with increase in annealing temperature, up to 700°C, which was dominantly due to improvement in crystallinity and grain size of the films. Figure 5(a), 5(b) and 5(c) shows the behavior of dielectric constant and dissipation factor, as a function of frequency, for BBT thin films annealed at 650, 700 and 750°C, respectively. The permittivity showed no appreciable dispersion with frequency up to about 1 MHz, indicating that the values were not masked by any surface layer effects or electrode barrier effects. As the frequency was increased above 1 MHz, the dielectric constant was found to decrease and the loss factor was found to increase with frequency. This behavior was found to be extrinsic in nature as similar behavior was observed at around the same frequency for thin films of other dielectric materials. At frequencies of the order of a few MHz, the stray inductance L of the contacts and wires and/or the presence of a finite resistance in series with the films, which may arise due to intrinsic or extrinsic sources, may cause such behavior. The high dielectric constant and low dielectric loss characteristics show the suitability of $BaBi_2Ta_2O_9$ thin films as the insulating dielectric layer for large value capacitors for various electronic devices.

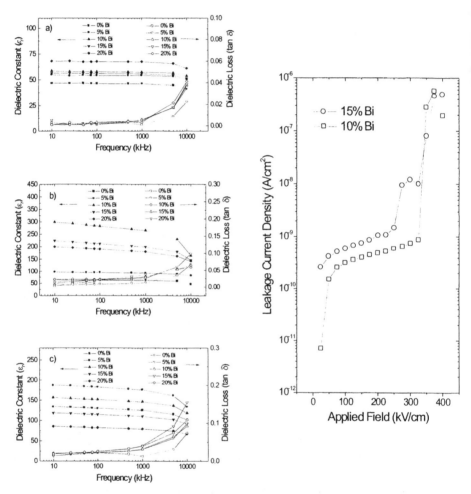

Figure 5: Dielectric constant and dissipation factor, as a function of frequency, for BBT thin films annealed at: (a) 650, (b) 700 and (c) 750°C.

Figure 6: Leakage Current density vs. Applied Electric Field

Dielectric Ceramic Materials

For DRAM applications, a material with high dielectric constant and good insulating characteristics is required. A limiting factor for any DRAM capacitor is leakage current. In a DRAM cell, the stored charge capacitor leaks off with time through various leakage mechanisms, which requires periodic refreshing of the stored charge. So the leakage current characteristics of the storage capacitor dielectric are very important for DRAM applications. Figure 6 shows the results, in terms of leakage current density vs. electric field characteristics of films with 10 and 15% excess bismuth content annealed at 700°C for 60 min. The leakage current densities of the film with 10% excess Bi was lower than 10^{-9} A/cm^2 up to an applied electric field of 300 kV/cm indicating good insulating characteristics. The charge storage density for this film was 38.4 fC/μm^2 at an applied electric field of 200 kV/cm. At this bias the leakage current density was of the order of 10^{-10} A/cm^2. These values appear to be consistent with the projected requirements for a DRAM cell capacitor with density beyond 64 Mbit.

CONCLUSION

In conclusion, the need for the development of high dielectric constant materials in DRAM capacitors continues to grow, and will be a necessity in Gb generation devices. The BBT film properties were found to be strongly dependent on the excess Bi content and annealing temperature. The best results were obtained for films with 10% excess Bi content annealed at 700°C. The film properties such as dielectric constant and dissipation factor at 100 kHz were 282 and 0.023, respectively, for films annealed at 700°C for 60 min. The leakage current density was lower than 10^{-9} A/cm^2 up to an applied field of 300 kV/cm. A charge storage density of 38.4 fC/μm^2 was obtained at an applied field of 200 kV/cm. The high dielectric constant, low dielectric loss, low leakage current density and high charge storage density suggest the suitability of BaBi$_2$Ta$_2$O$_9$ thin films as capacitor dielectric layer for microelectronic applications.

ACKNOWLEDGEMENTS

We would like to thank the support of CNPq - Brazil, process n$^{\underline{o}}$ 201036/93-2.

REFERENCES

[1] P. C. Fazan, Integrated Ferroelectrics, **4**, 247 (1994).

[2] M. Anthony, S. Summerfelt, and C. Teng, T.I. Technical Journal, Sept./Oct., 30 (1995).

[3] P. C. Fazan, H. C. Chan, V. K. Mathews, Semiconductor International, May, 108 (1992).

[4] C. A-Paz de Araujo, J. D. Cuchiaro, L. D. McMillan, M. C. Scott and J. F. Scott, Nature, **374**, 627 (1995).

[5] B. Aurivillius, Arkiv. för Kemi, **1**, 463 (1949).

[6] E. C. Subbarao, J. Phys. Chem. Solids, **23**, 665 (1962).

THE ROLE OF PROCESSING AND DOPANT ADDITIONS ON THE PHASE FORMATION AND HIGH FREQUENCY PROPERTIES OF ZIRCONIUM TIN TITANATE

Philip Pruna, James Wilson and Mohammed Megherhi
Ferro Corporation
1789 Transelco Drive
Penn Yan, PA 14527

INTRODUCTION

Ceramic microwave devices are playing an increasingly significant role in the wireless communications industry where they are used in a wide variety of filtering applications. These materials exhibit a range of permittivity from approximately 6 to 120 (ε'), very low dielectric losses (tan $\delta = 1/Q$) and excellent stability of coefficient of resonant frequency (τ_f) in the microwave region. Current industrial research efforts are focused on increasing ε' to reduce the physical dimensions of filtering devices and improve material Q values to allow greater channel-carrying capability while attempting to maintain near zero temperature stability.

One material of significant interest is Zirconium-Tin-Titanate (($Zr_{1-x} Sn_x)TiO_4$) which has been widely utilized for base station ring resonator applications. Rath first investigated the dielectric properties of $ZrTiO_4$ in 1941[1]. The crystal structure of $ZrTiO_4$ was identified by Newnham as having an orthorhombic symmetry in 1967[2]. Wakino and others studied the use of ($Zr_{1-x} Sn_x)TiO_4$ in microwave filtering applications during the 1970's[3-5]. Wolfram and Gobel identified a useful solid solution range in the ZrO_2-SnO_2-TiO_2 system in a 1981 publication[6]. Wakino, Minai and Tamura investigated the effects of transition doping of Zirconium Tin Titanate on microwave dielectric properties in the mid-1980's[7]. Additional work by Iddles, Bell and Moulson investigated the effects of ZnO, La_2O_3, and Nb_2O_5 dopants on the physical and electrical properties of ZST[8]. They reported an ε' of 38, Q of 13,000 at 3 GHz. and τ_f near 0 ppm/°C. Improvement in the dielectric properties were achieved using chemical synthesis routes in undoped ZST[9] and other researchers have studied the effects of donor/acceptor ion doping of this system throughout the late 1980's and into the 1990's[10-12].

While there has been a great deal published on the acceptable solid solutions range and dopant effects on the dielectric properties of Zirconium Tin Titanate, little has been reported on solid state processing/property relationships in this system. The present work is focused on studying the effects of calcination sequence and order of dopant additions on the densification and microwave dielectric properties of ($Zr_x Sn_{1-x})TiO_4$. Specifically, alternate calcination routes (e.g. pre-calcination of $ZrO_2 \cdot TiO_2$ and $SnO_2 \cdot TiO_2$ end members) the effects of adding dopants before and after ZST phase formation, and the use of partial dopant additions to enhance ZT endmember formation were investigated.

EXPERIMENTAL PROCEDURES

High Purity (>99%) ZrO_2, SnO_2, TiO_2, NiO, and ZnO raw materials were weighed in the appropriate amounts. The materials were milled in isopropyl alcohol for 3 hours and dried. The dried powders were calcined at various time/temperature profiles and X-rayed to evaluate phase formation. The calcined and uncalcined powders were then milled in distilled water and appropriate amounts of PVA binder and a polyelectrolyte dispersant were added. The slurries were spray dried. Discs were pressed and sintered over a range of temperatures, from 1325 °C to

1450 °C for 3 hours. The fired parts were X-rayed, machined to the desired geometry for high frequency measurements, and characterized for physical and electrical properties. The Courtney method was used to evaluate sample permittivity over the frequency range of 6.1 - 6.4 GHz. and the Cavity Perturbation method was used for both Q and τ_f determinations (4.1 - 4.5 GHz. resonant frequency range). Table 1 details the various calcination/firing conditions utilized in this experiment.

Table 1. Processing Routes

Process →	$ZrO_2 + SnO_2 + TiO_2 + 0.5$ wt. % ZnO + 0.5 wt. % NiO
Route 1	Calcined at 1100 °C/ 3 hrs.
	Sintered at 1350 °C - 1425 °C for 3 hrs.

Process →	$ZrO_2 + TiO_2$ Calcined at 1250 °C/ 2 hrs.
Route 2	Then recalcined at 1350 °C/ 4 hrs.
	$SnO_2 \cdot TiO_2$ calcined at 1325 °C/ 3 hrs.
	+ 0.5 wt. % ZnO + 0.5 wt. % NiO
	Sintered at 1350 °C - 1425 °C for 3 hrs.

Process →	$ZrO_2 + SnO_2 + TiO_2$ Calcined at 1350 °C/ 4 hrs.
Route 3	+ 0.5 wt. % ZnO + 0.5 wt. % NiO
	Sintered at 1325 °C - 1400 °C for 3 hrs.

Process →	$ZrO_2 \cdot TiO_2 + 0.25$ wt. % ZnO
Route 4	Calcined at 1200 °C/ 4 hrs.
	+ 0.25 wt. % ZnO + 0.5 wt. % NiO + $SnO_2 + TiO_2$
	Sintered at 1350 °C - 1425 °C for 3 hrs.

Process →	$ZrO_2 + TiO_2$ Calcined at 1250 °C/ 2 hrs.
Route 5	Then recalcined at 1350 °C/ 4 hrs.
	+ 0.5 wt. % ZnO + 0.5 wt. % NiO + $SnO_2 + TiO_2$
	Sintered at 1350 °C - 1425 °C for 3 hrs.

Process →	$ZrO_2 + SnO_2 + TiO_2$ Calcined at 1100 °C/ 3 hrs.
Route 6	+ 0.5 wt. % ZnO + 0.5 wt. % NiO
	Sintered at 1350 °C - 1425 °C for 3 hrs.

Process →	$ZrO_2 + SnO_2 + TiO_2 + 0.25$ wt. % ZnO
Route 7	Calcined at 1100 °C/ 3 hrs.
	+ 0.25 wt. % ZnO + 0.5 wt. % NiO
	Sintered at 1350 °C - 1400 °C for 3 hrs.

High temperature X-ray diffraction (XRD) was performed using a 60 kV Rigaku diffractometer with an Ultima+ Goniometer equipped with a 1400 °C furnace. Samples were heated to 1200 °C and X-rayed every 30 minutes over a 4 hour period.

Dielectric Ceramic Materials

A Shimadzu TMA-50 Thermomechanical Analyzer (TMA) was used to gather shrinkage data on all compositions. Samples were first heated at 10 °C/min to 400 °C and soaked for 30 minutes to remove volatiles/residual moisture. The materials were then heated to 1450 °C at 10 °C/min and held at temperature for 60 minutes.

An Amray 3300 Scanning Electron Microscope (SEM) equipped with an Oxford 300 Energy Dispersive Spectroscope (EDS) was used to perform microstructural and phase analysis on selected fired samples. Samples were etched at 1300 °C for 30 minutes prior to analysis.

RESULTS AND DISCUSSION

Room and In-Situ High Temperature XRD Results

Phase pure ZST was obtained upon sintering at 1325 °C/3 hrs. or higher, regardless of the process route selected. Processing Routes #1, 6 and 7 gave similar X-ray results, with Route #6 showing the highest percentage of unreacted constituents due to the absence of the ZnO sintering aid during calcination (See Figure 1).

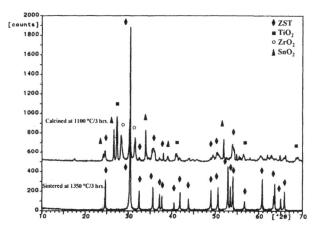

Figure 1 XRD patterns of Route #6 material calcined at 1100 °C/3 hrs. (top) and sintered at 1350 °C/3 hrs (bottom).

A higher calcination temperature of 1350 °C was required in Route #3 to produce the ZST major phase with minor amounts of TiO_2 and ZrO_2 observed. Upon adding the ZnO and NiO and sintering at 1350 °C - 1450 °C, phase pure ZST was achieved.

In Processing Route #2 nearly phase pure ZT was formed using a double calcine process with a peak temperature of 1350 °C. Due to the refractory nature of SnO_2-TiO_2 mixtures, no evidence of phase formation was observed after calcination at temperatures as high as 1400 °C which is consistent with the published phase equilibria data[13]. Upon adding the ZnO and NiO and sintering at 1350 °C - 1400 °C, phase pure ZST was achieved.

The dramatic effect of small additions of ZnO on the formation of the ZT phase at temperatures as low as 1200 °C was illustrated in Processing Route #4 (Figure 2). The XRD pattern of the calcined material revealed a higher degree of reaction than the material double calcined at 1250 °C and 1350 °C in Routes #2 and 5. To further demonstrate the effect of ZnO additions on ZT phase formation an in-situ high temperature X-ray diffraction study was performed on samples of ZrO_2 + TiO_2 with 0, 0.25 and 0.5 wt. % amounts of ZnO. The results of this investigation substantiate the effect of ZnO additions on ZT phase formation as illustrated by the plot of the peak intensity ratio of ZT (111) to TiO_2 (110) versus time at 1200 °C given in Figure 3.

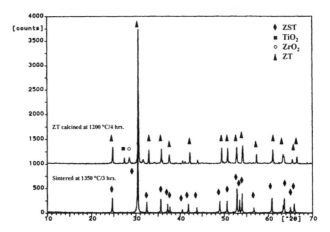

Figure 2 XRD patterns of Processing route #4: ZT calcined at 1200 °C/4 hrs. (top), and ZT + ZnO + NiO + SnO_2 + TiO_2 sintered at 1350 °C/3 hrs (bottom).

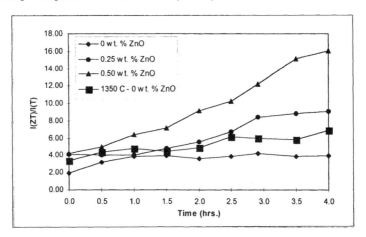

Figure 3 Plot of high temperature XRD I_{ZT} (111)/I_T (110) peak intensity ratios for ZT with and without ZnO at 1200 °C along with ZT without ZnO at 1350 °C.

Dielectric Ceramic Materials

Shrinkage Characteristics

Processing Route #2 exhibited the most refractory sintering profile as expected, due to the high calcination temperature necessary to form phase pure ZT (Figure 4). The addition of SnO_2+TiO_2 after calcining the ZT (Process Route #5) resulted in a higher shrinkage rate than Route #2. Processing Route #4 exhibited the most aggressive sintering profile. Again, this illustrates the effect of ZnO addition on phase formation of ZT.

Processing Route #3 illustrated the aggressive behavior of ZnO and NiO as sintering aids in the ZST system. Near complete phase formation prior to dopant addition and subsequent sintering results in the earliest onset of shrinkage and increased final shrinkage. The ZST intermediate processing routes #1, 6, and 7 resulted in similar shrinkage curves.

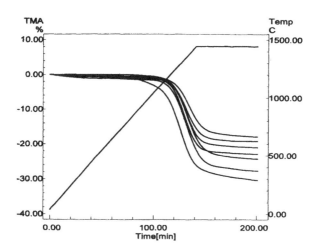

Figure 4   TMA Analysis of Processing Routes #1 - 7 conducted at a heating rate of 10 °C/min. to 1450 °C with a 60 minute soak. Shrinkage curves are in the following descending order: Processing route #2, 5, 6, 7, 1, 3 and 4.

SEM/EDS Analysis

Processing route #2 exhibited excessive microstructural porosity as expected, due to the refractory nature of the calcine procedure. Processing routes #1 and 6 resulted in comparable microstructures. This suggests that the role of the dopants is more that of sintering aid rather than a grain growth inhibitor. Processing routes #1, 3, and 6 showed evidence of a second phase, primarily ZST and concentrations of Ni. This second phase was not observed in route #2. This may be due to the poor densification obtained in the firing temperature range studied. The second phase was observed in fully developed microstructures. However, the firing temperature range where the second phase is observed is dependent on the processing route. This phenomenon has not been fully investigated. Figures 5 and 6 illustrate the presence of the second phase in the grain boundary regions. While Zn was detected in the first and second phases of routes #1, 2 and 6, no concentrations of Zn were observed. This may suggest that it is evenly distributed throughout the structure.

Dielectric Ceramic Materials   251

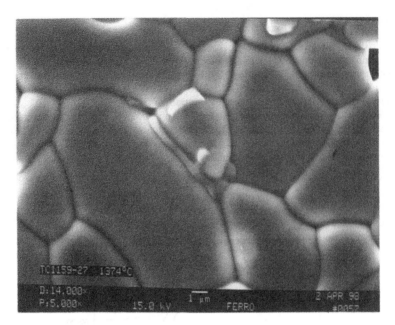

Figure 5 Processing Route #6 thermally etched fracture surface SEM micrograph (14,000x) of disc fired at 1375 °C for 3 hrs. The second phase can observed at the grain boundaries of the most centrally located grain.

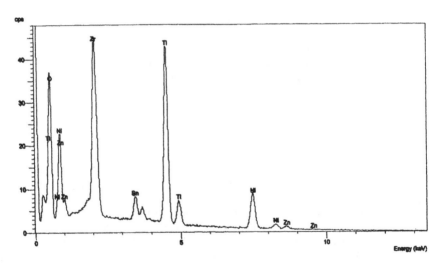

Figure 6 EDS analysis of second phase found in Processing route #6 sample seen in Figure 5.

Physical and Microwave Property Results

The effect of dopant addition sequence on ZST phase formation was studied in routes #1, 6 and 7. The inclusion of ZnO in the calcination step resulted in higher fired densities (See Table 2). The addition of the dopants to the ZST formed at 1350 °C (Process Route #3) resulted in similar densities to those obtained with the other routes. This illustrates the aggressive behavior of ZnO on densification. Processing route #2 exhibited the lowest fired densities.

All the above routes, regardless of the dopant addition sequence, resulted in permittivites greater than 38. Samples prepared via Route 1 exhibited Q*f's which ranged from 35,139 to 42,523 and τ_f's of -1.68 ppm/°C to -2.19 ppm/°C. Q maxima was obtained at 1375 °C. τ_f was observed to be consistent over the firing range. However, Q*f began to drop rather significantly at firing temperatures above 1375 °C. Processing Route #7 exhibited consistent τ_f values of -1.20 ppm/°C to -1.50 ppm/°C. However, the Q*f which ranged from 35,035 to 42,143, began to degrade at lower firing temperatures. Processing Route 3 exhibited similar behavior to Route 1. Q*f ranged from 33,450 at 1450 °C to 42,637 at 1325 °C. τ_f ranged from -1.34 ppm/°C at 1425 °C to -2.09 ppm/ °C at 1325 °C. Of particular note of interest to these researchers was the observation of a wide processing window with this route. This is despite the fact that the ZST phase was formed at the relatively high temperature of 1350 °C. This again suggests that the ZnO and NiO dopants are very effective sintering aids. Processing route #6 where dopants were added after an 1100 °C precalcination of ZrO_2 + SnO_2 + TiO_2 resulted in the highest Q*f products (40,193 to 48,644) among the routes studied. This suggests that post calcination dopant addition is preferred (versus Processing routes #1 and 7).

Processing routes 2, 4 and 5 using the $ZrTiO_4$ as an intermediate resulted in dissimilar physical properties. In processing route #2 both ZT and $SnO_2 \cdot TiO_2$ end members were precalcined at 1250 °C/1350 °C and 1325 °C, respectively, resulting in the lowest sintered densities obtained. In contrast, processing route #5 where the SnO_2 and the requisite amount of TiO_2 were added with the dopants to precalcined ZT, was able to achieve respectable sintered densities of greater than 5.0 g/cc. An addition of ZnO to $ZrO_2 \cdot TiO_2$ (Process route #4) allowed the formation of the ZT phase at 1200 °C. This resulted in the highest degree of sinterability and fired densities of the three ZT precursor routes studied.

No resonance frequency peaks were found for the Route 2 samples sintered at 1325 °C and 1350 °C. Q*f ranged from 8,847 to 36,678 at 1375 °C and 1450 °C, respectively. τ_f values of -1.74 ppm/°C to -2.97 ppm/°C were measured. Again, this trend in the data can be attributed to the densification kinetics of this processing route. Processing route #4, while possessing high sintered densities and permittivities, exhibited the lowest Q*f products of all routes studied. τ_f's were noted to be very consistent over the entire firing temperature range (-2.28 ppm/°C to -2.63 ppm/°C). Q values obtained from samples prepared via route #5 were some of the more moderate observed and ranged from 26,901 to 35,939. However, it was noted that there is an apparent positive shift in τ_f with this processing route and many of the values approach zero, -0.08 to -0.36 ppm/°C.

Dielectric Ceramic Materials 253

Table 2. Fired Density and High Frequency Properties

Processing Route/Firing Temperature (°C)	Fired Density (g/cm³)	ε'	Q*f (GHz)	τf (ppm/°C)
Route1				
1325	4.99	36.60	41,103	-2.19
1350	5.14	38.45	40,794	-2.00
1375	5.17	39.01	42,523	-1.83
1400	5.16	38.63	37,910	-1.69
1425	5.18	38.64	35,139	-1.74
1450	5.11	37.89	35,190	-1.68
Route 2				
1325	4.25	no freq.	no freq.	no freq.
1350	4.43	30.55	no freq.	no freq.
1375	4.64	33.06	8,847	-2.25
1400	4.79	34.43	36,557	-2.96
1425	4.93	35.90	32,966	-2.21
1450	4.96	36.33	36,678	-1.74
Route 3				
1325	4.92	35.95	42,637	-2.09
1350	5.05	37.29	41,039	-2.08
1375	5.14	38.44	40,858	-1.52
1400	5.15	38.58	38,910	-1.41
1425	5.16	38.56	34,204	-1.34
1450	5.10	37.92	33,450	-1.61
Route 4				
1325	5.17	38.38	28,893	-2.39
1350	5.18	38.58	27,324	-2.46
1375	5.18	38.47	26,893	-2.28
1400	5.08	38.08	25,210	-2.57
1450	5.07	37.14	22,809	-2.63
Route 5				
1325	4.73	33.46	28,787	-1.36
1350	5.01	36.83	35,939	-0.70
1375	5.10	37.80	35,328	-0.31
1400	5.15	38.42	31,919	-0.08
1450	5.10	37.82	26,901	-0.29
Route 6				
1350	4.97	37.45	48,644	-1.71
1375	5.13	38.54	46,949	-1.65
1400	5.14	38.51	44,282	-1.25
1425	5.16	38.52	40,193	-1.27
1450	5.08	37.91	40,193	-1.47
Route 7				
1350	5.17	38.71	42,143	-1.41
1375	5.18	38.71	39,886	-1.39
1400	5.17	38.60	35,035	-1.20
1425	5.16	38.40	35,527	-1.43
1450	5.08	37.31	37,360	-1.50

Dielectric Ceramic Materials

IV. CONCLUSIONS

Pre-reacting ZrO_2 + SnO_2 + TiO_2 resulted in the formation of sintered ceramics with the best electrical performance. Furthermore, higher Q's were observed by adding the dopants to the prereacted powders (Route #6). ZnO and NiO were observed to be very powerful sintering aids as evidenced by the results obtained with route #3. Even though high densities could be achieved with certain processing routes (e.g. Routes #4 and #5) where ZT was pre-formed, lower Q's were observed. It is believed that ZT may inhibit Sn diffusion into the crystal structure where it is thought to enhance lattice ordering and improve Q. The use of ZnO promotes the formation of ZT at significantly lower temperatures than are normally required. However, this does not result in any discernible improvement in those properties measured. Sinterability (as determined by TMA and fired density analysis) of powders produced from the various processing routes studied appears to have little effect on the resultant Q properties obtained.

VI. ACKNOWLEDGMENTS

The authors would like to thank Sarah Freeman for her assistance with the higher temperature XRD measurements, D. Gnizak for his help with the SEM/EDS analysis, Skip Quackenbush for his help with the TMA and XRD, Lana Garten and Richard Kizys for their assistance in material preparation and Howard Baube for his help with the high frequency material characterization.

REFERENCES

[1]W. Rath, Keram. Rdsch. 49 (1941), 137.

[2]R. E. Newnham, "Crystal Structure of $ZrTiO_4$", J. Am. Ceram. Soc., vol. 50 (1967) p. 216.

[3]K. Wakino, T. Nishikawa, S. Tamura, and Y. Ishikawa, Proc. IEEE MTT Symposium, 1975 p. 63.

[4]K. Wakino, T. Nishikawa, S. Tamura, and Y. Ishikawa, "Miniaturized band pass filters using half wavelength dielectric resonator with improved spurious response," IEEE MTT Symposium, 1978 pp. 230-232.

[5]K. Wakino, T. Nishikawa, S. Tamura, and Y. Ishikawa, Proc. IEEE MTT Symposium, 1979, p. 278.

[6]G. Wolfram and Gobel, "Existence range, structural and dielectric properties of $Zr_xTi_ySn_zO_4$ ceramics (x+y+z=2)," Mater. Res. Bull., vol. 16, No. 11, 1981 pp. 1455-1463.

[7]K. Wakino, K. Minai and H. Tamura, "Microwave characteristics of $(Zr,Sn)TiO_4$ and $BaO-PbO-Nd_2O_3-TiO_2$ dielectric resonators," J. Am. Ceram. Soc., vol. 67, No. 4, 1984, pp. 278-281.

[8]D. M. Iddles, A. J. Bell and A. J. Moulson, "Relationships between dopants, microstructure and the microwave dielectric properties of $ZrO_2-TiO_2-SnO_2$ ceramics", J. Mat. Sci. vol 27 (1992) p. 6303 - 6310.

[9]S-I Hirano, T. Hayashi and A. Hattori, "Chemical Processing and Microwave Characteristics of $(Zr,Sn)TiO_4$ Microwave Dielectrics", J. Am. Ceram. Soc. vol. 74 (1991) pp. 1320-1324.

[10]Y. C. Heiao, L. Wu, and C. C. Wei, "Microwave dielectric properties of $(ZrSn)TiO_4$ ceramic," Mat. Res. Bull., vol. 23, 1988 pp. 1687-1692.

[11]N. Michiura, T. Tatekawa, Y. Higuchi and H. Tamura, "Role of donor and acceptor ions in the dielectric loss tangent of $(Zr_{0.8}Sn_{0.2})TiO_4$ dielectric resonator material," J. Am. Ceram. Soc., vol. 78, No. 3, 1995, pp. 793-796.

[12]R. Christofferson, P. K. Davies and X. Wei, "Effect of Sn substitution on cation ordering in $(Zr_{1-x}Sn_x)TiO_4$ microwave dielectric ceramics," J. Am. Ceram. Soc., vol. 77, No. 6, 1994, pp. 1441-1445.

[13]N. N. Padurow, "System SnO_2-TiO_2; Subsolidus"; pp. 142 in Phase Diagrams for Ceramists, Edited by E. M. Levin, C.R. Robbins and H. F. McMurdle. Amercan Ceramic Society, Inc., 1964.

$BaO\text{-}Nd_2O_3\text{-}TiO_2$ MICROWAVE DIELECTRIC CERAMICS: EXPERIMENTAL AND COMPUTER MODELLING STUDIES

Feridoon Azough, Robert Freer, Paisan Setasuwon, Colin Leach and Paul Smith*
Materials Science Centre
University of Manchester/UMIST
Grosvenor Street
Manchester, M1 7HS UK

ABSTRACT

Ceramics of $Ba_{4.5}Nd_9Ti_{18}O_{54}$ have been prepared with additions of MgO (0.25-1.0wt%) and $MgO+TiO_2$ (1.0-4.0wt%) by the mixed oxide route. Specimens were sintered at temperatures in the range 1350-1450°C. Small additions of MgO (0.25wt%) led to the development of high density (5.67g cm^{-3}) ceramics with uniform microstructure, high relative permittivity ε_r (86) and enhanced Q.f product (9800). Higher levels of MgO or $MgO+TiO_2$ caused the formation of a second phase which degraded ε_r but stabilised Q.f at 5000-6000. Static simulation studies of $BaO\text{-}Nd_2O_3\text{-}TiO_2$ successfully reproduced the structure and lattice parameters to within 1-2%. A variety of interatomic potentials and starting configurations were explored. Predicted low frequency and high frequency ε_r values were in satisfactory agreement with experimental data.

INTRODUCTION

Ceramic solid solutions based on $BaO\text{-}Ln_2O_3\text{-}TiO_2$ are widely used as high permittivity microwave dielectrics [1,2]. The properties of most of the rare-earth analogues from La-Gd have been examined [2], but materials based on $BaO\text{-}Nd_2O_3\text{-}TiO_2$ have attracted most commercial interest. To maximise the dielectric properties additions of Pb and/or Bi are usually employed [1,3]. Detailed structural investigations by X-ray diffraction [4-7] and EXAFS [2] have gone a long way to explain how the dielectric properties depend upon the distribution of cations, and how additives improve performance.

In the present study, two distinct approaches have been adopted:(i) experimental, to explore the benefit of using MgO or $MgO+TiO_2$ as additives for the preparation of $Ba_{4.5}Nd_9Ti_{18}O_{54}$ ceramics, and (ii) computer modelling, to assess whether static simulation techniques can be employed to investigate the structure and dielectric properties of such complex, large volume unit cell materials as these $BaO.Nd_2O_3.4TiO_2$ (1-1-4) ceramics. Direct comparison is made between experimental results and predictions from simulations.

EXPERIMENTAL

The starting materials were high purity (>99.5%) powders of $BaCO_3$, Nd_2O_3 (from Fluka Chemicals, UK), TiO_2 and MgO (from Aldrich Chemicals, UK). Powders were mixed in the appropriate ratios to yield the base composition for $Ba_{4.5}Nd_9Ti_{18}O_{54}$. Batches of powder were ball-milled (wet, with zirconia media; typically 40g powder with equal amounts of milling media and 40 cc propan-2-ol) for 8 hours, calcined at 1275°C for 4 hours, and milled again for 8 hours. To this base powder, additions of either:

(i) MgO (up to 1 wt %) or

(ii) MgO+TiO_2 (up to 4 wt %)

were made to yield two series of samples. After the Mg-based addition was made the powder was milled again. The resulting slurries were dried and pressed (at 120 MPa) into pellets of 10mm diameter and length in the range 3-6mm, Pellets were sintered in oxygen at temperatures in the range 1350-1500°C for 4 hours and cooled at 120°C hour^{-1}.

X-ray diffraction analysis, using CuK_α radiation and a Philips diffractometer with horizontal goniometer, was used to investigate calcined powders and sintered specimens. Densities were determined by an immersion method. Product morphologies were examined by optical and scanning electron microscopy (Philips 505 and 525 instruments) on polished and etched surfaces.

Dielectric properties were determined at 3 GHz using a network analyser and sweep oscillator (Hakki and Coleman [8] method). Additional measurements of capacitance were made at 100 kHz over the range 20-100°C to yield values of temperature coefficient of capacitance (TCC).

RESULTS AND DISCUSSION

Physical Properties

X-ray diffraction analysis confirmed that the primary phase had an orthorhombic structure in all specimens. In samples prepared with MgO additions (up to 1 wt %) there was no evidence of second phase; for samples prepared with larger amounts of MgO+TiO_2 (up to 4 wt %) there was an indication of a second phase, but peaks were not strong enough to enable formal indexing.

Densities of sintered samples were generally in the range 5.0-5.67 g cm^{-3} (Figure 1). The highest densities (5.67 g cm^{-3}) were achieved with small additions of MgO (0.25 or 0.5 wt %) when sintered at 1400°C. These values are marginally higher than those achieved for $Ba_{3.75}Nd_{9.5}Ti_{18}O_{54}$ with additions of Bi_2O_3 [6]. Additions of $MgO+TiO_2$ were not as effective as MgO alone, and densities achieved with the former are highly sensitive to sintering temperature. For samples prepared with 2 wt % $MgO+TiO_2$ the density was 5.0 g cm^{-3} when sintered at 1350°C, and 5.4 g cm^{-3} when sintered at 1375°C. However, increasing the $MgO+TiO_2$ additions to 4 wt % did not degrade density.

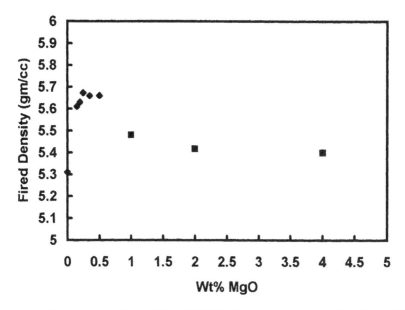

Figure 1. Sintered densities of $Ba_{4.5}Nd_9Ti_{18}O_{54}$ ceramics as a function of MgO (♦) or $MgO+TiO_2$ (■) content.

Microstructure

SEM micrographs of selected samples are shown in Figure 2. In general terms they exhibit the usual features expected for 1-1-4 type materials, rounded to elongated grains 2-10μm in length, but there are distinct differences which reflect the type and amount of additive used. Figure 2a shows the undoped 1-1-4 ceramic; the majority of grains are <5μm long and there is evidence of a few percent porosity. The addition of 0.25 wt % MgO (Figure 2b) improved densification; there is minimal evidence of porosity and no evidence of any second phase. However, increasing MgO additions to 1 wt % (Figure 2c) caused grain growth, and led to: the development of a second phase (rich in Mg and Ti), a reduction in density (due to the lower density of the second phase) and the appearance of isolated and interconnected porosity. In contrast, the preparation of samples with additions of MgO+TiO$_2$ always led to the formation of a second phase (rich in Mg and Ti). With increasing amounts of MgO+TiO$_2$ grain growth increased and the second phase grew in distinct isolated regions (Figure 2d). If this second phase is indeed MgTiO$_3$, then it may be expected to degrade the relative permittivity.

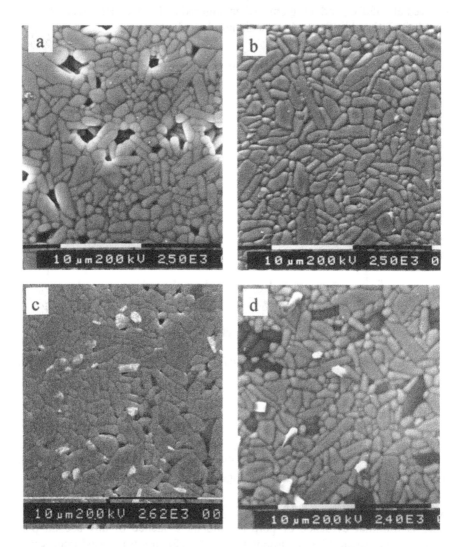

Figure 2. SEM micrographs of Ba$_{4.5}$Nd$_9$Ti$_{18}$O$_{54}$ ceramics prepared with (a) no additions, (b) 0.25 wt % MgO, (c) 1.0 wt % MgO and (d) 1.0 wt % MgO+TiO$_2$.

Dielectric Properties

Figure 3 shows the relative permittivity of $BaO-Nd_2O_3-TiO_2$ based ceramics at 3 GHz. In the high density samples prepared with 0.25 wt % MgO, the ε_r is high (86), but degrades to 79 with increasing additions (1 wt % MgO). The combined $MgO+TiO_2$ additions have a very similar effect, reducing the ε_r to 72, at the 2 wt % level, and to 66.7, at the 4 wt % level. As anticipated, the presence of the Mg-Ti rich second phase (generated by the presence of MgO or $MgO+TiO_2$) is detrimental to the relative permittivity of 1-1-4 ceramics; $MgTiO_3$ for example has a relative permittivity approximately one quarter of that of 1-1-4 materials.

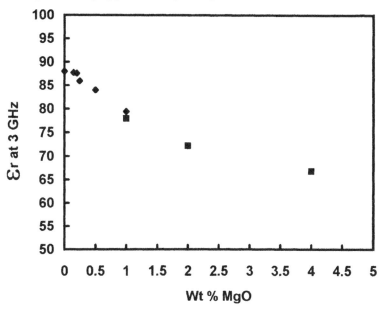

Figure 3 Relative permittivity of $Ba_{4.5}Nd_9Ti_{18}O_{54}$ ceramics as a function of MgO (♦) or $MgO+TiO_2$ (■) additions.

Figure 4 shows the Q.f product (dielectric Q value x frequency) for 1-1-4 ceramics prepared with MgO or $MgO+TiO_2$ additions. Whilst the majority of specimens exhibit Qf values in the range 5000-6000, which is typical for 1-1-4 materials without Bi or Pb additions, it is clear that small amounts of MgO (0.25 wt %) are beneficial, increasing the Qf value from 6200 (in the undoped ceramic) to 9800. Thus, these high density samples (Figure 2b) exhibit the highest ε_r and Qf values amongst all the doped ceramics. The mechanism, of enhancement is not known, but it is of interest to note that whilst large additions of MgO or

MgO+TiO$_2$ are detrimental to ε_r, the Qf product seems relatively unaffected, being approximately 6200 for high levels of MgO additions and 5000 for high levels of MgO+TiO$_2$ additions.

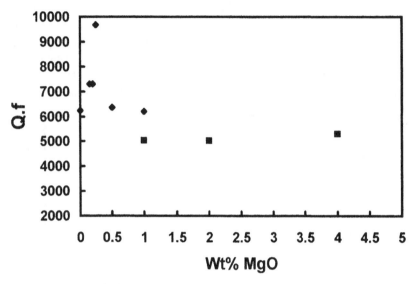

Figure 4. Product of dielectric Q value and frequency (Q.f) for Ba$_{4.5}$Nd$_9$Ti$_{18}$O$_{54}$ ceramics as a function of MgO (◆) or MgO+TiO$_2$ (■) additions.

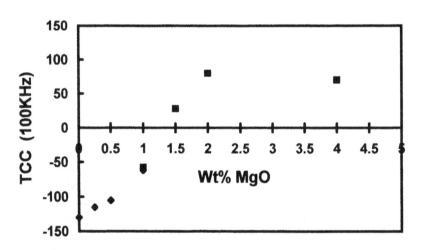

Figure 5. TCC (at 100kHz) of Ba$_{4.5}$Nd$_9$Ti$_{18}$O$_{54}$ ceramics as a function of MgO (◆) or MgO+TiO$_2$ (■) additions.

Dielectric Ceramic Materials

Figure 5 shows the TCC values (at 100kHz) for the 1-1-4 ceramics prepared with MgO or MgO+TiO_2 additions. For samples exhibiting high ε_r and high Qf products (eg prepared with ~0.25 wt % MgO) the TCC values are large (-116ppm/°C). However, it is noted that a zero TCC value is anticipated for MgO additions of approximately 1.3%. It will be of interest to learn how MgO, or perhaps MgO plus other additions, affect the TCC at microwave frequencies.

The benefit of Bi additions to 1-1-4 ceramics has long been known; it promotes densification, enhances ε_r (provided the solubility limit is not exceeded) and can adjust the temperature dependence of resonant frequency or TCC [9]. Table I compares the dielectric properties of 1-1-4 ceramics prepared with Bi_2O_3 additions [9] to those prepared with MgO or MgO+TiO_2 additions (this study, TS). Whilst the bismuth additions lead to high ε_r values, small additions of MgO (~0.25 wt %) are beneficial to both ε_r and Qf. Indeed, the MgO additions appear to enhance and stabilise the Qf more successfully than the Bi_2O_3 additions. It seems likely that the MgO reacts with TiO_2 to form a second phase, rich in Mg and Ti, which may be $MgTiO_3$. In small amounts this improves the dielectric properties, but with large amounts the ε_r steadily degrades. However, there appears to be scope for optimisation of 1-1-4 ceramics prepared with MgO additions.

Table I Dielectric properties of 1-1-4 type ceramics prepared with Bi_2O_3 or MgO additions.

addition	ε_r	Qf (GHz)	TCC/τ_f	ref
2.0 % Bi_2O_3	96	5600	27	9
2.45% Bi_2O_3	99	4631	16	9
3.0 % Bi_2O_3	92	3880	21	9
0.25% MgO	86	9800	-116	TS
0.5% MgO	84	6400	-110	TS
1.0% MgO	79	6300	-62	TS
1.0% MT	80	5000	-58	TS
1.5% MT	-	-	28	TS
2.0% MT	72	5000	80	TS
4.0% MT	67	5200	70	TS

MT=MgO+TiO_2

SIMULATION STUDIES

The structures of ceramics based on BaO-Ln_2O_3-TiO_2 have been the subject of investigation for more than two decades. For the Nd and Pr analogues there has been conflict over the space groups (several including Pbam, Pba2 and Pnam have been proposed [5,10,11]), and the detailed distribution of atoms between the four primary cation sites [6,7]. There is also a strong suggestion that the space group may vary with temperature in such materials, and this may help to explain apparent structural differences reported in the literature[6,11]. A further complication is that the refinement of large structures, such as those for BaO-Ln_2O_3-TiO_2 materials (~2000A°), depends critically upon the location of vacancies within the structure. It is anticipated that direct assistance will be provided by static simulation studies, both in terms of understanding the structure and the prediction of physical properties. Our work on the simulation of dielectrics is part of an extended programme; detailed results will be presented elsewhere. Here we outline the methodology and some of the achievements.

Static simulation studies were undertaken using the PARAPOCS code on a CRAY supercomputer at the Rutherford Appleton Laboratories (supported by the EPSRC, UK). The basis of atomistic simulation is that interatomic potentials are defined to describe the forces acting between pairs or groups of ions. Reliable sets of potential parameters are now available for the relevant cations (ie Ba, Nd, and Ti) and oxygen. Direct interactions between pairs of ions may be calculated in terms of Coulombic and short range potentials. It is also necessary to take account of polarization of oxygen ions (and in some cases cations) and co-ordination dependence of potentials for cations. For the simulation, the starting configuration for the location of ions and unit cell dimensions was that proposed by Matveeva et al [4]. A 'two-region' strategy was adopted for calculations: in the inner, region I, containing typically 200-300 ions, all interactions were determined explicitly; in the outer, region II (regarded as a continuum) the interactions were calculated by approximate techniques. By an interative procedure, the positions of ions were allowed to relax until a minimum energy configuration had been achieved. In this state, the equilibrium atom positions and physical properties (including piezoelectric parameters and relative permittivity) are predicted. Clearly the validity of the simulation is entirely dependent upon the reliability of the individual potentials and the overall potential model; convergence of the calculation is sensitive to the initial distribution of the atoms and any included vacancies. A wide variety of compositions, and cation distributions have been explored and examples are given below.

Starting with the 'comparatively' simple compound $Ba_6Nd_8Ti_{18}O_{54}$ there are in fact 45 ways in which the Ba^{2+} and Nd^{3+} may be distributed between sites (1), (2) and (3). All of these configurations were simulated; for many, the predicted lattice parameters in at least one crystal direction differed from the experimental value by more than 5%, but some reproduced the structure well. It is interesting to note that whilst the final lattice energies for 39 of the configurations exhibited little variation (2-3eV) the predicted relative permittivities varied from 7 to 137.

Considering next the composition $Ba_{4.5}Nd_9Ti_{18}O_{54}$ (which was the subject of the experimental part of this investigation) a variety of potential models, of differing complexity, were explored. The results are summarised in Table II.

Table II Simulated and experimental lattice parameters and dielectric properties for $Ba_{4.5}Nd_9Ti_{18}O_{54}$ (see text for details).

Model/data	a(A)	b(A)	c(A)	ε_r	ε (HF)
A	23.1	12.0	7.4	~30	3
B	22.9	12.2	7.6	<20	3
C	22.9	12.2	7.6	62, 49, 33	5.4
experimental	22.35	12.20	7.69	88	5.09

In model A, the potentials for the cations were taken from Bush et al [12]. The predicted lattice parameters were high for the a direction, but low for the b and c directions, and the predicted ε_r was low, less than 30. The simulated lattice parameters improved (Model B) by employing the potentials of Lewis [13] for Ti and O^{2-}, and those of Bush et al for Ba and Nd. However, the low frequency relative permittivity ε_r was still very low (<20). Finally, in model C, allowance was made for the polarisability of both oxygen and Ti ions. In this case there was good agreement between the experimentally determined lattice parameters [14] and the predicted values for the b and c directions; that for the a direction was only 2.5% high. The predicted high frequency relative permittivity ε(HF) is in moderately good agreement with the experimental value [15], and the predicted 'low frequency' ε_r values are now beginning to approach the typical value for ceramics (~88). Further refinement of the model is still required.

Preliminary work on the simulation of the solid solution $Ba_{12-2x}Nd_{16-4/3x}Ti_{36}O_{108}$ are most encouraging in terms of the structures and properties; there is a strong suggestion of anisotropy in the dielectric properties. Optimum structures obtained from the simulation studies should also be useful as input data for further structural refinements.

CONCLUSIONS

Good quality $Ba_{4.5}Nd_9Ti_{18}O_{54}$ ceramics have been prepared by the mixed oxide route with the aid of MgO or $MgO+TiO_2$ additions. Small amounts of MgO aid the development of high density, homogeneous ceramics having high ε_r values

(86) and enhanced Qf values (9800). Higher levels of additive cause the ε_r to degrade and the Qf to stabilise.

Static simulations have been successfully applied to BaO-Nd$_2$O$_3$-TiO$_2$ ceramics. Simulations tended to reproduce the structure and lattice parameters to within 1-2%. Predicted ε_r values are very sensitive to the potential models, but moderate agreement with experimental data was achieved. The techniques hold much promise for the investigation of dielectric materials.

ACKNOWLEDGEMENT. The support of the EPSCR through grant GR/L444218 is gratefully acknowledged.

REFERENCES

1. T. Negas and P.K.Davies, Ceramic Transactions vol 53, 179-196 (1995).
2. M. Valent, I Arcon, D Suvorov, A Kodre, T Negas and R Frahm, J Mater. Res., **12**, 799-804 (1997).
3. K Wakino, K Minai and H Tamura, J Am. Ceram. Soc.,**67**, 278-281 (1984)
4. R G Matveeva, M B Varfolomeev and L S Il'yushenko, Trans. From Zh. Neoirg Khimi., **29**, 31-34 (1984).
5. D Kolar, S Gabrscek and D Suvorov, in Proc. Third Euro-Ceramics (Faenza Editrice, Iberia S L, 1993) pp 229-234.
6. F. Azough, P Setasuwon and R Freer, Ceramic Transactions vol 53, 215-227 (1995).
7. C J Rawn, Thesis University of Arizona, Tucson Arizona (1995).
8. B W Hakki and R D Coleman, IER Trans on MTT, **MTT-8**, 402-410 (1960)
9. D Suvorov, M Valent and D Kolar, Ceramic Transactions vol 53, 197-207 (1995).
10. R S Roth, F Beach, A Santoro and K Davis, Abstract 07.9.9, 14th Int. Conf. On Crystallog., Perth, Australia August 1987.
11. F Azough, PE Champness and R Freer, J Appl. Cryst., **28**, 577-581 (1995).
12. T S Bush, J D Gale and CRA Catlow, J. Mater. Chem., **4**, 831 (1994).
13. GV Lewis, PhD Thesis University of London (1984).
14. H Ohsato, S Nishigaki and T Okuda, Jpn. J Appl. Phys. **31** 3136-3138 (1992).
15. K Fukada, R Kitoh and I Awai, J Mater. Res. **10**, 312 (1995).

* Dr Smith is at the IRC for Materials for High Performance Applications, University of Birmingham, Birmingham B15 2TT, UK

PYROELECTRIC AND DIELECTRIC PROPERTIES OF Pb(Sn$_{0.5}$Sb$_{0.5}$)O$_3$ -PbTiO$_3$-PbZrO$_3$ SYSTEM FOR IR SENSOR APPLICATION

Hak-in Hwang, Joon-shik Park,

KETI, PyungTaek, 451-860, Korea

Keun-ho Auh,

Han-yang Univ., Seoul, 133-791, Korea.

ABSTRACT

Pyroelectric and dielectric properties of $0.05(Pb_{1-x}La_x)(Sn_{0.5}Sb_{0.5})O_3 + yPbTiO_3 + z$ PbZrO$_3$+excess a wt%MnO$_2$ system were investigated. The peak pyroelectric coefficient of rhombohedral rich phase was larger than that of tetragonal rich phase. Pyroelectric coefficient was rapidly changed because of the lower rhombohedral phase transition at 97.3℃ in 0.05PSS+0.11PT+0.84PZ composition. As increasing MnO$_2$ content, dielectric constant and dielectric loss were decreased whereas pyroelectric coefficient was increased. The figures of merit, F$_v$ and F$_d$ were 7.2×10^{-9}C · cmJ^{-1} and 1.5×10^{-7}C · cmJ^{-1}, respectively in 0.05PSS+0.11PT+0.84PZ+0.4wt%MnO$_2$ composition. As increasing substitution content of La$_2$O$_3$, pyroelectric coefficient and dielectric constant were increased whereas dielectric loss was decreased. The figures of merit of $0.05(Pb_{0.990}$ -La$_{0.010}$)(Sn$_{0.5}$Sb$_{0.5}$)O$_3$+0.11PbTiO$_3$+0.84PbZrO$_3$+0.4wt% MnO$_2$ composition, F$_v$ and F$_d$ were 9.30×10^{-9}C · cmJ^{-1} and 1.88×10^{-7}C · cmJ^{-1}, respectively.

INTRODUCTION

Thermal detectors of long wavelength IR which convert the energy of the IR photons into heat, to be detected by a thermally sensitive device such as a thermopile, have been used for many years, and are low cost ambient temperature devices. Until the early-seventies, their performance restricted their range of application. However, research into pyroelectric materials over the last few years has produced a class of low cost, high performance thermal detectors which have brought IR intruder and fire alarms into many commercial and domestic premises, and has recently allowed the demonstration of ambient-temperature solid state thermal imagers with performance which compete with cooled technologies.[1] The good several attributes of a ceramic materials, namely ease of manufacture and processing, stability and uniformity, make this materials a particularly useful one for commercial IR detectors.[2] In this experiment, dielectric and pyroelectric properties of $0.05(Pb_{1-x}La_x)(Sn_{0.5}Sb_{0.5})O_3+ yPbTiO_3+ z PbZrO_3+$excess a wt%MnO_2, as a promising pyroelectric materials for IR sensor application were investigated.

EXPERIMENTAL PROCEDURE

Table 1 shows the fabricated compositions. In batch A series, the ratio of PZ/PT was changed from 0.45/0.50 to 0.84/0.11. In batch B series, Pb-ion was substituted by La-ion from 0.002 mole to 0.100 mole. In batch C series, excess MnO_2 as a sintering agent was added from 0 wt% to 0.60 wt%.

Figure 1 shows the fabrication process. Starting materials were weighed using electronic balance, and grinded and mixed by the 1st attrition milling using ZrO_2 ball for 2hr. The 1st milled powder was calcined at 850℃ for 2hrs, and milled again by the 2nd attrition milling for 2hrs. The average particle size after the 2nd attrition milling was 0.44μm. Particle size was measured using laser method (Mastersizer Micro and Microplus, Malvern Co.). Before sintering process, binder was burned out at 600℃ for 1hr. Then sintering process was at 1250℃ for 2hrs in PbO atmosphere. Microstructures were observed by Scanning Electron Microscope (SEM). Sintered samples were completely poled at 50kV$_{DC}$/cm and 140℃ in silicon oil bath. Dielectric properties were measured by impedance and gain phase analyzer (HP4194A).

Dielectric Ceramic Materials

Table I. Fabricated compositions

| Batch No. | Mole fraction (mole) | | | | | | additive | Compositions |
| | A-site | | B-site | | | | MnO$_2$ | |
	PbO	La$_2$O$_3$	SnO$_2$	Sb$_2$O$_3$	TiO$_2$	ZrO$_2$	(wt%)	
A1	1.000	-	0.025	0.025	0.110	0.840	0.40	Pb(Sn$_{0.025}$Sb$_{0.025}$Ti$_{0.11}$Zr$_{0.84}$)O$_3$+0.4wt%MnO$_2$
A2	↑	-	↑	↑	0.150	0.800	↑	Pb(Sn$_{0.025}$Sb$_{0.025}$Ti$_{0.15}$Zr$_{0.80}$)O$_3$+0.4wt%MnO$_2$
A3	↑	-	↑	↑	0.250	0.700	↑	Pb(Sn$_{0.025}$Sb$_{0.025}$Ti$_{0.25}$Zr$_{0.70}$)O$_3$+0.4wt%MnO$_2$
A4	↑	-	↑	↑	0.300	0.650	↑	Pb(Sn$_{0.025}$Sb$_{0.025}$Ti$_{0.30}$Zr$_{0.65}$)O$_3$+0.4wt%MnO$_2$
A5	↑	-	↑	↑	0.350	0.600	↑	Pb(Sn$_{0.025}$Sb$_{0.025}$Ti$_{0.35}$Zr$_{0.60}$)O$_3$+0.4wt%MnO$_2$
A6	↑	-	↑	↑	0.400	0.550	↑	Pb(Sn$_{0.025}$Sb$_{0.025}$Ti$_{0.40}$Zr$_{0.55}$)O$_3$+0.4wt%MnO$_2$
A7	↑	-	↑	↑	0.450	0.500	↑	Pb(Sn$_{0.025}$Sb$_{0.025}$Ti$_{0.45}$Zr$_{0.50}$)O$_3$+0.4wt%MnO$_2$
A8	↑	-	↑	↑	0.500	0.450	↑	Pb(Sn$_{0.025}$Sb$_{0.025}$Ti$_{0.50}$Zr$_{0.45}$)O$_3$+0.4wt%MnO$_2$
B1	0.999	0.001	↑	↑	0.110	0.840	↑	(Pb$_{0.998}$La$_{0.002}$)(Sn$_{0.025}$Sb$_{0.025}$Ti$_{0.11}$Zr$_{0.84}$)O$_3$+0.4wt%MnO$_2$
B2	0.997	0.003	↑	↑	↑	↑	↑	(Pb$_{0.994}$La$_{0.006}$)(Sn$_{0.025}$Sb$_{0.025}$Ti$_{0.11}$Zr$_{0.84}$)O$_3$+0.4wt%MnO$_2$
B3	0.995	0.005	↑	↑	↑	↑	↑	(Pb$_{0.990}$La$_{0.010}$)(Sn$_{0.025}$Sb$_{0.025}$Ti$_{0.11}$Zr$_{0.84}$)O$_3$+0.4wt%MnO$_2$
B4	0.993	0.007	↑	↑	↑	↑	↑	(Pb$_{0.986}$La$_{0.014}$)(Sn$_{0.025}$Sb$_{0.025}$Ti$_{0.11}$Zr$_{0.84}$)O$_3$+0.4wt%MnO$_2$
B5	0.990	0.010	↑	↑	↑	↑	↑	(Pb$_{0.980}$La$_{0.020}$)(Sn$_{0.025}$Sb$_{0.025}$Ti$_{0.11}$Zr$_{0.84}$)O$_3$+0.4wt%MnO$_2$
B6	0.970	0.030	↑	↑	↑	↑	↑	(Pb$_{0.940}$La$_{0.060}$)(Sn$_{0.025}$Sb$_{0.025}$Ti$_{0.11}$Zr$_{0.84}$)O$_3$+0.4wt%MnO$_2$
B7	0.950	0.050	↑	↑	↑	↑	↑	(Pb$_{0.900}$La$_{0.100}$)(Sn$_{0.025}$Sb$_{0.025}$Ti$_{0.11}$Zr$_{0.84}$)O$_3$+0.4wt%MnO$_2$
C1	1.000	-	↑	↑	↑	↑	0.00	Pb(Sn$_{0.025}$Sb$_{0.025}$Ti$_{0.11}$Zr$_{0.84}$)O$_3$+0.0wt%MnO$_2$
C2	↑	-	↑	↑	↑	↑	0.15	Pb(Sn$_{0.025}$Sb$_{0.025}$Ti$_{0.11}$Zr$_{0.84}$)O$_3$+0.15wt%MnO$_2$
C3	↑	-	↑	↑	↑	↑	0.25	Pb(Sn$_{0.025}$Sb$_{0.025}$Ti$_{0.11}$Zr$_{0.84}$)O$_3$+0.25wt%MnO$_2$
C4	↑	-	↑	↑	↑	↑	0.30	Pb(Sn$_{0.025}$Sb$_{0.025}$Ti$_{0.11}$Zr$_{0.84}$)O$_3$+0.30wt%MnO$_2$
C5	↑	-	↑	↑	↑	↑	0.40	Pb(Sn$_{0.025}$Sb$_{0.025}$Ti$_{0.11}$Zr$_{0.84}$)O$_3$+0.40wt%MnO$_2$
C6	↑	-	↑	↑	↑	↑	0.50	Pb(Sn$_{0.025}$Sb$_{0.025}$Ti$_{0.11}$Zr$_{0.84}$)O$_3$+0.50wt%MnO$_2$
C7	↑	-	↑	↑	↑	↑	0.60	Pb(Sn$_{0.025}$Sb$_{0.025}$Ti$_{0.11}$Zr$_{0.84}$)O$_3$+0.60wt%MnO$_2$

And pyroelectric current was measured by static method using computer controlled system.[3] And pyroelectric coefficient, spontaneous polarization and phase transition temperature were obtained from measured results.

RESULTS AND DISCUSSIONS

1. The properties of compositions with the ratio of PT to PZ (Batch A series)

The dielectric and pyroelectric properties of $0.05(Pb_{1-x}La_x)(Sn_{0.5}Sb_{0.5})O_3$+ yPbTiO$_3$+ z PbZrO$_3$+excess α wt%MnO$_2$ system with x=0, $0.11 \leq y \leq 0.50$ and $0.45 \leq z \leq 0.84$, and α =0.4 were investigated. Figure 2 shows the dielectric constant and dielectric loss at 1kHz of batch A series. In the rhombohedral rich phase, the dielectric constant was decreased. Dielectric loss was lower than 0.5% in the all compositions.

Dielectric Ceramic Materials

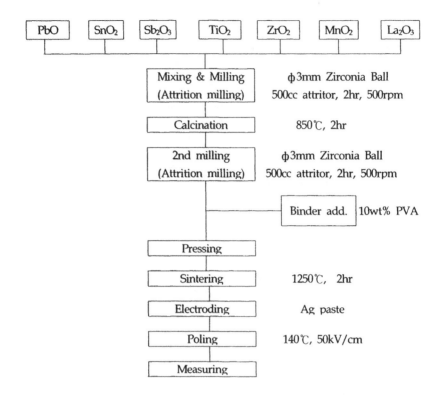

Fig. 1 Fabrication process

Figure 3 shows pyroelectric coefficient and spontaneous polarization with temperature. The temperature of phase transition was decreased and pyroelectric coefficient was increased with increasing of PZ content. The peak of pyroelectric coefficient were increased from $0.0258C/m^2K$ at PT/PZ=40/55 to $0.0541C/m^2K$ at PT/PZ=11/84 because of large variation of spontaneous polarization in the rhombohedral phase region. At the composition of PT/PZ=0.11/0.84, pyroelectric coefficient was $0.001911C/m^2K$ at 97.3℃ of lower transition. Similar results were reported by other reports.[4,5] It is possible to use these figures of merit as a rough guide to the selection of materials. The voltage response for a pyroelectric element feeding into a high input impendence, unity gain amplifier (such as a source follower FET) is proportional to F_v, where $F_v = p/(C \varepsilon \varepsilon_0)$. And the signal to noise ratio for a

Dielectric Ceramic Materials

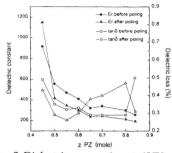
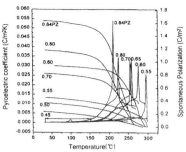

Fig. 2 Dielectric properties at 1HKz
with z of 0.05PSS-yPT-zPZ
+0.4wt%MnO₂

Fig. 3 Pyroelectric properties with z
of 0.05PSS-yPT-zPZ+0.4wt%MnO₂

voltage detector depends on the nature of dominant noise source. It is usually found that the noise is dominated by the Johnson noise by the lossy capacitance of the detector element. In this case the detectivity is proportional to F_d, where $F_d = p/(C \varepsilon \varepsilon_0 \tan \delta)$.[6] Figure 4 shows the calculated F_v and F_d from measured pyroelectric coefficient (p), volume heat capacity (C), dielectric constant (ε) and dielectric loss ($\tan \delta$). The F_v and F_d of rhombohedral rich composition were higher than that of the tetragonal rich composition. So, the composition of rhombohedral rich phase such as 0.80 PZ was good as a pyroelectric IR sensor materials.

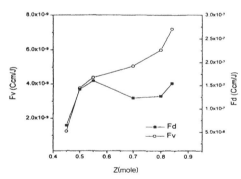

Fig. 4 F_v and F_d with z in 0.05PSS-yPT-zPZ+0.4wt%MnO₂

2. The properties of compositions with MnO₂ additive (Batch C series)

Microstructures, dielectric and pyroelectric properties of $0.05(Pb_{1-x}La_x)$ $(Sn_{0.5}Sb_{0.5})O_3+ y$

PbTiO$_3$+ zPbZrO$_3$+excess α wt%MnO$_2$ system with x=0, y=0.11, z=0.84, 0.0$\leq \alpha \leq$0.6 were investigated. Figure 5 shows SEM Microstructures of fracture surface of MnO$_2$ added Pb(Sn$_{0.025}$Sb$_{0.025}$Ti$_{0.11}$Zr$_{0.84}$)O$_3$+(0.0~0.6)wt%MnO$_2$ system sintered at 1250℃ for 2hrs. As adding MnO$_2$, the liquid phase sintering was observed and grain size was increased. In case of composition over 0.5wt% MnO$_2$ content, liquid phase was existed at grain boundary. Figure 6 shows dielectric constant and dielectric loss of the batch C series sintered at 1250℃ for 2hrs in PbO atmosphere. Dielectric constant was decreased with MnO$_2$ addition and was the lowest value at 0.5wt%MnO$_2$ content. Oxygen vacancy is formed by substitution by Mn^{+2} or Mn^{+3} for Pb^{+2} to compensate for charge valance. Dielectric constant and loss were decreased by the role of domain stabilizer in sintered body by the formed Mn-V$_o$ bonding.[8]

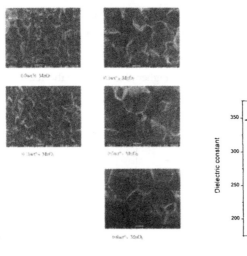

Fig. 5 Microstructures of 0.05PSS   Fig. 6 Dielectric properties of 0.05PSS
　　+0.11PT+0.84PZ with MnO$_2$.　　　　 -0.11PT-0.84PZ with MnO$_2$

Figure 7 shows pyroelectric coefficient and spontaneous polarization with temperature. The temperature of lower rhombohedral phase transition was decreased and the peak pyroelectric coefficient became higher with increasing of MnO$_2$ content. Because of the high peak pyroelectric coefficient in the case of 0.5wt%MnO$_2$ content the

Dielectric Ceramic Materials

pyroelectric coefficient in the room temperature was increased. The spontaneous polarization was the highest value at 0.5wt%MnO$_2$ content and rapidly decreased at 0.6wt% MnO$_2$ because of second phase at grain boundary. Figure 8 shows the calculated results of F$_v$ and F$_d$. F$_v$ was increased with MnO$_2$ content, F$_d$ was increased to 0.4wt%MnO$_2$, and then decreased. From these results, excess 0.4wt%MnO$_2$ added Pb(Sn$_{0.025}$Sb$_{0.025}$Ti$_{0.11}$Zr$_{0.84}$)O$_3$ system was good as a pyroelectric IR sensor materials.

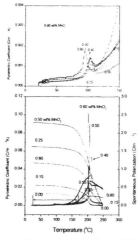

Fig. 7 Pyroelectric properties of 0.05PSS -0.11PT-0.84PZ with MnO$_2$

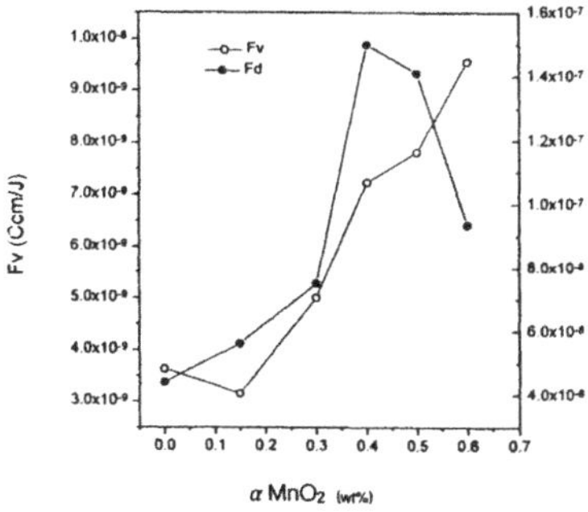

Fig. 8 F$_v$ and F$_d$ of 0.05PSS-0.11PT -0.84PZ with MnO$_2$.

3. The properties of compositions doped with La$_2$O$_3$ (Batch B series)

Microstructures, dielectric and pyroelectric properties of 0.05(Pb$_{1-x}$La$_x$)(Sn$_{0.5}$Sb$_{0.5}$)O$_3$+ yPbTiO$_3$+ zPbZrO$_3$+excess α wt%MnO$_2$ system with 0.002≤x≤0.100, y=0.11, z=0.84, α=0.40 system were investigated. Figure 9 shows SEM microstructures of batch B series sintered at 1250℃ for 2hrs. Grain size was decreased with La-ion substitution. Figure 10 shows the dielectric constant and dielectric loss at 1kHz. Dielectric constant was increased and dielectric loss was decreased with La-ion substitution. It is explained by charge neutrality and the grain size effect. The cation deficiency of Sb^{+3} in B site of ABO$_3$ structure was compensated by La^{+3} in A site.

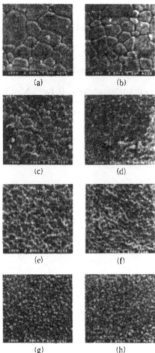

Fig. 9 Microstructures of 0.05(Pb$_{1-x}$La$_x$)
(Sn$_{0.5}$Sb$_{0.5}$)O$_3$-0.11PT-0.84PZ
+0.4wt%MnO$_2$ with x.

(a) x=0.000 (b) x=0.002 (c) x=0.006
(d) x=0.010 (e) x=0.014 (f) x=0.020
(g) x=0.060 (h) x=0.100

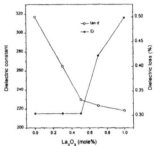

Fig. 10 Dielectric properties of 0.05(Pb$_{1-x}$La$_x$)
(Sn$_{0.5}$Sb$_{0.5}$)O$_3$-0.11PT-0.84PZ+0.4wt%
MnO$_2$ with La$_2$O$_3$

Figure 11 shows pyroelectric coefficient with temperature. The temperature of phase transition was decreased and pyroelectric coefficient was decreased with La-ion substitution over 0.3 mole% La$_2$O$_3$. The rhombohedral phase transition temperature was increased with La-ion substitution. And the peak pyroelectric coefficient was decreased with La$_2$O$_3$ and the peak was disappeared more than 1.0mole%La$_2$O$_3$. The peak pyroelectric coefficient was the largest value at 0.7mole% La$_2$O$_3$. Figure 12 shows the calculated results of F$_v$ and F$_d$. The F$_v$ and F$_d$ were increased with La$_2$O$_3$ and were the highest values at 0.5mole%La$_2$O$_3$.

Dielectric Ceramic Materials

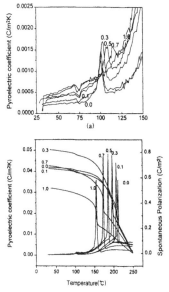

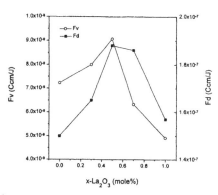

Fig. 11 Pyroelectric properties of $0.05(Pb_{1-x}La_x)$ $(Sn_{0.5}Sb_{0.5})O_3$-0.11PT-0.84PZ+0.4wt%MnO$_2$ with La$_2$O$_3$

Fig. 12 F_v and F_d of $0.05(Pb_{1-x}La_x)$- $(Sn_{0.5}Sb_{0.5})O_3$+0.11PT+0.84PZ +0.4wt%MnO$_2$ with La$_2$O$_3$

CONCLUSIONS

The relationship between microstructures, dielectric properties and pyroelectric properties was studied on the $0.05(Pb_{1-x}La_{x=0.0-0.05})$ $(Sn_{0.5}Sb_{0.5})O_3$ + (0.11-0.50) PbTiO$_3$ +(0.84-0.45)PbZrO$_3$+(0.0-0.6)wt%MnO$_2$ system expected to be a good pyroelectrics. In order to analyze the effects of MnO$_2$ as a sintering agent and La$_2$O$_3$ as a softner, the microstructure was investigated by scanning electron microscope, and the dielectric and pyroelectric properties were measured. It was found that rhombohedral phase was observed, when the PT/PZ mole ratio of specimen was lower than 0.43/0.52. The pyroelectric coefficient of rhombohedral-rich phase was larger than that of tetragonal-rich phase and the pyroelectric coefficient was rapidly changed by the phase transition at 97.3℃ in the 0.05PSS+0.11PT+0.84PZ composition. As increasing MnO$_2$ content, dielectric constant and dielectric loss were decreased whereas pyroelectric coefficient was increased. However, the phase transition temperature was not affected by the concentration of MnO$_2$. The figures of merit F_v and F_d in the

0.05PSS+0.11PT +0.84PZ+0.4wt%MnO$_2$composition was 7.2×10^{-9} C $\cdot$ cmJ^{-1} and 1.5×10^{-7}C $\cdot$ cmJ^{-1}, respectively. As increasing substitution content of La$_2$O$_3$, the pyroelectric coefficient and the dielectric constant were increased whereas dielectric loss and phase transition temperature were decreased. The figures of merit F$_v$ and F$_d$ in the 0.05(Pb$_{0.995}$La$_{0.005}$)(Sn$_{0.5}$Sb$_{0.5}$)O$_3$+0.11PbTiO$_3$+0.84PbZrO$_3$+0.4 wt%MnO$_2$ composition are 9.03×10^{-9}C $\cdot$ cmJ^{-1} and 1.88×10^{-7}C $\cdot$ cmJ^{-1}, respectively.

REFERENCES

[1]R.W. Whatmore, "Pyroelectric Ceramics and Devices for Thermal Infra-red Detection and Imaging", Ferroelectrics, 1991, vol.118, pp.241-259.

[2]R.W. Whatmore, P.C. Osbond and N.M. Shorrocks, "Ferroelectric Materials for Thermal IR Detectors", Ferroelectrics, 1987, vol.76, pp.351-367.

[3]M.Shimhony, A.Shaulov, "Measurement of the Pyroelectric Coefficient and Permittivity from the Pyroelectric Response to Step Radiation Singnals in Ferroelectrics", Appl. Phys. Lett., Vol. 21, pp. 375-377, 1972.

[4]M.Toyoda, M.Ishida and Y.Hirose, "Temperature Dependence of Pyroelectric Responsivity in Barium Titanate and Barium Strontium Titanate Ceramics", Memories of the Fac. Eng. Kobe Univ., 1977, vol.23, pp.189-193.

[5]R. R. Zeyfabg, W.H.Sehr and K.V.Kiehl, "Enhanced Pyroelectric Properties at a F.E.-F.E. Phase Transition", Ferroelectrics, 1976, vol.11, pp.355-358.

[6]R.W. Whatmore, "Ferroelectric Materials for Thermal IR Sensors, State Of The Art and Perspectives", Ferroelctrics, 1990. Vol.104, pp. 269-283

[7]R.Clark, A.M.Glazer, "Phase Transitions in Lead Zirconate-Titanate and Their Application in Thermal Detectors", Ferroelectrics, 1976, vol.11, pp.359-364.

[8]R.B.Atkin, R.M. Fulrath, "Point Defects and Sintering of Lead Zirconate-Titanate", Journal of the American Ceramamic Society, 1971, vol.54, pp.265-270.

INFLUENCE OF THE REACTIVITY OF INORGANIC FILLERS TO A GLASS MATRIX ON THE PROPERTIES OF GLASS-CERAMIC COMPOSITES

H.Mizutani*, M.Sato, H.Yokoi, K.Ohbayashi, and S.Iio
R&D Center
NGK Spark Plug Co., Ltd.
2808, Iwasaki, Komaki, Aichi 485-8510, Japan

ABSTRACT

Glass-ceramic composites were prepared from an alkaline and lead-free glass and various inorganic fillers. The purpose of this study was to reveal the influence of the chemical reactions between the glass matrix and the inorganic fillers during sintering on the final dielectric properties.

When TiO_2 (rutile) was used as a filler, no chemical reaction occurred and the relative permittivity and the temperature coefficient of resonant frequency simply depended on the nature and the volume fraction of the filler. The use of $SrTiO_3$ as a filler resulted in a decrease of the dielectric loss of composites. This phenomenon was attributed to crystallization of the residual glass due to the chemical reaction during sintering. The particle size of fillers was an important factor to control the dielectric properties.

INTRODUCTION

The properties of glass-ceramic composites can be controlled by selecting of a glass and fillers. The properties of composite materials can be approximated by a mixing rule when the properties of the component materials have been predetermined[1]. It has been reported that the thermal coefficient of expansion of glass-ceramic composites can be controlled by changing the glass/ceramic mixture ratio without the secondary crystal formation.[2]

The glass-ceramic composites are widely used as low temperature co-fired substrates which have been developed by several companies aiming at the application to microelectronics packages[3-5]. The low temperature co-fired materials have excellent features: a low sintering temperature below 1000°C to permit co-firing with a highly conductive metal such as Ag and Cu, a low thermal coefficient of expansion to match that of silicon, and a low dielectric constant. These glass-ceramic composites have been also used as the multilayered devices for the communication system according to the requirements of miniaturization. For this electronic components, the temperature coefficient of resonant frequency (TCF) of the materials is a very important factor because the resonant frequency changes depending on temperature from the practical point of view.

This paper describes the influence of the reactivity of inorganic fillers to a glass matrix on the dielectric properties and the development of a glass-ceramic composite which satisfies the requirement of TCF (within ±10 ppm/°C).

EXPERIMENTAL PROCEDURE

1. Sample Preparation

The commercially available powders of $CaCO_3$, $Al(OH)_3$, SiO_2, $MgCO_3$, H_3BO_3, and ZrO_2 were weighted and mixed. The mixture was melted in a platinum crucible at 1450°C for 5 hours and then, dropped into cold water to form a frit. The obtained frit was crushed to a powder of 3 μm in the average size with an alumina ball mill in ethanol. The resulting slurry was dried.

Next, the prepared glass powder of the CaO-Al_2O_3-SiO_2-MgO-B_2O_3-ZrO_2 system and commercially available inorganic fillers were mixed in ethanol for 5 hours. After drying, the mixtures were pressed into green compacts with dimensions of 20mm in diameter and 12mm in thickness or with dimensions of 40mm $\times$ 50mm $\times$ 40 mm under a compaction pressure of 80MPa and CIPed under a pressure of 150MPa. These green compacts were sintered at 850~950°C for 30min in air. The both end surfaces of the disk samples were ground and the rectangular samples were ground into dimensions of 30mm×40mm×35 mm using a #200 diamond wheel.

2. Characterization of Glass-Ceramic Composites

The relative permittivity, ε_r and the $Q \cdot f$ value were measured in the TE_{011} resonant mode by means of Hakki-Coleman's dielectric resonator method.[6] The temperature coefficient of resonant frequency was measured in the temperature range of $30 \sim 80°C$. The crystalline phases were identified by X-ray powder diffraction (XRD) using Cu-Kα radiation. The flexural strength was evaluated by a three-point bending method. Thermal coefficient of expansion (TCE) was measured with a dilatometer in the range of $30 \sim 400°C$ using a fused silica as a standard. The dimensions of the typical thermal expansion specimen were 3mm × 3mm × 20mm. The microstructures of the glass-ceramic composites were observed by scanning electron microscopy (SEM).

RESULTS AND DISCUSSION

1. Addition of the Al_2O_3 Filler

Figure 1 shows the effect of Al_2O_3 addition on the flexural strength and the thermal coefficient of expansion. With an increase of the Al_2O_3 content, the thermal coefficient of expansion increased as expected from the mixture rule. The flexural strength also increased up to 300MPa because the dispersed ceramic particles (Al_2O_3 in this case) restrained the size of Griffith crack.[7] The addition exceeding 30vol% resulted in the flexural strength deterioration. This was attributed to an increase of number of internal pores from the SEM observation.

Figure 2 shows the microstructure of the glass-Al_2O_3 composite fired at 900°C. The Al_2O_3 particles were uniformly dispersed in the glass matrix. The crystalline phases, Al_2O_3 and anorthite ($CaAl_2Si_2O_8$) resulted from the crystallization of the glass matrix during sintering were identified by XRD. The SEM observation and XRD analysis revealed that no chemical reaction occurred between the glass matrix and Al_2O_3. The glass-ceramic composite had a simple

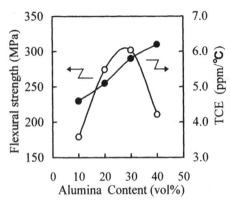
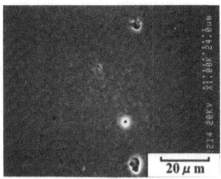

Fig.1 Effect of the Al₂O₃ content on
flexural strength and TCE

Fig.2 Microstructure of the sintered
glass - alumina composite

physical mixing state.

2.Addition of Inorganic Fillers Having Positive TCF

Because the glass-Al₂O₃ composite had a negative TCF, -70ppm/°C, it was not suitable for the application to electronic components. Therefore, the various fillers having a positive TCF were added to the glass matrix to improve its TCF. To estimate the TCF of glass-ceramic composites consisting of the filler particles and the glass matrix, the following simple mixture rule was applied:

$$\tau_{fc} = V_m \cdot \tau_{fm} + V_p \cdot \tau_{fp} \qquad \cdots\cdots\cdots \quad (1)$$

where τ_f is the TCF, V is the volume fraction and the subscripts c, m, and p represent composite, matrix, and filler particles, respectively.

Figure 3 shows that the effect of addition of the various secondary fillers to the glass-Al₂O₃ composite on TCF, where the total filler content of Al₂O₃ and the secondary was fixed at 20vol%. The solid lines indicate the values calculated from the equation (1) for each filler. When TiO₂ (τ_f = +450) or Bi₄Ti₃O₄ (+70) was used as a secondary filler, the experimental results showed good agreement with the calculated. On the other hand, when CaTiO₃ (+800), SrTiO₃ (+1670) or SrSnO₃ (+1850) was used as a secondary filler, the experimentally obtained value was much lower than expected one from the calculation.

Dielectric Ceramic Materials

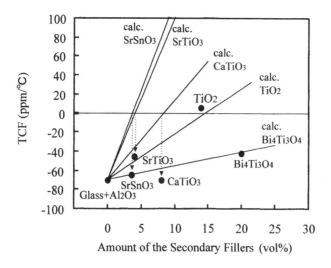

Fig.3 Effects of the various secondary fillers on the TCF

Figure 4 shows the microstructures of the glass-ceramic composites containing TiO_2 or $CaTiO_3$ as a secondary filler.

When Al_2O_3 and TiO_2 were used as fillers, both particles remained. In contrast, when Al_2O_3 and $CaTiO_3$ were used as fillers, only the Al_2O_3 particles remained. These results suggested that the fillers such as $CaTiO_3$, $SrTiO_3$, and $SrSnO_3$, of which the addition resulted in large deterioration of TCF from the calculation, reacted with the glass matrix and did not remain in the composites.

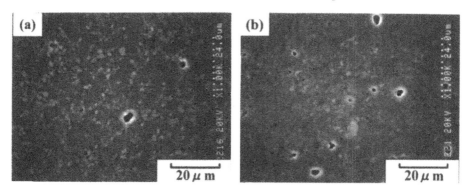

Fig.4 Microstructures of the glass-ceramic composites:(a) Addition of Al_2O_3 and TiO_2 , and (b) Addition of Al_2O_3 and $CaTiO_3$

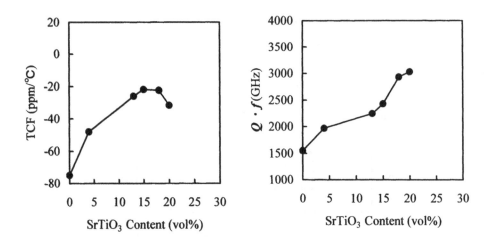

Fig.5 Effect of SrTiO₃ content on the TCF and the $Q \cdot f$ value

3. Influence of the Particle Size of SrTiO₃ Fillers on the Dielectric Properties

To clarify the influence of the particle size of fillers on dielectric properties, two types of SrTiO₃ were tested ; one was as received SrTiO₃ (D_{50}=1.73 μm) and the other was thermally treated SrTiO₃ (D_{50}=15.15 μm).

Figure 5 shows that the effects of as received SrTiO₃ addition on TCF and $Q \cdot f$ value of the sintered glass-ceramic composites. With an increase of the SrTiO₃ content, the TCF and the $Q \cdot f$ value were improved up to -20ppm/°C and 3000 GHz, respectively.

When the thermally treated SrTiO₃ was used as a filler, TCF dramatically increased with an increase of SrTiO₃ content (Fig.6). This was attributed to the remains of SrTiO₃ particles in the glass-ceramic composite as shown in Fig. 7. The $Q \cdot f$ value of the glass-ceramic composite used the thermally treated SrTiO₃ as a filler was lower than that of the glass-composite used as received SrTiO₃. These results indicated that the chemical reaction between the glass matrix and the SrTiO₃ particles was not completed during sintering.

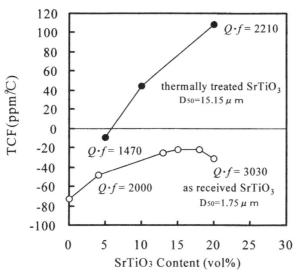

Fig.6 Influence of particle sizes of $SrTiO_3$ on TCF

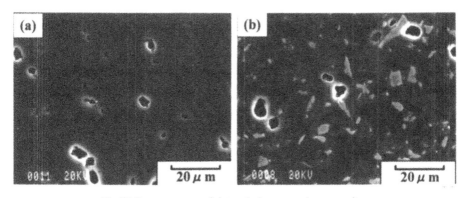

Fig.7 Microstructures of sintered glass-ceramic composites:
(a) Addition of as received $SrTiO_3$ and (b) Addition of thermally treated $SrTiO_3$

Figure 8 shows XRD patterns of the sintered glass-ceramic composites prepared using two types of $SrTiO_3$ as a filler. When as received $SrTiO_3$ was used as a filler, $CaAl_2Si_2O_8$, $SrAl_2Si_2O_8$, and $MgTi_2O_5$ formed and no $SrTiO_3$ remained in the glass-ceramic composite. On the other hand, when thermally treated $SrTiO_3$ was used as a filler, $SrTiO_3$ remained in the glass-ceramic composite.

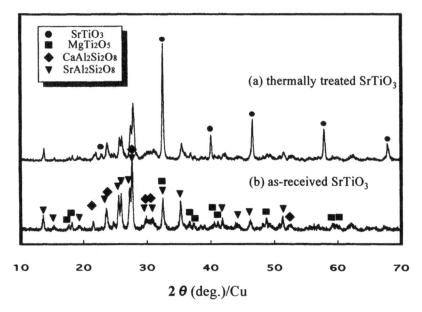

2 θ (deg.)/Cu

Fig.8 XRD patterns of the sintered glass-ceramic composites:
(a) Addition of thermally treated SrTiO$_3$ and (b) Addition of as received SrTiO$_3$

These results implied that the improvement of $Q \cdot f$ value was attributed to the crystallization of the residual glass (the formation of SrAl$_2$Si$_2$O$_8$) due to the reaction between the glass matrix and SrTiO$_3$ particles during sintering.

4. Controlling TCF of the Glass-Ceramic Composite

The TCF of the glass ceramic composite used as received SrTiO$_3$ was improved up to -20ppm/°C but not enough for the application to electronic components. Figure 9 shows effects of TiO$_2$ content on the dielectric properties of glass-ceramic composites where the total filler content of as received SrTiO$_3$ and TiO$_2$ was fixed at 20vol%. With an increase of TiO$_2$ content, the relative permittivity increased and at the TiO$_2$ content between 2vol% and 3vol%, the TCF could be controlled within ±10ppm/°C. The obtained glass-ceramic composites are suitable to the application of electronic components because of their excellent $Q \cdot f$ values and TCF.

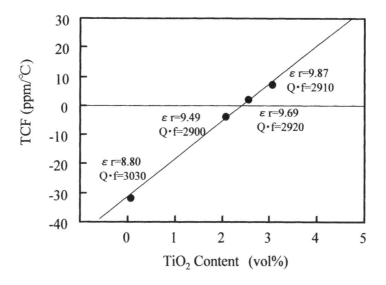

Fig.9 Effect of TiO₂ content on the dielectric properties

CONCLUSIONS

The influence of the reactivity of inorganic fillers to the glass matrix on the properties of glass-ceramic composites was investigated.

When TiO_2 was used as a filler, no chemical reaction occurred and the dielectric properties of glass-ceramic composites simply depended on the nature and the volume fraction of the filler. The use of $SrTiO_3$ as a filler resulted in the improvement of $Q \cdot f$ value up to 3000GHz. This phenomenon was attributed to the crystallization of the residual glass (the formation of $SrAl_2Si_2O_8$) due to the chemical reaction between the glass matrix and the filler during sintering.

It was found that when the materials of which reactivities were high such as $SrTiO_3$ were added as a filler, the dielectric properties of the glass-ceramic composite are greatly influenced by the particle size of the fillers.

The glass-ceramic composites which were available for the application to electronic components were obtained by simultaneously using $SrTiO_3$ and TiO_2 as fillers.

REFERENCES

[1] Imanaka. Y, Aoki S, Kamahara N, and Niwa K, "Crystallization of Low Temperature Fired Glass-Ceramic Composite, "Yogyo-Kyokaishi, **95** [11] 1119-1121 (1987)

[2] Niwa K, Kamahara N, Yokouchi K, and Imanaka Y, "Multilayer Ceramic Circuit Board With a Copper Conductor," Adv. Ceram. Mater., **2** [4] 832-835 (1987)

[3] R.R. Tummala, "Ceramic and Glass-Ceramic Packaging in the 1990s, "J. Am. Ceram. Soc., **74** [5] 895-908 (1991)

[4] P.C. Donohue, S. Gallo and A. Dow, "High Reliability Copper MCM System," ISHM Proceeding, 607-612 (1992)

[5] Nishigaki S., "Low Temperature Co-fireable Ceramics for Hibrid IC and Role of Glass," New Glass, Vol.4, No2, 37-49 (1989)

[6] B.W. Hakki and P.D. Coleman, "A Dielectric Resonator Method of Measuring Inductive Capacitors in the Millimeter Range," IRE Trans. Microwave Theory Tech., MTT-8, 402-410 (1960)

[7] A.A. Griffith, "The Phenomena of Rupture and Flow in Solids," Philos. Trans. Roy. Soc. Lond., Vol.A221, 163 (1920)

STRONTIUM-LEAD TITANATE BASED COMPOSITE THERMISTOR MATERIALS

Zhilun Gui, Dejun Wang, Jun Qiu, and Longtu Li,
Dept. of Mater. Sci.&Eng., State Key Lab. of New Ceramics & Fine Processing,
Tsinghua University, Beijing 100084, CHINA

Yttrium-doped $(Sr_{0.45}Pb_{0.55})TiO_3$ ceramics were prepared by liquid chemical method and the conduction mechanism was studied by the complex impedance analysis. The materials exhibited a significantly large negative temperature coefficient of resistivity below Tc in addition to the ordinary PTC characteristics above Tc. The minimum resistivity was at the magnitude of $10^2 \Omega \cdot cm$ and the negative temperature coefficient of resistivity was better than $-3\%^\circ C^{-1}$. Complex impedance analysis indicated that the strong NTC effect was not originated from the deep energy level of donor (bulk behavior), but from the electrical behavior of the grain boundary. Furthermore, the NTC behavior in the materials was depended on the composition and microstructure of the grain boundary. The NTC-PTC ceramics was a grain boundary controlled materials.

I. INTRODUCTION

It is well known that perovskite PTC materials consist of semiconducting barium titanate and its solid solutions with strontium titanate and lead titanate,[1] which exhibit a rapidly increasing resistivity when heated above Tc and an almost constant resistivity below Tc. $(Sr,Pb)TiO_3$ ceramics is a new perovskite ferroelectric semiconductor with NTC–PTC (V-type PTC) characteristics,[2,3] which can fulfill multiple functions such as temperature control and overcurrent protection etc. Therefore its research has great practical significance,[4] meanwhile, it has aroused great interest in interpreting the strong NTC effect of $(Sr,Pb)TiO_3$ materials.

In the previous literature,[2] it was suggested that the NTC effect was related to the depth of donor energy level and the strong NTC effect was induced by the deep donor energy level, however, it is lack of enough experimental evidence. In this paper, the bulk and grain boundary resistances of the V-type PTC material have been investigated by complex impedance analysis which is an efficient method to study the grain boundary conduction mechanism of the polycrystalline ceramics.

II EXPEPERIMENT PROCEDURE

1. Sample Preparation

The samples with formulation of $(Sr_{0.45}Pb_{0.55})TiO_3$ were prepared from strontium nitrate (>99.5%), lead (II) nitrate (>99%), yttrium nitrate (>99%), and tetrabutyl titanate $(Ti(OC_4H_9)_4,>98\%)$. The weighed nitrates and the tetrabutyl titanate were dissolved in a $CH_3COOH-C_2H_5OH-H_2O$ system. The coprecipitates were obtained by adding an excess amount of oxalic acid dihydrate (>99.8%) into the solution. The chemical reaction was describrd by the following equation:[5]

$$0.45Sr(NO_3)_2 + 0.55Pb(NO_3)_2 + Ti(OC_4H_9)_4 + 2H_2C_2O_4 \cdot 2H_2O + H_2O \rightarrow$$

$$(Sr_{0.45}Pb_{0.55})TiO(C_2O_4)_2 \cdot 4H_2O \downarrow + 4C_4H_9OH + 2HNO_3 \qquad (1)$$

The precipitates were washed, dried, and calcined at 700°C for 45 min in air to obtain $(Sr_{0.45}Pb_{0.55})TiO_3$ powders.

Small amount of ultrafine SiO_2 particles were coated on the surface of $(Sr_{0.45}Pb_{0.55})TiO_3$ powders by surface modification method[6] in order to obtain $SiO_2-(Sr_{0.45}Pb_{0.55})TiO_3$ composite powders.

The $SiO_2-(Sr_{0.45}Pb_{0.55})TiO_3$ composite powders were then pressed into pellets (10mm in diameter and 1mm in thickness) and sintered at 1080 °C~1240°C for 10min~180min to prepare the glass-ceramics composite materials.

2. Measuring Procedure

The surfaces of the $(Sr,Pb)TiO_3$ specimen were coated with In-Ga alloy and the resistance-temperature property was measured with a DC resistance-temperature measuring system, while the applied voltage was generally about 0.5V. The temperature range was 15°C~450°C, the temperature interval of the measurement was 1°C, and the heating rate was 100°C/h.

Complex impedance analysis was measured with a LF impedance analyzer (HP4192A). The frequency range was 5Hz~13MHz, the temperature range was 20°C ~250°C, and the heating rate was 1°C/min.

The microstructure and composition of the materials was analyzed by H9000NAR transmission electron microscope and EDAX/9100 system respectively.

Thermal mechanical analysis was measured by SETARAM92-24 TMA analyzer (SETARAM, France). The atmosphere was oxygen and the heating rate was 10°C/min.

III RESULTS AND DISCUSSION

1 Resistance-temperature Characteristics

A series of (Sr,Pb)TiO₃ materials with various curie temperatures can be obtained by modifying the Sr/Pb ratio. The relation between Sr/Pb ratio and curie temperature is shown in Fig.1. It is can be seen that due to the Pb loss in the sintering process, the experimental results are slightly lower than the theoretical values.

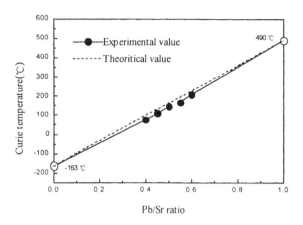

Fig.1 The dependence of curie temperature on Pb/Sr ratio

In the compostion design of (Sr,Pb)TiO₃ material, it is found that both the dopant Y and the additive SiO₂ have remarkable effects on the Resistance-temperature characteristics of (Sr,Pb)TiO₃ ceramics. It is an essential condition to fabricate the V-PTCR material that the compostion of Si and Y must be higher than those of the typical PTCR material.

By modifying the process parameters, the typical R-T curve of V-PTCR materials is obtained, shown in Fig.2 as solid line. The room temperature resistance was 7.8kΩ, the lowest resistance was 83.95Ω and the highest resistance was 3.83MΩ. In the PTC range, the resistance jump was 4.7 orders of magnitude and the temperature coefficient of resistance was +6.61%/°C. In the NTC range, the temperature coefficient of resistance was -3.54%/°C. Besides strong PTC effect above Tc, the sample also presented strong NTC effect below Tc.

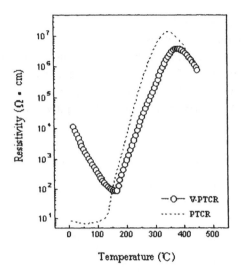

Fig. 2 Resistance-temperature curves of (Sr,Pb)TiO$_3$ ceramics.

2 Bulk and Grain Boundary Resistance

For V-PTCR elements, the simplified equivalent circuit and the Cole-Cole plot are shown in Fig.3 and the corresponding impedance equation (2) is as below:

$$Z' = R_b + \frac{R_{gb}}{1 + \omega^2 \tau_{gb}^2} \; ; \qquad Z'' = \frac{\omega \tau_{gb} R_{gb}}{1 + \omega^2 \tau_{gb}^2} \tag{2}$$

where Z', Z'' represent the real and imaginary part of the impedance respectively; R_b and R_{gb} represent bulk and grain boundary resistance respectively; ω represents angular frequency; τ_{gb} represents relaxation time of the grain boundary. When $\omega \to 0$, $Z' = R_b + R_{gb}$; $\omega \to \infty$, $Z' = R_b$, so we can obtain R_b and R_{gb} from the intercept of Z' of the Cole-Cole plot .

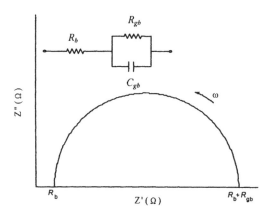

Fig. 3 Simplified equivalent circuit and its Cole-Cole plot.

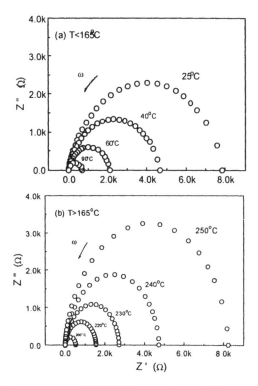

Fig.4 Cole-Cole patterns of V-PTCR ceramics at different temperature.

The Cole-Cole plots of the sample at different temperatures are shown in Fig.4. It can be seen that the variation of the left intercept Rb at the real axis Z' of the plot was small, and the variation of the right intercept $R_{gb}+R_b$ was large, which indicated the large variation of R_{gb}. The right intercept $R_{gb}+R_b$ decreased below Tc (Tc=165°C) and increased rapidly above Tc with the temperature increasing. According to the impedance diagrams of the specimen at different temperatures, the variations of grain resistance R_b and grain boundary resistance R_{gb} with the temperature were obtained in Fig.5. For V-PTC material, grain boundary resistance was large and presented V-type PTC characteristics with the variation of the temperature. When the temperature was below Tc, grain boundary resistance decreased with the temperature increasing and the temperature coefficient is -3.54%/°C; when the temperature was above Tc, grain boundary resistance elevated rapidly with the temperature increasing and the temperature coefficient is +6.61%/°C. It is known that the overall resistance consists of grain resistance and grain boundary resistance. Because grain resistance was small, the overall resistance R was approximately equal to grain boundary resistance R_{gb}. Therefore, It was concluded that V-PTC effect was originated from the electrical behavior of grain boundary.

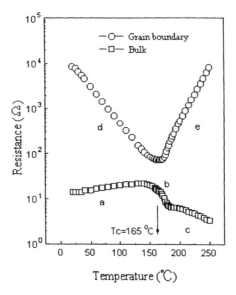

Fig.5 Dependence of Rb and Rgb on temperature

3 Microstructure and Composition

TEM micrographs clearly indicated that the glass phase existed in the grain boundary, which was shown in Fig.6. EDAX analysis of the bulk and grain boundary phase , which was shown in Fig.7, also showed that element Si and Y were rich in glass phase. Therefore, The NTC behavior was closely related to the composition of the grain boundary phase.

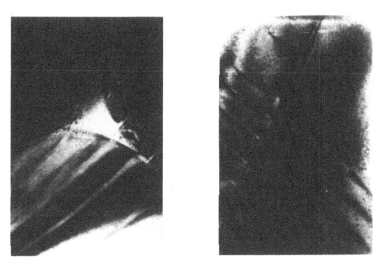

Fig.6 TEM microscopies of $(Sr,Pb)TiO_3$ composite thermistor material

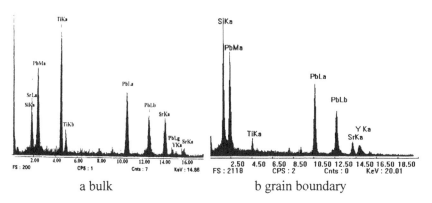

a bulk   b grain boundary

Fig.7 EDAX analysis of bulk and grain boundary phase

4 Thermal Mechanical Analysis

Thermal mechanical analysis also provided an implemental evidence for the above therory. Fig.8 are the TMA analysis results of (Sr,Pb)TiO$_3$ materials with SiO$_2$ additive of 10mol% and 0.3mol% respectively. Unlike the material with single phase structure, V-PTCR material expand in the final stage of the sintering process above 1160°C due to the existence of the glass phase, meanwhile the previous works indicate that the sintering temperature for the V-PTCR material was above 1170°C. Therefore it can be concluded that V-PTCR characteristics was closely related to the formation of the glass phase in the material.

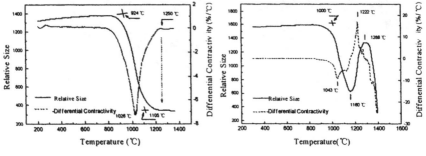

a (Sr,Pb)TiO$_3$ with 0.3mol% SiO$_2$ b (Sr,Pb)TiO$_3$ with 0.3mol% SiO$_2$

Fig.8 Thermal mechanical analysis of (Sr,Pb)TiO$_3$ material

V. CONCLUSIONS

In summary, our study indicates that V-PTC effect was a grain boundary effect, which overturn the theoretical model of deep donor energy level in the previous literature. The NTC-PTC ceramics was a grain boundary controlled materials.

REFERENCES

[1]B.M. Kulwicki, "PTC Materials Technology, 1955-1980"; p.138-54 in Advances in Ceramics, Vol.1, Grain Boundary Phenomena in Electronic Ceramics. Edited by L.M. Levinson. American Ceramic Society, Columbus, OH, 1981.

[2]S. Iwaya, H. Masumura, H. Taguchi, and M. Hamada, "V-type Characteristics of PTC Materials," Electron. Ceram. Jpn., 19, 33 (1988).

[3]C. Lee, I-N Lin, and C.-T. Hu, "Evolution of Microstructure and V-Shaped Positive Temperature Coefficient of Resistivity of (Pb$_{0.6}$Sr$_{0.4}$)TiO$_3$ Materials," J. Am. Ceram. Soc., 77, 1340 (1994).

[4]Longtu Li, Shan Wan, Shiping Zhou, and Zhilun Gui, "Preparation and

Characteristics of (Sr,Pb)TiO$_3$ Ultrafine Particles," Chin. J. Mater. Res., **Suppl.**, 148 (1994).

[5]Dejun Wang, J. Zhou, Z. Gui, and L. Li, "Preparation and Characteristics of (Sr,Pb)TiO$_3$ Ultrafine Particles," Wuji Cailiao Xuebao / Journal of Inorganic Materials, **12**, 231 (1997).

[6]Y. Azuma and K. Nogami, "Coating of Ferric Oxide Particles with Silica by Hydrolysis of TEOS," J. Ceram. Soc. Jpn., **10**, 646 (1992)

INTERFACIAL PROPERTIES OF PZT THIN FILMS WITH BOTTOM ELECTRODES (Pt, Ir, IrO$_2$)

Hee-Soo Lee, Keun-Ho Auh, Kwang-Bo Shim, Min-Seok Jeon, Woo-Sik Um*, and In-Shik Lee*
Dept. of Ceramic Eng., College of Engineering, Ceramic Processing Research Center, Hanyang University, Seoul 133-791, Korea
*Material Analysis Team, Korea Testing Laboratory, Div. of KITECH, Seoul 152-053, Korea

ABSTRACT
Polycrystalline PZT thin films with columnar structure and random orientation were reactively sputtered on Pt, Ir, and IrO$_2$. When PbTiO$_3$ buffer layer were inserted between PZT and bottom electrodes, PZT had a high (100) peak and showed improvement in the electrical properties such as hyteresis loops and dielectric constant. Electrical properties of PZT on Ir were better than that of Pt, which is attributed to the enhancement in the interface properties and crystallinity of PZT films. Ir and IrO$_2$ electrodes were annealed to increase their crystallinity and modify the surface in Ar and O$_2$ ambient, respectively. High quality PZT films were obtained on IrO$_2$. There was no appreciable roughening in the PZT/IrO$_2$ interface respective to the PZT/Ir. In case of PZT/Ir, IrO$_2$ phase was also formed by the oxidation of Ir electrode during PZT deposition process. Electrical properties of PZT thin films using IrO$_2$ electrode were better than that of Ir electrode and, especially, PZT on IrO$_2$ showed no fatigue up to 10^{11} cycles. PZT/Ir and PZT/Pt showed fatigue after 10^9 and 10^7 polarization reversals, respectively.

INTRODUCTION
Ferroelectric thin films are being actively studied for non-volatile random access memories(NVRAMs)[1] as well as for dynamic random access memories(DRAMs)[2] applications. Ferroelectric thin film capacitors using PZT as ferroelectric materials have been investigated for memory applications[3]. These devices undergo a large amount of read/write cycles($10^{11} \sim 10^{13}$ cycles for commercial use) when they are

used in a destructive readout mode in order to dynamically retrieve and store information and, thus, they are required to have long term reliability under various operating conditions. The application of ferroelectric memory devices requires an improvement in the degradation such as remanent polarization, coercive field, retention, imprint and fatigue. Fatigue is defined as a loss of switchable polarization with read/write cycles. Logic state retention and imprint are also very important and consequently have received most of attention for NVRAMs application. To improve these properties, many studies have been carried out on crystalline quality, microstructure, electrode materials and so on[4-6]. The electrical characteristics such as fatigue, retention, imprint properties have mostly depended on interfacial properties, i.e., electrode materials.

To improve interfacial properties, IrO_2 electrode which is a conductive oxide as a new electrode material is studied in comparison with Pt and Ir for the application to FRAM

EXPERIMENTS

PZT thin films were deposited at 550℃ by dc reactive sputtering using multiple 3-inch metal targets. The Pt, Ir, and IrO_2 bottom electrodes were deposited by sputtering method on p-type Si wafer with a SiO_2 layer(100nm). $PbTiO_3$ buffer layer was deposited to increase the crystallinity of PZT films. The Ir and IrO_2 electrodes were annealed at 250~750℃ by RTA system to investigate the relation between the ferroelectric properties of the PZT thin films and thermally modified surface of the bottom electrodes, i.e., the interface. Top Pt electrodes were deposited by dc sputtering at room temperature and patterned with a metal shadow mask (200μm in diameter). The deposition conditions of the PZT films are summarized in Table 1. The crystallinity of the films was examined using an X-ray diffractometer (MAC science) and the chemical composition of the films was determined by wavelength dispersive X-ray spectroscopy (WDX, Microscope). The surface morphology and surface roughness of the bottom electrodes and the PZT films were obtained from scanning electron microscopy (SEM) and atomic force microscopy (AFM). The depth profiles of each elements and the interface were analyzed by auger electron spectroscopy (AES) and transmission electron microscopy (TEM), respectively. Ferroelectric properties and reliability characteristics of the PZT thin films were investigated using a RT66A standard ferroelectric tester (Radiant Technologies). All the measurements were carried out at room temperature.

Dielectric Ceramic Materials

Table 1. The sputtering conditions of PZT thin films

Base pressure		$2\sim3\times10^{-6}$ Torr
Target		3-inch Pb, Zr, Ti metal
Substrate temperature		550℃
Working pressure		10mTorr
Ar/O₂ ratio		95/5
RF power	Pb	12~15W
	Zr	36~45W
	Ti	90~105W

RESULTS AND DISCUSSION

Figure 1 shows the X-ray diffraction patterns without and with $PbTiO_3$ buffer layer. All the films exhibited only a perovskite single phase without other second phases. In case of PZT(Zr/Ti=52/48) on Ir electrode, IrO_2 phase was formed by oxidation of Ir electrode by reaction with oxygen during PZT deposition process. It could be also confirmed by TEM. By using the buffer layer, polycrystalline films changed to highly (100) oriented films at the deposition temperature of 550℃.

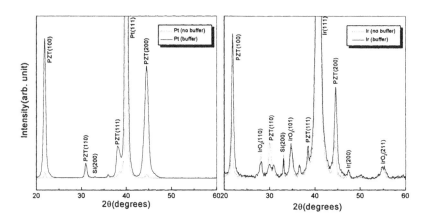

Fig. 1 X-ray diffraction patterns of PZT films on Pt and Ir.

Polarization-electric field (P-E) hysteresis loops of PZT thin films, as illustrated in Fig. 2 (a), were measured at virtual ground mode using a RT66A ferroelectric tester. The hysteresis property of the films improved toward having a good squareness by using the buffer layer. Remanent polarization of the films increased and coercive field

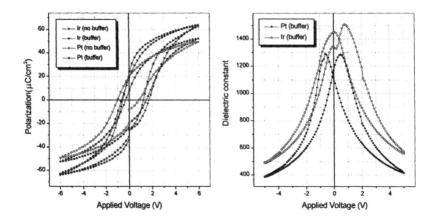

Fig. 2 P-E hysteresis loops and dielectric constants of PZT films on
Pt and Ir

of them decreased. Hysteresis properties of the PZT films on Ir were
better as compared to those on Pt. The measurement results of
dielectric constant are shown in Fig. 2 (b). Dielectric constant of film
on Ir was also higher than that of Pt with buffer layer. This is due
to the higher crystallinity of PZT films on Ir than on Pt, as shown in
Fig. 1. From these results, improvement in P-E hysteresis and
dielectric constant was attributed to $PbTiO_3$ buffer layer. It is
suggested that this buffer layer is apt to increase the crystallinity of
PZT films. Moreover, Pb-deficient disturbed layer can be formed at
the interface between PZT and Pt as reported[7,8], previously and this
interfacial layer degrades the ferroelectric properties of PZT films.

In order to improve the degradation of PZT films for application to
FRAM, IrO_2 electrode which is a conductive oxide as a new electrode
material was adopted for the bottom electrode in comparison with Pt
and Ir. Ir and IrO_2 electrodes were annealed to increase their crystallinity
and modify the surface topology of the bottom electrodes. Ir and IrO_2
were annealed in Ar and O_2 atmosphere, respectively. Fig. 3 shows the
change in crystallinity of Ir and IrO_2 with various annealing temperatures.
With the increase of annealing temperature, the crystallinity of both
electrodes increased. In case of the IrO_2, it was crystallized above 350℃.
The surface roughness of the bottom electrodes are illustrated in Fig. 4.
The roughness of Ir and IrO_2 increased with the annealing temperatures
and Ir had larger roughness than IrO_2 at the same annealing temperature.

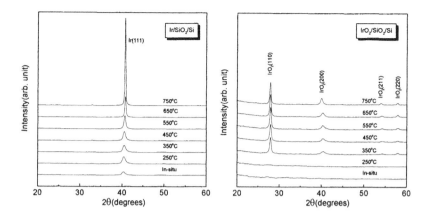

Fig. 3 X-ray diffraction patterns of Ir and IrO₂ with various annealing temperatures.

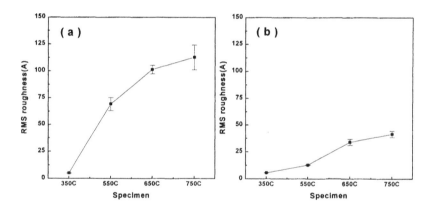

Fig. 4 Surface roughness of the bottom electrodes; (a) Ir and (b) IrO₂

This was observed by cross-sectional TEM images.

Figure 5 shows the P-E hysteresis loops of the PZT films on Ir and IrO₂ annealed at various temperatures. The hysteresis properties in both cases appeared better at 650℃ than at other annealing temperatures. From AFM results, the surface roughness of the bottom electrodes increased and this can affect the phase formation and crystallinity of the PZT films. Generally, the increase in roughness enhance the crystallinity of PZT films. Moreover, as the bottom

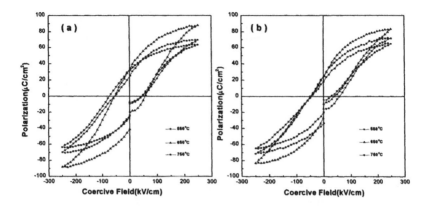

Fig. 5 P-E hysteresis loops of PZT films on (a) IrO$_2$ and (b) Ir.

electrodes were annealed at higher temperature, the grain size of PZT films increased from the observation of SEM. In case of 750℃, the slight decrease in P$_r$ and P$_s$ may be due to the degradation in interface by high surface roughness of the electrode.

The fatigue test was conducted for the PZT thin films deposited at identical conditions on IrO$_2$ and Ir annealed at 650℃ and these results were compared with the PZT/Pt(Fig. 6) The alternating square pulse with 5V high was applied in the test. It is thought that the PZT/IrO$_2$ and PZT/Ir thin film can be applied for the capacitor

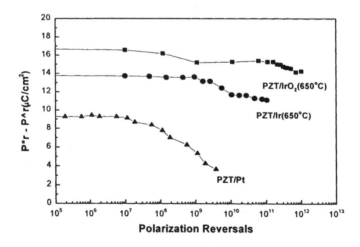

Fig. 6 Fatigue characteristics of PZT films on various electrode.

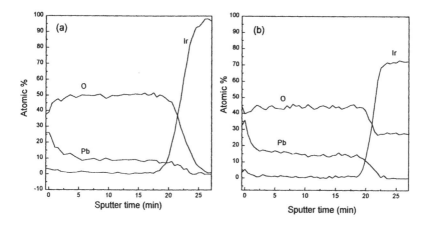

Fig. 7 AES depth profiles of PZT films on (a) Ir and (b) IrO$_2$.

cell of FRAM from the high value of $P^*_r - P^{\wedge}_r$. The PZT/Ir film showed some fatigue after 10^9 cyclings, whereas the PZT/IrO$_2$ film did not show nearly any fatigue until 10^{11} cyclings. This difference in fatigue result is thought to be induced by interfacial properties. In case of the PZT/Ir, Pb diffused into the bottom electrode in spite of the formation of IrO$_2$ phase between PZT and Ir, while PZT/IrO$_2$ did not showed a diffusion of Pb(Fig. 7). This Pb diffusion in PZT/Ir films may be due to the irregular formation of IrO$_2$. Pb diffusion seems to form Pb-deficient layer at the interface and this layer resulted in the decrease of $P^*_r - P^{\wedge}_r$ in PZT/Ir. In addition, the high surface roughness of Ir electrode may also affect the fatigue characteristics.

CONCLUSIONS
    The ferroelectric characteristics of PZT films were studied to investigate interfacial properties by inserting PbTiO$_3$ buffer layer and applying various bottom electrodes, especially IrO$_2$ which is conductive oxide as a new electrode material for the application to FRAM.
    PZT/Pt and PZT/Ir with PbTiO$_3$ buffer layer showed improvements in the electrical properties such as hysteresis loops and dielectric constant. The enhancement in electrical properties of PZT on Ir respective to Pt is attributed to the enhancement in the interface properties and the crystallinity of PZT films. The improvement of crystallinity and modification of surface topology were conducted by annealing Ir and IrO$_2$. There was no appreciable roughening of the interface between the PZT

and the IrO_2 electrode respective to the PZT films on Ir by TEM. In case of PZT/Ir, irregular IrO_2 phase was also formed by oxidation of Ir electrode during PZT deposition process. PZT thin films on IrO_2 showed better electrical properties than on Ir and, especially, PZT/IrO_2 films showed no fatigue up to 10^{11} cycles. From the depth profiles of the PZT/Ir and IrO_2, Pb diffusion was observed in PZT/Ir opposite to PZT/IrO_2.

REFERENCES

[1] J. F. Scott and C. A. Paz de Araujo, Science, 246, 1400 (1989).

[2] W. A. Geideman, IEEE Trans. Ultrasonics, Ferroelectrics and Freq. Control, 38, 704 (1990).

[3] S. S. Eaton, D. B. Butler, M. Paris, D. Wilson, and H. Mcnellie, Digest of Technical Papers of the IEEE International Solid-State Circuits Conference No. 31 (IEEE New York, 1988), p. 130.

[4] D. P. Vijay, C. K. Kwok, W. Pan, I. K. Yoo, and S. B. Desu, Proc. 8th Int. Symp. on Applications of Ferroelectrics, Greenville, SC, 408 (1992).

[5] J. Lee, L. Johnson, and A. Safari, Appl. Phys. Lett., 63, 27 (1993).

[6] W. L. Warren, D. Dimos, B. A. Tuttle, R. D. Nasby, and G. E. Pike, Appl. Phys. Lett., 65, 1018 (1994).

[7] K. Aoki, Y. Fukuda, K. Numata, and A. Nishimura, Jpn. J. Appl. Phys., 33, (1994) 5155.

[8] M. Shimizu, M. Sugiyama, H. Fusisawa, and T. Shiosaki, Jpn. J. Appl. Phys., 33, (1994) 5167.

DIELECTRIC PROPERTIES OF TIN SUBSTTUTED $Ca(Li_{1/3}Nb_{2/3})O_{3-\delta}$ CERAMICS

Ji-Won Choi, Seok-Jin Yoon, Chong-Yun Kang, Sergey Kucheiko, Yong-Wook Park, Hyun-Jai Kim, and Hyung-Jin Jung
Thin Film Technology Research Center, Korea Institute of Science and Technology, P.O.Box 131, Cheongryang, Seoul 130-650, Korea

ABSTRACT

The dielectric properties of $Ca[(Li_{1/3}Nb_{2/3})_{(1-x)}Sn_x]O_{3-\delta}$ $(0 \leq x \leq 0.3)$ ceramics have been investigated. The ceramic specimens were multiphase materials. However, single phase specimens having orthorhombic perovskite structure similar to $CaTiO_3$ could be obtained in the range of $0.2 \leq x \leq 0.3$. As Sn concentration increased, the dielectric constant (ε_r) decreased, the quality factor Q increased till the composition of x = 0.15 and then slightly decreased, and the temperature coefficient of resonant frequency (τ_f) changed from negative to positive direction till the composition of x = 0.1 and then changed to the negative direction. The $Q \cdot f_0$ value and ε_r for the composition range of $0.1 \leq x \leq 0.3$ were found to be $52100 \sim 46000$ GHz and $25.4 \sim 23.5$, respectively

INTRODUCTION

Recently, dielectric materials which use microwave frequency range have been developed for the purpose of adapting in the fields of mobile telephone, satellite broadcast and communication. The application of high frequency dielectric ceramics such as resonator, band pass (stop) filter and duplexer has been increased for practical use of a new communication media[1,2] and they can reduce the size and weight of the components by a factor of $\sqrt{\varepsilon_r}$.

It is known that the materials for microwave resonators and filters are required to excel in three dielectric characteristics[3] as follows: the first characteristic is high dielectric constant ε_r, the second characteristic, high Q, which is the inverse of the dielectric loss (tanδ), is required for achieving high frequency selectivity and stability in microwave transmitter components, and the third characteristic is stable temperature coefficient of the resonant frequency τ_f.

There are a number of microwave dielectric materials which were developed such as (Ba,Pb)-Nd_2O_3-TiO_2[4] and (Pb,Ca)ZrO_3[5] which have the dielectric constant over 70, (Zr,Sn)TiO_4[6], $Ba_2Ti_9O_{20}$[7] and $CaTiO_3$-$Ca(Al_{1/2}Ta_{1/2})O_3$[8] which have the dielectric constant range of 30~60 and $Ba(Zn,Ta)O_3$[9,10] which have the dielectric constant below 30 and all these materials have sintering temperature range of 1300~1400 °C.

In this paper, the effect of Sn^{4+} substitution for $[Li_{1/3}Nb_{2/3}]^{3.67+}$ in the $Ca(Li_{1/3}Nb_{2/3})O_{3-\delta}$[11] system was investigated to improve the Q value and preserve the low sintering temperature of the system.

EXPERIMENTAL PROCEDURE

$Ca[(Li_{1/3}Nb_{2/3})_{(1-x)}Sn_x]O_{3-\delta}$ powder compositions were synthesized by the conventional solid-state reaction method and the starting materials were high purity (>99.9%, Aldrich co.ltd.) $CaCO_3$, Li_2CO_3, Nb_2O_5 and SnO_2.

These powders were mixed in a ball mill with ethyl alcohol to prevent dissolution of Li_2CO_3, then dried and calcined at 750~900 °C for 2 h in air. The calcined powder was milled again with a 5 wt% polyvinyl alcohol as a binder. The dried powders were pressed into rods of 10 mm in diameter and 6 mm in thickness under a pressure of 140 MPa. The pellets were placed in a closed Pt box and sintered at 1100~1300 °C for 1~5 h in air.

X-ray diffraction (Cu Kα radiation) was used to study the phase of the specimens. Pure Si was utilized as an internal standard. The density was determined by the Archimedes method. The fractured surfaces of the specimens were investigated by scanning electron microscopy (SEM). The dielectric constant (ε_r) and the unloaded Q value were measured at 11-13 GHz using the parallel-plate method combined with a network analyzer (HP8720C)[12]. The temperature coefficient of resonant frequency (τ_f) was measured in the range of -20~+80 °C.

RESULTS AND DISCUSSION

The properties of ceramics investigated are presented in Table I. All ceramics were good sintered at 1150 °C for 3-5 h in air. In general, the sintering temperature of perovskite type dielectric materials having dielectric constant range of 20-50 is high temperature which is range of 1300-1600 °C. But in this composition, good sintered specimens were achieved at low temperature of 1150 °C by using nonstoichiometry $Ca(Li_{1/3}Nb_{2/3})O_{3-\delta}$ Ceramic.

The bulk density of $Ca[(Li_{1/3}Nb_{2/3})_{(1-x)}Sn_x]O_{3-\delta}$ specimens sintered at 1150 °C for 5h as a function of x increased from 4.17 to 4.44 g/cm³ with increasing Sn content from 0 to 0.3 mol as shown in figure 1.

Figure 2 shows powder x-ray diffraction patterns for $Ca[(Li_{1/3}Nb_{2/3})_{(1-x)}Sn_x]O_{3-\delta}$ ceramics sintered at 1150 °C for 5 h as a function of x. The sintered

Dielectric Ceramic Materials

Table I. Dielectric properties of $Ca[(Li_{1/3}Nb_{2/3})_{(1-x)}Sn_x]O_{3-\delta}$ ceramics sintered at 1150 °C.

x	Sintering time (hr)	ε_r	$Q \cdot f_o$ (GHz)	τ_f (ppm/°C)	Density (g/cm³)
0	3	29.6	40000	-21	4.17
0.1		25.2	48200	-14	
0.15		24.8	49100	-25	4.28
0.2		23.3	50600	-30	
0.3		22.6	46300	-39	4.32
0.1	5	25.4	48300	-15	4.22
0.15		24.9	52100	-29	4.24
0.2		23.9	51000	-32	4.31
0.3		23.5	46000	-40	4.44

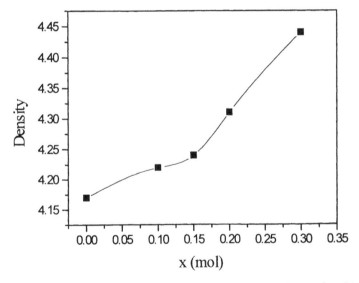

Fig. 1. Bulk density of $Ca[(Li_{1/3}Nb_{2/3})_{(1-x)}Sn_x]O_{3-\delta}$ ceramic sintered at 1150 °C for 5 h as a function of x.

$Ca(Li_{1/3}Nb_{2/3})O_{3-\delta}$ (x = 0) specimen was multiphase. However, the stability of the perovskite phase increased with increasing in Sn content and the single phase

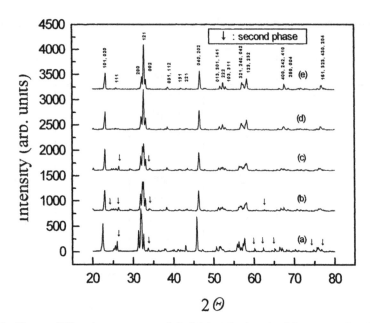

Fig. 2. X-ray diffraction pattern of $Ca[(Li_{1/3}Nb_{2/3})_{(1-x)}Sn_x]O_{3-\delta}$ ceramic sintered at 1150 °C for 5 h : (a) x = 0, (b) x = 0.1, (c) x = 0.15, (d) x = 0.2, (e) x = 0.3.

specimens were obtained in the range of $0.2 \leq x \leq 0.3$. The x-ray diffraction peaks for $Ca[(Li_{1/3}Nb_{2/3})_{1-x}Sn_x]O_{3-\delta}$ could be indexed based on the $CaTiO_3$-type orthorhombic perovskite structure.

Figure 3 shows fractured scanning electron microscopy(SEM) photographs of the $Ca[(Li_{1/3}Nb_{2/3})_{(1-x)}Sn_x]O_{3-\delta}$ ceramics sintered at 1150 °C for 5 h as a function of x. The fractured photographs couldn't measure the exact grain size but it was reasonable that the grain size slightly decreased with increasing Sn content from 0.1 to 0.3 mol and crystalline status was good.

Figure 4 shows dielectric constant (ε_r) of the $Ca[(Li_{1/3}Nb_{2/3})_{(1-x)}Sn_x]O_{3-\delta}$ ceramics sintered at 1150 °C for 5 h as a function of x. The ε_r decreased slightly from 29.6 to 23.5 with increasing Sn content from 0 to 0.3 mol. This result is reasonable when we think over that Sn content only slightly affected the dielectric constant as reported by Kucheiko et.al.'s.[13]

It is generally known that, in the narrow microwave region, the Q value varies inversely with frequency ; hence the product of Q and f_o is constant. Therefore, we use $Q \cdot f_o$ for the evaluation of the dielectric loss instead of the Q value.

Figure 5 shows $Q \cdot f_o$ values of the $Ca[(Li_{1/3}Nb_{2/3})_{(1-x)}Sn_x]O_{3-\delta}$ ceramics sintered

Dielectric Ceramic Materials

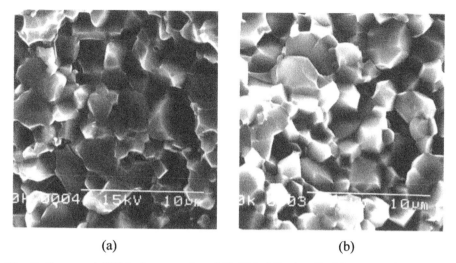

(a)   (b)

Fig. 3. Fractured SEM photographs of $Ca[(Li_{1/3}Nb_{2/3})_{(1-x)}Sn_x]O_{3-\delta}$ ceramic sintered at 1150 °C for 5 h : (a) x = 0.1, (b) x = 0.3.

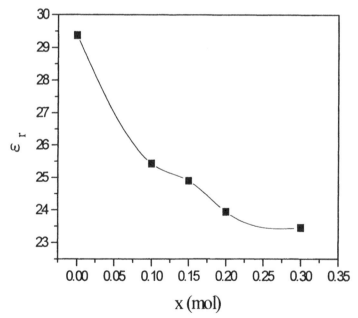

Fig. 4. Dielectric constant of $Ca[(Li_{1/3}Nb_{2/3})_{(1-x)}Sn_x]O_{3-\delta}$ ceramic sintered at 1150 °C for 5 h as a function of x.

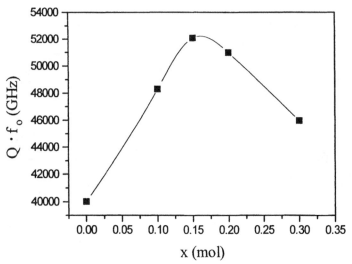

Fig. 5. The Q·f value of $Ca[(Li_{1/3}Nb_{2/3})_{(1-x)}Sn_x]O_{3-\delta}$ ceramic sintered at 1150 °C for 5 h as a function of x.

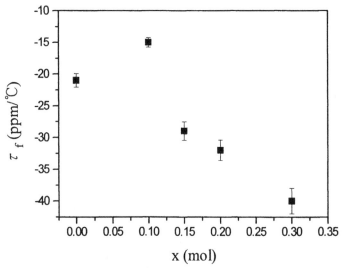

Fig. 6. Temperature coefficient of resonant frequency of $Ca[(Li_{1/3}Nb_{2/3})_{(1-x)}Sn_x]O_{3-\delta}$ ceramic sintered at 1150 °C for 5 h as a function of x.

at 1150 °C for 5 h as a function of x. The $Q \cdot f_o$ value increased from 40000 to 52100 GHz with increasing x from 0 to 0.15 and then decreased from 52100 to 46000 GHz with increasing x from 0.15 to 0.3. It is worth noting that Sn substitution improves significantly the quality factor and the improvement is within the limited Sn consentration.

Figure 6 shows temperature coefficient of resonant frequency (τ_f) of the $Ca[(Li_{1/3}Nb_{2/3})_{(1-x)}Sn_x]O_{3-\delta}$ ceramics sintered at 1150 °C for 5 h as a function of x. The τ_f value increased from -21 to -15 ppm/°C with an increase in Sn content from 0 to 0.1 mol and then decreased from -15 to -40 ppm/°C with an increase of Sn consentration over 0.1 mol.

CONCLUSIONS

The microwave dielectric properties of $Ca[(Li_{1/3}Nb_{2/3})_{(1-x)}Sn_x]O_{3-\delta}$ ($0 \leq x \leq 0.3$) ceramics were investigated. These ceramics can be sintered at a relatively low temperature of 1150 °C for 5 h. A single phase ceramic with an orthorhombic perovskite structure similar to $CaTiO_3$ was obtained in the range of $0.2 \leq x \leq 0.3$. An increase of Sn concentration from 0 to 0.3 mol in the system $Ca[(Li_{1/3}Nb_{2/3})_{(1-x)}Sn_x]O_{3-\delta}$ only slightly affected the dielectric constant (ε_r) and improved the Q values.

REFERENCES

[1]K.Wakino, "Recent Development of Dielectric Resonator Materials and Filters in Japan", *Ferroelectrics*, **91**, 68-86 (1989).

[2]K.Pobl and G.Wolfram, "Dielectric Resonators, New Components for Microwave Circuits", *Siemens-Components*, **17**, 14-18 (1982).

[3]W.Wersing, "High Frequency Ceramic Dielectrics and Their Application for Microwave Components"; pp67-119 in *Electronic Ceramics* Edited by B.C.H.Steele, Elsevier Applied Science, London, U.K., 1991.

[4]K.Wakino, K.Minai and H.Tamura, "Microwave Characteristics of (Zr,Sn)TiO$_4$ and BaO-PbO-Nd$_2$O$_3$-TiO$_2$ Dielectric Resonators", *Journal of the American Ceramic Society*, **67**(4), 278-81 (1984).

[5]J.Kato, H.Kagata and K.Nishimoto, "Dielectric Properties of Lead Alkaline-Earth Zirconate at Microwave Frequencies", *Japanese Journal of Applied Physics*, **30**, 2343-46 (1991).

[6]Y.C.Heiao, L.Wu and C.C.Wei, "Microwave Dielectric Properties of (ZrSn)TiO$_4$ Ceramic", *Material Research Bulletin*, **23**, 1687-92 (1988).

[7]J.K.Plourde, D.F.Linn, H.M.O'Bryan and J.Thomson, "Ba$_2$Ti$_9$O$_{20}$ as a Microwave Dielectric Resonator", *Journal of the American Ceramic Society*, **58** (9-10), 418-20 (1975).

[8]S.Kucheiko, J.W.Choi, H.J.Kim and H.J.Jung, "Microwave Dielectric Properties of $CaTiO_3$-$Ca(Al_{1/2}Ta_{1/2})O_3$ Ceramics", *Journal of the American Ceramic Society*, **79** [10], 2739-43 (1996).

[9]S.Kawashima, M.Nishada, I.Uedo and H.Ouchi, "$BaZn_{1/3}Ta_{2/3}O_3$ Ceramics with Low Dielectric Loss at Microwave Frequencies", *Journal of the American Ceramic Society*, **66** [6], 421-23 (1983).

[10]S.B.Desu and H.M.O'Bryan, "Microwave Loss Quality of $BaZn_{1/3}Ta_{2/3}O_3$ Ceramics", *Journal of the American Ceramic Society*, **68** [10], 546-51 (1985).

[11]J.W.Choi, S.Kucheiko, S.J.Yoon, H.J.Kim and H.J.Jung, "Dielectric Properties of Ti Substituted $Ca(Li_{1/3}Nb_{2/3})O_{3-\delta}$ Ceramics", *Journal of the Korean Physical Society*, **32**, S334-36 (1998).

[12]B.W.Hakki and P.D.Coleman, "A Dielectric Resonator Method of Measuring Inductive Capacities in the Millimeter Range", *IRE Transaction Microwave Theory & Technology*, **8**, 402-10 (1960).

[13]S.Kucheiko, J.W.Choi, H.J.Kim, S.J.Yoon and H.J.Jung, "Microwave Characteristics of $(Pb,Ca)(Fe,Nb,Sn)O_3$ Dielectric Materials", *Journal of the American Ceramic Society*, **80** [11], 2937-40 (1997).

Mechanical Anisotropy of PZT Piezoelectric Ceramics

S. Wan, W. F. Shelley II, and K. J. Bowman
School of Materials Engineering, Purdue University, West Lafayette, IN 47907

ABSTRACT

Preferred crystal orientation of piezoelectric materials can be induced by poling with an applied electric field. This texture of poled piezoelectric ceramics can be eliminated by thermal depoling. Mechanical stress can also lead to texture in PZT. In this investigation fracture anisotropy from Vickers indentation was observed in mechanically stressed Navy VI PZT. Surprisingly, elastic isotropy was found from ultrasound measurement on the same material. The corresponding crystal orientation was characterized by X-ray diffraction and it showed that the applied compressive stress changes the surface texture state and not the bulk ceramic sample. The mechanical anisotropy induced by compression was apparently a surface effect. Complete mechanical poling may be difficult to achieve due to the limitation of the compressive strength.

I. INTRODUCTION

Electrical properties, chemistry, and processing of lead zirconate titanate (PZT) have been thoroughly investigated[1]. With wide application as actuators, their mechanical properties and related performance have received significant attention in recent years. A number of investigations have shown that electrical poling can induce both electrical and mechanical anisotropy of PZT. Grekov and Kramarov showed the anisotropy of mechanical strength of ferroelectric ceramics[2]. Pisarenko et al. have demonstrated fracture toughness anisotropy of $BaTiO_3$ and PZT[3]. Recent results from Wan and Bowman show the changes in fracture anisotropy with various thermal treatments[4]. Park and Sun reported that fracture of PZT materials is dependent on applied electric field[5].

Mechanical stress can also introduce the preferred domain orientation in piezoelectric ceramics. Grinding and abrasion can change the 90° domain orientation of surface layer and this surface texture can be eliminated by fine polishing [6,7]. Karun and Virkar have shown a compressive stress effect on domain switching in PZT materials[8]. Kim et al. also observed compressive stress effects on the domain configuration of $BaTiO_3$[9]. The application of mechanical stress to induce domain motion can be called mechanical poling. Mechanical poling effects on the mechanical anisotropy of piezoelectric ceramics are not well known. Therefore, the objectives of this work are to probe the effect of mechanical stress on anisotropy of piezoelectric materials and to reveal differences between electrical poling and mechanical poling.

II. EXPERIMENTAL PROCEDURE

a)Thermal depoling

Poled and unpoled Navy VI PZT samples were obtained from Channel Industries, CA. Electrically poled samples were thermally depoled by annealing over 400°C for two hours.

b) Mechanical poling

Unpoled PZT samples were cut with dimensions 8mm ×6mm×12mm. Samples were finely polished using 0.1μm alumina to eliminate the effect of grinding. The samples then were subjected to compression loads of 100 MPa, 200 MPa, 300 MPa and 400 MPa at room temperature with a loading rate of 3.6 Mpa/min (Fig.1). Once the compressive stress was reached, the load was released.

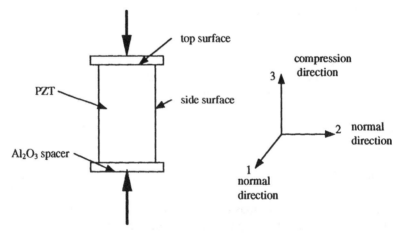

Fig.1 Schematic of the mechanical poling and corresponding sample coordinate.

c) Measurement of fracture toughness and elastic constants

The fracture toughness of unpoled, electrically poled, thermally depoled, and mechanically poled PZT samples was measured by Vickers indentation with a load of 5 kg. By measuring the residual fracture length c and the half diagonal length of the indent a, the fracture toughness K_c can be calculated by [10]:

$$K_c = 0.016 \left(\frac{E}{H} \right)^{1/2} \frac{P}{c^{3/2}}$$ (1)

where E is Young's modulus, P is the applied load, and H is the hardness, which is determined as:

$$H = \frac{P}{2a^2}$$ (2)

Then elastic stiffnesses were measured using ultrasound at a frequency of 40 MHz. Sound wave velocity v was determined by the sample thickness and dividing the pathlength by the time of flight. The stiffness constant, C, was calculated by the equation

$$C = \rho v^2$$ (3)

where ρ is the density of material and v is the normal soundwave velocity.

d) X-ray characterization

Unpoled, poled, thermally depoled and mechanically poled PZT samples were examined using X-ray diffraction (Cu Kα) while focusing on 002/200 peaks. After measuring the fracture toughness and elastic constants, mechanically poled samples were polished again to remove approximately by 400 µm, then were characterized by X-ray once more.

III. RESULTS AND DISCUSSIONS

a) Domain orientations in poled and thermally depoled PZT samples

The crystal orientation of unpoled, fully electrically poled and thermally depoled Navy VI PZT are shown in the X-ray diffractogram of Fig.2. For an unpoled sample with random crystal orientation, the intensity ratio of the (002) peak and the (200) peak is nearly 0.5. As expected, the poling process produces preferred orientation in the PZT. The (002) planes in most crystals become oriented normal to the poling direction, as indicated by the change of intensity ratio of the (002) peak relative to the (200) peak on the side surface.

The Curie temperature of Navy VI PZT is 200°C. Thermally depoling at 400°C for two hours effectively eliminates the preferred orientation of piezoelectric

ceramics. Usually, the <001> direction represents the polarization direction for 90°
domains. As shown in Fig.2, after thermal depoling the 90° domain orientation
distribution is almost the same as the state of the unpoled sample. Thus the sample
returns to a random state after thermal depoling. Table I lists elastic constant and
fracture toughness data of PZT materials at three orientation states. Mechanical
property anisotropy is exhibited by the poled material wherein $C_{33} > C_{11}$ and $K^3 >
K^1$. The other two materials have isotropic mechanical properties.

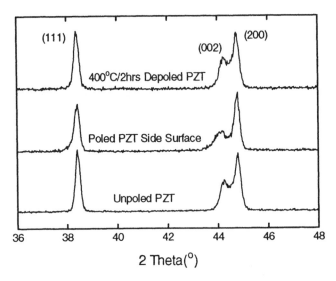

Fig.2. XRD of unpoled, electrically poled, and thermally depoled Navy VI PZT.

Table I. Elastic stiffness and fracture toughness of unpoled, poled and
thermally depoled Navy VI PZT.

	C_{11} (GPa)	C_{33} (GPa)	K^1 (MPa.m$^{1/2}$)	K^3 (MPa.m$^{1/2}$)
Unpoled PZT	131±2	131±2	0.7±0.1	0.7±0.1
Poled PZT	128±2	156±2	0.5±0.1	1.1±0.1
Depoled PZT	130±2	130±2	0.7±0.1	0.7±0.1

For the poled sample 3 is the direction parallel to and 1 is the direction normal to the poling
direction

b) Mechanical anisotropy of PZT under various compressive stresses

Mechanical stress can also change the domain orientation in piezoelectric
ceramics. Fig.3 shows fracture toughness anisotropy from Vickers indentation with
increasing compressive stress. XRD results from side surfaces of samples after

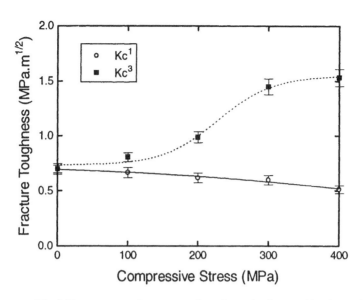

Fig.3 Fracture toughness as a function of prior application of compressive stress in PZT materials.

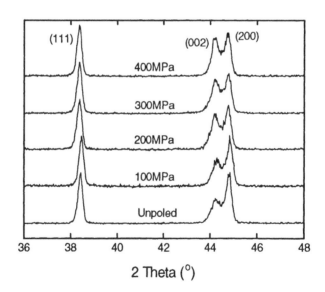

Fig.4 XRD from side surface of mechanically poled samples.

compression indicate domain orientation (Fig.4). The compressive stress leads to 90° domain switching of the (002) planes parallel to the compression direction.

With this result, it is easy to expect a similar anisotropy of elastic constants in compressed samples. Fig. 5 shows the elastic stiffness as a function of compressive stress. In contrast to the obvious anisotropy of fracture toughness from Vickers indentation, there is no apparent difference between the normal stiffness in the compression direction and the normal direction. The stiffness values are 130±2 GPa which are very close to those of the unpoled random sample. This is distinctly different from the electrically poled PZT for which fracture anisotropy and elastic anisotropy were induced simultaneously. At first glance, the elastic stiffness isotropy and the fracture anisotropy are not consistent.

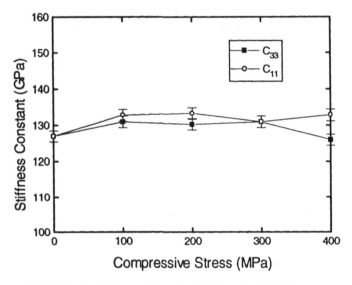

Fig.5 Elastic stiffness of mechanically poled PZT versus prior application of compressive stress

The inconsistency in mechanical properties is related to the measurement methods for fracture toughness and elastic stiffness. If the sample has a uniform distribution of microstructure and orientation, the value from the Vickers indentation can represent the material properties. However, for the mechanically poled sample, this assumption is questionable. Moreover, the Vickers indentation technique may exaggerate the anisotropy of fracture toughness [11]. Alternatively, the elastic constants measured by ultrasound are bulk properties since the ultrasound wave travels from one side to the other side in pulse/echo

Dielectric Ceramic Materials

measurements. The difference between the fracture anisotropy and the elastic anisotropy is possibly from domain orientation confined to near the surface. Therefore, to remove the surface effect, each sample was thinned by 400 μm and examined by X-ray diffraction again. Fig.6 shows the orientation state in the interior of compressed samples. Apparently, the intensity of the (002) peak decreases and the intensity ratios of the (002) peak to the (200) peak for all the samples are close to 0.5. For the sample compressed by 400 MPa, this ratio is 0.82 before surface removal and 0.57 afterwards. This suggests that domain orientations in the interior of compressed samples are similar to the unpoled random state. Thus the mechanical poling by compression up to 400 MPa only changes the orientation near the surface and the orientation state in the sample interior remains random.

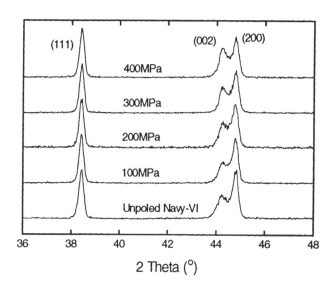

Fig.6 XRD of thinned side surface of mechanically poled samples

The surface switching may be explained by recognizing the different constraints on domains near the surface and in the interior of the sample. Domains in the sample interior are subjected to the compressive stress and the constraints of adjoining grains, whereas the material on the surface is free of constraints normal to the surface. Consequently, domains at the surface may be easier to switch and are less likely to switch back after unloading. With sufficient compressive load, all favorably oriented domains should undergo domain switching. Complete mechanical poling is difficult since stresses higher than 400 MPa often lead to fracture.

The 90° domain switching in PZT materials usually leads to nonlinear deformation and strain hysteresis under a compressive stress[12] and a tensile stress[13]. If 90° domain switching occurs near the sample surface, it is expected that a stress-strain hysteresis will be observed during loading and unloading cycles. And the magnitude of the hysteresis should be related to the volume fraction of the switched 90° domains. Stress-strain analysis will be included in future work to further understand the localized surface effect in mechanical poling.

IV. SUMMARY

The domain orientation state in piezoelectric PZT ceramics can be changed by electrical poling and thermal depoling. Apparent fracture anisotropy and elastic anisotropy were induced by electrical poling, then returned to an isotropic state by thermal depoling. Similarly, mechanical stress can change the orientation state of piezoelectric ceramics. X-ray diffraction results showed 90° domain switching after compression. Fracture toughness anisotropy by Vickers indentation was detected after compression in unpoled PZT samples, but the elastic constant measured by ultrasound remained isotropic. X-ray diffraction on the surface of each sample after surface removal indicated the compression can only cause a change in orientation state near the surface. The orientation state at the interior remained random even after compression. Complete mechanical poling is difficult to achieve due to the limitation of compressive strength.

V. ACKNOWLEDGEMENT

This research is supported by the National Science Foundation under Grant No. DMR-9528928.

REFERENCES

[1] B. Jaffe, W. R. Cook and H. Jaffe, *Piezoelectric Ceramics*, Academic Press, London and New York, 1971

[2] A. Grekov and S. O. Kramarov, "Mechanical Strength of Ferroelectric Ceramics", *Ferroelectrics*, **18**, 249-255 (1978)

[3] G. Pisarenko, V. M. Chushko and S. P. Kovalev, "Anisotropy of Fracture Toughness of Piezoelectric ceramics", *J. Am. Ceram. Soc.*, **68**[5], 259-265 (1985)

[4] S. Wan and K. J. Bowman, "Thermal Depoling Effects on Anisotropy of Lead Zirconate Titanate materials", submitted to *J. Am. Ceram. Soc.*

[5] Seungbae Park and Chin-Teh Sun, "Fracture Criteria for Piezoelectric ceramics", *J. Am. Ceram. Soc.*, **78**[6], 1475-1480 (1995)

[6] S. Cheng, I. K. Lioyd, and M. Kahn, "Modification of Surface Texture by Grinding and Polishing Lead Zirconate Titanate Ceramics", *J. Am. Ceram. Soc.*, 75[8], 2293-2296 (1992)

[7] C. I. Cheon, and H. G. Kim, "Effects of Abrasion on the Phase Coexistence Region of Tetragonal and Rhombohedral Phase in Pb(Zr,Ti)O$_3$ ceramics", *Ferroelectrics*, 110, 227-234 (1990)

[8] Karun Mehta and Anil V. Virkar, "Fracture Mechanisms in Ferroelectric-Ferroelastic Lead Zirconate Titanate (Zr:Ti = 0.54:0.46) Ceramics, *J. Am. Ceram. Soc.*, 73[3], 567-574 (1990)

[9] S. B. Kim, T. J. Chung and D. Y. Kim, "Effect of External Compressive Stress on the Domain Configuration of Barium Titanate Ceramics", *J. European Ceramics Society*, 147-151(1993)

[10] G. Anstis, P. Chantikul, B. Lawn and D. Marshall, "A Critical Evaluation of Indentation Techniques for Measuring Fracture Toughness: I. Direct Crack Measurements", *J. Am. Ceram. Soc.*, 64[9], 533-538(1981)

[11] J. M. Calderon-Moreno, F. Guiu, M. Meredith and M. J. Reece, "Fracture Toughness Anisotropy of PZT", *Mater. Sci. & Eng.* : A234-236, 1062-1066(1997)

[12] H. Cao, and G. Evans, "Nonlinear Deformation of Ferroelectric Ceramics", *J. Am. Ceram. Soc.*, 76, 890-896(1993)

[13] T. Tanimoto, K. Okazaki, and K. Yamamoto, "Tensile Stress-Strain Behavior of Piezoelectric ceramics", *Jpn. J. Appl. Phys.*, 32, 4233-4236(1993)

SINTERING AND HUMIDITY SENSITIVE RESISTIVITY OF SPINEL Ni_2GeO_4

M. J. Hogan*, A. W. Brinkman, T. Hashemi
Dept. of Physics, University of Durham, DH1 3LE, U.K.
*email: m.j.hogan@durham.ac.uk

ABSTRACT

The spinel ceramic material nickel germanate, Ni_2GeO_4, has been prepared by the solid state reaction of nickel (II) oxide and germanium (IV) oxide in air. Formation of the monophase compound has been investigated in detail by powder x-ray diffraction (XRD) and infra-red spectroscopy (FTIR) over the temperature range from 973 K to 1673 K. These investigations suggest formation was complete in 12 hrs at 1473 K. The humidity dependant resistivity of the sintered pellets has been measured using d.c. and a.c. techniques over an extensive range of humidity. Pressed pellets, fired at 1573 K, have a d.c. resisitivity which varied from $\sim 10^9$ Ωcm to $\sim 10^5$ Ωcm as the ambient humidity was changed from 10 %R_H to 90 %R_H at 293 K. The impedance modulus of the pellets at 100 Hz varied from $\sim 10^8$ Ω to $\sim 10^4$ Ω over the humidity range 20 %R_H to 80 %R_H at 298 K. A.c. impedance results were analysed in terms of complex plane plots, as a function of humidity.

INTRODUCTION

In recent years, many investigations have taken place into the use of porous ceramic materials as humidity sensors [1,2], since they have shown advantages over polymer thin film sensors in terms of resistance to chemical attack, and their thermal and physical stability [3]. Nickel germanate, Ni_2GeO_4, is a polycrystalline ceramic material, with a spinel-type crystal structure, which has been found by us to demonstrate humidity dependant impedance [4], and is therefore a suitable candidate for use as a humidity sensor. Previous investigations of this material have been limited to its formation [5-8], optical properties [9,10], and crystal structure [11].

EXPERIMENTAL

High purity (>99.99%) powders of nickel (II) oxide and germanium (IV) oxide, in the molar ratio 2:1, were intimately mixed in a pestle and mortar to obtain a precursor material. This mixture was placed in a high purity alumina boat (Multilab Alsint 99,7), and calcined at varying temperatures between 700 °C and 1400 °C in a tube furnace (Lenton Thermal Designs) open to the atmosphere. At each stage in the fabrication, X-ray diffraction (Philips PW1130 diffractometer with $\lambda = 1.5406$ Å), infra-red spectroscopy (Bruker IFS48, samples mounted in potassium bromide discs) and scanning electron microscopy (JEOL JSM IC848) were employed to monitor the reaction progress, and to identify the optimum calcination temperature.

Monophasic Ni_2GeO_4 powder formed at 1200 °C was then reground, pressed at 2 tons cm^{-2} into 13 mm diameter pellets of approximately 1 mm thickness, and subsequently fired at 1573 K. Gold contacts were sputtered onto the faces of these pellets, using an argon plasma, in order to provide electrical connection to the sample.

The d.c. resistance of the samples was measured by a two-point probe technique using a current source and digital electrometer (Keithley 617), with the sample humidity and temperature controlled by a computer-driven chamber, over the range 20 %R_H to 98 %R_H at a constant temperature of 293 K, as measured by a Rotronic AM3-HP100A polymer thick film sensor.

The ambient humidity and temperature of the samples for a.c. measurements was controlled in a Hereaus environmental chamber, utilising dew-point mirror control of humidity over the range 20 %R_H to 98 %R_H at a constant temperature of 298 K. The a.c. impedance measurements were carried out using a Solartron 1296 Impedance Analyser, interfaced to the sample through a Solartron 1295 Dielectric Interface, over the frequency range 10 mHz to 10 MHz.

RESULTS AND DISCUSSION
Calcination

Quantitative phase analysis has been performed on the x-ray results obtained from samples of the (nickel oxide + germanium oxide) precursor fired at different temperatures, as shown in figure 1. The concentration shown was calculated according to equation 1 below :

$$I_1 / I_S = const. \; x_1 \qquad (1)$$

where the ratio I_1 / I_S is the ratio of the intensity of component 1 to the internal standard, which in this case was 50 wt.% Si, and x_1 is the molar fraction of component 1 in the mixture [12].

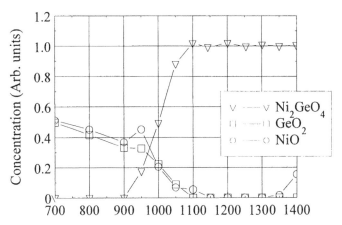

Calcination temperature (°C)

Figure 1   Quantitative phase analysis of XRD
of the reaction ($2NiO + GeO_2 \rightarrow Ni_2GeO_4$)

From figure 1, we see that detectable production of Ni_2GeO_4 started at 950 °C, monophase material was formed by 1150 °C, and that decomposition occured from 1350 °C upwards involving evaporation of GeO_2 and production of NiO.

These results were confirmed by FTIR analysis of the powders, as shown in figure 2. FTIR spectra of four samples are shown, the precursor powder, and specimens of the precursor powder which had been fired at 700 °C, 900 °C, and 1200 °C. In the trace corresponding to the as-prepared precursor powder, there are two main features, a strong absorption at $v = \sim 880 \; cm^{-1}$ due to a GeO_2 bond vibration, and a wide absorption band between $v = \sim 600 \; cm^{-1}$ and $v = \sim 400 \; cm^{-1}$, of which GeO_2 is responsible for absorption between $v = \sim 600 \; cm^{-1}$ and $v = \sim 500 \; cm^{-1}$, and NiO is responsible for absorption between $v = \sim 500 \; cm^{-1}$ and $v = \sim 400 \; cm^{-1}$. The top trace, which is due to pure Ni_2GeO_4, shows two strong absorptions, one at $v = \sim 670 \; cm^{-1}$, and another at $v = \sim 440 \; cm^{-1}$. However, for the purposes of this analysis, only the absorption at $v = \sim 670 \; cm^{-1}$ was considered, since it does not correspond to any absorptions of the precursor mixture.

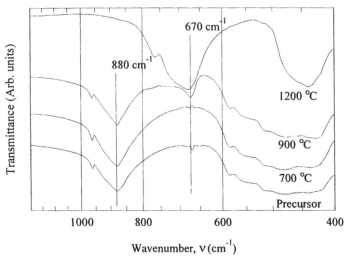

Figure 2   FTIR Spectra of precursor powder calcined at 1200 °C, 900 °C, 700 °C and as prepared

The FTIR trace produced by the sample calcined at 700 °C was essentially identical to that produced by the precursor material, showing that no conversion into the desired product had taken place. However, at 900 °C, there was a clearly visible absorption at $\nu = \sim 670$ cm^{-1}, which is indicative of Ni_2GeO_4 formation. Once 1200 °C has been reached, there were no signs of the absorptions due to GeO_2 or NiO, indicating that monophase material had been formed.

Nickel germanate formation is detected at a lower temperature using the FTIR technique, 900°C with FTIR as compared to 950 °C with XRD, due to the higher sensitivity of FTIR to small concentrations of material.

Sintering

Upon firing at 1300 °C, the pellets underwent a lateral shrinkage of (3.5±0.5)%, indicating that sintering of the ceramic grains had taken place. Density measurements of the sintered pellets revealed an average density of 3.5×10^3 kg m^{-3}, compared to a theoretical single crystal density of 6.1×10^3 kg m^{-3} [13]. This equates to a total porosity (open and closed) of 43%.

X-ray powder diffraction performed on the sintered pellets confirmed that the material was still monophasic, as shown in figure 3.

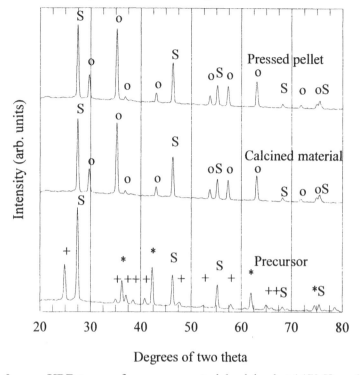

Degrees of two theta

Figure 3   XRD traces of precursor, material calcined at 1473 K, and pressed pellets sintered at 1573 K (S = Si, + = GeO_2, * = NiO, o = Ni_2GeO_4)

The microstructure of the pellets was investigated by scanning electron microscopy (SEM) and nitrogen adsorption isotherms, and showed these samples to be highly porous. The specific surface area of these samples, as measured by nitrogen adsorption isotherms via the B.E.T. method, was $(1.00\pm0.1)m^2g^{-1}$. The mean grain size of the material was measured from SEM micrographs using the line-intercept method, giving a value of $(0.5\pm0.2)\mu m$.

Electrical Properties

The d.c. resistivity of the pellet samples was observed to change with humidity as per figure 4.

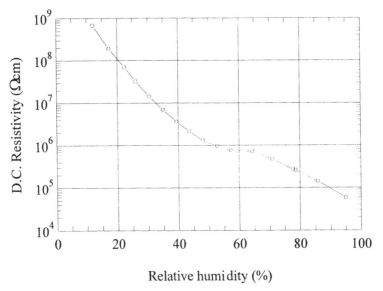

Relative humidity (%)

<u>Figure 4   D.C. resistivity vs. humidity of pellet sample at 293 K.</u>

The resistivity of the sample varied from $\sim 10^9$ Ωcm to $\sim 10^5$ Ωcm as the ambient humidity was changed from 10 %R_H to 90 %R_H at 293 K. The presently accepted model for this change of resistivity is that as the humidity is increased, water is adsorbed onto the surface of the ceramic grains, providing a path for conduction via protonic means [14]. However, the shape of the characteristic implies that at least two mechanisms of conduction are present in the system, one of which is dominant below 60 %R_H.

A.c. impedance spectra have been found to provide more explicit information about the nature of the processes responsible for conduction.

We have found previously that the impedance modulus of the pellets at 100 Hz varied from $\sim 10^8$ Ω to $\sim 10^4$ Ω over the humidity range 20 %R_H to 80 %R_H at 298 K [4]. At $R_H = 98$ %, the impedance-frequency characterstic showed three distinct regions at high ($> 10^6$ Hz), medium and low (< 10 Hz) frequency. At high frequencies, the impedance characteristic showed the usual roll-off due to capacitative effects. Over the middle part of the frequency spectrum where bulk charge transport processes were taking place, the impedance was largely independant of the frequency and strongly dependant on the humidity. At low frequencies, the impedance began to increase with decreasing frequency, due to charge transfer processes at the electrodes.

Cole-Cole complex plane plots of the impedance are presented for selected humidities between 22 %R_H and 98 %R_H in figure 5.

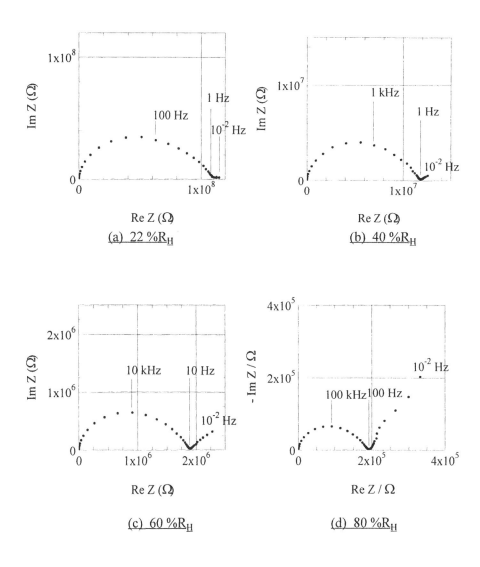

(a) 22 %R_H

(b) 40 %R_H

(c) 60 %R_H

(d) 80 %R_H

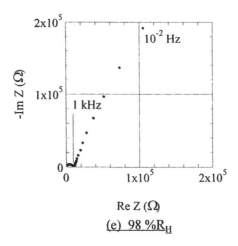

Figure 5   Cole-Cole plots of the impedance at humidities
between (a)22 %R_H and (e)98 %R_H

In all of the graphs presented in figure 5, two features are present in
varying degrees. There is a distorted semicircle, depressed slightly below the
axis, at the high frequency end of the graph (the end nearest the origin), and a
straight line tail-up at the low frequency end of the graph.

The semicircle corresponds to a parallel combination of a non-dispersive
d.c. resistance and a slightly dispersive capacitance [4]. The dispersion in the
capacitance accounts for the depressed and distorted shape of the semicircle. The
resistance corresponds to charge transport in the sample via surface conduction,
and the capacitance represents the geometrical and artifactual capacitances of the
measurement combined with modifications caused by the dielectric nature of the
adsorbed water. The inclined straight line is due to diffusion of ionic species
through the interfacial water layer formed at the gold electrodes, and other charge
diffusion processes occurring in the sample.

Observation of the graphs with respect to the change in humidity shows
that the size of the semicircle decreases with increasing humidity, which means
that the bulk resistivity of the sample decreased with increasing humidity.
However, the imaginary part of the impedance of the inclined straight line stays
essentially the same, and thus at high levels of humidity starts to dominate the
characteristic at increasingly higher frequencies. However the real part increased
with decreasing humidity suggesting that the phase angle is humidity dependant.
The full implications of this have still to be analysed.

                         Dielectric Ceramic Materials

At very low levels of humidity (< 20 %R_H), there are isolated water molecules present on the surface of the ceramic grains which dissociate into a proton and a hydroxyl group. The hydroxyl group bonds to the surface of the ceramic, whilst the proton is free to migrate across the surface by hopping from site to site [15]. At higher levels of humidity, layers of surface water are formed through which conduction may occur via diffusive motion of protons. As the humidity is increased, water is progressively adsorbed, increasing the water layer thickness. This allows an easier conduction path for the carriers currently present in the sample, but also increases the dielectric constant of the material, a consequence of which is an increased surface charge, which promotes the dissociation of the water leading to yet more carriers.

CONCLUSION

In summary, we have investigated the temperature of formation, and the electrical properties with respect to humidity, of the spinel ceramic material nickel germanate. Monophase material may be produced by firing an intimately mixed powder consisting of NiO and GeO_2 in the respective molar ratio 2:1, at 1200 °C for 12 hours.

The resistivity, and consequently the impedance, of pellet samples was noted to vary by several orders of magnitude as the humidity was changed from ~ 10 %R_H to ~ 90 %R_H, which is ascribed to a mechanism by which the ceramic grains become coated with an increasingly thick layer of water as the humidity is increased, thus increasing the opportunity for conduction by protonic means. It is proposed that at low humidities, charge transport is via a protonic surface hopping mechanism, which becomes superceeded by a diffusive motion of protons in the absorbed water layers at high humidity.

ACKNOWLEDGEMENTS

The authors would like to thank Elmwood Sensors UK Ltd. for their support. M. J. Hogan acknowledges the support of the UK Engineering and Physical Sciences Research Council.

REFERENCES

[1] G. Gusmano, G. Montesperelli, P. Nunziante *et al.*, "The electrical behaviour of $MgAl_2O_4$ pellets as a function of relative humidity", Mater. Eng., **3**, 417-434 (1992).

[2] S.S. Pingale, S.F. Patil, M.P. Vinod *et al.*, "Mechanism of humidity sensing of Ti-doped $MgCr_2O_4$ ceramics", Mater. Chem. Phys., **46**, 72-76 (1996).

[3] E. Traversa, "Ceramic sensors for humidity detection: the state-of-the-art and future developments", Sensors and Actuators B, **23**, 135-156 (1995).

[4] M. J. Hogan, A. W. Brinkman, and T. Hashemi, "Humidity dependant impedance in porous spinel nickel germanate ceramic", App. Phys. Lett., *In Press*, (1998).

[5] T. Hashemi, A.W. Brinkman, and M.J. Wilson, "Preparation, sintering and electrical behaviour of cobalt, nickel and zinc germanate", J. Mater. Sci., **28**, 2084-2088 (1993).

[6] A. Navrotsky and O.J. Kleppa, "Thermodynamics of formation of simple spinels", J. Inorg. Nucl. Chem, **30**, 479-498 (1968).

[7] A. Navrotsky, "Thermodynamics of formation of the silicates and germanates of some divalent transition metals and of magnesium", J. Inorg. Nucl. Chem., **33**, 4035-4050 (1971).

[8] A. Navrotsky, "Thermodynamic relations among olivine, spinel, and phenacite structures in silicates and germanates: II. The systems $NiO-ZnO-GeO_2$ and $CoO-ZnO-GeO_2$", J. Solid State Chem., **6**, 42-47 (1973).

[9] J. Preudhomme, "Correlations entre spectre infrarouge et cristallochimie des spinelles", Ann. Chim., **9**, 31-41 (1974).

[10] M. Lenglet and C.K. Jorgensen, "Optical spectra of Ni(II) in Ni_2GeO_4 and germanate spinels", Chem. Phys. Lett., **185** (1,2), 111-116 (1991).

[11] K. Hirota, T. Inoue, N. Mochida *et al.*, "Study of germanium spinels (Part 3)", J. Ceram. Soc. Japan, **98** (9), 976-986 (1990).

[12] A Koller, *Structure and Properties of Ceramics* (Elsevier, Amsterdam, 1994).

[13] JCDPS Card 10-266, , 1960.

[14] H. T. Sun, M. T. Wu, P. Li *et al.*, "Porosity control of humidity-sensitive ceramics and theoretical-model of humidity-sensitive characteristics", Sensors and Actuators, **19** (1), 61-70 (1989).

[15] J.H. Anderson and G.A. Parks, "The electrical conductivity of silica gel in the presence of adsorbed water", J. Phys. Chem, **72**, 3662-3668 (1968).

PROPERTIES OF DONOR DOPED BaTi(Mn)O$_3$ + SiO$_2$ SINTERED IN REDUCING ATMOSPHERES

David I. Spang and Ahmad Safari
Rutgers University
Department of Ceramic Science and Engineering
Piscataway, NJ 08855

Ian Burn
Degussa Corporation
3900 South Clinton Avenue
South Plainfield, NJ 07080

ABSTRACT

Studies into the effect of trivalent (Ho, Nd, Y) and pentavalent (Nb, V) donor dopants on the material properties of BaTi(Mn)O$_3$ + SiO$_2$ sintered under reducing conditions are presented. The effects of acceptor (Mn) concentration are also discussed.

INTRODUCTION

The goal of engineering base metal compatible dielectrics for multi-layer capacitors (MLC's) has long been present in both industry and academia. Dielectric formulations that are properly designed can prevent the material from taking on a semiconducting nature as a result of oxygen loss during firing. A substantial cost saving can then be achieved by utilizing Ni electrodes as internal metallization over the more expensive precious metals such as Pd or Ag/Pd. While some success has been achieved in designing Ni compatible dielectrics for high K applications such as Z5U and Y5V [1,2], the design of mid-K dielectrics for X7R capacitors is still quite empirical [3,4].

With the advent of chemical coating techniques, various donor and acceptor dopants can be uniformly applied to BaTiO$_3$ powders to impart reduction resistance and temperature stability to the formulations. Base metal formulations must have physical and electrical properties comparable to those of conventional air-fired noble metal compatible dielectrics. Reduction resistant formulations, however, do not always exhibit acceptable DC bias performance and it has been reported in the literature that various donor dopants can significantly improve life performance. Sakabe[4] reported properties of Ni MLC's utilizing Nb as a donor. Hitomi et.al. [5] have introduced a hypothesis on the effect of rare earth donor dopants such as Dy, Ho, and Y on the degradation mechanism. Kobayashi et.al. [6] have reported Ni MLC properties utilizing Y and V as donors. On the other hand, these various studies have been largely empirical and a general consensus as to the mechanism by which the donor modifications affect the life characteristics has not yet been reached.

BACKGROUND

Ni/NiO Equilibrium

MLC's utilizing Ni electrodes must be fired in an atmosphere with an oxygen partial pressure below that for Ni/NiO equilibrium. The equilibrium standard free energy (in calories) can be determined for the reaction NiO $\Leftrightarrow$ Ni + $1/2O_2$ at a particular temperature according to the relation $\Delta G = A + BT \log T + CT$ with A = 58450, B = 0, and C = -23.55 for the temperature range 298-1725K [7]. The equilibrium constant for this reaction is calculated according to the relation $\Delta G = -RT \ln K$ which converts to $\Delta G = -4.576 T \log K$ with ΔG in calories. The partial pressure of oxygen (pO_2) necessary to attain Ni/NiO equilibrium can then be determined from the fact that 2 log K equals log pO_2. Values of pO_2 necessary for various firing temperatures with Ni electrodes are given in Table I.

Table I. Ni/NiO Equilibrium pO_2 values

Temp (°C)	pO_2 (atm)
1150	2.19×10^{-8}
1250	3.30×10^{-7}
1350	3.57×10^{-6}

However, to prevent excessive interaction between the dielectric and the Ni electrodes in an MLC it is usually necessary to fire several orders of magnitude below the Ni/NiO equilibrium [1]. Therefore, firing atmospheres containing oxygen partial pressures of 10^{-11} atm were investigated in the current study.

Gas System

The reducing atmosphere was achieved by mixing two gas streams consisting of 5% H_2 (nominal) in N_2, and CO_2. This gas system was chosen due to

Dielectric Ceramic Materials

its ability to give the desired reducing potential as well as to avoid handling CO gas, which is toxic. The latter factor is important when considering a large scale production process.

Thermodynamic calculations facilitate the determination of K_p for the CO $+ 1/2O_2 = CO_2$ and $H_2 + 1/2O_2 = H_2O$ gas reactions which in turn can be used to calculate K_p for the $CO + H_2O = CO_2 + H_2$ reaction. From these relations the theoretical CO_2/H_2 ratios necessary to obtain a desired oxygen partial pressure can be calculated. Values of CO_2/H_2 ratios for various firing temperatures and pO_2 are tabulated in Table II below.

Table II. Theoretical CO_2/H_2 ratios

Temp (°C)	pO_2 (atm)		
	10^{-7}	10^{-9}	10^{-11}
1150	216.4	22.1	2.6
1250	45.9	5.1	0.79
1350	12.4	1.7	0.29

Oxygen Sensor

While, in principle, a predetermined oxygen partial pressure can be provided by a suitable mixture of CO_2 and H_2 gases, it is important that the actual furnace atmosphere can be accurately measured. In this work the pO_2 in the hot zone of the furnace was measured using a solid electrolyte sensor. The sensor consists of a closed end yttrium stabilized zirconia tube with a platinum lead on the outside of the tip of the tube exposed to the furnace atmosphere. Another platinum lead inside the tube is exposed to pure oxygen. The conduction of oxygen through the tube at firing temperatures produces an EMF that can be measured utilizing the platinum leads and a high impedance voltmeter. See Figure 1 for a schematic of the oxygen sensor.

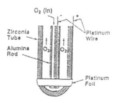

Figure 1. Schematic of Oxygen Sensor

The resultant voltage can be translated into an oxygen partial pressure according to the relation E (volts) = - 4.96 x 10^{-5} T log pO_2 [8]. A listing of voltage values for various firing temperatures and pO_2 are given in Table III.

Table III. Oxygen Sensor Voltage Readings For Varying Temperature and pO_2

Temp (°C)	pO_2 (atm)		
	10^{-7}	10^{-9}	10^{-11}
1150	0.49	0.64	0.78
1250	0.53	0.68	0.83
1350	0.56	0.72	0.89

The actual CO_2/H_2 ratio needed to produce the desired voltage can be compared to the theoretical ratio as described above to determine the integrity of the furnace seal. Even if a lower ratio is needed, indicating some air intrusion into the furnace, the voltage can be used to confirm that the desired atmosphere has actually been achieved.

EXPERIMENTAL

In the present study, high purity $BaTiO_3$ (HPBT-1, Fuji Titanium Co.) was coated with MnO and SiO_2 together with dopants selected from trivalent ions Ho $^{3+}$, Nd $^{3+}$, and Y $^{3+}$ or pentavalent ions Nb $^{5+}$ and V $^{5+}$. The MnO and SiO_2 comprised the base coating and were present to impart reduction resistance to the formulation and to act as a sintering aid, respectively. Table IV illustrates the typical properties of the starting $BaTiO_3$ powder.

Table IV. Physical Characteristics of the starting $BaTiO_3$ Powder

	Fuji HPBT-1
SA (m^2/g)	2.90 - 3.90
d90 (μm)	1.10 - 1.60
d50 (μm)	0.50 - 0.70
d10 (μm)	0.25 - 0.40
A/B	0.994 - 997

The initial base composition, 0.05 weight % MnO (0.16 mole %) and 0.1 weight % SiO_2 (0.38 mole %), was modified by high speed blending with a coating solution containing 0.103 mole % of Ho $^{3+}$, Nd $^{3+}$, Y $^{3+}$, Nb $^{5+}$, V $^{5+}$, each in separate coatings. This allowed the comparison of modifiers on a molar basis. The powders were coated and calcined in air at 7.5 °C/min to 450°C for 2 hours to evolve any residuals from the coating process and to aid in the conversion of metals to oxides.

In a second set of experiments, the base composition was modified to contain 0.1 weight % MnO (0.32 mole %) , i.e. twice the amount of MnO, and the same amount of SiO_2 (0.38 mole %) and donor ions (0.103 mole %).

The dielectric formulations produced were milled in an acrylic binder system using a two stage milling procedure. The milling consisted of the powder, solvent, and dispersant in the initial stage, milled for approximately 14 hours, and the binder and plasticizer in the second stage, milled for approximately 4 hours. The slurries were then tape cast using a Wallace laboratory tape caster on a mylar substrate. The tape was then removed from the substrate before stacking and laminating. Laminating conditions consisted of stacking 29 layers of dry cut tape and pressing at 50°C under 3200lb/in² for 2 minutes. Green thickness was approximately 0.5mm and the laminates were then cut into squares 7.3mm x 7.3mm. The laminates ,or K-squares, then underwent binder burn out in air by heating at a rate of 1°C/min to 300°C for 5 hours.

The laminates were fired in an alundum boat in a laboratory CM Inc. tube furnace at 1225-1250°C for 2 hours under an oxygen partial pressure of 10^{-11} atm (approximately 10^{-12} Mpa). The laminates were inserted into a mullite tube and pushed into the hot zone at the rate of 1cm per minute utilizing an alumina push rod with a platinum tie wire. The furnace temperature and atmosphere were adjusted prior to insertion of the boat into the hot zone. This procedure kept the firing profile of the laminates consistent between runs. The temperature of the hot zone inside the mullite tube was accurately measured using a type R thermocouple placed just above the tip of the oxygen sensor. The EMF produced by the oxygen sensor was measured using the high impedance voltmeter function of a HP34401A Multimeter (>10Gohms). Figure 2 below shows a schematic of the firing apparatus.

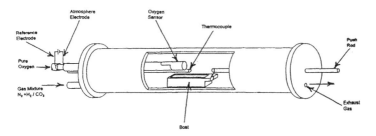

Figure 2. Schematic of the Firing Apparatus

Silver electrode paste (Dupont 1188D) was then applied to the surfaces and termination fired in air under the conditions of 15°C/min to 700°C for 10 minutes. The laminates were then leaded and tested for electrical properties such as capacitance (cap), dissipation factor (df), insulation resistance (IR), and temperature coefficient of capacitance (TCC). This procedure allowed for the quick determination of dielectric properties which were essentially independent of dielectric-electrode interactions.

When desired, multilayer capacitors (MLC's) with Ni internal electrodes were produced by a procedure similar to that for the laminates as described above. Screen printed Ni electrodes were applied to six of the internal layers to produce five active dielectric layers. The Ni pastes used were development products from Degussa Corporation (South Plainfield, NJ). The MLC's were cut to the dimensions of approximately 3.17mm x 3.68mm and processed with the same burn out and firing procedures as described for the laminates. The same electrical properties, capacitance, df, , IR and TCC were measured.

RESULTS

Table V shows the electrical results for the base composition of 0.05 weight % MnO (0.16 mole %) and 0.10 weight % SiO_2 (0.38 mole %) with the indicated dopant modification at levels of 0.103 mole %.

It can be seen from the Table that all formulations at these dopant levels reduced, as indicated by the zero values of IR at 25°C and 125 °C. All dissipation factor values (df) are 100% except for the V^{5+} doped composition which was 27.84% and exhibited a dielectric constant (K) of 2642. No K values are given for the other formulations as the high df (100%) produced artificially high capacitance values. The extremes of the TCC values, as compared to the EIA X7R specification can also be seen in Table V. The TCC behavior is more completely described graphically in Figure 3.

Table V. Laminate Properties of dopant modified 0.05 weight % MnO, 0.1 weight % SiO_2 composition

Dopant	cap (pF)	K	df (%)	IR (25°C) (Ohms)	IR (125°C) (Ohms)	% Dev (-55°C)	% Dev (125°C)
Nd^{3+}	27042	-	100	0	0	-41.1	45.6
Ho^{3+}	48383	-	100	0	0	-67.4	-43.7
Y^{3+}	13608	-	100	0	0	-21.3	94.8
Nb^{5+}	51931	-	100	0	0	-64.3	-55.1
V^{5+}	1083	2642	27.84	0	0	-39.6	94.4

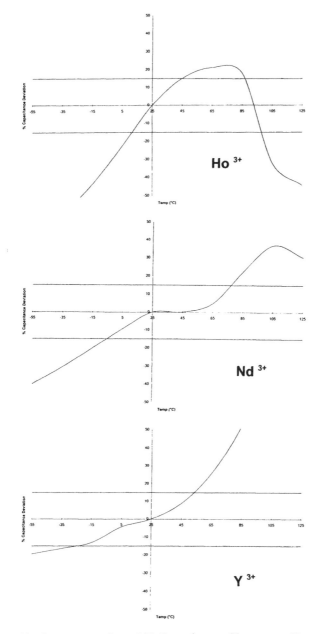

Figure 3. Graphical representation of % Capacitance Change vs. Temperature for the Formulations in Table V

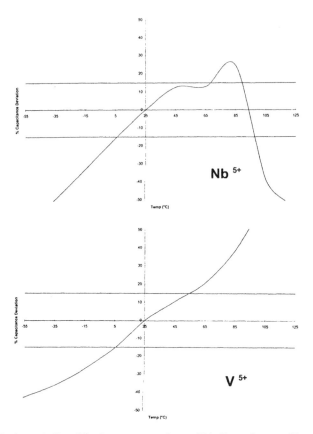

Figure 3. (cont.) Graphical representation of % Capacitance Change vs.
Temperature for the Formulations in Table V

In this Figure, it can be seen that there is little difference between the TCC behavior of the Y^{3+} and V^{5+} containing compositions. Both show essentially no suppression of the characteristic curie peak for $BaTiO_3$ at approximately 120°C. The Ho^{3+}, Nd^{3+}, and Nb^{5+} compositions, however, all show some suppression and shifting to lower temperatures of the curie peak of $BaTiO_3$. This information is potentially useful for attempting to ascertain the mechanism by which each of

Dielectric Ceramic Materials

these dopants alters the behavior of the formulation and when attempting to further modify these formulations to meet the EIA X7R specification.

Table VI below shows the electrical results for the indicated dopant modification, at a level of 0.103 mole %, to the composition containing 0.10 weight % MnO (0.32 mole %) and 0.1 weight % SiO_2 (0.38 mole %) which now has twice the MnO level that was previously discussed.

It can be seen from Table VI that now none of the formulations have reduced, as indicated by the IR values at 25°C and 125 °C. The dissipation factor values (df) are now much lower than in the previous formulation, with the highest value exhibited by Nd $^{3+}$ at 2.11%. The highest capacitance value was observed for Y $^{3+}$ while the lowest value was observed for Nb $^{5+}$. The extremes of the TCC values, as compared to the EIA X7R specification can be seen in Table VI. The TCC behavior for these formulations are more completely described in Figure 4.

Table VI. Laminate Properties of dopant modified 0.10 weight % MnO, 0.1 weight % SiO_2 composition

Dopant	cap (pF)	K	df (%)	IR (25°c) Ohms	IR (125°C) Ohms	% Dev (-55°C)	% Dev (125°C)
Nd $^{3+}$	911	2223	2.11	777	160	-37.0	76.5
Ho $^{3+}$	742	1810	1.42	485	118	-31.8	98.5
Y $^{3+}$	920	2245	1.55	774	132	-34.2	100.0
Nb $^{5+}$	660	1610	1.06	768	123	-34.5	70.0
V $^{5+}$	820	2000	1.06	619	98	-31.2	72.4

In Figure 4, it can be seen that there is essentially no difference in the TCC behavior between the different compositions. All the compositions show essentially the same TCC behavior with no suppression of the characteristic curie peak. This result suggests that, at these dopant levels, the TCC behavior is dominated by the base composition. While the higher MnO level is necessary to prevent reduction, it overcomes any TCC effects imparted to the formulation by the trivalent or pentavalent additions.

It is possible that the formulations of Table V and VI can be stoichiometrically modified to compensate for the trivalent or pentavalent dopant to give low df and reduction resistant material. Toward this aim, the initial base composition of 0.05 weight % MnO (0.16 mole %) and 0.1 weight % SiO_2 (0.38 mole %) had a BaO addition made at a level of 0.388 weight % BaO (0.58 mole %). This level of BaO was determined empirically to give the best electrical properties and will be used as a control for future modifications to the above mentioned formulations. Table VII shows the electrical properties for the stoichiometrically adjusted composition for both a laminate and a 5 layer MLC.

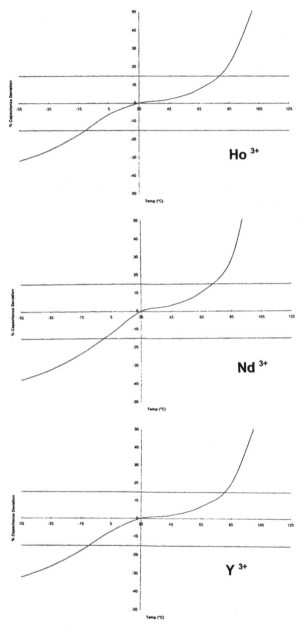

Figure 4. Graphical representation of % Capacitance Change vs. Temperature for the Formulations in Table VI

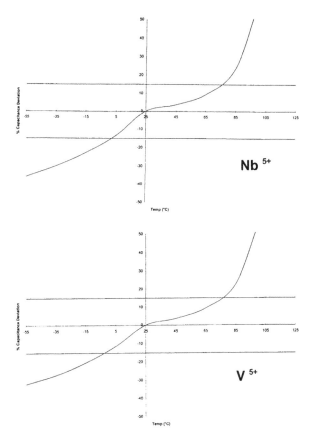

Figure 4. (cont.) Graphical representation of % Capacitance Change vs. Temperature for the Formulations in Table VI

Table VII. Electrical Properties of the stoichiometrically adjusted control composition

(0.05 weight % MnO, 0.10 weight % SiO_2, 0.388 weight % BaO)

	cap (pF)	K	df (%)	IR (25°c)	IR (125°c)	% Dev (-55°C)	% Dev (125°C)
Laminate	963	2350	0.98	777 Ohms	69 Ohms	-6.8	100.0
MLC	31700	1593	1.7	115 GOhm	8.8 GOhm	0	4.0

The TCC behavior for the laminate and MLC of this composition is graphically represented in Figure 5, again compared to the EIA X7R specification. The TCC curve for the laminate shows fairly stable behavior at temperatures below 105°C and exhibits the characteristically sharp curie peak at approximately 120°C. The TCC behavior for the MLC of this composition is graphically represented below in Figure 5, again compared to the EIA X7R specification. The curie peak appears to have been suppressed in the MLC as compared to the laminate possibly due to electrode-dielectric and/or termination-dielectric interactions.

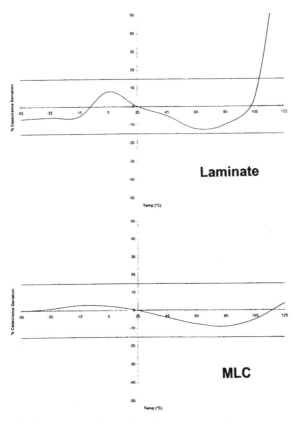

Figure 5. Graphical representation of % Capacitance Change vs. Temperature for the Laminate and MLC described in Table VII.

Dielectric Ceramic Materials

Figure 6 below shows a fracture surface of an MLC characteristic of the data presented in Table VII. The MLC shows reasonable density and acceptable dielectric-electrode compatibility.

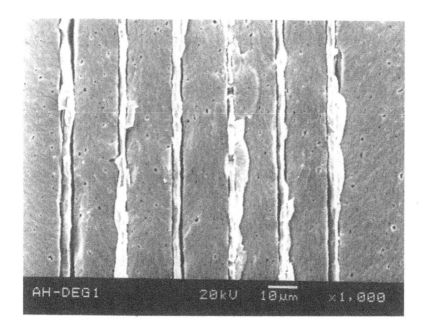

Figure 6. Fracture Surface of Base Metal MLC

FUTURE WORK
Future work includes stoichiometrically adjusting the compositions in Tables V and VI to see if properties comparable to those of Table VII can be achieved, both for laminates and MLC's. The formulations will also be life tested to see the effect of the various donors on the reliability of the base metal compositions for both laminates and MLC's.

ACKNOWLEDGMENTS
The authors would like to thank Degussa Corporation for their cooperation and enthusiasm in supporting this work as well as Rutgers University, The Malcom G. McLaren Center for Ceramic Research. The authors would also like to thank Mr. Mel Mehta of Degussa Corporation and Dr. Hyun Park of Kemet Corporation for providing benchmarking electrical and microstructural characterization of the MLC's.

REFERENCES

[1] I. Burn and G.H.Maher, "High resistivity $BaTiO_3$ ceramics sintered in $CO-CO_2$ atmospheres," J.Mater.Sci., 10 (1975) 633-640.

[2] Y. Sakabe , "Dielectric materials for base-metal multilayer ceramic capacitors," Ceram. Bull., 66(9) (1987) 1338-41.

[3] H. Kishi, Y. Okino, and N. Yamaoka, "Electrical properties and reliability study of multilayer ceramic capacitors with nickel electrodes," Program Summary and Extended Abstracts of The Seventh US-Japan Seminar on Dielectric and Piezoelectric Ceramics, pp.255-260 (1995).

[4] T. Nomura et al., "Aging behavior under DC field of Ni-electrode multilayer ceramic capacitors with X7R characteristics," Program Summary and Extended Abstracts of The Seventh US-Japan Seminar on Dielectric and Piezoelectric Ceramics, pp.265-268 (1995).

[5] A.Hitomi, et.al., "Hypothesis on Rare Earth Doping of $BaTiO_3$ Ceramic Capacitors", Proceedings of the 8th US-Japan Seminar on Dielectric and Piezoelectric Ceramics, p.44-47 (1997)

[6] Ryo Kobayashi, "Multilayer Ceramic Capacitors Incorporating Nickel Internal Electrodes and Ultra Thin Dielectric Layers", Proceedings of the 1st IEMT/IMC Symposium , p.372-377 (1997)

[7] O. Kubashewski, E.L. Evans and C.B. Alcock, "Metallurgical Thermochemistry," 4th Ed., Pergamon Press, p. 426 (1967).

[8] M. Sato, "Electrochemical Measurements and Control of Oxygen Fugacity and Other Gaseous Fugacities with Solid Electrolyte Systems," in *Research Techniques for High Pressure and High Temperature*, G.C.Ulmer, editor. Springer-Verlag, pp. 43-99 (1971).

PROCESS-STRUCTURE-PROPERTY RELATIONSHIPS IN RECRYSTALLIZING CaO-B$_2$O$_3$-SiO$_2$ LOW TEMPERATURE COFIRED CERAMIC FOR MICROELECTRONIC PACKAGING

Andrew A. Shapiro
Department of Chemical and Biochemical Engineering and Materials Science
Henry P. Lee
Department of Electrical and Computer Engineering
Martha L. Mecartney
Department of Chemical and Biochemical Engineering and Materials Science
University of California, Irvine
Irvine, CA 92607-2575

ABSTRACT

Microelectronic packaging for high frequency applications strongly depends on the selection of packaging materials due to interactions with the electric field. Low temperature cofired ceramic (LTCC) of CaO-B$_2$O$_3$-SiO$_2$ offers low loss dielectric characteristics and compatibility with Ag as the best high frequency conductor. Variability in loss properties has been a significant concern in application of this system. Four crystalline phases (β-CaO-SiO$_2$, CaO-B$_2$O$_3$, α-CaO-SiO$_2$ and 3CaO-SiO$_2$) and a glass phase were identified in varying quantities as a function of processing conditions. High frequency loss was correlated to the presence of the glass phase. A series of microstrip lines were measured from 0.5-18GHz with the conductor/dielectric fired from times of 6-60 min. at temperatures from 750-950°C. A first order theoretical model, based on changes in loss tangent, was developed that fit the data reasonably well over the frequency range.

BACKGROUND

Some of the most important microelectronic packaging applications are in the high frequency transmission market segment. Applications such as satellites, point-to-point communications, PCS and radars operating from L to K band require low loss circuitry for improved performance. The packaging is critical in these applications because the signal resides primarily outside the conductor line and interacts with the surrounding air and dielectrics.[1]

Key characteristics for microelectronic packaging in this environment include: low signal loss at high frequency; coefficient of thermal expansion; ability to tolerate harsh environments; low DC resistance; strength or toughness and good heat transfer.[2],[3] This study is focussed on the first requirement on the list, low signal loss at high frequency.

A low temperature cofired ceramic (LTCC) system consisting of a recrystallizing CaO-B$_2$O$_3$-SiO$_2$ glass with Ag conductors has been used in electronic packaging for more than ten years. Low temperature (<1100°C) firing offers the advantages of being able to use high conductivity metals such as Au, Ag or Cu. The cofiring of many layers simultaneously reduces process cost and variability due to a gradient of structure properties as layers are progressively fired[4]. Significant variability, however, was still observed[5]. The aim of this study was to identify the sources of the variability in loss and to determine the contribution of each source.

EXPERIMENTAL DESIGN

A general process flow is depicted in Figure 1. Starting from the left (A), via holes are punched in sheets of a green (unfired) tape. The flexible tape is generally a few mils thick consisting (in this system) of glass $CaO-B_2O_3-SiO_2$ particles, a few microns in diameter, incorporated in an organic binder. The contacts are made (B) by filling the vias and printing the conductor lines with Ag ink (Ag metal and glass frit). The sheets are then aligned and laminated together (C) in a heated press. The laminate is then processed through a burnout (D) and sintering (E) profile both in air. The burnout is to remove the organic binders completely without delamination or vaporization which causes blistering. The higher temperature firing allows sintering of both the metal in the vias and lines as well as the recrystallizing glass dielectric.

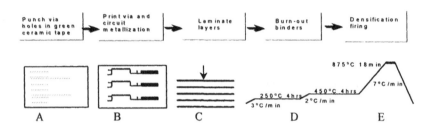

Figure 1. Simplified LTCC process flow.

Statistical design of experiments was used as a screen to determine the significant ranges of the variables being examined. The control variables were initially sintering time (6-60min), temperature (800-950°C) and heating ramp rate (1°C/min-10°C/min). Ramp rate was not found to be a significant variable. Traditional single variable experiments were then performed at the various times and temperatures in order to determine the optimal process parameters (Table I). The process experiments covered essentially the entire usable range of the $CaO-B_2O_3-SiO_2$ system with Ag. Below the low end temperature, the glass would not recrystallize or densify and the Ag would not sinter well. Above the high-end temperature the Ag conductors would melt. Single variable linear experiments were then performed to extend the range for time at temperature.

Additionally, an attempt at statistical variation reduction was employed. Several parts were fired at the same time to assess variation across the firing furnace. Parts were made from different lots of material to observe lot to lot variation. Identical firing profiles were repeated more than a year apart for general process variation evaluation. Table I shows replicated conditions as two X's.

The measured response variables for this system were: the presence of phases determined by x-ray diffraction; the amount of phases determined by x-ray diffraction; and rf loss.

Table I. Experimental design

		Temp.	(C°)	
		800	875	950
	6	XX	XX	XX
Time	12	X	XX	X
(min.)	18	X	X	XX
	24	X	X	X
	60			X

A twenty centimeters (eight in) microstrip was used as a measurement vehicle for rf loss. The critical dimensions used for the calculations are shown in Figure 2, where w and t are the height and thickness of the Ag conductor and h is the thickness of the dielectric.

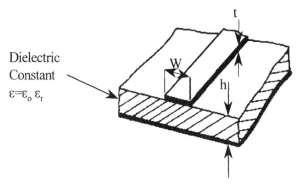

Dielectric
Constant
$\varepsilon = \varepsilon_o \, \varepsilon_r$

Figure 2. Microstrip critical dimensions [6]

The test set-up used a network analyzer (HP8510) for making rf measurements. Measurements were made from 0.5-18GHz. Fixturing was used for holding the microstrip for measurement with the network analyzer. A clamp was used to ensure a good contact and launch from the SMA connector to the circuit.

RESULTS AND DISCUSSION

The x-ray diffraction pattern in Figure 3 shows a typical profile from the center of the experimental matrix. The two primary phases are identified as $CaO-B_2O_3$ and $\beta-CaO-SiO_2$ and appear in roughly equal amounts are. Two additional phases of $\alpha-CaO-SiO_2$ and $3CaO-SiO_2$ were present in much smaller amounts and appeared to be transient appearing at low temperatures and shorter times and disappearing at longer times and higher temperatures. The two peaks marked with the asterisks were identified as isolated peaks for quantitative comparison of the two main phases. The relative peak values were measured using powder standards of the pure materials and mixed materials as weight percentages. They were found to follow a linear function, meaning that the proportional areas of the two peaks reflected the proportional weight percentage of each phase. In addition, the presence of some residual glassy phase can be seen by the amorphous background at lower angles.

Dielectric Ceramic Materials   349

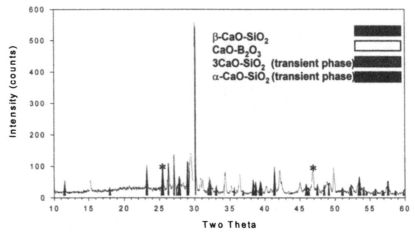

Figure 3. Typical (875°C and 12 min.) x-ray diffraction profile with phases identified.

The X-ray diffraction sequence for most of the experimental matrix is shown in Figure 4, with longer times with higher temperature at the end. The diffraction patterns were normalized to a quartz standard run both before and after the samples. The first few conditions at 800°C give primarily glass, and the final conditions at 950°C are nearly completely crystalline with $CaO-B_2O_3$ and β-$CaO-SiO_2$ appearing in approximately equal amounts.

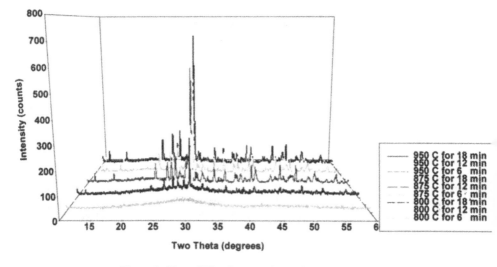

Figure 4. X-ray diffraction experimental sequence.

Dielectric Ceramic Materials

Figure 5 shows the ternary phase diagram[7] depicting the CaO-B$_2$O$_3$-SiO$_2$ system. The dotted line between the two primary phases shows the region for the recrystallization. An asterisk marks the approximate starting composition near the eutectic composition with a melting temperature of 1017°C. After crystallization, the composition should phase partition into CaO-B$_2$O$_3$ and β-CaO-SiO$_2$ in approximately equal amounts by the lever rule. This agrees with the x-ray diffraction experimental results. At lower temperatures phases such as 3CaO-SiO$_2$ were obtained, but it can be seen by the phase diagram that this is a non-equilibrium phase and not within any of the phase triangles including CaO-SiO$_2$ and CaO-B$_2$O$_3$. α-CaO-SiO$_2$ is a polymorph of β-CaO-SiO$_2$ but β-CaO-SiO$_2$ is most stable at high temperature. An attempt was made to quantify the glass content using a Fourier analysis of the x-ray diffraction patterns. The relatively short times and lower energy of the Cu target made the x-ray diffraction analysis inconclusive.

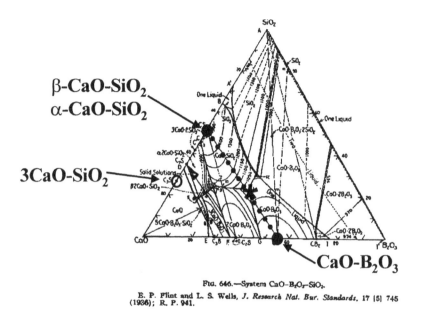

Figure 5. Ternary phase diagram for the CaO-B$_2$O$_3$-SiO$_2$ system[7].

Thus, electron dispersive spectroscopy (EDS) in the SEM was used for identification and quantification of phases. Figure 6 shows typical x-ray dot maps with a backscattered electron image. The dot maps are for Si, Ca and O. B is too light to be identified by the EDS system. The dark regions on the Si map show CaO-B$_2$O$_3$ (no presence of Si). Bright areas on the backscattered image show higher density β-CaO-SiO$_2$, confirmed by the Ca map which has a high yield (brighter region) for β-CaO-SiO$_2$ than for CaO-B$_2$O$_3$. Glass pockets appear as Si rich and Ca deficient areas. The glassy nature of these pockets was confirmed by TEM electron diffraction. The percentage of each phase was determined by a histogram of the pixel count on each map. The amount of glass could be determined by the EDS/SEM analysis. Figure 7 shows a surface of the percentage of glass calculated from the EDS/SEM maps as a function of the two primary process variables, time and temperature. Even under the longest firing time (60 minutes) and highest temperature (950°C) about 10% glass still remains. Both the $CaO-B_2O_3$ and β-$CaO-SiO_2$ crystallize in nearly equal amounts increasing in total volume going from lower temperatures (800°C) to higher temperatures (950°C).

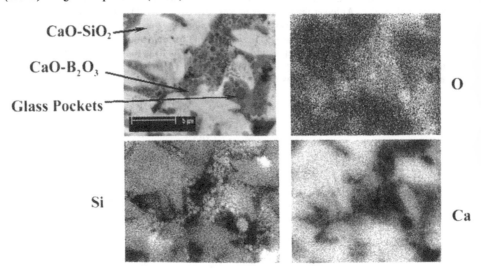

Figure 6. BSE and x-ray dot maps of a 950°C, 18 minute fired sample.

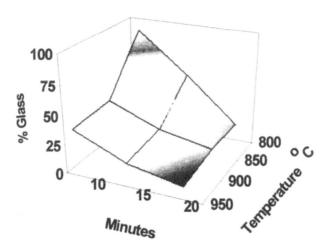

Figure 7. Percentage of glass as a function of process variables.

The transmission loss parameter S_{21} was measured using the meander microstrip line shown in Figure 2. The plot in Figure 8 shows the transmission loss in dB/cm for two different firings 800°C and 950°C at 6 minutes. These samples had 90% glass and 25% glass respectively. The determination of which phases are more responsible for the loss was made in the following manner. The relative amounts of the two primary phases $CaO-B_2O_3$ and $\beta-CaO-SiO_2$ were determined by x-ray diffraction and it was found that these two phases were present in nearly equal amounts for all processing conditions with only 5-10% of the transient phases present. High loss is seen at low temperatures and short times. From the percentage of glass shown in Figure 7 and the loss plot shown in Figure 8 it is clear that loss is correlated with the percentage of glass. Higher amounts of residual glass had higher losses. Samples with lower temperature firings had higher porosity, which may give the dielectric a slight decrease in dielectric constant, but the small amount of porosity (perhaps 5%) is not proportional to the loss seen.

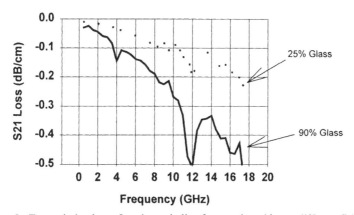

Figure 8. Transmission loss of a microstrip line for samples with two different firings.

ANALYSIS AND MODEL

A transmission line loss model was developed using a combination of models for conductor attenuation and dielectric attenuation. The dielectric attenuation model used is based on the work of Wheeler[8] and further developed by Hammerstad and Bekkadal[9]. The conductor attenuation is based on a model by Bahl[10]. The parameter list and values used for the calculations are found in Appendix I.

Equations were developed as first order approximations. Equation (1) calculates an effective dielectric constant, ε_{eff}, based on sample geometry. It assumes that the dielectric constant is constant with frequency (only true to a first order). Equation (2) is used to calculate the characteristic impedance, Z_0, of the microstrip. The conductor attenuation, α_c determined by equations (3) and (4), is calculated using the impedance as a function of the square root of frequency. As with the previous equations all other variables are assumed to be constant with frequency. The permittivity is assumed to be 1 (the material is not magnetic). Radiation losses are assumed to be negligible.[1]

$$\varepsilon_{eff} = \frac{W}{h}\frac{\varepsilon_r+1}{2} + \frac{\varepsilon_r-1}{2}\left[\left(1+12\cdot\frac{h}{W}\right)^{-.5} + 0.04\left(1-\frac{W}{h}\right)^2\right] \quad (1)$$

$$Z_o = \frac{120\cdot\dfrac{\pi}{\sqrt{\varepsilon_{efff}}}}{\dfrac{W}{h}+1.393+(0.660)\cdot\ell n\left(\dfrac{W}{h}+1.444\right)} \quad (2)$$

Conductor Loss $\quad \alpha_c = -6.1\cdot10^{-5}\cdot\left[\dfrac{R_s\cdot Z_o\cdot\varepsilon_{eff}}{h}\cdot\left[\dfrac{W_e}{h}+\dfrac{0.667\cdot\dfrac{W_e}{h}}{\dfrac{W_e}{h}+1.444}\right]\right]\cdot B \quad (3)$

Where $\quad B=\left[1+\dfrac{h}{W_e}\cdot\left(1.25\cdot\dfrac{t}{\pi\cdot W}+\dfrac{1.25}{\pi}\cdot\ell n\left(\dfrac{4\pi\cdot W}{t}\right)\right)\right] \quad (4)$

Equation (5) calculates the attenuation loss from the dielectric, α_d and is a linear function of loss tangent and frequency with all other variables held constant.

Dielectric Loss $\quad \alpha_d = -27.3\cdot\dfrac{\varepsilon_r\cdot(\varepsilon_{eff}-1)\cdot\tan\delta\cdot f\cdot10^9}{\sqrt{\varepsilon_{eff}}\cdot(\varepsilon_r-1)\cdot c} \quad (5)$

The total attenuation, α_T, can be estimated by a linear sum of the two attenuations in equation 6.

$$\alpha_T = \alpha_d + \alpha_c \quad (6)$$

The primary variable, in addition to frequency, that influences the model is the loss tangent of the material. The loss tangent for the well-fired material with a high amount of crystalline phase was found to be 0.0013, (see Xu[11] et al). Values for glasses range between 0.1 and 0.01.[12] A value of 0.05 was found to fit the loss data for the glass the best.

The loss model from equation (6) is plotted against measured data from the microstrip meander line in Figure 9. The loss tangent was chosen to correspond to the appropriate mixture of crystalline phases and glass.

The dominance of either dielectric or conductor in the loss is of general interest. According to this first order model, for well-fired substrates with a dielectric loss tangent of 0.006, the model predicts 78% of the loss is due to the conductor. A glassy substrate with a loss tangent of 0.05 has the dielectric responsible for 70% of the loss.

Dielectric Ceramic Materials

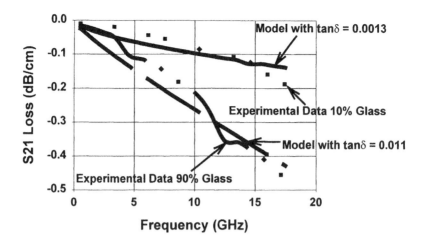

Figure 9. Measured and modelled data for microstrip loss.

This simple model shows the interesting phenomena that in a narrow range of firing conditions, the system switches from dominance by the dielectric to dominance by the conductor. The dielectric does not change composition significantly with longer firing times at high temperature, however, longer times at the high firing temperature of 950°C would undoubtedly show diffusion of the Ag conductor into the dielectric. The further diffusion will add to the roughness, from the electric field's point of view, so one would expect longer firing times (beyond those examined) to show greater loss. At first the increased loss would be due to the increased roughness and eventually due to the complete dispersion of the Ag into the dielectric.

SUMMARY

- Variability in the loss properties can be accounted for by changes in the phases of the recrystallizing glass dielectric. Specifically, the amount of glass present correlates to the rf loss.

- A first order model was developed using the sum of dielectric and conductor attenuations. The change in dielectric was modeled as a change in the loss tangent in a range comparable to known similar materials. The model fit the experimental data well.

- The loss dominance was shifted from the dielectric to the conductor as a result of longer processing times and higher processing temperatures.

- Future work is underway to improve the model by including second order functions of frequency.

REFERENCES

1. R. Sturdivant, C. Ly, J. Benson, and M. Hauhe, *Design and Performance of a High Density 3D Microwave Module*. preprint for *IEEE MTT-S*, IEEE (1997).
2. J.H. Alexander, S.K. Muralidhar, G.J. Roberts, T.J. Vlach., *A Low Temperature Cofiring Tape System Based on a Crystallizing Glass*. in *Proceedings ISHM International Conference on Microelectronics*. San Francisco, California: International Society for Hybrid Microelectronics 414-417 (1991).
3. D.I Amey, *Overview of MCM Technologies; MCM-C*. in *Proceedings ISHM International Symposium on Microelectronics*. San Francisco, California: International Society for Hybrid Microelectronics, 225-234 (1992).
4. M.F Bender, F.K. Patterson, E.A. Kemp, and J.E. Gantzhorn Jr., *Low-Temperature Cofired Ceramic Tape System Meets Industry Needs*, in *Hybrid Circuit Technology*. 23-26 (1989).
5. D.F Elwell and A.A. Shapiro, *CMI Tools in LTCC Component Development*. in *Proceedings, 3'd Annual Variability Reduction Symposium*. El Segundo California: Hughes Aircraft, 77-82 (1991).
6. T. Edwards, "Static-TEM Parameters and Designs at Low Frequencies"; pp. 52-54 *Foundations for Microstrip Circuit Design*. John Wiley & Sons, West Sussex 1991.
7. E.P. Flint and L.S. Wells, *The System Lime-Boric Oxide-Silica*. Journal of Research for the National Bureau of Standards, **17**[5] 727-752 1936.
8. H.A. Wheeler, *Transmission Line Properties of a Strip on a Dielectric Sheet on a Plane*. IEEE Transactions on Microwave Theory and Technology, **MTT-25** 631-647 (1977).
9. E.O. Hammerstad and F. Bekkadal, *Microstrip Handbook*, p.8, NTNF The Royal Norwegian Council for Scientific and Industrial Research: Trondheim 1975.
10. I. Bahl and P. Bhartia, "Transmission Lines and Lumped Elements"; p. 18 in *Microwave Solid State Circuit Design*. John Wiley & Sons, New York, 1988.
11. M. Xu, *Microstrip Losses at Microwave Frequencies*, Thesis Defense, J. Dougherty, L. Carpenter, J.S. Jang, and R.E. Newnham committee members. Center for Dielectric Studies at The Penn State University 1992.
12. A. von Hippel, "Dielectric Data"; pp. 301-407 in *Dielectric Materials and Applications*, Boston: Artech House 1995.

APPENDIX I

Parameters used for loss model.

f=0.5-18	Frequency (GHz)	ε_{eff}	Effective Dielectric Constant
$\sigma=6.17 \times 10^7$S/m	Conductivity of Ag	Z_0	Impedance
$\varepsilon_0=8.85 \times 10^{-12}$F/m	Permeability	α_c	Conductor Attenuation
$\varepsilon_r=5.6$	Relative Permeability	α_d	Dielectric Attenuation
μ	Permittivity	α_T	Total Attenuation
t=12μm	Conductor Thickness	c	The Speed of Light
W=1015μm	Width of Conductor	tanδ	Loss Tangent
h=625μm	Height of Dielectric	$\eta_0=376.73\Omega$	Free Space Impedance

Dielectric Ceramic Materials

PREPARATION AND ELECTRICAL PROPERTIES OF SrTiO₃-BICRYSTALS

Noboru Ichinose, Masanobu Nomura, Katsuhiko Yamaji*
Hajime Haneda** and Junzo Tanaka**
School of Science and Enjineering, Waseda University
3-4-1 Ohkubo Shinjuku-ku, Tokyo 169-8555, Japan
*National Institute of Materials and Chemical Research
1-1 Higashi, Tukuba, Ibaraki 305-0047, Japan
**National Institute for Research in Inorganic Materials
1-1 Namiki, Tukuba, Ibaraki 305-0047, Japan

ABSTRACT

In order to make model interfaces for $SrTiO_3$ grain boundaries, $SrTiO_3$ bicrystals with twisted and tilted angles were prepared. In the case of twisted angles, two Nb-doped $SrTiO_3$ single crystals were contacted by HIP in air. Annealed HIP-bounded samples in air showed nonlinear resistive properties depending on both of the twist angle and annealed temperature. Nonlinear coefficient (α) had the maximum value of 7.3 at $45°$ of twist angle and annealed temperature of 1673K. On the other hand, tilted angles were prepared by hot-pressing method. Nonlinear properties for tilted angles were different from twisted ones, but they are not almost depending on the tilted angle and equal to 8. A molecular dynamics calculation was applied in order to understand the detailed interface structure. The periodicity of the $\Sigma 5$ twisted angle was calculated to be 1.58 times longer than the lattice constant of $SrTiO_3$ along the [310] axis, and agreed with the periodicity observed by a high resolution transmission electron microscope, i.e. 1.57.

INTRODUCTION

Grain boundaries play an important role in physical and electrical properties in many ceramics.[1] For example, $SrTiO_3$ and ZnO ceramics exhibit nonlinear current-voltage characteristics which are influenced by structure, impurity-segregation etc. of grain boundaries.[2] Fujimoto et al.[3] formed $SrTiO_3$ bicrystals joined by a thin film of molten Bi_2O_3 and evalued their electrical properties and structures. These experiments indicated that bicrystals were useful for the elucidation of the structure and properties of the grain boundary.

In this paper, $SrTiO_3$ bicrystals with twisted or tilted angles were prepared for the purpose of making model interface for $SrTiO_3$ grain boundaries. Furthermore, the interface structures of the $\Sigma 5$ twisted and $\Sigma 13$ tilted boundaries in the concidence site lattice (CSL) notation,[4] as shown in Fig.1 were investigated with a high resolution transmission electron microscope (HRTEM) and molecular dynamics (MD) calculation.

EXPERIMENTS

Samples

Strontium titanate [001] single crystals with 0.05 mol% Nb_2O_5 and 99.99% purity were supplied by Furuya Materials Japan. One side of the crystal sheets was mirror finished by polishing with diamond paste (0.3 μm). The polished sample was cleaned in the ultrasonic cleaner in acetone for 2 min. To form twisted interfaces rotated around the [001] axis, the crystals were fixed at 0° for a $\Sigma 1$ boundary, at 36.9° for a $\Sigma 5$ boundary and at 22.6° for a $\Sigma 13$ boundary using Pt wires. The crystals were fired at 1873K for 2 hours under 50MPa in Ar by a hot isostatic press (HIP) method. Finally the crystals were annealed at 1573K for 2 hours in air.

In the case of tilted boundaries, all samples were prepared at 1623K for 5 hours under 0.6MPa in air by hot pressing. For checking current-voltage characteristics, post annealing in air or Ar at 1573K for 2 hours and diffusion of Bi_2O_3 using Bi pasted material near boundary at 1373K for 1 hour were tried.

Measuring Method

Contacting interfaces were investigated by a scanning electron microscope (SEM) and TEM. Specimens for TEM observation were prepared by ion milling with an Ar^+ ion beam of 4keV and a grazing incidence angle of 12°. The specimens were observed at 800kV by HRTEM (HITACHI, H-1500, JAPAN).

The current-voltage (I-V) characteristics of the bicrystals obtained were measured by using In-Ga alloy electrodes. The I-V relation was measured by the Curve Tracer with 50Hz at room temperature. The capacitance under the constant DC bias was measured to check the interfacial barrier using Impedance Analyzer at room temperature.

Molecular Dynamics Calculation Method

MD calculations were carried out for various $SrTiO_3$ interfaces using the following potential function :

$$u_{ij} = \frac{Z_i Z_j e^2}{r_{ij}} + f_0(b_i b_j)\exp\left(\frac{a_i + a_j - r_{ij}}{b_i + b_j}\right) - \frac{c_i c_i}{r_{ij}^6}$$

Here, Z_i and Z_j are the charges of the i-th and j-th ions, respectively, and r_{ij} is their distance. $a_{i,j}$ and $b_{i,j}$ are the potential parameters corresponding to ionic radius and ionic stiffness, respectively. $c_{i,j}$ correspondings to the *van der waals* potential related to a non-symmetric electronic distribution. e is an electron charge and f_0 is a constant.

As given by the above equation, the potential function consists of three terms : the first term is an ionic two-body potential, i.e. coulombic interaction, the second term Born-Mayer type repulsive interaction, and the third term *van der waals* attractive interaction. The parameters used in the present paper are given in Table 1. The program used for the MD calculation was MXDORTO written by Hirao et al. [5] and running on a DEC Alpha Server 2100 5/250.

The equations of motion were intergrated using a predictor method based on the verlet algorithm with a time step of $\Delta t = 2.5$ fs. All the calculations were performed at constant temperature and pressure.

RESULTS AND DISCUSSION

Interfacial structure of a SrTiO₃ bicrystal observed by TEM

Figure 2 shows the TEM image of a Σ 5 twisted boundary in a $SrTiO_3$ bicrystal. This image was observed along the [310] axis and the interface was well joined even at atomic scale. At the Σ 5 twisted boundary, the lattice constant along the c-axis adjacent to the interface is 1.3 times longer than that of the original $SrTiO_3$ single crystal, and the periodic length along the interface is 1.57 times longer than that of the $SrTiO_3$ crystal. The periodicity of $1.57a_0$ corresponds to half of the length of the Σ 5 consident site lattice when the bicrystal is viewed along the [310] axis.

Electrical properties of SrTiO₃ bicrystals

Figure 3 shows the current-voltage (I-V) characteristics of $SrTiO_3$ twisted bicrystals measured along the c-axis. The I-V characteristics are nonlinear, suggesting that an interfacial barrier formed at the bicrystal interface. From Fig.3, it is found that the nonlinearity of I-V characteristics increases in the order of twisted angle. It is thus considered that a higher electrical barrier forms at a twisted boundary with more disordered atomic configuration can probably induce interfacial states which are separated from the conduction band in energy.

For the tilted bicrystals, they show almost same nonlinear coefficient (α) value of around 8 not depending on the tilted angle as shown in Fig.4. For the annealed samples in Ar atmosphere, they did not show the nonlinear characteristics. However, nonlinear characteristics were appeared after annealing samples in air again. Diffusion of Bi_2O_3 in tilted boundary is also effective for appearance of non-linear conduction. Capacitance is decreasing with increase of voltage for nonlinear resistive samples as found in Fig.5. From above experimental facts, double Schottky barrier and interfacial state may be induced by adsorbed oxygen in boundaries. For investigating the density of state in the tilted boundary, DV-X_α calculation was applied for model clusters as shown in Fig.6. It is found from these calculations that molecular orbitals consisting mainly of O-2p orbital are existing in the band gap as shown in Fig.7. These are corresponding to interfacial state.

Mechanism for nonlinear resistivity

Figure 8 represents the model for mechanism of nonlinear resistivity. The adsorbed oxygen plays an important role to make the formation of the energy barrier as reported in [6]. It is found in Fig.8 that the interfacial state is formed by electrons coming from donor level.

CONCLUSIONS

$SrTiO_3$ bicrystals with twisted and tilted boundary were prepared by a HIP and hot pressing method, respectively. The bicrystals exhibited nonlinear I-V characteristics. In the case of twisted boundary, they are depending on the twisted angles. However, they are not depending on angles for tilted boundary. The maximum value of nonlinear coefficient is around 8 for both cases. C-V characteristics suggest that there is interfacial state in the bicrystal interface.

A MO calculation was applied in order to understand the detailed interface structure. The periodicity of the $\Sigma 5$ twisted boundary was calculated to be 1.58 times longer than the lattice constant of $SrTiO_3$ along the [310] axis and agree with the periodicity observed by HRTEM, i.e. 1.57.

DV-X_α calculation using cluster model for $SrTiO_3$ bulk, surface and interface shows that interfacial state formed by electron coming from donor level exsists in the $(TiO_5)^{-6} + O^{-2}$ cluster. This may be origin of mechanism for nonlinear resistivity.

ACKNOWLEGMENT
This work was supported by " Frontier Ceramics " project from Science and Technology Agency of Japan.

REFERENCES
[1]L. M. Levinson : " Zinc Oxide Varistors-A Review " Ceram. Bull. **68** (1989) 866
[2]M. Fujimoto : " Potassium Grain Boundary Segregation and Site Occupancy in SrTiO$_3$ Ceramics " Jpn. J. Appl. Phys. **26** (1987) L2065
[3]M. Fujimoto, N. Yamamoka and S. Shirasaki : " The Electrical Properties of SrTiO$_3$ Bicrystal Joined by Thin Film of a Molten Bi$_2$O$_3$ " Jpn. J. Appl. Phys. **26** (1987) 1594
[4]W. Bollman : " Crystal Defects and Crystalline Interfaces " Berlin, Springer-Verlag (1970)
[5]K. Hirao and U. Kawamura : " Material Design by Personal Computer " (1994)
[6]S. Fujitsu, H. Toyoda and H. Yanagida : " The Enhanced Diffusion of Oxygen in ZnO Varistor " J. Ceram. Soc. Japan **96** (1988) 119

Table 1. Potential parameters used for the MD calculation.
The meanings of z, a, b and c are given in the text.

Parameter	z	a	b	c
O	−1.331	1.629	0.1113	20.0
Ti	2.662	1.235	0.0850	0.0
Sr	1.331	1.632	0.0850	15.0

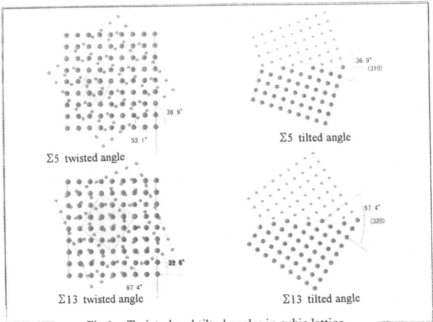

Fig.1　Twisted and tilted angles in cubic lattice

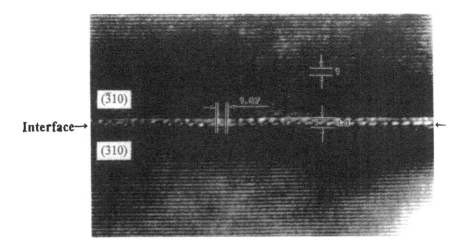

Fig.2　Σ 5 twisted boundary of SrTiO₃ bicrystals
observed by HRTEM along the [310]-axis

　　　　　　　　　　　　　　Dielectric Ceramic Materials

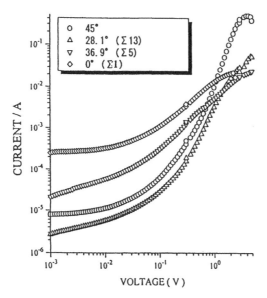

Fig.3   I-V characteristics of bicrystals

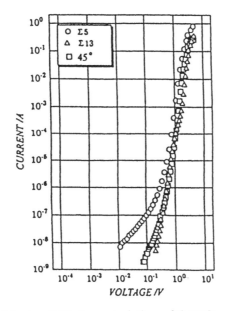

Fig.4   I-V characteristics of SrTiO₃ bicrystals with tilted angle

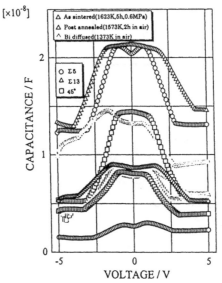

Fig.5   C-V characteristics of SrTiO$_3$ bicrystals
with twisted or tilted angle

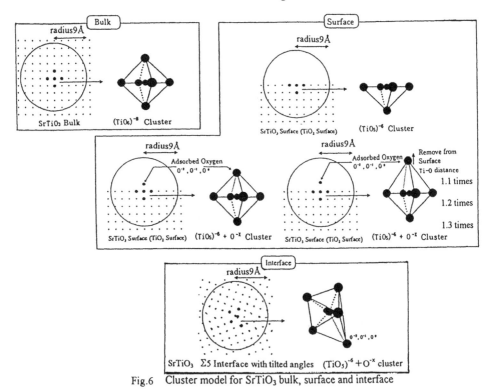

Fig.6   Cluster model for SrTiO$_3$ bulk, surface and interface

Dielectric Ceramic Materials

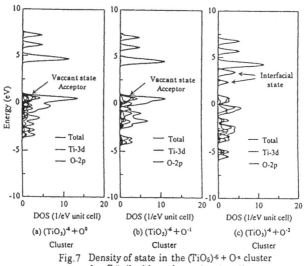

Fig.7 Density of state in the $(TiO_5)^{-6} + O^{-x}$ cluster
for $\Sigma 5$ tilted boundary $(x = 0, 1, 2)$

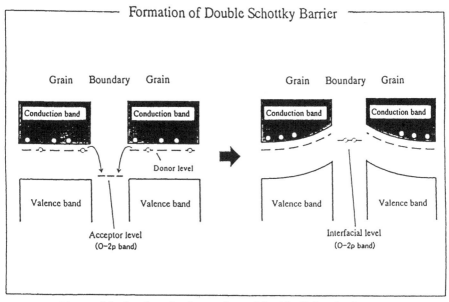

Fig.8 Mechanism for formation of double Schottky barrier

ELECTRICAL CHARACTERIZATION OF SOL-GEL DERIVED Pb(Zr,Ti)O$_3$ CERAMICS

Ming Dong and Rosario A. Gerhardt
School of Materials Science and Engineering, Georgia Institute of Technology
778 Atlantic Drive, Atlanta, GA 30332-0245, USA

ABSTRACT

PbZr$_{0.53}$Ti$_{0.47}$O$_3$ fine powders were synthesized by an air-stable sol-gel route. These fine powders allow sintering PZT ceramics at much lower temperature than 1200°C. The specimens sintered at 900°C attained about 97% of the theoretical density and had a grain size of about 2μm. The frequency and temperature dependence of the dielectric properties were determined from room temperature to 500°C at frequencies ranging from 5 to 10^7 Hz. The PZT ceramic shows a Curie temperature of 400°C and a large frequency dependence at T$_c$ due to highly conductive grain boundaries. At room temperature, the dielectric constant k and the dissipation factor tanδ are 820 and 0.025 respectively at 1kHz. The ac and dc resistivity measurements coincide and show an activation energy of 1.10eV. The bulk conductivity of the PZT ceramic is about 10^{-12} Sm^{-1} at 40°C.

INTRODUCTION

Lead-based perovskite ferroelectric ceramics, such as Pb(Zr$_x$Ti$_{1-x}$)O$_3$, have been the subject of extensive research for many decades because of their exceptional piezoelectric, ferroelectric, dielectric, and electro-optic properties. The quality of the sintered PZT ceramics strongly depends on the chemical homogeneity, particle size, and morphology of the starting powder. Synthesis and sintering processing always play major roles in determining properties of the sintered ceramics. These ceramics are normally sintered at temperatures higher than 1200°C. Loss of PbO, which leads to the degradation of electric properties, is inevitable at these temperatures because of its high volatility. Typically, packing powders are used to prevent or suppress the PbO loss during sintering [1,2]. Many researchers have used low temperature sintering techniques to eliminate the PbO loss and control the microstructure. The different techniques so far reported for the low-temperature sintering of PZT are (1) cationic substitutions in the compositions [3,4], (2) addition of low melting glass, oxide and oxyfluoride [5-7], (3) hot pressing process[8], and (4) using very fine and chemically active powders prepared by sol-gel process [9-12]. The intrinsic nature of sol-gel process offers some advantages over other methods, such as high purity, molecular homogeneity and low-temperature synthesis processing [13].

Our interests in PZT are twofold. First, we want to obtain low defect and high-density ceramics by low-temperature sol-gel process. Second, we wish to carry out an ac impedance study of PZT ceramics in order to understand better the influence of microstructure on the electrical properties of this type of ferroelectric material.

EXPERIMENTIAL

The starting reagents for PZT (53/47) precursor solution synthesis were lead acetate trihydrate (Alfa Co. Grade 1), titanium (diispropoxide) bis (2, 4-pentanedionate) (75% in isopropanel, Alfa Co.), zirconium(IV) acetylacetonate (Fluka Chemika) and 1,3 -propanediol (Aldrich Co.). The details of the sol preparation procedure are described in reference [13]; sols were prepared with 10% excess lead acetate trihydrate to compensate for PbO losses during firing. A stock solution (about $1.3g/cm^{3)}$ was synthesized by the above route. This solution was then concentrated to bulk gel by vacuum evaporation at 120°C. The dry gel was transferred into a box furnace set at 350°C for 30 minutes and then transferred into a box furnace set at 700°C for 30 minutes. X-ray diffraction analysis of obtained powders showed that the samples were single pseudocubic "perovskite" PZT phase (Figure 1).

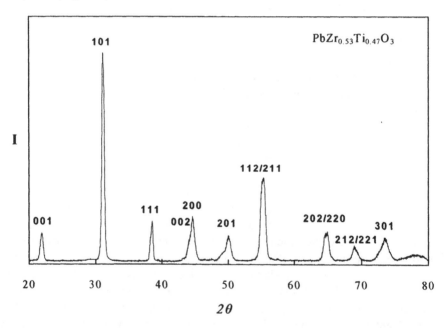

Figure 1. X-ray spectrum for $PbZr_{0.53}Ti_{0.47}O_3$ powder and ceramic.

Ceramics were then densified by sintering cold-pressed disks at 900°C in air for 2 hours. The shrinkage is greater than 14%. The relative density is higher than 97%. A new XRD analysis on finely pulverized ceramic powder showed that the ceramic is the same phase as the synthesized powder. For SEM analysis, the ceramic was polished to 0.3 μm finish and thermal etched at 850°C for 30 minutes. Silver-paste electrodes were deposited on opposite faces of the ceramic disks. The electrical measurements were carried out from room temperature to 500°C in dry air. A preliminary heating-cooling cycle was carried out to ensure measurement reproducibility. Frequency sweep from 1 to 10^6 Hz with a 500mV nominal applied voltage was carried out using a HP 4192a LF Impedance Analyzer. Dc conductivity measurements were performed with a 500 mV constant dc voltage source using Keithley 617 Electrometer/Source. For both ac and dc measurements, the temperature was maintained at each measurement temperature point for 30 minutes with an accuracy of ±0.5°C before collecting the data.

RESULTS AND DISCUSSION

The SEM image of the PZT ceramics indicates that the grain size is about 2μm and the thickness of grain boundaries are less than 0.05 μm (Figure 2).

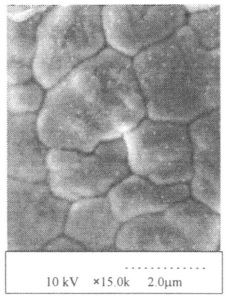

10 kV   ×15.0k   2.0μm

Figure 2. SEM image for PZT ceramic.

Data directly measured as impedance Z^* were converted into complex resistivity ρ^*, electric modulus M^* and complex relative permittivity ε^*, using the relations:

$$\rho^* = Z^*k = Z^*(S/d), \quad M^* = j\omega C_o Z^*, \quad \varepsilon^* = 1/(j\omega C_o Z^*) = \varepsilon_r' - j\varepsilon_r''$$

where S is the area of the silver electrode, d is the thickness of the ceramic sample, C_0 is the vacuum capacitance and $j = (-1)^{0.5}$.

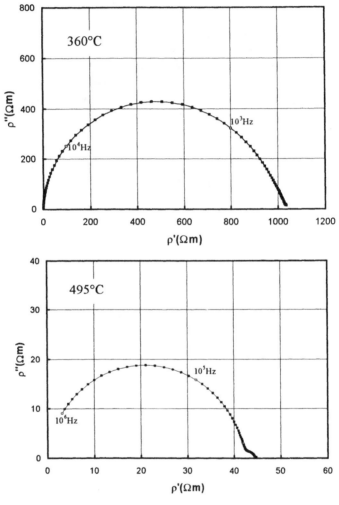

Figure 3. Typical complex resistivity plots of the PZT Ceramics.

Dielectric Ceramic Materials

Typical complex resistivity ρ^* plots are shown in Figure 3. At temperatures below 400°C, similar ρ^* plots with only one depressed arc were obtained. At temperatures higher than 400°C, a second arc became visible at low frequencies. In these plots each arc can be represented by a parallel RC component which corresponds to the individual active region in the ceramic. The characteristic relaxation time τ of each RC element is given by the product of R and C, i.e. $\tau=RC$. RC elements are theoretically separable due to relationship, $\omega_{max}RC = 1$, which holds at the frequency of maximum loss, ω_{max}, in impedance spectrum. At least two electrical active regions should exist in the ceramic sample. At low temperature the time constant τ of the two regions are very close, the two arcs are overlapped; as the temperature increases, they differ from each other. The two arcs are then separable. The resistance and the capacitance of each RC component can be obtained by CNLS fitting [14]. The relative capacitance values for the high frequency arcs are about $10^{-10}\sim10^{-9}$ F/cm, which correspond to the bulk ferroelectric near Curie temperature; while those for the low frequency arcs are about 10^{-11}-10^{-8} F/cm, which correspond to the grain boundary [15]. Fitting the high frequency arcs with CNLS program, the bulk conductivity σ could be obtained. Figure 4 exhibits the Arrhenius plot of σ obtained by CNLS fitting. For comparison, the dc conductivity measurement data are also presented in the same figure. Because the instrument's limit, the ac conductivity data could only be

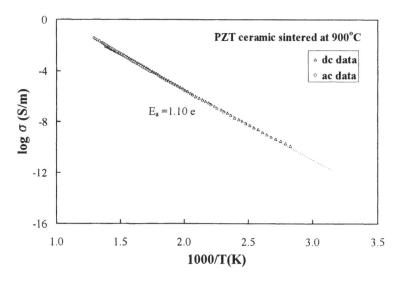

Figure 4. Arrhenius plot of σ of PZT ceramic.

obtained at elevated temperature. We notice that the ac and dc data match well, both of them are related to the same process thermally activated with an activation energy of about 1.10eV. This value is in general corresponding to an ionic conduction process. From Figure 4, we can estimate that the ionic conductivity of the PZT bulk is about $10^{-12}Sm^{-1}$ at 40°C, which is in agreement with a previously reported value [16]. As the ceramic has high density, the dimension of grain boundary are very small, or the resistance of the grain boundary is very small (the conductance is very high), the contribution of bulk grains to the resistance are thus more prominent.

Figure 5 shows the temperature dependence of dielectric constant k (dielectric constant k = relative permittivity ε_r') of the PZT ceramic at 8 frequencies. The k maximum appears at about 400°C, which is the ferroelectric-

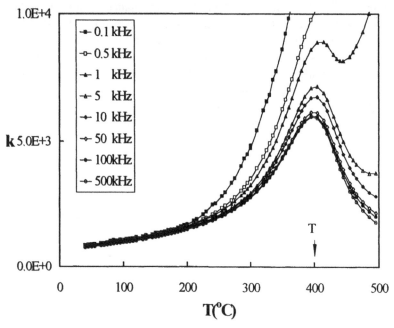

Figure 5. Temperature dependence of the dielectric constant k
of PZT Ceramic at 8 different frequencies.

paraelectric phase transition point or Curie temperature T_c. The dielectric constant is about 800 at room temperature, but it is larger than 5900 (at 500kHz) at T_c. The low frequency dispersion are noticeable at temperatures above 200°C. As the frequency gets lower than 0.5kHz, the dielectric dispersions are very strong that the "dielectric anomaly" are overshadowed and a peak is no longer discernible.

This results from the ionic conduction that occurs in the ceramic [17-20]. The ionic conduction may result from the oxygen vacancies in the PZT crystal caused by the volatilization of PbO. Ionic conduction in ferroelectric materials will cause a low frequency dielectric dispersion as the charge carriers (O^{2-}) will be blocked at the ceramic-silver electrode interface region under low-frequency electrical stimulus. Very commonly, dielectric constant k is obtained at a single frequency only, often 1 or 10 kHz and then plotted either directly against temperature or as Curie-Weiss plots of reciprocal dielectric constant against temperature. Curie-Weiss plots are shown in Figure 6 at eight fixed frequencies for the PZT ceramic. The high frequency (higher than 50kHz) plots are almost linear at paraelectric state (at temperatures higher than T_c). However, at lower frequency, gross departures from linearity occur. The deviations from linearity in Figure 6 could be explained by the frequency dependence of the dielectric properties.

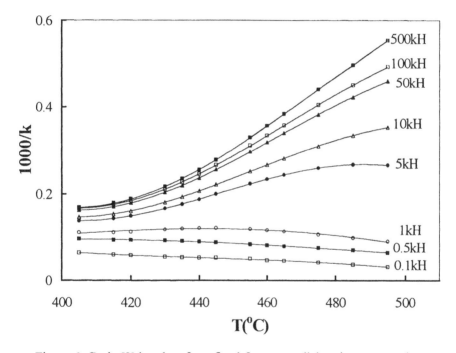

Figure 6. Curie-Weiss plots from fixed-frequency dielectric constant data.

Figure 7 shows the frequency dependence of the dielectric constant at four temperatures around T_c. In the frequency range higher than 50kHz, the dielectric constant approaches the maximum at T_c, as temperature deviates (increasing or

Dielectric Ceramic Materials   373

decreasing) from T_c, the dielectric constant decreases. This is due to the ferroelectric grain or bulk behavior. At lower frequencies, the low-frequency dispersion which results from the ionic conduction and the grain boundary effect is very strong, the higher the temperature, the higher the dielectric constant. The ferroelectric grain or bulk behavior are masked.

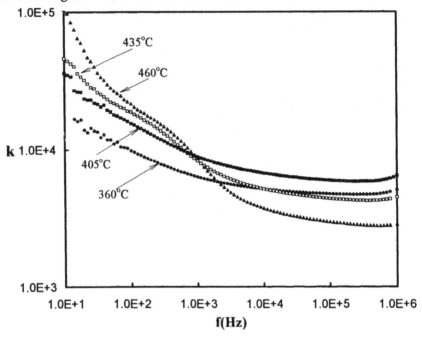

Figure 7. Frequency dependence of the dielectric constant of PZT ceramic.

CONCLUSIONS

Low-temperature sintering of PZT ceramic at 900°C has been successfully attained with sol-gel derived $PbZr_{0.53}Ti_{0.47}O_3$ powder. The high-density (97%) PZT ceramic has high dielectric constant of 800 and low dissipation factor tanδ of 0.025 at room temperature. The Curie temperature T_c is found to be 400°C.

Based on dc and ac electrical analysis, we were able to distinguish two electrical active regions in the ceramic: the ferroelectric grain or bulk region, which has a higher characteristic relaxation frequency; and the grain boundary, which is more conductive and has a relaxation at low frequency.

Dielectric Ceramic Materials

The ionic conductivity of the PZT grain or bulk is about 10^{-12}S/m at 40°C and has an activation energy of 1.10eV. The ionic conduction probably results from the oxygen vacancies in the PZT crystal caused by the volatilization of PbO. The frequency dispersion due to conductive grain boundaries affect the dielectric properties even at high frequency in the PZT ceramic. In conclusion, fixed frequency dielectric measurement at frequencies lower than 100kHz could not accurately represent the ferroelectric bulk behavior of the PZT ceramic sintered at 900°C because of the substantial contribution of the grain boundaries to the dielectric reponse.

REFERENCES
[1] K. Okazaki, I. Ohtsubo and K. Toda, "Electrical, optical and acoustic properties of PLZT ceramics by two-stage processing", *Ferroelectrics*, **10**, 195-97 (1976).
[2] S. Chiang, M. Nishioka, R. M. Fulrath and J. A. Pask, "Effect of processing on microstructure and properties of PZT ceramics", *Am. Ceram. Soc. Bull.*, **60** [4], 484-89 (1981).
[3] P. G. Lucuta, F. Constantinescu and D. Barb, "Structure dependence on sintering temperature of Lead Zirconate-Titanate solid solutions", *J. Am. Ceram. Soc.*, **68** [10], 533-37(1985).
[4] R. B. Atkin, R. L. Holman and R. M. Fulrath, "Substitution of Bi and Nb ions in Lead Zirconate Titanate", *J. Am. Ceram. Soc.*, **54** [2] , 113-15(1971).
[5] S. Takahashi, 'Sintering Pb(Zr,Ti)O$_3$ ceramic at low temperature", *Jpn. J. Appl. Phys.*, **19**, 771-72 (1980).
[6] D. E. Wittmer and R. C. Buchanan, "Low-temperature densification of Lead Zirconate-Titanate with Vanadium Pentoxide additive", *J. Am. Ceram. Soc.*, **64** [8], 485-90(1981).
[7] O. Ohtaka, R. Von der Mühll and J. Ravez, "Low-temperature sintering of Pb(Zr, Ti)O3 ceramics with the aid of Oxyfluride addition: x-ray diffraction and dielectric studies", *J. Am. Ceram. Soc*, **78** [3], 805-08 (1995).
[8] N. D. Patal and P. S. Nicholson, " Comparison of piezoelectric properties of hot-pressed and sintered PZT", *Am. Ceram. Soc. Bull.*, **65** [5], 783-87 (1986).
[9] L. F. Francis, Y-J. Oh,and D. A. Payne, "Sol-gel processing and properties of Lead Magnesium Niobate Powder and Thin Films", *J. Mater. Sci.*, **25**, 5007-16 (1989).
[10] I. Kato and T. Yoshimoto, "Low-sintering technique of ceramics using alkoxides as raw materials", in *Multilayer Ceramic Capacitor*, Edited by K. Okazaki et al, (Gakkensya, Tokyo, 1988), pp 87-93.
[11] R. Lal, N. M. Gokhale, R. Krishnan and P. Ramakrishnan, "Effect of sintering

parameter on the microstructure and properties of Strontium modified PZT ceramics prepared using spray-dried powders", *J. Mater. Sci.*, **24**, 2911-16 (1989).

[12] T. Yamamoto, "Optimum preparation methods for piezoelectric ceramics and their evaluation", *Am. Ceram. Soc. Bull.*, **71**[6], 978-85 (1992).

[13] Y. L. Tu and S. J. Milne, "Characterization of single layer PZT films prepared from an air-stable sol-gel route", *J. Mater. Res.* **10** [12], 3222-31(1995).

[14] J. R. Macdonald, LEVM program V7.0 (1997).

[15] L. T. S. Irvine, D. C. Sinclair and A. R. West, "Electroceramics: characterization by impedance spectroscopy", *Adv. Mater.*, **2 [3]**, 132-38 (1990).

[16] V. M. Gurevich, in " Electric conductivity of ferroelectrics", pp145, (Israel Program for Scientific Translations Ltd., Jerusalem, 1971).

[17] V. Andriamampianina, J. Ravez, M. Dong and J.-M. Reau, " Préparation et Caractérisation de céramiques ferroélectriques dérivées de $Ba_3TiNb_4O_{15}$", *C. R. Acad. Sci. Paris*, t.321, Serie IIb, 467-469 (1995)

[18] M. Dong, J.-M. Reau and J. Ravez, "Ac impedance analysis of $Ba_5Li_2Ti_2Nb_8O_{30}$ ferroelectric ceramics", *Solid State Ionics*, **91**, 183-190 (1997).

[19] V. Andriamampianina, J. Ravez, M. Dong and J.-M. Reau, "Ferroelectricity, electronic and ionic conductivity in $Ba_3TiNb_4O_{15}$ ceramics", *Ferroelectrics*, **196**[1-4], 39-42 (1997).

[20] R. A. Gerhardt, "Cause of dielectric dispersion in ferroelectric materials", *Ceram. Trans.* **88**, 41-60 (1998).

EFFECTS OF DONOR DISTRIBUTIONS ON DOMAIN STRUCTURES IN PZT FERROELECTRICS

Qi Tan, Z. Xu and Dwight Viehland
Department of Materials Science and Engineering, and Seitz Materials Research Laboratory, University of Illinois at Urbana-Champaign, Urbana, IL 61801

ABSTRACT
    Decreasing higher valent A-site substituent radii resulted in higher coercive fields, less decrease in transition temperatures and disappearance of relaxor characteristics with respect to largest A-site substituents La^{3+}. These smaller substituents have enhanced tendency to exchange onto B-sites, resulting in lower valent substituent characteristics, "wavy" domains and "hard" PZT properties.

INTRODUCTION
    La^{3+} has been found to be effective in causing significant change in the microstructural property relations[1,2,3]. Previous transmission electron microscopy investigations have revealed that increasing La^{3+} concentration in PZT results in a decrease in domain size and an increase in ferroelectric disordering effects[4,5]. Other higher valent substituents in PZT have not been reported to induce important changes such as relaxor ferroelectric characteristics, large electrostrictive effects, or high quadratic electro-optic coefficients. The reason for La^{3+} being superior to other higher valent A-site substituents in causing these changes is unknown.
    Recent investigations have demonstrated that substituent distributions[6-9] play an important role in changes in domain stability and properties. Substituents and associated defects which are mobile until temperatures below that of the ferroelectric transformation result in fine "wavy" domains, domain pinning effects, and "hard" ferroelectric behavior. In general, higher valent substituents (La^{3+}) and associated defects are randomly quenched-in from temperatures above that of the ferroelectric phase transformation, whereas lower valent substituents (K^{1+}) and associated defects remain mobile until temperatures below that of the transition and develop preferential distributions near domain boundaries.
    The role of La^{3+} and other lanthanide distributions in PZT has been investigated. Hardtl et al[10] reported co-distribution of vacancies on A and B sites in La-modified PZT (PLZT). Haertling et al[1] found that La^{3+} and associated vacancies in PLZT have an 85% occupancy on the A-sites for PZ and a 25% on the B-site for PT. Gonnard et al[11] proposed that trivalent substituents can occupy both divalent A-sites and tetravalent B-sites. The occupancies on the two types of sites was found to depend on valence and ionic size differences. Studies of A-site lanthanide

substituents in PZT have shown evidence that some of the substituents undergo a site exchange to the B-sites[12-14]. These investigations clearly demonstrate the importance of substituent distributions in PZT. However the effect of subsitutent distribution on dielectric properties and domain stability remains unknown.

In this work, the dependence of domain stability and macroscopic properties on substituent distribution between A and B-site occupancies was investigated. It was anticipated that smaller lanthanide substitutents might undergo a site occupancy exchange to the B-sites, resulting in differences in structure property relationships between various higher valent modified PZTs. The results will demonstrate the importance of A-site occupation in inducing relaxor ferroelectric characteristics and polar nanodomain state, whereas exchange to B-site occupancy results in "hard" ferroelectric characteristics.

EXPERIMENTAL

Rhombohedral-structured PZT ceramics modified La^{3+}, Bi^{3+}, and Dy^{3+} were fabricated by a conventional mixed oxide method according to the formula $Pb_yM_{1-y}(Zr_{0.65}Ti_{0.35})_{1-y/4}O_3$, (M=La, Bi, Dy and Y). These compositional sequence are designated as PLZT 100y/65/35, PBZT 100y/65/35, PDZT 8/65/35 and PYZT 8/65/35 respectively. Dense specimens were prepared by hot-pressing at 1100°C for 2 hours followed by an annealing at 1300°C for 2 hours. Detailed materials processing can be found elsewhere[7]. Similar processing methods were used to fabricate PLZT 5/65/35 compositions co-modified with various lanthanide ions. The lanthanide ions studied included Nd^{3+}, Er^{3+} and Lu^{3+}. These co-modified compositional sequences are designated as PLLnZT 5/3/65/35. For the substituents investigated, the ionic radii as a function of atomic number is given in Figure 1 according to Shannon[15]. It can be seen in this figure that La^{3+} is the largest of these trivalent substituents.

The complex dielectric permittivity was measured using a Hewlett-Packard 4284A inductance-capacitance-resistance (LCR) meter which can cover a frequency range between 20 and 10^6 Hz and an ac driving field range between 0.005 and 20V. The P-E behavior was characterized with a computer-controlled, modified Sawyer-Tower circuit using a measurement frequency of 50 Hz. X-ray diffraction

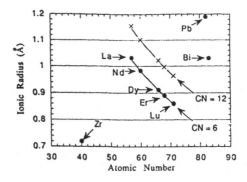

Figure 1. Comparisonof substitution ionic radius with those of Pb^{2+} and Zr^{4+}.

measurements were taken on a Rigaku D-MAX diffractometer with CuKα radiation at 45 kV and 20 mA. The TEM studies were done on a EDS-attached Phillips EM-420 microscope operating at an accelerating voltage of 120 kV.

COMMONALITY OF HIGHER VALENT SUBSTITUENTS ON FERROELECTRIC ORDER

Effects of smaller trivalent substituents on Pb-sites

La^{3+} substitution in PZT results in significant changes in the phase transformation characteristics[5,16]. Dramatic changes in the dielectric response occur with increasing La^{3+} concentration. For concentrations above 8 at.%, relaxor ferroelectric behavior results, as shown in Figure 2. Increase in the La^{3+} concentration results in a lowering of T_{max} and stronger relaxor characteristics.

The dramatic change in phase transition behavior of PZT is believed to be due to a weakened coupling between BO_6 octahedra[17]. To systematically understand substituent modification effects on phase transitions, other trivalent substituents with smaller ionic sizes were investigated. Bi^{3+} was chosen as higher valent substituents on the A-site. The ionic radius Bi^{3+} is smaller than that of La^{3+}. For PBZT 100y/65/35, the dielectric constant was found to be less influenced with increasing y than for La^{3+} modification, as shown in Figure 2(a). Increasing Bi^{3+} concentration results in very small decreases in T_{max}. For example, the dielectric maximum shifted to 305°C for y=0.04. However, with increasing y to 0.2, no additional decrease in T_{max} was observed; although further decreases in the peak height were found. Since Bi^{3+} has a high solubility in PZT as revealed by X-ray diffraction, the limited peak shift indicates less ferroelectric disordering effects. A frequency dependence of dielectric response near T_{max} was obvious. However, no

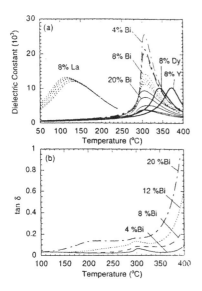

Figure 2. (a) Dielectric constant as a function of temperature for PLZT 8/65/35, PBZT 4/65/35, PBZT 8/65/35, PBZT 20/6535, PDZT 8/6535 and PYZT 8/65/35 compositions. The data for PBZT were taken using frequencies of 10^2, 10^3, 10^4, 10^5 and 5×10^5 Hz from the top to the bottom curves, and 10^5 for PDZT and PYZT. (b) Loss factors as a function of temperature for various PBZT 100y/65/35 compositions at a measurement frequency of 10^3 Hz.

Dielectric Ceramic Materials

shift in T_{max} was found with frequency as for relaxor PLZTs. Measurements of the dielectric loss factor for various compositions are shown in Figure 2(b). These data reveal an enhanced dielectric loss with increasing Bi^{3+} concentration. For example, $\tan\delta$ increased by more than three times at T>150°C for 0.04<y<0.2.

For smaller substituents Dy^{3+} and Y^{3+} ions, the dielectric response was influenced even less with increasing y than that for Bi^{3+} modification. In fact, the dielectric behavior resembles that of undoped PZT, as shown in Figure 2(a). XRD studies reveal a small amount of a second phase due to the insolubility of 8 at.% Dy and Y ions. Although this insolubility of Dy and Y in PZT may be one of the reason for the decreased changes with increasing y, significant changes appear in the phase transformational characteristics with decreasing substituent radii for La, Bi, Dy and Y ions. Cations with smaller ionic radii may be more easily exchanged onto B-sites, resulting in less disruption of ferroelectric order.

It was found that for PZT ceramics with randomly distributed La ions, an obvious ac drive dependence of the dielectric constant was observed[6]. For Bi modified PZT, the ac drive effect was found to be weaker, as shown in Figure 3(a). In addition to the slight decrease in transition temperature, the nonlinearity and the frequency dispersion in PBZT 4/65/35 are smaller than those in PLZT 100y/65/35. DC bias effect on the dielectric responses turned out to be similar to that in PLZT, but the highest DC bias that the specimens can withstand is considerably reduced especially for high Bi concentration (Figures 3(a) and (b)). This difference may be associated with the higher mobility of induced defects in PBZT specimens.

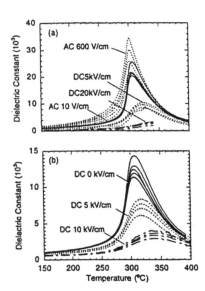

Figure 3. Electric field dependence of dielectric responses for various PBZT 100y/65/35 compositions. (a) 4/65/35, ac and dc fields, and (b) 8/65/35 dc fields. The measurement frequencies used were 10^2, 10^3, 10^4, 10^5 and 5×10^5 Hz from the top to the bottom curves.

Figure 4 illustrates the domain structures of PZT modified by Bi, Dy and Y. For PBZT 4/65/35 (Figure 4(a)), regular micron sized domains were observed throughout the entire specimen, similar to that for PLZT 4/65/35. With increasing Bi concentration to 8 at.%, the domain size was decreased, although micron-scale morphologies persisted (Figure 4(b)). For PBZT 20/65/35, the domain structures were decreased even further in size, however nanodomains were not observed (Figure 4(c)). It should be noted that a small waviness in the domain boundaries appear for higher Bi concentrations. This suggests an interaction of Bi^{3+} induced vacancies with domain boundaries, as was discussed by Tan et al[18,19]. More pronounced domain structural distortions were observed in Dy modified PZT ceramics (Figure 4(d)), where submicron sized domains were found for PDZT 8/65/35. An obvious "wavy" domain structure was observed, indicating that domain boundaries are becoming pinned. Similarly, coarse domain structures were also found for Y modified PZT, as shown in Figure 4(e).

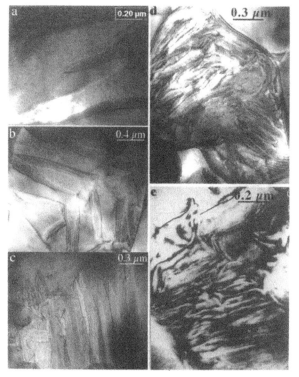

Figure 4. Bright-field images for (a) PBZT 4/65/35, (b) 8/65/35, (c) PBZT 20/65/35, (d) PDZT 8/65/35 and (e) PYZT 8/65/35

P-E behaviors are measured as shown in Figure 5 for PLZT 8/65/35, PBZT 8/65/35, PDZT 8/65/35 and PYZT 8/65/35. In general, higher electric fields were needed to completely switch the polarization with decreasing cation radii from La^{3+}, Bi^{3+}, Dy^{3+} to Y^{3+}. The increase in coercive fields and the decrease in remanent polarization suggest a pinning of domain boundaries and/or polarization with decreasing radii. This suggests that with decreasing radii, cation substituents on the A-site are exchanged onto the B-site. Clearly PZT modified with trivalent "donors" can behave differently depending on the substituent species and distribution.

The inexistence of tweed-like structure and nanodomains demonstrates that relaxor behavior can not be induced by trivalent modifications on the A-site. La^{3+} has a similar ionic radii as Pb^{2+}. Substitution of Pb^{2+} by La^{3+} unavoidably disrupts long-range ferroelectric ordering resulting in the nanodomains, relaxor behavior and a pronounced decrease in T_{max}. A plausible explanation for the effect of Bi^{3+}, Dy^{3+} and Y^{3+} can be made in terms of ionic radii. These ions are smaller than La^{3+} and Pb^{2+}, and have lower electronegativities, therefore they have a higher tendency to occupy B-sites and cluster into extended defect structures. The enhanced occupancy of B-sites results in less disruption of long-range ferrolectric order. The occupation of B-sites by "donors" is also consistent with the observation of the increased dielectric losses for the PBZT and PDZT compositions. The conclusion that smaller ions favor the B-sites is also consistent with previous theoretical calculations and experimental observations[11,12,14].

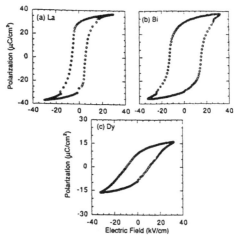

Figure 5. Room-temperature Sawyer-Tower polarization (P-E) curves for (a) PLZT 8/65/35, (c) PBZT 8/65/35 and (d) PDZT 8/65/35.

Influence of Lanthanide Distributions on Phase Transitions and Domain Structures
   The redistribution of "donors" postulated above was testified by co-substitution of lanthanide ions in PLZT 5/65/35. PLZT 5/65/35 modified by an additional 3

Dielectric Ceramic Materials

at.% La has relaxor characteristics, as shown in Figure 6. However, when PLZT 5/65/35 is further modified by a lanthanide donor smaller than La^{3+}, apparent differences in dielectric responses was observed. By doping Nd^{3+}, only a small shift in T_{max} towards higher temperatures was observed. However, smaller substituents such as Er, Lu and In resulted in dramatic differences. Not only was T_{max} significantly increased, but the frequency dispersion also disappeared. These results demonstrate with decreasing lanthanide ionic radius that long-range-ferroelectric order is enhanced.

Previous calculations of unit cell volumes and measured densities for PLZT co-modified by lanthanide ions[14] showed that the substitution site of lanthanide ions was changed from A to B sites with decreasing lanthanide ionic radius. The observation that the dielectric response is nearly unchanged for small substituents indicates that this transition from A to B site occupancy is correct as, for example, In^{3+} radii is close to that of Ti/Zr ions and consequently may primarily substitute for Ti/Zr ions. In this case, so called "donors" are actually acceptors because B-sites occupancy requires oxygen vacancies for charge neutrality.

Figure 7 shows the dielectric loss factors for Nd^{3+}, Er^{3+}, Lu^{3+} and In^{3+} modified PLZTs. It can be seen that $\tan\delta$ increases with decreasing lanthanide radii. In addition, significant frequency relaxation was obvious. These results clearly demonstrate enhanced electrical transport characteristics with decreasing radii.

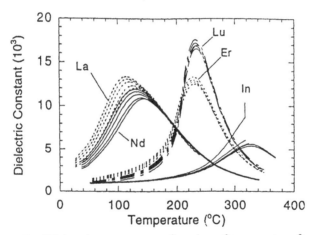

Figure 6   Dielectric constant as a function of temperature for various PLLnZT 5/3/65/35 compositions. The data was taken using frequencies of 10^2, 10^3, 10^4, 10^5 and 5×10^5 Hz from the top to the bottom curves.

Domain Morphologies of Lanthanide Modified PZT

Transmission electron microscopy studies revealed a progressive evolution of domain morphologies in PLZT 5/65/35 co-modified with lanthanide ions on the A-site. Figure 8(a) reveals the presence of polar nanodomains morphology for

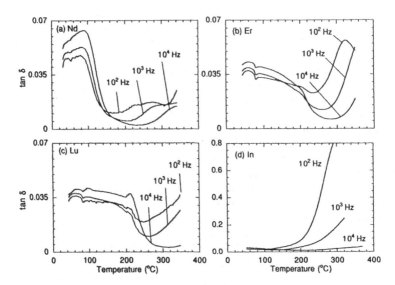

Figure 7. Loss factors as a function of temperature for various PLLnZT 5/3/65/35 compositions. (a) PLNdZT, (b) PLErZT, (c) PLLuZT and PLInZT.

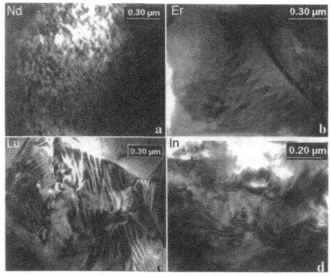

Figure 8. Bright-field images for (a) PLNdZT 5/3/65/35, (b) PLErZT 5/3/65/35, (c) PLLuZT 5/3/65/35 and (d) PLInZT 5/3/65/35, showing the evolution of domain structures toward coarse but irregular feature.

Dielectric Ceramic Materials

PLNdZT 5/3/65/35. Similar features were observed for the corresponding PLZT 8/65/35 composition. Figure 8(b) shows the effect of Er^{3+} on domain morphology. An obvious difference with Figure 8(a) is that the domain size is increased and the domain shape is more regular. Correspondingly, normal ferroelectric behavior was observed in the dielectric response (Figure 6). With Lu^{3+} substituents, the PLZT specimen possessed a more dense domain structure. The length of the domains were submicron and their widths were much smaller than 100 nm. In addition, a small degree of "waviness" was observed in the domain boundaries. More pronounced "wavy" domains were observed for In modified PLZT 5/3/65/35 as shown in Figure 8(d). Polarization studies, shown in Figure 9, revealed an increase in the coercive field with decreasing ionic radii of the lanthanide substituents, indicating enhanced domain pinning effects.

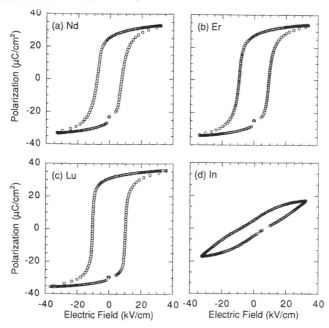

Figure 9.   Room-temperature Sawyer-Tower polarization (P-E) curves for (a) PLNdZT, (b) PLErZT, (c) PLLuZT and (d) PLInZT 865/35.

CONCLUSION
    A-site donor modification of PZT ceramics were studied by dielectric spectroscopy, TEM and polarization switching characteristics. Much different dielectric responses and domain structures with respect to those of PLZT compositions were observed. Unlike La substitution inducing the easy switching characteristics and dramatic domain refinement till nanometer size, Bi and lanthanide ions originally substituting for Pb^{2+} result in relatively larger coercive fields, less ferroelectric disordering and the perseverance of submicron domains,

similar to the role of lower valent substituents. Moreover, domain waviness was revealed in these small "donors" modified PZT and PLZT ferroelectrics. These results show the pinning characters of these "donors" on domain boundaries and/or polarization. A transition of "donors" from A sites to B sites inducing the creation of oxygen vacancies was believed to be responsible for these phenomena. In comparison with other smaller trivalent cations, La^{3+} due to its largest trivalent cations would primarily locate at A-sites, which maybe the essential cause resulting in greatest changes of ferroelectric properties. Consequently, understanding and well control of the donor character and distribution at A and B sites will ensure the observation of relaxor characteristics, desired dielectric properties and domain structures.

ACKNOWLEDGMENTS
This work was supported by the Office of Naval Research (ONR) under contract No. N00014-95-1-0805 and by Naval UnderSea Warfare Center contract No. N66604-95-C-1536. The use of the facilities in the Center for Microanalysis in Materials Research Laboratory at the University of Illinois at Urbana-Champaign is gratefully acknowledged.

REFERENCES
[1] G.H. Haertling and C.E. Land, J.Am.Ceram.Soc. **54**, 1 (1971).

[2] N. Setter, and L.E. Cross, J. Appl. Phys. **51**, 4356 (1980).

[3] H. Ito, K. Nagatsuma, H. Takeuchi and S. Jyomura, J.Appl.Phys. **52**, 4479 (1981).

[4] C.A. Randall, D.J. Barber, R.W. Whatmore and P. Groves, J. Mater. Sci. **21**, 4456 (1987).

[5] J.F. Li, X.H. Dai, A. Chow and D. Viehland, J.Mater. Res. **10**, 926 (1995).

[6] Q. Tan and D. Viehland, Phys.Rev.B**53**, 14103 (1996).

[7] Q. Tan, Z. Xu, J.F. Li and D. Viehland, J. Appl. Phys. **80**, 5866 (1996).

[8] Q. Tan and D. Viehland, Ferroelectrics. in press (1998).

[9] Q. Tan, and D. Viehland, Phil. Mag. B.**76,** 59 (1997).

[10] K.H. Hardtl and D. Hennings, J.Am.Ceram.Soc. **55**, 230 (1972).

[11] P. Gonnard and M. Troccaz, J. of Solid State Chemistry **23**, 321 (1978).

[12] H.D. Sharma, A. Govindan, T.C. Goel, P.K.C. Pillai, J. Mater. Sci. Lett. **15**, 1424 (1996).

[13] S.J. Hong and A.V. Virkar, J.Am.Ceram.Soc. **78**, 433 (1995).

[14] H.-B. Park, C.Y. Park, Y.S. Hong, K. Kim and S.-J. Kim, J.Am.Ceramic Soc., to be published (1998)

[15] R.D. Shannon, Acta Cryst., A**23**, 751 (1976).

[16] X.H. Dai, Z. Xu and D. Viehland, Phil.Mag.B **70**, 33 (1994).

[17] N.W. Thomas, J. Phys. Chem. Solids **51**, 1419 (1990).

[18] Q. Tan and D. Viehland, J. Mater.Res. accepted (1998).

[19] Q. Tan and D. Viehland, J. Am.Ceram.Soc.**81**, 328 (1998).

RELAXOR DIELECTRIC CHARACTERISTICS IN POLAR DIELECTRIC BARIUM MAGNESIUM NIOBATE

S. M. Gupta*, E, Furman, E. V. Colla, Z. Xu, and D. Viehland,
Department of Materials Science & Engineering,
University of Illinois at Urbana Champaign, Urbana, Illinois-61801, USA.

ABSTRACT

Relaxor-like dielectric behavior, analogous to the well-known ferroelectric material Lead Magnesium Niobate (PMN), has been induced in the polar dielectric Barium Magnesium Niobate (BMN). The dielectric constant was increased and the temperature of the maximum dielectric constant was shifted to lower temperature with decrease in measurement frequency for BMN containing A-site vacancies. The frequency dispersion of the permittivity maximum showed good agreement with the Arhenius relationship. An activation energy of 0.15 eV and a pre-exponential factor 10^{14} sec.$^{-1}$ were determined. No frequency dispersion in the imaginary part of the permittivity was found below the temperature of the dielectric maximum, analogous to PMN. A fictitious freezing temperature, calculated using the Vogel-Fulcher relationship, was estimated to be near 0 K. The Curie constant, calculated using the Curie-Weiss law, was 1.77 x10^5. Phase analysis revealed a single phase perovskite structure with a hexagonal unit cell with a = 5.77 Å and c = 7.08 Å. A linear polarization-electric field dependence was observed at room temperature. A polarization of 0.3 $\mu C/cm^2$ was found under a field of 75 kV/cm at 25 °C. <110> selected area electron diffraction patterns revealed superlattice reflections along the 1/3<111> direction. Low temperature polarization and microstructural studies were also performed.

INTRODUCTION

Barium magnesium niobate is a polar dielectric which has a relative permittivity (ε) of 32 and a Q-factor (1/Tan δ) of 5600 at 10 GHz [1]. On the basis of ionic radii differences between B-site cations, BMN was conjecture to posses no B-site cations ordering, but in actuality it has tripled unit cell due to three B-site cation sublattice oriented along the <111> [2-3]. Stoichiometric ordering influences the Q-factor. Due to the high Q-factor and dielectric constant barium based perovskite materials are very suitable microwave dielectric resonators for commercial wireless technologies.

Recently this material has been explored for the optimization of its dielectric properties via chemical substitution, crystal structure and microstructure [4]. Viehland et al. [5] studied the ordering behavior and Thomas et al. [6] studied the transformation from relaxor to polar dielectric behavior in Ba-doped Lead magnesium niobate. Here we report the relaxor-like dielectric characteristics in the polar dielectric BMN containing A-site vacancies. The frequency dispersion in the dielectric constant and the dissipation factor are analogous to that of PMN. The relaxor behavior in BMN is explained by Skanavi model [7]. Phase analysis, microstructural and field induced polarization are also studied.

EXPERIMENTAL

The specimens studied in this investigation were fabricated according to the formula Ba$_{1-2x}$ $_{2x}$ (Mg$_{1/3}$ Nb$_{2/3}$)$_{1+x}$ O$_3$ (BMN-x) for 0<x<0.25. The purity of the starting raw materials was 99.9% and the specimens were prepared by the columbite precursor method [8]. The columbite precursor ($MgNb_2O_6$) was first prepared by mixing predetermined amounts of MgO and Nb_2O_5 in isopropanol and ball milling for 5 hours using a ZrO_2 grinding media. The slurry was dried at 80 $^{\circ}$C and the powder was calcined at 1100 $^{\circ}$C for 4 hours. Single phase formation of $MgNb_2O_6$ was confirmed by x-ray diffraction (XRD). The columbite precursor was then mixed and ball milled with predetermined amounts of $BaCO_3$ powders and calcined at 1300 $^{\circ}$C for 2 hours. The calcined powders were mixed again and pressed into cylinders using polyvinyl alcohol (PVA) as a binder at a hydrostatic pressure of ~ 172 MPa. These cylinders were fired at 1500 $^{\circ}$C for 4 hours in air after binder burnout at 450 $^{\circ}$C for 2 hours. The densities of all sintered pellets were measured volumetrically and determined to be above 95% of the theoretical [4]. The calcined and sintered specimens were analyzed by Rigaku x-ray diffractometer for second phase formation. The sintered blocks were then cut into thin disks and polished on different grades of emery papers to obtain parallel surfaces. The polished surfaces were ultrasonically cleaned to remove dust particles and then annealed at 650 $^{\circ}$C for half an hour to remove surface strains during polishing. Sample discs were electroded using sputtered on gold, followed by a thin coating of air drying silver paste to insure good electrical contact.

The dielectric response was measured using a Hewlett-Packard 4284A inductance-capacitance-resistance (LCR) bridge which can cover a frequency range from 20 to 10^6 Hz. For low temperature measurements, the samples were placed in a Delta Design 9023 test chamber, which can be operated between -180 and +250 $^{\circ}$C. The temperature was measured using a HP 34401A multimeter via a platinum resistance thermocouple mounted directly on the ground electrode of the sample fixture. The LCR, test chamber and multimeter were interfaced with computer to collect data while cooling at a rate of 4 $^{\circ}$C/minute at ten frequencies between 10^2 and 10^6 Hz.

Polarization electric field (P-E) measurements were made using a modified Sawyer-Tower circuit at 10 Hz. The voltage from the polarization change was fed into a voltage amplifier and then digitized by the computer.

Scanning electron micrographs were taken from the fractured surfaces. The fractured surfaces were sputtered by gold and palladium. The fractured surfaces were used for grain size and morphology determination. A qualitative study was carried out using energy dispersive X-ray spectroscopy (Link EDS System, Oxford X-ray System with detector resolution 133 eV) equipped with Zeiss DSM-960 scanning electron microscope (SEM).

TEM specimens were prepared by ultrasonically drilling 3-mm discs which were mechanically polished to ~100 μm. The center portions of these discs were then further ground by a dimpler to ~10 μm and argon ion-milled to perforation. Specimens were coated with carbon before examination. The TEM studies were done on a Phillips EM-420 microscope operating at an accelerating voltage of 120 kV. Scanning transmission electron microscope (STEM) studies were performed using model HB501 (resolution 10 Å) at an accelerating voltage 100 kV.. Quantitative EDX and line scans were carried out using Link EDS, Oxford X-ray System with Si Li detector (resolution 136 eV) equipped with oxford virtual standards.

RESULTS AND DISCUSSION

Microstructural studies:

Figures 1(a)-(d) show the SEM pictures of the fracture surfaces for sintered $MgNb_2O_6$, BMN-0, BMN-10 and BMN-25, respectively. Intergranular fracture and unidirectional grains (approximately 10 μm in length) can be seen in Figure 1(a). However, Figures 1(b)-(d) show intragranular fracture and well defined grains. The average grain size was approximately 1 μm. No unreacted starting elements or second phases were observed in these samples. Relaxor-like characteristics has previously been reported in Bi_2O_3 doped $SrTiO_3$ [9]. In these samples bulk grains were depleted and grain boundaries were richer in Bi^{3+} ions. To reveal the chemistry of grains and grain boundaries in BMN-x samples, STEM study was performed. The resolution of the STEM was approximately 10 Å which was sufficient to measure the compositions at and near grain boundaries of 1μm grains. The concentration of elements present at and near GB for BMN-10 are presented in table 1. It should be noticed from this table that grain boundaries and grains are of nearly the same composition. A STEM picture of BMN-10 is shown in Figure 3. The points at which quantitative analysis were performed are marked in this figure.

Table 1. Compositional study of grain and grain boundaries for BMN-10

Positions	Ba (at.%)	Mg (at.%)	Nb (at.%)
Grain 1	50.7	11.8	37.5
Grain 2	50.3	12.3	37.4
Grain 3	50.0	13.4	36.6
Grain boundary	52.1	11.0	36.9

To analyze the elemental gradient through the grain boundary between two adjacent grains, line scanning of Ba, Mg and Nb elements was performed. Four lines scan were taken at different positions on BMN-10 samples. One of the line scans is shown in Figure 3 for BMN-10. No change in the concentrations of Ba, Mg and Nb elements was observed. All four line scans revealed that there was no segregation of starting elements at the grain boundaries and the grain compositions were nearly equivalent.

Qualitative energy x-ray dispersive analysis (EDX) of $MgNb_2O_6$, BMN-0, BMN-10 and BMN-25 are shown in Figures 4(a)-(d), respectively. All the peaks in these spectra are identified. EDX spectra taken from the top of intragranular fractured surfaces and from well defined grains were identical. This clearly reveals that grain boundaries are free from segregation of starting reagents or from second phase formation.

Selected Area Electron Diffraction

The 1:2 ordering of BMN was studied using SAED. Figures 5(a)-(b) show <110> SAED patterns taken at room temperature for BMN-0 and BMN-10, respectively. 1/3<111> superlattice reflections are clearly visible in both specimens, as marked by an arrow. These superlattice reflections demonstrate an 1:2 ordering of B-site cations. No changes were observed in the superlattice reflections during cooling or heating in the temperature range between -180-50 °C. The presence of superlattice reflections in BMN-10 reveals that A-site vacancies have no effect on the 1:2 ordering. 1/3<111> Superlattice reflections were also present in BMN-25.

Phase analysis

X-ray diffraction was used to examine the phases present and the unit cell structure. Figures 6(a)-(e) show XRD patterns for BMN-x for 0<x<25. All the major x-ray diffraction lines of BMN were identified with the reported JCPDS file [10]. The perovskite structure of BMN has a hexagonal unit cell with a = 5.77 Å and c = 7.08 Å. Our results revealed that the cell dimensions remain unchanged with increasing A-site vacancy concentration. Additional peaks were identified as magnesium niobate. The intensity of these peaks were found to increase with increasing A-site vacancy concentration, as shown in Figures 6 (c)-(e). The concentration of magnesium niobate was less than 5% in BMN-x (x>10) specimens studied. It should be noticed from x-ray data that no superlattice lines are present for the BMN-x specimens. The absence of superlattice lines indicates that the 1:2 ordered domains are small. Recently, Davies et al. [4] found superlattice lines in pure BMN samples. Differences in ordered domain sizes can be attributed to variations in processing conditions. The sintering temperature was 1640 $^{\circ}$C for 10 hours in Davies samples, whereas, our specimens were sintered at 1500 $^{\circ}$C for 4 hours .

PROPERTIES STUDIES
Dielectric properties:
    The dielectric responses of BMN-x for 0< x< 25 are shown in Figures 7 (a)-(d). The dielectric constant increased with decreasing temperature for BMN-0, as shown in Figure 7(a). No maxima in permittivity was observed for BMN-5. However, typical relaxor-like dielectric properties were observed for BMN-10, BMN-15 and BMN-25 specimens as can be seen in Figures 7 (b)-(d). The dissipation factor was lower for lower frequency, analogous to that observed in PMN. Interestingly only very small changes in the dielectric constant (< 6) were observed between 100 $^{\circ}$C and T_c (temperature of ε_{max}). The dielectric permittivity at T_{max} of the BMN-15 ceramic was ~63 (at 1KHz), which was higher than that of BMN-0 (ε = 33 at 1 KHz) or pure magnesium niobate (MN) (ε = 23 at 1 KHz). The maximum dielectric constant for BMN-15 was ~33 which is close to that expected from the dielectric mixing rules (parallel, series or logarithmic) for a BMN containing 15 at.% magnesium niobate. The room temperature DC-resistivity of MN (2×10^{12} ohm-cm) is comparable to BMN-0 (2.5×10^{13} ohm-cm), thus the Brick-wall model and space charge contributions to the low frequency dielectric constant are not reasonable. The temperature of ε_{max} was nearly equal for BMN-10, BMN-15 and BMN-25, whereas, the dielectric constant increased with increasing Ba-vacancy concentration.
    The frequency dispersion of the permittivity maximum revealed close agreements to an Arhenius relationship, as can be seen in Figure 8(a). An activation energy of 0.15 eV and pre-exponential factor of 10^{14} sec^{-1} were calculated for BMN-10. No frequency dispersion in the imaginary part of permittivity, below the temperature of the dielectric maximum was observed. A fictitious freezing temperature, calculated using Vogel-Fulcher relationship [11], was estimated to be near 0 K as shown in Figure 8(b). The Curie constant for BMN-10, calculated using the Curie-Weiss law, was 1.77 x10^5.
    The dipole relaxation in BMN-15 and BMN-0 was studied at room and low temperatures and these data are shown in Figures 9(a)-(d). The value of ε' remains unchanged up to a frequency of 10^5 Hz, but it decreased at higher frequencies for BMN-15. This decrease in ε' occurred sharply at lower temperatures as can be seen in Figure 9(a). No such relaxation was found for BMN-0 (Figures 9(c)-(d)). On comparing the data for BMN-15 and BMN-0, it is evident that higher concentrations of A-site vacancies result in dielectric relaxation at higher frequencies. The higher dielectric constant in BMN-15 can be attributed to contributions from a relaxational polarization due to Ba-

vacancy. The origin of this relaxational polarization can be explained using the Skanavi model [7], in which weakly bound ions displace from equilibrium positions.

Field Induced Polarization :
Electric field induced polarizations were measured for MN and BMN-x (0< x< 25). A linear P-E curve was observed for all the samples, as shown in Figure 10. The polarization under a field of 30 kV/cm was found to increase with increasing Ba-vacancy concentration. It is plausible that A-site vacancies in BMN have defect dipole polarization contributions.

CONCLUSIONS

Relaxor-like dielectric characteristics have been induced in the polar dielectric BMN by creating A-site vacancies. Scanning transmission electron microscopy revealed homogeneous compositions in bulk grains and near grain boundaries. No segregation of secondary phases was observed near grain boundaries by SEM, whereas a small peak of secondary phase was found by x-ray diffraction. An activation energy of 0.15 eV, a pre-exponential factor 10^{14} sec^{-1}, a fictitious freezing temperature near 0 K, and a Curie constant of 1.77 x10^5 were calculated for BMN-10. A linear P-E curve was observed at room temperature and the polarization was found to increase with increasing A-site vacancy concentration. The origin of relaxational polarization in BMN is explained using the Skanavi model.

REFERENCES

[1] Nomura S., "Ceramics for microwave dielectrics resonators", Ferroelectrics, **49**, 61-70 (1983).

[2] F. Galasso and J. Pyle, "Ordering of the compounds of the $A(B'_{0.33} Ta_{0.67})O_3$ type", Inorg. Chem. **2**, 482-84 (1963).

[3] F. Galasso and J. Pyle, "Preparation and study of ordering in $A(B'_{0.33} Nb_{0.67})O_3$ perovskite type compounds", J. Phys. Chem. **67**, 1561-62 (1962).

[4] M. A. Akbas and P. K. Davies, "Ordering-induced microstructures and microwave dielectric properties of the $Ba(Mg_{1/3}Nb_{2/3})O_3$-$BaZrO_3$ system", J. Am. Ceram. Soc., **81**, 670-76 (1998).

[5] D. Viehland, N. Kim, Z. Xu and D. A. Payne, "Structural studies of ordering in the $Pb_{1-x}Ba_x$ $(Mg_{1/3}Nb_{2/3})O_3$ crystalline solution series", J. Am. Ceram. Soc., **78**, 2481-89 (1995).

[6] S. J. Butcher and N. W. Thomas, "Ferroelectricity in the system $Pb_{1-x}Ba_x$ $(Mg_{1/3}Nb_{2/3})O_3$ ", J. Phys. Chem. Solids, **52**, 595-601 (1991).

[7] G. I. Skanavi, IA. M. Ksendzov, V. A. Trigubenko and V. G. Prokhvatilov, "Relaxation polarization and losses in non ferroelectric dielectrics with high dielectric constant", Sov. Phys. JETP, **33**, 250-59 (1958).

[8] S. L. Swartz and T. R. Shrout, "Fabrication of perovskite Lead Magnesium Niobate", Mater. Res. Bull., 17, 1245-50 (1982).

[9] A. K. Mehrotra, "Microstructure and dielectric behavior for pure and bismuth doped strontium titanate polycrystalline ceramics" Ph.D. Thesis, University of Illinois at Urbana Champaign, 1983.

[10] JCPDS file Number 17-173.

[11] D. Viehland, S. J. Jang, M. Wuttig and L. E. Cross, "Local polar configurations in lead magnesium niobate relaxors", J. Appl. Phys., **69**, 414 -20(1991).

Dielectric Ceramic Materials

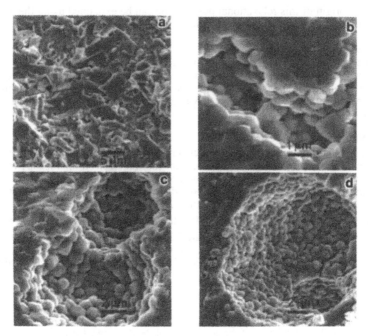

Figure 1 SEM pictures of fractured surface (a) $MgNb_2O_6$, (b) BMN-0,
(c) BMN-10 and (d) BMN-25.

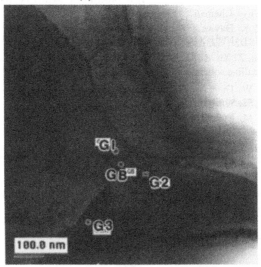

Figure 2. STEM picture of BMN-10 for compositional studies. (Bar = 100 nm).
Grain1, Grain2 , Grain3 and Grain boundary GB are marked.

Dielectric Ceramic Materials

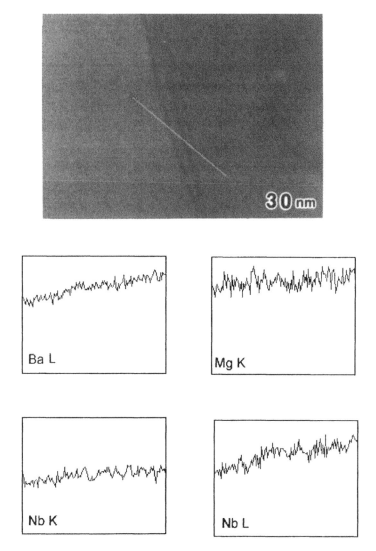

Figure 3. Line scanning of BMN-10 for Ba, Mg and Nb elements. Length of the line is 30 nm. K lines were used for Mg and L lines were used for Ba and Nb.

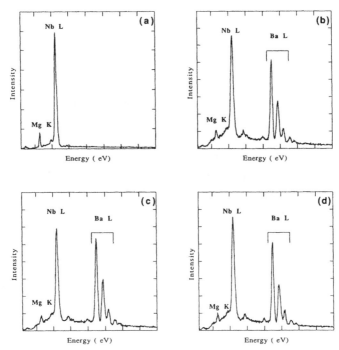

Figure 4- Energy dispersive x-ray spectra of (a) MgNb$_2$O$_6$, (b) BMN-0, (c) BMN-10 and (d) BMN-25.

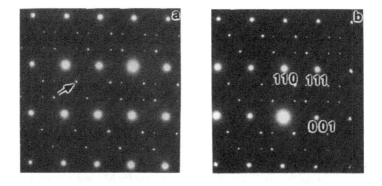

Figure 5. Selected area electron diffraction along <110> unit axis for (a) BMN-0 and (b) BMN-10. Superlattice reflections along 1/3<111> direction are marked by an arrow head.

Dielectric Ceramic Materials

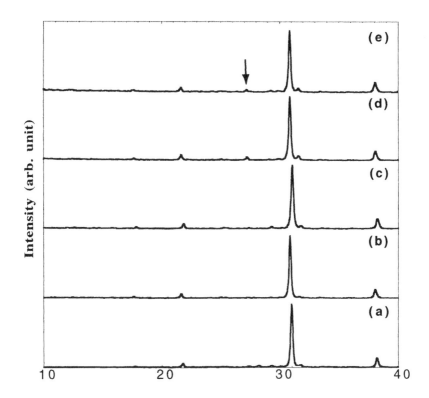

Figure 6. Comparison of x-ray diffraction patterns for (a) BMN-0, (b) BMN-5, (c) BMN-10, (d) BMN-15, and (e) BMN-25. Additional phase peak is marked by an arrow.

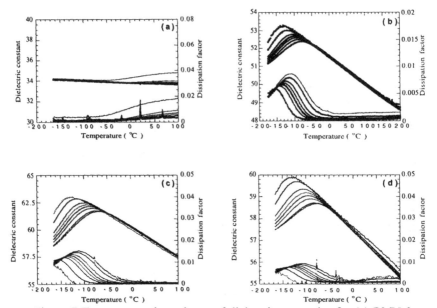

Figure 7 Temperature dependence of dielectric properties for (a) BMN-0, (b) BMN-10, (c) BMN-15 and (d) BMN-25. Frequencies used (from top to bottom) are 100 Hz, 1, 10, 25, 50, 100, 300, and 500 KHz.

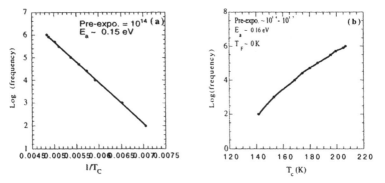

Figure 8. (a) Arhenius, (b) Vogel-Fulcher fitting of frequency dispersion of permittivity maximum for BMN-10.

Dielectric Ceramic Materials

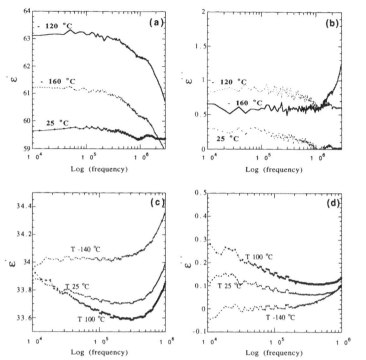

Figure 9. Frequency dependence of real and imaginary part of relative permittivity for (a&b) BMN-15 and (c&d) BMN-0.

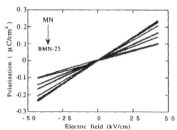

Figure 10. Comparison of P-E curves for $MgNb_2O_6$ and BMN-x where x varies from 0 to 25.

MONOBLOCK λ/4 DIELECTRIC FILTER USING HIGH K MATERIAL

C.Y. Kang, S.J. Yoon[o], J.W. Choi[o], H.J. Kim[o], Y.S. Ko[oo], C.D. An[oo], C.Y. Park
Electrical Eng. Yonsei Univ. Seoul 120-749, Korea
[o]Thin Film Technology Research Center, KIST, Seoul 130-650, Korea
[oo]Korea First Co., LTD. Ichonshi, Kyunggido, 467-860, Korea

ABSTRACT

Monoblock λ/4 dielectric band pass filter (BPF) for 900 MHz band portable telephone terminal is presented. This BPF is a kind of combline filter. Computer-aided design (CAD) was used in modeling procedure. Equivalent circuit of the monoblock BPF is represented by transmission lines and lumped elements based on Z_{oe} and Z_{oo}. BPF model was designed for surface mounted device (SMD) type. The simulations of the equivalent circuit and BPF structure had been performed to optimize the filter design. A newly developed high K dielectric material having K=86, $Q \cdot f_0$=8600 GHz, and τ_f=+0.5 ppm/℃ was used for BPF fabrication. Experimental results of the fabricated device were in a good agreement with simulation.

INTRODUCTION

Recently, with the rapid development and demand for compactness of portable communications, the requirement for compact and low-cost filter is increasing.

Dielectric Ceramic Materials

For previous conventional coaxial type dielectric BPF, dielectric substrates were used for coupling between adjacent resonators and additional input and output ports were needed[1]. These raised the cost of fabrication and restricted to reduce the size. Monoblock dielectric filter which has holes in a single body has the benefits of low cost and small size, because additional coupling elements are not needed. Thus, The design and manufacture for this kind of filter have been studied intensively[2]. However, the structure design procedures for this kind of filter are not easily achieved, because of the existence of coupling mode in the filter block.

This paper describes a structure and design procedure for 900 MHz band monoblock dielectric BPF by using software tools having high accuracy. A circuit analysis using even and odd mode impedances is described. Then structure simulations are performed to accord with circuit design in structure design.

STRUCTURE OF MONOBLOCK λ/4 DIELECTRIC BPF

Figure 1 shows the structure of Monoblock λ/4 dielectric BPF which is composed one dielectric monoblock resonator, a pair of electrodes for coupling resonator holes, and two ports to external loads. The block includes two metallized cylindrical holes which act as inner conductors of a coaxial resonator. The surfaces of the block, except two surfaces, are fully metallized. One nonmetallized upper surface acts as open end on which a pair of electrodes is made up to couple resonators. Input and output ports are constituted on the other lateral surface. The length of the resonator is designed to be a quarter-wavelength of the fundamental mode. The dielectric materials used for manufacturing are shown in Table I . The materials are relatively high quality factor and small temperature coefficient, thus improves unloaded Q of the BPF and restricts to the shift of center frequency according to temperature. The physical dimension of the

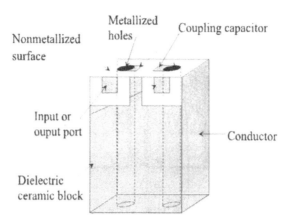

Fig. 1. Structure of monoblock λ/4 dielectric BPF

Table I. Characteristics of dielectric materials

Composition	$(Pb_{0.45}Ca_{0.55})[(Fe_{0.5}Nb_{0.5})_{0.9}Sn_{0.1}]O_3$
K	86
$Q \cdot f_0$	8600 GHz
τ_f	+0.5 ppm/℃

Table II. Physical dimension of the designed filter (mm)

Filter size (l × w × h)	6 × 3 × 8.77
Diameter of hole	1
Distance between the centers of hole	3
Port size	1 × 1

filter is set as Tables II.

Dielectric Ceramic Materials   401

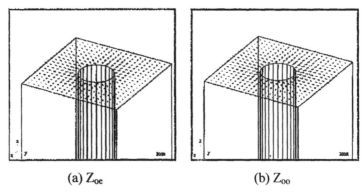

(a) Z_{oe} (b) Z_{oo}

Fig. 2. FEM analysis for Z_{oe} and Z_{oo}

DESIGN

Coupling Analysis

This BPF structure is a kind of combline filter[3]. In this structure, two different field distributions can exist between two holes depending on the relative current flow in the holes[4]. The current flow in each hole is in the same direction and the field distribution is known as even mode. The flow is in the opposite direction and the distribution as odd mode. The even mode axis of symmetry is known as magnetic wall and the odd mode as electric wall. The characteristic impedances for either mode will be different, and the even and odd mode impedaces are denoted by Z_{oe} and Z_{oo}. To design the equivalent circuit of this BPF structure, firstly, Z_{oe} and Z_{oo} of the structure must be defined. In this paper, the finite elements method which is simple and versatile was applied to solve Z_{oe} and Z_{oo}. Figure 2 shows electric field distribution of half of the structure for solving Z_{oe} and Z_{oo}. The even mode symmetric plane is defined as perfect magnetic boundary and the odd mode symmetric plane as perfect electric boundary. As a result, in this structure, Z_{oe} is 8.5401 and Z_{oo} is 7.5822.

Dielectric Ceramic Materials

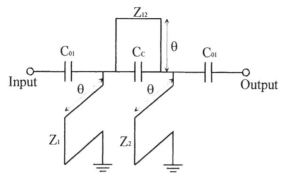

Fig. 3. The equivalent circuit configuration for the monoblock λ/4 dielectric BPF

Table III. Target characteristics of filter

Center frequency (f_0)	914.5 MHz
3 dB bandwidth	20 MHz
Insertion loss	3 dB max
Attenuation (at $f_0 \pm 44$ MHz)	24 dB min

The coupling coefficient k is obtained with

$$k\,[\text{dB}] = 20\log\left(\frac{Z_{oe} - Z_{oo}}{Z_{oe} + Z_{oo}}\right) \qquad (1)$$

Equivalent Circuit

Target characteristics of filter to be fabricated are shown in Table III. Based on target characteristics, prototype g-parameters and the admittance inverters were derived[5]. The equivalent circuit for the monoblock λ/4 dielectric BPF was

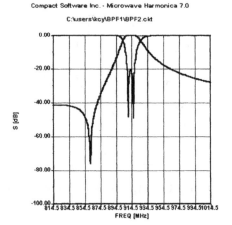

Fig. 4. Equivalent circuit simulation result

derived from these parameters. Figure 3 shows the equivalent circuit configuration for the monoblock $\lambda/4$ dielectric BPF. Input and output coupling capacitors C_{01} are connected in series between input and output ports. A coupling capacitor C_c is connected between resonators.

In this configuration, Z_1, Z_2, and Z_{12} are found as follows,

$$Z_1 = Z_2 = Z_{oe}$$

$$Z_{12} = \frac{2Z_{oe}Z_{oo}}{Z_{oe} - Z_{oo}} \qquad (2)$$

Also, C_{01} can be calculated as the procedure described in [5]. However, C_c should be computed as considered Z_{12}.

The simulation results of this equivalent circuit are shown in figure 4. C_c makes

Dielectric Ceramic Materials

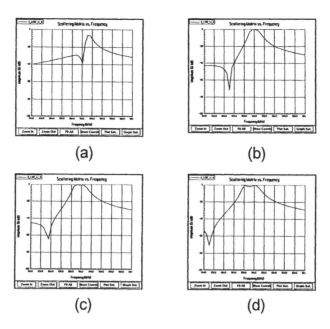

(a)   (b)

(c)   (d)

Fig. 5. Structure simulation results

attenuation pole at lower passband (f_0-50 MHz) to improve attenuation characteristics. However, C_{01} does not affect position of attenuation pole.

Structure Simulation

Structure simulations were performed to verify the structure design for monoblock $\lambda/4$ dielectric BPF by High Frequency Structure Simulator (HFSS). The simulation results are shown in figure 5. The results show transmission responses according to the gap of electrodes for coupling between holes, which are on the open ends. Relatively, (a), (b), (c), and (d) are 2mm, 1mm, 0.8mm, and 0.6mm of the gap. As the gap decreases, coupling capacitance value increases. In case that the gap of electrodes for coupling between resonators was 1mm, the result was a concord with the circuit simulation result.

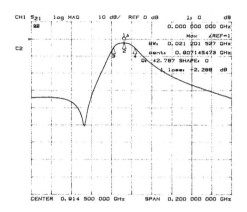

Fig. 6. Electrical characteristics of the filter

PERFORMANCE

The electrical characteristics of the BPF are shown in Figure 6. This measurement was completed with a network analyzer. The results show a center frequency of 907.1 MHz, 3dB-bandwidth of 21.2 MHz, and insertion loss of 2.286dB. Also, attenuation pole locates at f_0-42 MHz. The results show 0.8% error of center frequency compared with simulation results.

CONCLUSIONS

Monoblock $\lambda/4$ dielectric band pass filter was developed. This filter is preferred in mobile communication systems because it has a number of merits. It exhibits small size, simple structure, and low cost. This paper presented a design technique using software tools in designing monoblock BPF. A newly developed high K dielectric material having K=86, Q·f_0=8600 GHz, and τ_f=+0.5 ppm/℃ was used for BPF fabrication. The experimental results show good agreement with design.

REFERENCES

[1] T.Nishikawa, "Miniaturized Duplexer Using Rectangular Coaxial Dielectric Resonators for Cellular Portable Telephone," *IEICE TRANSACTIONS*, **E74** [5] 1214-1220, 1991

[2] Cheng-Chyi You, "Single-Block Ceramic Microwave Bandpass Filters," *Microwave J.* 24-35 November 1994

[3] G. L. Matthaei, "Comb-line Band-pass Filters of Narrow or Moderate Bandwidth," *Microwave J.* 82-91, August 1963

[4] E. G. Cristal "Coupled Circular Cylindrical Rods Between Parallel Ground Planes," *IEEE Trans. on Microwave Theory and Techniques*, **MTT-12**, 428-439, 1964

[5] G. L. Matthaei, L.Young, and E.M.T.Jones, *Microwave Filters, Impedance-Matching Networks, and Coupling Structures*, McGraw-Hill, New York,1964

VOLUME FRACTION GRADIENT PIEZOELECTRIC CERAMIC /POLYMER COMPOSITES FOR ULTRASONIC MEDICAL IMAGING TRANSDUCERS

Rajesh K. Panda, Andrei Kholkin, Stephen C. Danforth and Ahmad Safari
Department of Ceramic and Materials Engineering, Rutgers University, Piscataway, NJ 08855-0909

ABSTRACT

Sanders Prototyping (SP), a Solid Freeform Fabrication (SFF) technique, was used to fabricate a variety of novel piezoelectric ceramic/polymer composite transducers. The composites were processed using a modified lost mold route. A variety of novel ceramic structures such as annular, octagonal rods in a hexagonal pattern and PZT plate composites with and without volume fraction gradients (VFG) were fabricated using this process. Different mathematical distributions including, linear, exponential and gaussian functions, decreasing from the center towards the edges have been used to describe the gradient. In this paper, the fabrication, electrical properties, and the vibration amplitude profiles of the different VFG composite designs have been discussed.

INTRODUCTION

Ultrasonic imaging is one of the most important and still growing diagnostic tools in use today. It is a very popular modality for imaging structures within the human body. In this type of imaging a piezoelectric transducer is used to generate ultrasonic waves and transmit it into the body. These waves are reflected from impedance discontinuities associated with organ boundaries and other features within the body. The resultant echo is usually received by the same transducer and processed to produce the images of the body structures [1-3].

The ultrasonic transducers utilize a piezoelectric material to generate the transmitted ultrasonic pulse. There are a variety of materials which can be used for this application, including piezoceramics, piezopolymers or piezoelectric ceramic/polymer composites. A piezocomposite consists of an active piezoceramic embedded in an inactive polymer. These composites provide many advantages for ultrasonic transducers, including low acoustic impedance, high electromechanical coupling coefficients, formability into complex shapes, suppression of lateral modes and tailorable electromechanical properties [4,5].

In ultrasonic imaging it is desirable to have peak pressure output along the axial direction. However, apart from the main lobe, grating and side lobes are also generated in an ultrasonic beam. Grating lobes and to a lesser extent side lobes give off axis sensitivity to a transducer used as both, a source and receiver of acoustic waves. A decrease in the amplitude of the grating lobes is desirable, as they are potential sources of ambiguity in determining the direction of the echoes returned to the transducer. One of the techniques utilized to reduce the intensity of the grating lobes is by varying the piezocomposite properties within the same transducer. It is known that the peak pressure output for a composite transducer is proportional to the ceramic volume fraction. A gradual lowering of the ceramic content from the center towards the edges leads to a gradual and smooth decrease in the sensitivity towards the ends. This apodization or shading would lead to a reduction in the out of plane artifacts [6,7]

Although many possible solutions to reduce the grating lobe intensity have been proposed, it has proved difficult to design and fabricate complex and fine scale piezoceramic structures by conventional ceramic processing techniques. At Rutgers University, piezoelectric, bio-ceramic and structural components have been processed using various SFF techniques [8-10]. In this project the SP technique was used to produce a variety of novel lead zirconate titanate (PZT) composite transducers, including composites with a volume fraction gradient the center towards the edges. The processing method, design, fabrication and the electromechanical properties of these composites have been discussed in the following sections.

EXPERIMENTAL

The piezocomposites developed in this study were manufactured using an indirect SFF route. This fabrication route utilizes a lost mold technique. In this process, sacrificial molds having a negative of the desired structures were manufactured using the Sanders Prototype (SP) Model-Maker system MM-6PRO (SPI Inc., Wilton, NH). The MM-6PRO is a liquid to solid inkjet plotter, which deposits the polymer on a movable Z-platform layer by layer. The SP machine deposits a build wax for the mold, and support wax to maintain any overhangs in the mold structure. This particular feature makes it possible to fabricate many complex mold designs. The support wax can later be dissolved in an organic solvent without any damage to the build polymer. Another advantage of using the SP technique includes a very high resolution, and ability to get mold wall thickness as low as 75 μm with a good surface finish. This is possible because of the cutting of excess material after building each layer, making it possible to control the individual layer thickness and surface flatness for the deposition of the next layer of polymer.

A high solids loading aqueous PZT ceramic slurry, with a low viscosity was specially developed to infiltrate the polymer molds with openings as small as 75-100 μm. The high PZT loading of 52 vol.% was also helpful in avoiding cracking in the sample during drying and the subsequent binder burn out process. After fabrication, the green structures were heat treated to burn out the binder and then sinter the ceramic. The sintered samples were embedded in a standard Spurr epoxy (Ernest F. Fullam Inc., Latham, NY) and cured in an oven at 70°C for 12 hours. The samples were then cut, polished, electroded and poled using a corona poling technique. The electrical properties of the composite were measured after aging the specimens for at least one day. The capacitance (C_p), dielectric loss factor (tanδ) and d_{33} coefficient were measured, before and after poling, at 1kHz. Other electromechanical properties including volume percent of ceramic, dielectric constant, and thickness and planar coupling coefficients were also calculated. The mold, ceramic and composite architectures were observed under a scanning electron microscope (Amray 1400, Amray Corporation) to observe the VFG and wall dimensions.

The vibration amplitude profiles of the composite surfaces were measured using a laser probe technique, by applying a 28 V AC voltage to the composites and using a probe with a diameter of 3mm. A frequency of 40 Hz was chosen to avoid any interference from the line current frequency (60Hz) and its harmonics. The surfaces of 2-2 VFG composites were scanned from the center towards the edges, with readings taken at intervals of 1mm each. The probe readings were allowed to stabilize for 30 seconds before acquiring the data.

RESULTS AND DISCUSSION

Before fabricating the structures it was crucial to understand the available features and limitations of the Sanders machine used to produce the molds. In order to optimize the process, some of the factors considered while fabrication of the molds included build time, smallest feature sizes, and mold wall surface finish. Figure 1 (a) shows successive layers of build polymer deposited by the SP machine in a regular build pattern. It can be observed that each layer is flat with a thickness of ~ 50 μm. However, the surface finish of this structure is very poor with jagged edges protruding from the walls. Although this build pattern is fast and saves machine time, it would be difficult to make ceramic structures with fine features, such as fine scale ceramic walls (~100 μm) separated by similar distances. The jagged edges would lead to uneven wall widths, causing poor poling of the structure and hence show inferior electromechanical properties. In fabricating fine scale piezoceramic / polymer composite structures for this study, a fine build strategy was used where the layer thickness was much smaller. Figure 1 (b) taken at a magnification three times higher than Figure 1 (a), shows a mold wall with 15 μm thick individual layers and a much better surface finish.

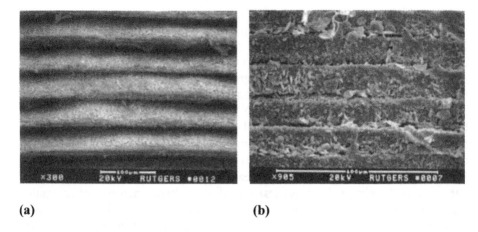

(a)   **(b)**

Figure 1: SEM photographs of Sanders Prototyping (SP) build wax mold wall fabricated using, (a) regular build option and (b) the fine build option.

Many polygon structures were fabricated by the lost mold method using the SP machine in the fine build mode. Figure 2 (a) shows a SEM photograph of a sintered PZT-5H regular polygon structure with a ceramic wall thickness of 225 μm separated by a spacing of 225 μm. This structure, with 51 vol. % ceramic, has more than 20 concentric polygons in a specimen 2.4 cm in diameter. Figure 2 (b) shows the surface of a polished PZT / spurr epoxy polygon composite with a volume fraction gradient (VFG). The ceramic volume percent linearly decreases from 45 % at the center to 22 % at the edges. The thickness coupling constant for this structure was measured to be ~ 68% as compared to 69 % for the structure in Figure 2 (a). Thus, inspite of the introduction of a VFG, the coupling constant stays high because of a high height/width ratio of 3.33 for the PZT rings.

Another novel annular composite transducer design fabricated by the indirect SFF process was the hexagonal pattern structure shown in Figure 3 (a). This structure consisted of solid rods of PZT arranged in a concentric hexagonal pattern. These rods themselves were ~ 500 μm in width and octagonal in shape as shown in Figure 3 (b). In the VFG hexagonal structure shown in Figure 3 (a), the ceramic content changed from 28 vol.% at the center to about 16 vol.% at the edges. This structure had a clean thickness mode resonance with a k_t of about 67%. The hexagon pattern structure without a VFG and the same PZT rod dimensions had a 14 vol. % ceramic content with a k_t of 69 %. Thus by proper control of the volume fraction gradient and the ceramic rod aspect ratio, the VFG composites can be tailored to show very high coupling constants.

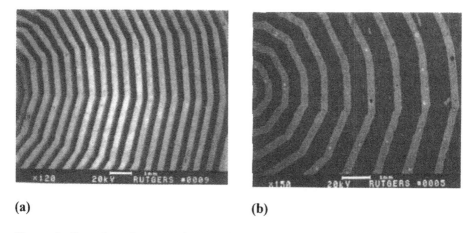

(a)   **(b)**

Figure 2: Scanning electron micrographs of (a) a sintered regular polygon PZT-5H structure and (b) an polygon volume fraction gradient PZT /spurr composite with decreasing ceramic content from the center to the edges.

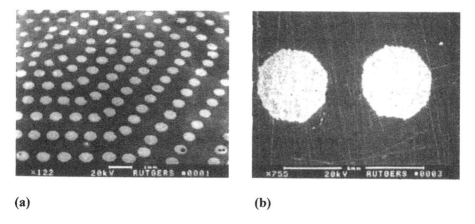

(a)   **(b)**

Figure 3: SEM micrographs of (a) the top surface of a volume fraction gradient hexagonal pattern composite and (b) each element of the pattern made up of octagonal shaped rods.

Composites with 2-2 connectivity are widely used in medical imaging. Hence, volume fraction gradient (VFG) piezoelectric 2-2 composites, comprising of parallel piezoelectric ceramic sheets separated by an inactive spurr epoxy were fabricated using the indirect SFF route. Many mathematical functions including gaussian, linear and exponential gradient were used to describe the gradient. Figure 4 shows the variation of volume percent ceramic with distance from the center for the three functions. In order to achieve a high pressure output at the center, all the distributions were designed to have ~ 60 vol. % ceramic in that region. The ceramic content gradually decreases to 20 % at the edges, falling at different rates depending on the function. For example, the initial drop would be maximum for the two exponential gradients while it would decrease slowly for the gaussian distribution.

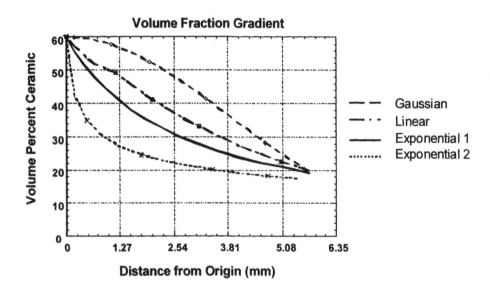

Figure 4: Variation of vol.% ceramic with distance from the center in a 2-2 PZT /polymer composite design.

Pro Engineer™ software was used to design the molds with a negative of the desired 2-2 structure. In order to obtain the required VFG distributions, the ceramic shrinkage during sintering was also taken into consideration while designing the molds. Figure 5 shows two of the sacrificial molds obtained using the Sanders Prototyping technique. Both the molds have an air gap of 135 μm between the walls throughout the structure. However, the mold wall thickness increases from 95 μm at the center to 575 μm at the edges, following the respective mathematical function. The Sanders wax molds were very accurate with an X-Y deposition error of only ± 5 μm.

Dielectric Ceramic Materials

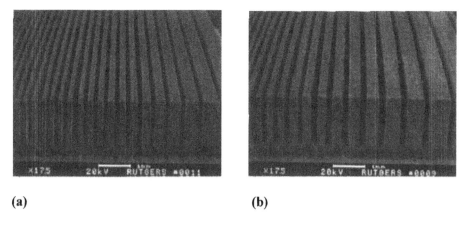

(a)   (b)

Figure 5: Scanning electron microscope (SEM) micrographs of sacrificial wax molds with a (a) gaussian distribution and (b) linear volume fraction gradient.

After infiltrating PZT slurry into the molds and heat treatment, the sintered structures were embedded in spurr epoxy. Figure 6 shows the top surface of the gaussian and linear VFG composites. From left to right, the SEM photographs show different regions in the composite; center and 2, 3, and 5 mm away in a sample with a total width of 12.5 mm. The increase in the spacing between ceramic walls from the center to the edges for the two distributions follow distinctly different functions. As shown in Table I, the average ceramic wall width was ~115 μm with the spacing changing as designed in Figure 4. Incidentally, the total ceramic content was 34 vol. % for the gaussian distribution as compared to 28 % for the linear VFG composite. Both the structures had a clean thickness mode resonance with high values of thickness coupling coefficient as shown in Table I.

The vibration profiles of the composites were determined by the laser probe technique, as described earlier. Figure 7 shows the vibration profiles of the 2-2 gaussian and linear volume fraction gradient composites. The probe had 2.3 mm. internal diameter, hence the observed displacement values were an average of the vibration of various sheets of PZT ceramic contained in that area (Figure 6). It can be seen from Figure 7 that the vibration amplitude of the gaussian composite decayed much slowly and was always higher than the linear composite. They seemed to follow the ceramic volume percent profile. Hence by properly designing the composite gradient, the output acoustic pressure distribution and hence the axial intensity of the beam can be easily tailored.

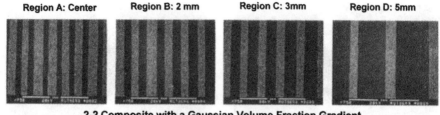

Region A: Center   Region B: 2 mm   Region C: 3mm   Region D: 5mm

2-2 Composite with a Gaussian Volume Fraction Gradient

2-2 Composite with a Linear Volume Fraction Gradient

Figure 6: SEM photographs of the top surface of the gaussian and linear VFG composites showing the central region and regions 2, 3, and 5 mm away from it. In this images the brighter lines are the ceramic and the dark regions are spurr epoxy.

Table I: The Macrostructural and Electrical Properties of the Gaussian and Linear VFG Composites.

Type	Ceramic wall width (μm)	Spurr epoxy width (μm)		Vol. % ceramic	K	k_t (%)	k_p (%)
		center	edges				
Gaussian VFG	115	85	500	34	900	68	26
Linear VFG	115	95	550	28	825	66	28

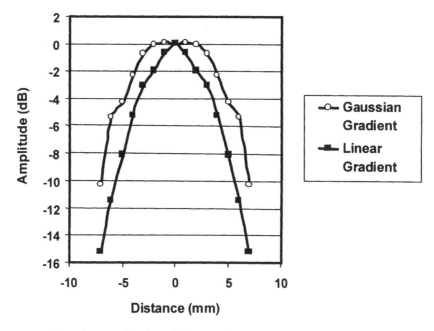

Figure 7: Vibration amplitudes of the gaussian and linear volume fraction gradient 2-2 composite specimens as a function of the distance from the center.

SUMMARY AND CONCLUSIONS

In this work, the Sanders Prototyping technique was used to fabricate a variety of complex shaped molds to produce different piezoelectric composite transducers, including polygon, hexagonal pattern with octagonal rods and 2-2 plate composites with and without ceramic volume fraction gradients. The normal and VFG polygon and hexagonal distribution composites showed very similar coupling coefficients inspite of the introduction of the gradients. The 2-2 gaussian and linear VFG composites were also made with high design accuracy by using the lost mold technique. The vibration amplitudes of these composites seemed to follow their VFG distribution. Hence, the ultrasonic output pressure distribution can be controlled by changing the arrangement of the ceramic elements.

ACKNOWLEDGMENTS

The authors would like to acknowledge ONR for financial support under grant #00014-94-1-0588. They are also grateful to Dr. T. R. Gururaja for his valuable advice and help on the design and electromechanical measurements of these composites.

REFERENCES
1. J. A. Zagzebski, "Physics of Ultrasound", pp. 1-92 in Textbook of Diagnostic Ultrasonigraphy ed. S. L. Hagen-Ansert, The C. V. Mosby Company, St. Louis, 1989
2. T. R. Gururaja, " Piezoelectric Transducers for Medical Ultrasonic Imaging", Am. Ceram. Soc. Bull., 73[5] 50-55, 1994
3. K. K. Shung and M. Zipparo, "Ultrasonic Transducers and Arrays", 20-30 (1996)
4. W. A. Smith, "New Opportunities in Ultrasonic Transducers Emerging from Innovation in Piezoelectric Materials" pp. 3-?? in SPIE International Symp. on New Development in Ultrasonic Transducers and Transducer Systems, ed. F. Lizzi, SPIE, Bellingham, 1992
5. V. F. Janas and A. Safari, " Overview of Fine-Scale Piezoelectric Ceramic/Polymer Composite Processing," J. Am. Ceram. Soc., 78 [11] 2945-55, 1995
6. D. A. Christensen, Ultrasonic Bioinstrumentation, Chapter 5, pp. 69-121, John Wiley & Sons, New York, 1988
7. V. Murray, G. Hayward, and J. Hossack, " Spatial Apodization Using Modular Composite Structures", pp. 767-770 in Proceedings of IEEE Ultrasonics Symposium, , IEEE, Piscataway, NJ, 1989
8. A. Bandyopadhyay, R. K. Panda, V. F. Janas, S. C. Danforth, and A. Safari, " Processing of Piezocomposites by Fused Deposition Technique", J. Am. Ceram. Soc., 80[6] 1366-72, 1997
9. R. K. Panda, P. Teung, S. C. Danforth, and A. Safari, " Solid Freeform Fabrication of Hydroxyapatite Bioceramics with a Controlled Pore Network" submitted to Biomaterials, February 1998
10. M. K. Agrawala, A. Bandyopadhyay, R. vanWeeren, A. Safari, S. C. Danforth, N. A. Langrana, V. K. Jamalabad, and P. J. Whalen, " FDC, Rapid Fabrication of Structural Components" Am. Ceram. Soc. Bull., 75 [11] 60-65, 1996

TiO$_2$-BASED GAS SENSORS AS THICK OR THIN FILMS: AN EVALUATION OF THE MICROSTRUCTURE

P. I. Gouma, S. Banerjee, and M. J. Mills
Center for Industrial Sensors and Measurements (CISM)
The Ohio State University, Columbus, OH 43210

ABSTRACT

Multiple phase systems based on titania that exhibit gas sensing behavior have been characterized by means of electron microscopy (SEM, TEM) and chemical analysis techniques (EDX). The materials examined have been either thick (thickness > 10 μm) or thin films (thickness < few μm). These films have been analyzed prior to, as well as after gas sensing tests were performed. The morphological, structural and chemical changes observed in each case are presented and discussed.

INTRODUCTION

The surface conductance of semiconducting oxides is affected by the concentration of the ambient gases [1]. Resistive gas sensors are based on this principle and are used to detect and quantify the gases present in the operating environment. The nature of the sensing mechanism is related to the electrical response of gas sensors to reactive gases, and the latter is considered to be a two-step process. Initially, oxygen from the atmosphere adsorbs on the surface of the sensor material, extracts electrons from its valence band and creates a depletion layer in the near-surface region [2]. The sensor surface, is therefore, very resistive initially. When the sensor comes in contact with a combustible gas, the gas molecules react with the adsorbed oxygen and inject electrons back to the depletion region. The change in sensor's resistance is used to measure the gas concentration[1].

The important parameters for a sensor are sensitivity, selectivity, and response time. Sensitivity is defined as the percentage change in resistance when exposed to the reactive gas. Furthermore, for a sensor to be useful, it is necessary that it is selective to a particular gas with negligible cross-sensitivity to other gases. These parameters can be improved vastly by additions of various second phase oxides and/or catalysts [3].

Dielectric Ceramic Materials

Traditionally, resistive gas sensors utilizing semiconducting oxides as sensing elements have been produced in the form of thick films[4]. These are usually printed on Al_2O_3 substrates with electrodes pasted on them [5]. Thick films usually have thicknesses that range from about 10 μm to 100 μm or even more [6]. Recently, there has been a trend towards developing planar sensors based on thin-film oxides with vapor-deposited catalytic activators.

Thin films have enhanced compatibility with microfabrication techniques and have lower power consumption. They can also be produced in arrays, by including various "chemically tuned" elements to introduce a level of kinetic selectivity into the operation of each element [7]. Thin films have thicknesses that range from a few ångstroms to about 0.1 μm. In both thick and thin film sensors, the fundamental sensing process is the same, i.e., oxygen chemisorption and its reaction with the reducing gases. In thin films, the changes in electrical conductivity is due to changes occuring in the external region of the layer in contact with the gases, while, in sintered layers, it is due to changes happening through the thickness of the material, on all grains whose surface comes in contact with the gases.

Most of the developments in resistive gas sensors have been based on tin oxide (SnO_2). Tin oxide shows favorable response to different gases under various combinations of doping and operating temperature [4]. Work on SnO_2-based thick film sensors has shown that sensitivity increases with decreasing crystallite size [8]. This is related to the Debye length (depletion layer thickness) of the material. With decreasing crystallite size, the Debye length becomes a larger fraction of the total grain volume, and hence the effect of decreasing or increasing Debye length on resistance change is more significant.

Similar results have been obtained for SnO_2 thin film sensors by Mochida et al [9]. Their work suggested that in addition to crystallite size, surface roughness and porosity are other important factors which determine sensitivity. Sensitivity enhancement by decreasing grain size has also been demonstrated in other oxide systems like In_2O_3 [10]. On the other hand, recently it has been suggested that r.f sputtered epitaxial thin films with single-crystal orientation and few grain boundaries should produce more adsorption-desorption so that the conducting response is more homogeneous [11].

Studies of the effect of film thickness on sensing behavior by Bruno et al [12] have shown that sensitivity increases with decreasing film thickness in SnO_2 CVD films. They suggested that CVD films are formed by piling up of grains which are connected to each other by necks or grain boundaries. Electrical conduction, thus, takes place through preferential percolation paths where the resistivity is the lowest. The thinnest films are more disordered and discontinuous and there are very few percolation paths, hence their initial resistance is very high and they are more sensitive to gas absorption effects.

From the above it is apparent that several different factors need to be considered in order to optimize the processing and microstructural design steps so as to obtain a satisfactory and reproducible sensing behavior. Furthermore, all of the above leads us to conclude that there is a large demand for microstructural characterization and structure-property correlation studies which can help us formulate suitable models to explain various aspects of the sensing behavior.

The present study focusses on the characterization of the structural and chemical characteristics of TiO_2-based gas sensors in thick and thin film designs, intended for use in hostile environments (e.g. for monitoring automotive exhaust gases). Representative samples of the TiO_2-based system have been studied using electron microscopy techniques. Most of the gas sensing work within CISM has been performed on thick TiO_2-based films. Recently, it has been extended to TiO_2-based thin films prepared by spin-coating. The ways in which the microstructure affects the sensing behavior of the material are discussed.

EXPERIMENTAL TECHNIQUES

Thick Films

Anatase-based thick films of titania were prepared by first dry-mixing commercial powders and subsequently ball-milling them to achieve homogeneity. Thick slurries were then prepared in 1-Heptanol solvent and screen-printed on alumina substrates. These films were heated in air to burn off the solvent.

Thin Films

Anatasematerial for the thin films of titania was prepared via a sol-gel route [13]. Thin films were produced by spin coating sol-gel droplets on: (i) alumina substrates with gold electrodes painted on them for sensing purposes or (ii) on sodium chloride single crystals and thin glass slides for microstructural characterization. Subsequently the films were annealed at 600°C for 2 hrs in air.

Sample Preparation for Characterization

Specimens for SEM examination were prepared in planar and cross-sectional views of the thick and thin films. The microscope used for this work was a Philips XL30 FEG SEM equipped with analytical capabilities.

TEM samples were prepared from the thick films by removing some of the powder material, suspending it in methanol and depositing it on holey carbon copper grids. TEM samples were also prepared from the thin films deposited on unheated substrates of sodium chloride. The films were removed from the substrate by floating them off in water and were deposited on copper grids. The microscopes used for this work were the Philips CM200 and Philips CM300 FEG STEM.

RESULTS AND DISCUSSIONS

Thick Films
Anatase-TiO₂-10wt%La₂O₃-2wt%CuO powder

Rutile is the stable and most widely used form of titanium dioxide. Anatase is a metastable form of titania (TiO_2) and transforms to rutile at temperatures that are a function of grain size, impurities, etc.[14]. Earlier work in CISM has shown that anatase-based systems could be used for gas sensing applications at relatively high temperatures, provided that the anatase form is stabilized at the operating temperatures[15]. Rare-earth oxide additions have been used for that purpose[16]. Both anatase-yttria and anatase-lanthana systems have been successfully tested for CO sensing[17]. Addition of CuO was shown to improve the sensing behavior of the systems tested [17].

The material studied in this work is also a complex system with three components, each of which is expected to have a different effect on the physio-chemical behavior of the material. The morphology of the system is shown in figure 1, which is a secondary image scanning electron micrograph. The thickness of the film, as calculated from cross-sectional analysis of the sample (not shown herein) was found to vary between 7-15μm. The surface porosity of the film (area fraction) has been calculated using image analysis and it was found to be 20% (±2.3).

The technique used to study the uniformity of the distribution of the various components was backscattered electron imaging (BEI), based on the different atomic numbers of Ti, La, and Cu. In this way, each phase was identified and studied sepa-

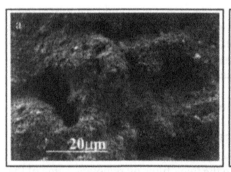

Figure 1: (a) SI SEM micrograph showing the morphology of the surface of the thick film of the anatase-based system; note the large pores in the sample; (b) BEI SEM micrograph showing phase contrast, in which the different components have been labelled with help from EDX analysis.

rately. Figure 1b is a BEI micrograph which illustrates the relative distribution of the various constituents (the background grey phase is the TiO_2 matrix and the lighter phases correspond to the secondary and ternary components). Imaging based on atomic number contrast alone was not sufficient to differentiate between La_2O_3 and CuO, so EDX analysis was used to complement the study. The large "bright"-contrast features in figure 1b are La_2O_3 particles, ($\geq 0.5\mu m$ long) and the smaller ones are clusters of very fine CuO particles, on different sites in the TiO_2 matrix.

Characterization of the material using TEM revealed a variation in the size of the TiO_2 phase particles, ranging between 70-200nm (see figure 2a). All of the titania particles examined had anatase-type structures. Figure 2b shows a electron diffraction pattern of the anatase type grains.

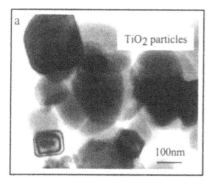 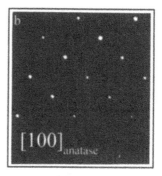

Figure 2: (a) BF TEM micrograph showing the particles of the anatase matrix; (b) Electron diffraction pattern of the [100] zone axis of anatase.

Further examination of the sensing material using the TEM has revealed the presence of fine clusters of particles on some TiO_2-phase powders (see figure 3a). EDX analysis performed on these clusters contain Cu–see figure 3b (a significantly more intense peak than the background Cu signal from the support grid) which indicates that these might be CuO particles.

High Resolution electron microscopy (at 300kV) was employed to study the distribution of secondary particles on TiO_2 in more detail. Only anatase-type TiO_2 particles were detected. Most of these particles were surrounded by a continuous amorphous layer. Often, anatase particles were also seen having fine features attached to their surfaces. Figure 4 shows a CuO particle attached on anatase. The CuO structure was deduced from direct measurements of the lattice spacings. Copper oxide was not found to be reduced, as one would expect for a catalytic component exposed to CO. However, in-situ TEM studies are required to confirm the behavior of copper oxide particles during sensing.

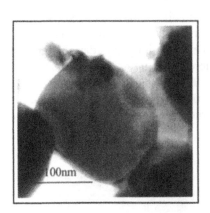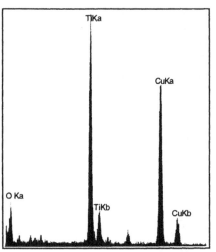

Figure 3: (a) BF TEM micrograph showing the presence of "dark" fine features on a titania matrix particle; (b) EDX spectrum revealing the presence of Cu in these features, along with the Ti and Oxygen peaks.

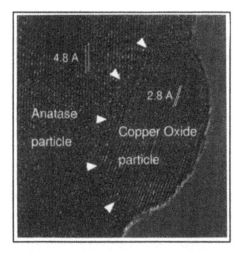

Figure 4: High Resolution TEM micrograph showing the side of an anatase particle where a CuO particle is attached.

Rutile-based TiO₂-2wt%CuO

The rutile-type titania powders used for the preparation of this material were produced by heat treating commercial anatase powder at 1000°C until it completely transformed to rutile. Figure 5a shows a BEI micrograph with rutile particles of different sizes and morphologies, ranging from rod-like to platelet structures. The average rutile particle size was found to be 1µm. There was a small percentage of Cu-based clusters dispersed in the matrix (average cluster size ~2µm). The porosity of the film was interconnected. The films thicknesses varied between 16µm and 25µm. The surface porosity was also measured to be 16% (±1.1).

Figure 5b is a TEM micrograph of a thick rutile particle. The electron diffraction pattern also shown in the figure corresponds to the [110] rutile zone axis. The rutile particles in this thick film were found to be larger in size and heavily faulted, compared to those of anatase. Twinning of rutile particles occurred during the heat treatment given to the starting powder. There were also many Cu-based particles seen in the vicinity of the faulted areas of the rutile particles.

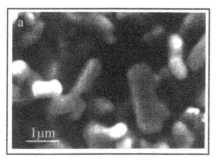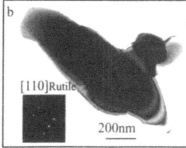

Figure 5: (a) BEI SEM micrograph showing rutile matrix particles; (b) BF TEM micrograph of a thick rutile particle and SAD pattern of the [110] zone axis of rutile.

The difference in the particle size and morphology of the matrix material in the anatase-based versus the rutile-based sensor can have a significant effect on the sensing response of these systems. The efficiency of electronic conduction depends not only on the surface characteristics but also on the number of contacts (boundaries, etc.) formed among the particles (grains) of the sensing element; (grain boundaries typically act as barriers to electronic conduction). The longer rutile grains with the high aspect ratio are expected to form fewer contacts than the smaller, round anatase particles. Furthermore, the heavily faulted surface of the rutile particles might enhance oxygen and gas adsorption.

Thin Films of TiO$_2$ Sol-gels

Sol-gel powder
Drying the sol-gel of titania resulted in the formation of large chunks of powder. Examination of the powder supported on a holey carbon copper grid in the TEM revealed fine clusters of nanometer size particles, having the anatase-type structure. An increase in particle size was observed with time (from 3.5nm to about 30nm, within an hour of TEM operation at 200kV)-as determined by the electron diffraction patterns (see figure 6). It is believed that the exposure of the powder to the electron beam induced the crystallization and growth of the sol-gel powder.

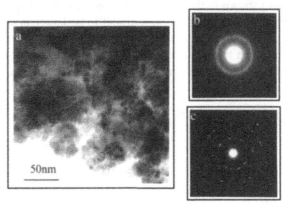

Figure 6: (a) BF TEM micrograph showing the typical microstructure of the titania sol-gel; (b) and (c) are SAD patterns of the same region taken 1 hr apart, showing the crystallization and growth of the anatase nanometer size grains.

Thin Films
The thin films deposited on sodium chloride substrates were found to consist of nanocrystalline, equiaxed grains (see Figure 7a). The electron diffraction patterns showed rings corresponding to the anatase-type structure. The average grain size, as determined from the ring width was about 8 nm (see figure 7). Multiple coatings were used to make sure that the film would cover the electrodes of the substrate for the sensing tests, resulting in a film thickness close to 1μm.

The films that were annealed at 600° C for two hours have also been studied in the TEM. They consisted of overlapping layers (due to the multiple coatings) with granular structures and interconnected porosity (see figure 8). Individual grains having the rutile structure, were identified by electron diffraction. The morphology of the rutile particles tended towards the rod-like and plate-like shapes, similar to the transformed commercial powders found in thicker films.

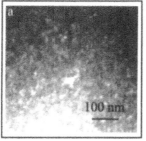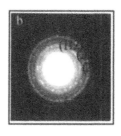

Figure 7: (a) DF TEM micrograph of a planar view of the as-deposited thin film showing nanocrystals of the anatase type with bright contrast; (b) SAD pattern of the region showing rings and some spots.

Comparing the above findings with published work in the literature on thin films of titania, it might be argued that the transformation from anatase nanocrystals to rutile grains during sintering was achieved at lower temperatures than it would be expected. Nagpal [18] et al spin-coated suspensions of sub-100 nm TiO_2 particles in an organic solution, on quartz and Si substrates. Sintering of these films resulted in the formation of anatase films at 600^0 C. At 900^0 C, these films were still predominantly anatase with some fine rutile crystals forming. Film thicknesses, in their study, were a few hundred nanometers. Another study [19] of sol-gel films of TiO_2 spin-coated on fused silica, Si (100) and rutile (100), involving film thicknesses around 0.3-0.4 microns for single coatings has indicated 650°C as the crystallization temperature for anatase. The conversion to rutile was substrate dependent. It appears that the onset of the transformation depends on the grain size of the starting material. Further study is currently underway to understand this polymorphic reaction.

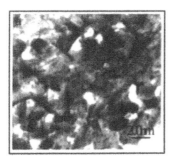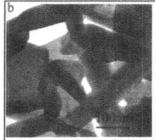

Figure 8: (a) BF TEM micrograph of a planar view of the annealed thin film; (b) higher magnification micrograph showing the rutile crystals in more detail.

CONCLUSIONS

Thick and thin films of titania-based systems for gas sensing applications have been characterized by means of electron microscopy and microanalysis techniques. The microstructures of the thick films varied significantly with the type of titania used (i.e. anatase vs. rutile). The structural and morphological differences observed are expected to determine the variation in the sensing behavior of these systems. The fine particle sizes of the thin films, on the other hand, were found to induce phase transformation at lower temperatures than the thick films. Although the processing route followed was different in the thick vs thin films, the rutile crystals grown had similar morphologies.

REFERENCES:

[1]M. J. Madou and S. R. Morrison, " Chemical Sensing with Solid State Devices", Academic Press, San Diego,(1989).

[2] S.R. Morrison, *Sensors and Actuators*, 11, 283-287, (1987).

[3] D. Kohl, *Sensors and Actuators B*, 1, 158-165 (1990).

[4]J Watson, *Sensors and Actuators*, 5, 29-42 (1984).

[5]W. Gopel and G. Reinhardt, "Sensors Update", 1, VCH Publication, (1996).

[6]R. E. Loehman, "Characterization of Ceramics", Butterworth Heinemann,(1993).

[7]G.E. Poirier, R.E. Cavicchi and S. Semancik, *J. Vac. Sci. & Tech. A*, 12, 2149-2153 (1994)

[8]Chaonan Xu, Jun Tamaki, Norio Miura and Noboru Yamazoe, *Chemistry Letters*, 441-444, (1990).

[9]Tadashi Mochida, Kei Kikuchi, Takehiko Kondo. Hironobo Ueno, Yoshinobu Matsuura, *Sensor and Actuators B*, 24-25 (1995).

[10]A. Gurlo, M Ivanonvskaya, N Barsau, M Schweizer-Berberich, U. Weimar, W Gopel and A. Diegnez, *Sensors and Actuators B*, 44, 327-333, (1997).

[11] S. Semancik and R.E. Cavicchi, *Thin Solid Films*, 206, 81-87, (1991).

[12]L. Bruno, C. Pijolat and R. Lalauze, *Sensors and Actuators B*, 18-19, 195-199, (1994).

[13] Qunyin Xu and M.A Anderson, *J. Mater. Res.*, 6, 1073-1081, (1991).

[14]X-Z. Ding, X-H. Hu, Y-Z. He, *J. Mat. Sci. Lett.*, 15, 789-791, (1996).

[15]L. Younkman, "Development and Characterization of Ceramic-Based Carbon Monoxide Sensors, Ph.D. Thesis, The Ohio State University, (1994).

[16]S. Hishita, I. Mutoh, K. Koumoto and H. Yanagida, *Ceramics International*, 9, 61-67, (1983).

[17]CISM, Unpublished Work (1997 - present).

[18] V. J. Nagpal, R. M. Davis and S. B. Desu, *J. Mater. Res.*, 10, 3068-3078, (1995).

[19]U. Selvaraj, A. V Prasadarao, S. Komarneni and R. Roy, *J. Am. Ceram. Soc.*, 75, 1167-1170. (1992).

THICK FILM TECHNOLOGY: AN HISTORICAL PERSPECTIVE

Robert J. Bouchard, Photopolymer and Electronic Materials, Dupont Experimental Station, Wilmington, Delaware 19880

INTRODUCTION

Screen printing is a relatively mature technology . Yet, the materials are continually being improved, and ever finer resolution of features is being developed by advances in patterning technology. This renewal process is required to accommodate the demands of miniaturization, circuit complexity, multilayer assemblies, higher frequencies, reduced toxicity, thinner layers, tighter specifications, reduced cost and higher yields. Because of the pivotal role played by thick film materials in this age of electronics, it is interesting to follow it's surprisingly rich historical path that has led to the current era.

DESCRIPTION OF THE TECHNOLOGY

Thick film technology is a method whereby resistive, conductive and dielectric pastes are typically applied to ceramic substrates by screen printing or some other patterning method. The circuits thus defined are fired, typically at 850 deg C, to fuse the films to the substrates to produce the desired functions.

If these interconnect patterns are used to attach or package semiconductor devices, the result is a so-called *thick film hybrid microcircuit*. If no active device is involved, the result is a single or multiple resistor or capacitor network. Discrete, surface mounted resistors or capacitors are called *chips*.

"Thick film" technology is a term that was coined in the early `60`s to distinguish it from "thin films", typically applied by a vapor deposition process. Thick films are typically 5-10 microns and thicker, while thin films are typically on the order of 1 micron.

THE EARLY DAYS

In the beginning (1930), Dupont acquired R&H Chemicals, which had a product line for decorating and coloring chinaware, including liquid bright gold. The decorative product line was expanded in Perth Amboy to include liquid bright Pt, Pd, glazes, stains for chinaware, enamel pigments, and glass enamels.

In the 1940's, screen printed silver electrodes replaced Al foil for stacked mica capacitors. Proximity fuses in WWII used screened Ag conductors and C-phenolic resistors. $BaTiO_3$ was developed for disk caps with Ag electrodes, and replaced stacked mica capacitors for the military. Dupont Ag conductors were also used for RC networks. The first monolithic capacitors with glass dielectrics and Ag electrodes were produced.

In the 1950's, Dupont abandoned the production of capacitors to make materials only. Television, disc capacitors, and Ag paste grew rapidly. In the ceramic area, ceramic labels for beverage containers and porcelain enamels for metals grew rapidly. Pd/Ag "fired on" resistors were developed to replace carbon phenolics.

THE LAST FORTY YEARS

The 1960's were characterized by the vigorous growth of the microcircuit business, as hybrids entered the world of semiconductor technology. This was the beginning of the thick film hybrid circuit industry as we know it today. The first "fired on" resistors were commercialized, based on Pd, Ag, PdO combined with frit, and were adopted by the military for the Minuteman missile and by IBM for the System 360. IBM was one of the first to recognize the value of thick film hybrid technology, and their influence had an enormous effect upon creating an awareness throughout the rest of the electronic industry. Electrodes and terminations were developed for emerging multilayer capacitors. At the end of the decade, the first Birox® resistors were developed by Dupont, based on oxides of ruthenium.

In the 1970's, membrane touch switches and thick film defoggers for autos were developed. Strain gauges based on the piezo-resistive properties of thick film resistors were designed. Fodel® photosensitive gold conductors were invented, but had little commercial utility at this time. Perhaps the major factor in this decade was the vigorous growth of the thick film business with the expansion of hybrid microcircuit technology in the '75-'80 period. At the same time, silver pastes for disc capacitors declined.

The 1980's was another decade of solid growth, but it brought with it a period of competitive turbulence. Plessey and Cermaloy were acquired by Heraeus, Thick Film Systems by Ferro, AVX Materials by Johnson Matthey, and EMCA by Rohm & Haas. In the meantime, Shoei emerged as a leader in Japan. At the same time, new or improved materials were developed for resistor systems, copper multilayers, gold pastes for microcircuits, silvers for defoggers, and membrane touch switches. Custom formulations for major customers were much more common. Expansion of manufacturing sites occurred into other geographical areas across the world.

In this last decade of the 1990's, we have seen the limits of screen printing technology pushed to finer and finer features. Diffusion patterning allows a 50% reduction in via size vs traditional thick films from 10-20 mils to 4-6 mils, which permits higher interconnect density when combined with fine line printing. New thick film material systems have been developed for flat panel display applications. Fodel® photo-formable technology has produced resolution of 40 micron lines and 30 micron spaces, required for microwave applications. Improvements in processing of multilayers and in multilayer materials has resulted in significant cost savings.

Recent achievements of screen printing technology that can be reached by a determined manufacturer is perhaps best demonstrated by the production of large area plasma display panels, where fine pixel displays require exceedingly fine resolution and uniformity. Recently, Noritake (1) has developed equipment that achieves precise screen printing over a 42 inch AC plasma display panel. This example demonstrates what can be accomplished by a determined manufacturer to re-define and extend state-of-the-art thick film technology.

On the materials side, Cd and Pb have been eliminated from many material systems. In the dielectrics area, lead-free multilayer systems are now available with excellent aged adhesion and compatibility with a range of conductors. Dielectric constant of dielectrics is gradually decreasing for higher and higher speed applications. Better understanding of the interactions between resistor materials and terminations, and the dependence of the conductive microstructure on processing conditions, have reduced resistor process sensitivity.

New fine particle gold powder developments have increased the density and reduced the line width possible with conventional thick film printing, and the high frequency performance is equivalent to thin film deposition.

In addition, we are seeing advances in new applications and types of thick films, e.g. polymer thick film inner layers, buried resistors and capacitors, humidity and smoke sensors, piezoelectric thick films.

MATERIALS TECHNOLOGY (2-5)

RESISTORS

The development of thick film resistor materials in the late '60's was critical to establishing thick film as a viable technology. Earliest materials were based on mixtures of Pd, Ag and glass. Electricals were controlled by the ratios of Pd/Ag solid solution, PdO, and glass in the fired film. These materials required extensive process control. They had to be fired at<760 deg. C to remain below 800 deg. C, where PdO, formed during firing, thermally decomposed back to Pd. At the same time, the amount of PdO formed was also influenced by the amount of Ag.

For this chemically dynamic system, reproducibility could only be obtained by careful control of the firing time-temperature profile. In addition, even slightly reducing conditions could markedly change electricals. As a result, only a limited range of resistivities (500-10,000 ohms) were available and the TCR's were in the 250-500 ppm/degree range, an order of magnitude higher than modern materials.

In the late 60's, Dupont's Central Research Department discovered a new series of precious metal oxides based on ruthenium with the pyrochlore crystal structure having the general formula $A_2B_2O_{6-7}$. These materials were interesting for several reasons:

1. Some of them were excellent electrical conductors, with conductivities among the highest known for oxides, rivaling those of metals.

2. They could have *intrinsically* very large or very small TCR's, which could be varied over a wide range by the large number of chemical substitutions into the pyrochlore formula (TCR or temperature coefficient of resistivity is a critical performance parameter of resistor material).

3. As required for all thick film materials, they were able to be fired at 850 deg C in air with aggressive Pb-containing frits without chemical decomposition.

Typical compounds are $Bi_2Ru_2O_7$ and $Cu_{0.5}Bi_{1.5}Ru_2O_{6.5}$. This development expanded the range of available resistivity down to below 10 ohms to above 10 megohms with TCR's below 100 ppm/degree.

As these materials became more widely used, understanding of glass chemistry and its influence on resistor composite physical and electrical properties increased (6,7). The influence of glass in conductive phase sintering and dispersion became better understood. Small amounts of transition metal oxides, presumably dissolved in the glassy phase, on electrical characteristics were found to be useful for adjusting TCR, either in the positive or negative direction.

Controlling TCE of the resistor composite *vs* the alumina substrate increased stability to laser trimming. (Laser trimming, where a significant part of the resistor is removed, is always required for tight control of resistivity. This is because of the inability to control the volume of paste to the high tolerance required for resistivity (<1%). Fortunately, laser trimmers are capable of trimming to very close resistivity values in a fraction of a second.) Higher conductive loadings across a broad range of resistivity increased the power handling capability, especially at high resistivity.

Dielectric Ceramic Materials

At about the same time, RuO_2, which is the critical chemical building block of the new pyrochlore systems, came into its own as a conductor. Although the range of possible chemical substitution is not as large for the rutile structure of RuO_2 as for the pyrochlore structure of $Pb_2Ru_2O_6$, it can be made in a wide range of particle size and surface area. It also has intrinsically a higher conductivity, which makes it suitable for lower resistivities. However, it's high conductivity can be detrimental, since highly resistive films (>10Kohm) require only a small amount of RuO_2 to achieve resistivity values. As a result, it's ability to handle voltage is diminished, since the small number of conductive chains must each carry more current.

The ability of RuO_2 at low resistivities overlaps and complements the ability of the ruthenates at higher resistivities. Therefore, most commercial systems today consist of RuO_2-based low end members blending into $Pb_2Ru_2O_6$ high end members. This assures adequate power handling and low TCR over six or seven orders of magnitude in resistivity. At the very low resistivity end, Pd/Ag alloys/glass composites have been developed further for surge protection applications. They are typically formulated for the 0.1 to 1 ohm range, and the TCR's are controlled by the ratio of Pd to Ag in the alloy.

A major change in the past decade is the increase in the number of chip resistors. Chips still use thick film technology, but are prepared off-line in a separate manufacturing process. They are then precisely deposited onto substrates with pick-and-place equipment and soldered on as surface-mounted devices. They dominate the resistor materials used today. The complexity is demonstrated in Figure 1, where it can be seen that resistor material is only a small part of the chip resistor. Hundreds of billions of these parts are made today, mostly in Japan, and size has been decreasing steadily. Dimensions of 0.5X0.5 mm are common. Even with this complicated design, typical selling price is <one cent each!

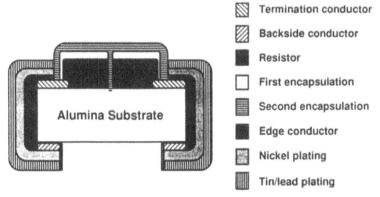

Figure 1. Schematic of a modern chip resistor

The development of resistors over time has been impressive. The high stability and tight specifications to which modern resistors are manufactured are illustrated in the table below:

PARAMETER	TYPICAL SPECIFICATION
Wide range of R	1 ohm-10 megohm
Low TCR (-55->+125 C)	+/- 50 ppm/deg C
Good power handling	5 watts/5 sec
Electrostatic Discharge	<0.5% change, 5X5KV pulses
Low process sensitivity	1 ppm/deg TCR, 1%/deg R
High laser trim stability	<0.5%/1000 hrs/85 C/85 RH
No changes	<1% over lifetime

DIELECTRICS

Thirty years ago, dielectrics were mainly single-component, low-melting glasses that were used to isolate the top layer conductor from the rest of the circuit below. Because of undesirable interaction between the glass phase and the overlying conductor, crystallizable glasses with low dielectric constants appeared. The crystalline phase strengthens the matrix glass, and retards its flow during subsequent firing. These were much more stable, and are still used today in simple circuits.

As miniaturization demands required smaller circuit area, multilayers were developed, and for the simpler two or three layer multilayers, crystallizable dielectrics remained adequate. However, as layer count increased, factors like TCE mismatch between dielectric and substrate became more critical, since bowing had to be minimized. As a result, filled dielectrics were developed from higher softening point glasses combined with inorganic fillers, to fine tune the expansion between dielectric and substrate. In addition, these filled systems formed more adherent interfaces with screen printable conductors.

The early multilayer conductor was usually gold to insure reliability, since silver had a tendency to migrate through the glassy phase. However, as cost became more of an issue, the drive to silver was too powerful to resist. That required the development of filled crystallizable glasses, combining the characteristic hermaticity and density of crystallizable glass with the ability to fine tune the dielectric properties by suitable filler additives. The crystallizable glasses are typically calcium zinc aluminosilicates, which precipitate celsian or hardystonite

Dielectric Ceramic Materials

on firing. Alumina can be added for toughness. The remnant vitreous phase after crystallization can be designed to inhibit ionic migration, and this combination leads to a highly reliable, silver-based system.

CONDUCTOR MATERIALS

Conductors generally constitute the largest quantity of thick film material used in a hybrid microcircuit. Because of precious metal content, they also often represent the most expensive materials in the circuit. Conductors are passive elements that connect and transmit signals from one circuit element to another. The choice of conductor depends on a number of factors, e.g. low resistivity, solderability, solder leach resistance, aged adhesion, cost, etc. Because different applications value these factors differently, a number of metallurgies exist to optimize utility. As cost has become more critical, for example, the use of gold conductors has steadily decreased, as well as the amount of Pd in Ag/Pd conductors. The demands on conductors has also increased as multilayer circuits become more and more complicated. Often the conductor has to withstand a number of firing steps while the other constituents of the circuit are built up in repetitive firing cycles.

Copper

Copper was considered as a potential hybrid conductor in the mid-1970's, because of its much lower cost, high conductivity (second only to silver), excellent solder leach resistance, and lower tendency to migrate in dielectrics. Its primary limitation was that it must obviously be fired in an inert atmosphere like nitrogen. Dielectric materials used with copper were similar to air fired systems, although burnout of the organics is more difficult and oxidants had to be added in small amounts. Because of the higher processing costs, copper tended to be used in the higher value applications, like the military. A better understanding of the interactions at the conductor/dielectric interface under inert or reducing atmospheres was required to minimize blistering and voids. Highly adherent copper with excellent solderability and solder leach resistance was eventually developed, aided by close control of powder morphology.

By the early 1980's, resistor systems compatible with copper processing were developed. The challenge was to develop a resistor system of new conductives and glasses stable in a low oxygen environment, and with suitable burnout characteristics. Traditional ruthenate systems are reduced at high temperatures under nitrogen. Systems based on lanthanum boride, indium oxide, strontium ruthenate, and metal silicides have been developed to meet these needs. In general, it is more difficult to develop a system that spans a large range of resistivity without the flexibility of firing in air. And encapsulation of the fired resistor may be required for long term stability.

Dielectric Ceramic Materials   435

LOW TEMPERATURE COFIRED CERAMIC (LTCC)

Cofired ceramic tape technology based on alumina and refractory metals like tungsten or molybdenum have been in use for many years. Perhaps the most complex multilayer of this type is the IBM Thermal Conduction Module developed in 1980. However, because it requires a large capital investment and complex processing, including firing with atmosphere control at 1500 deg C, it is not widely practiced. Extension of the technology in the form of a low temperature, cofired tape system made it accessible to all thick film circuit manufacturers in the early 1980's. LTCC then, is a combination of the best features of thick film and high temperature cofired alumina technologies. It provides an alternative approach to standard print and fire techniques to fabricate complex, high layer count, multilayer circuits.

As with screen-printable dielectrics, tape dielectrics were originally formulated from filled glass systems, but crystallizable glass systems were also developed. In many cases, tape material systems are very similar to their thick film dielectric counterparts. Early tape systems were designed to have TCE matched to alumina, but later systems have lower TCE's to allow direct attachment of silicon chips. Although the strength of LTCC's started out much less than their high-fired alumina counterparts, improvements in materials technology and tape processing have produced almost equivalent properties. For low layer count multilayers, tape can be laminated to traditional alumina substrates to build up the circuit. In this way, many of the shortcomings of tape, such as low strength, low thermal conductivity, high shrinkage, etc. can be markedly reduced.

As the trend toward higher speed integrated circuit technology continues, the dielectric constant of the tape dielectric becomes more important, and the preferred conductors are copper and gold. Dielectric constants are gradually decreasing to accommodate the increase in operating frequency (8), and TCE's matched closely with silicon will be required.

PATTERNING TECHNOLOGY

New patterning techniques have evolved as circuits increased in complexity from single-sided hybrids to double-sided, to multilayers with finer lines/spaces and smaller vias. Screen printing is clearly the technology of choice because of its simplicity and low cost. As a result, much of the work has centered around improving the paste rheological characteristics to allow the printing ever finer features. Organic vehicle development which allows better design of the viscosity at low and high shear rates, combined with multiple solvents and various additives to control thixotropy, has resulted in marked improvements. Newer screen designs can now produce lines as thin as 3 or 4 mils. Ultimately, these modifications reached a plateau, and other approaches were developed.

One approach was to combine the screen printing with traditional etching technology. Circuits Processing Technology (Oceanside, CA) has developed a photoresist-defined etching process on conventionally screen printed gold conductor. This combination produces high density, reliable circuits that can operate in the many applications that must use frequencies in the 1-100 GHz range, e.g. satellite uplink and downlink, cellular, PCs and paging base stations, etc. Historically, more costly thin film metallizing was required. The CPS technology allows 0.001 in. lines and spaces to be patterned on substrates as large as 4 X 4 inches. The screen printing/etching process yields conductor traces with smooth, flat surface topology, well-defined edges, and near vertical walls–all key requirements for low attenuation and dielectric loss at frequencies above 1 GHz. .

Photoimaging is another technology which has been available since the early 1970's, although not widely practiced. Instead of a screen, the circuit is defined by a photo mask, where circuit elements are illuminated through transparent sections of the mask. Although pastes are still required, the screen printing is very simple, since the paste is printed in an open pattern, covering the entire substrate surface uniformly. However, the pastes are chemically more complex, since they require photosensitive polymers, monomers and other additives.

Most photopatterning is subtractive, in that the unexposed areas are washed out in a development process, and this is undesirable from a materials cost standpoint. Therefore it can be justified only where resolution requirements are critical. However, this technology has undergone a resurgence in recent years for use in flat panel displays, where fine line conductors over a very large area are required.

Some of the principal patterning technology types are reviewed below.

FODEL® PATTERNING

Photoimageable thick film processes have been developed that are capable of producing .001" lines with .002" spaces and .003" diameter vias. In this process, a photoactive paste is applied to a substrate, dried, and exposed to UV light through a mask to define lines or vias. Those area exposed to the light are crosslinked into insoluble polymers. The materials are then developed with an aqueous carbonate process, and then fired using conventional thick film methods.

A variety of application methods can be used to apply the material, e.g. spraying, doctor knife coating, and screen printing. Since the final features are not defined by the screening pattern, these material systems are formulated as relatively low viscosity, Newtonian pastes. As a result, dried and fired surfaces are smoother than those obtained by conventional screen printing, which is desirable for multilayers. Both copper and gold metallizations up to ten layer circuits have been demonstrated. The basic principles are demonstrated in Figure 2:

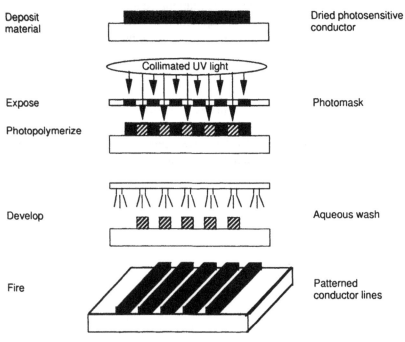

Figure 2. The sequence of operations in *Fodel® patterning*.

PHOTOFORMED CONDUCTORS

PCS (Photoformed Conductor System) is a photoimageable, thick film-compatible process for high resolution and high definition conductors on ceramic substrates. The process (Figure 3) consists of coating a substrate with light-sensitive organic material, which becomes tacky when exposed to UV light through a photo mask. The surface is then covered with metal powder "toner", which sticks to the tacky areas. They can then be fired using conventional thick film equipment and profiles. The basic process is shown in Figure 3. The resultant metal pattern has excellent wire bonding properties, and has been demonstrated with Au, Ag, Cu, Pt, and Ni metallizations, and in multilayer connections.

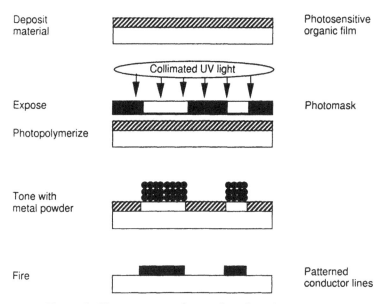

Figure 3. The sequence of operations in *PCS patterning.*

DIFFUSION PATTERNING

Vias or holes in dielectric layers are critical to allow interconnection between circuits above and below the layer. Screen printing vias is troublesome, since it is difficult to keep paste from re-filling them after printing. This limits the size of vias that can be formed with the confidence required for high yields. One of the patterning techniques that overcomes some of the limitations of conventional screen printing is called *diffusion patterning*. This technique involves printing and drying an unpatterned dielectric layer, then overprinting with a pattern of dots corresponding to the desired vias. Each dot contains an organic reagent that diffuses into the corresponding dielectric layer below and converts it, as the paste is dried, to a column of material which is soluble in water. The vias are generated by washing the layer with water.

With this technique, vias as small as six mils across with defect levels as low as one per hundred thousand are routinely achieved. Costs are lower because of higher yields and throughput, and the higher interconnect density made possible by smaller via size. The process is outlined schematically in Figure 4.

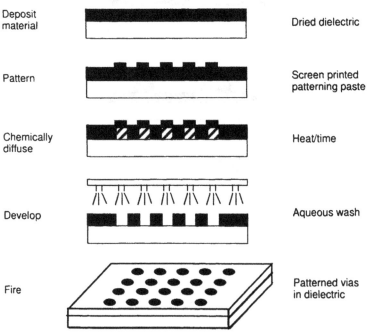

Figure 4. The sequence of operations in *diffusion patterning*.

Dielectric Ceramic Materials

TRENDS

ECONOMIC

With the increasing importance of worldwide competition, there is ever increasing pressure to provide materials to tighter specifications and that are robust to processing variability. Both of these attributes contribute to lower cost and higher yields. Conductor applications that originally were based on Pt, Pd and Au have been displaced largely by Ag except in special military applications. Lowering the cost of precious metals has been a major driver for lower firing temperature thick films, e.g. MLC electrodes, dielectrics that are stable to silver migration, copper thick film technology, etc.

POLITICAL/ENVIRONMENTAL

Cd-based materials are much less common today, and will gradually be phased out completely. There is no evidence of thick film toxicity problems with lead-based thick films in electronic applications. However the concern for child safety with lead-based paints, lead-based plumbing, and atmospheric quality with tetraethyl lead has carried over into electronic materials. The trend to elimination of heavy metals is unlikely to diminish, regardless of the dubious environmental significance for electronic materials. It is fortuitous that in many cases, substitution can be achieved with a minimum penalty in performance, e.g. Zn can usually replace Cd as an oxide bonding promotor in conductor thick films; and Pb-based frits can be replaced by Bi, Zn and alkaline earth aluminosilicates and aluminoborosilicates in dielectrics. For resistor thick films, it is more difficult. Although RuO_2 can often replace $PbRuO_3$, for high ohm resistor applications that require excellent power handling, no effective substitute has yet been found.

CONCLUSIONS

Since its earlier days, there has been marked change in thick film technology: the shift from PdO to Ru-based resistors, use of cheap silver conductors, extensive use of screen printed multilayers, viable patterning alternatives to screen printing, and low temperature cofired tape. Failure mechanisms are better understood. These changes have been driven by reliability, low cost, and increased circuit density for greater functionality per unit size. These technical and economic drivers will continue to impact thick film materials development to test it's competitiveness with other not-in-kind technologies like printed circuit boards and thin films.

Perhaps the best example of the recent accomplishments of this technology is in the complicated multichip modules with improved dimensional, electrical and thermal performance and reduced cost. Particularly impressive is the development of large area (10X10") multilayer (36 layer) circuits with line width/spaces down to 0.002" and via sizes down to 0.003".

Thick film hybrid technology has been applied in the electronics industry in military and commercial applications for more than three decades. Predictions of its decline and demise have been made through almost the same period. Yet the technology applications are growing and thick film techniques still offer significant advantages vs thin films and printed circuit boards for high-performance circuitry.

REFERENCES

1. M. Hiroshima et al, "Noritake's New Equipment for Manufacturing Large PDP's, Information Display, pp. 12-14, Nov. 1997.

2. Much of the material presented here has been based on previous extensive reviews by others in Dupont, mainly Don DeCoursey and John Larry, in various ISHM talks, esp. J.R. Larry, "The past 25 Years: Advances in Thick Film Materials", Inside ISHM, 19, #6, 8-15 (1992).

3. W. Borland, "Thick Film Hybrids", Volume 1, Electronic Materials Handbook, ASM International, 1989.

4. J. R. Larry et al, "Thick Film Technology: An Introduction to the Materials", IEEE Trans., CHMT-3, #2, June 1980.

5. R.J. Bouchard, "Materials Aspects of Thick Film Resistors", Ceramic Transactions 33, 391-404 (1993).

6. R.W. Vest and B.S. Lee, "A model for sheet resistivity of RuO2 Thick Film Resistors", IEEE-CHMT 14, #2, 396-406 (1991).

7. T. Pfeiffer and R.J. Bouchard, "Modeling of Thick Film Resistors", Ceramic Transactions 33, 405-418 (1993).

8. D. I. Amey et al, "Beyond the 1 GHz Barrier with Thick-Film Ceramic", Microwave Journal, 122-132, November, 1997.

RESONANT STRAIN LEVELS IN MODERN CERAMIC PLATE ACTUATORS

Arthur Ballato
US Army Communications-Electronics Command
Fort Monmouth, NJ 07703-5201, USA

ABSTRACT
Ceramic actuators driven piezoelectrically are used in a wide variety of modern applications such as in optical modulators, microtransducers, and certain types of resonant acoustic sensors and cellular radio control elements. For these uses it is necessary or desirable to know the maximum mechanical amplitudes obtainable, and how these are limited by the various internal dissipation mechanisms such as internal conduction, dielectric loss, and acoustic viscosity. For poled ceramics, the displacement amplitude may exceed 100 Å per volt at VHF frequencies. The analysis also shows that the unavoidable interfacial stresses existing between the plate surfaces and the exciting electrodes can become exceedingly large at the frequencies commonly used for frequency control of cellular communications.

INTRODUCTION
The resonant amplitude levels of piezoelectrically driven ceramic resonators are of interest in a variety of modern applications [1]. For simplicity of illustration, we consider here a plate executing simple thickness motion – that is, motion that depends solely on the thickness coordinate, and is independent of the lateral dimensions. Also for simplicity, the discussion is limited to a single mode of motion, which may be either extensional or shear. These restrictions permit the essence of the solution to be described in simple terms; more general cases are considerably more complex [2-5].

We will apply the formalism to be given in the next sections to a single configuration and posited material parameters, viz., a 250 MHz resonator with 1 mm^2 electrodes. The input quantities are given in Table I.

THE PLATE IMPEDANCE MATRIX
The plate is considered to have three ports: port one is the 'left' surface; port two is the 'right' surface; and port three is the electrical port, consisting of the connection to the two electrodes. The variables at port 3 are electrical voltage and current; those at mechanical ports 1 and 2 are, correspondingly, force (surface stress times electrode area) and particle

velocity (circular frequency times particle displacement for time-harmonic motion). Subscripts refer by number to these ports.

Table I. Input constants for ceramic transducer

Quantity	Unit	Numerical Value	
Electrode area, A	m^2	1.00 E-6	(1 mm^2)
Plate thickness, t	m	8.80 E-6	
Mass density, ρ	kg/m^3	7.52 E+3	
Acoustic velocity, v	m/s	4.40 E+3	
Piezoelectric coupling, k	dimensionless	0.51	
Dielectric permittivity, ε	F/m	1.723 E-9	(1.723 nF/m)
Conductivity, σ	S/m	1.00 E-7	
Motional time const., τ_1	s	1.00 E-12	(1 ps)
Static time constant, τ_0	s	$\tau_0 = \varepsilon/\sigma = 17.23$ E-3	
Acoustic viscosity, η	Pa-s	$\eta = c \cdot \tau_1 = 0.146$	
Mechanical frequency, f_0	hertz	$f_0 = v/2t = 250$ MHz	

The following definitions are used in the sequel: acoustic impedance, $Z_0 = A \cdot \rho \cdot v$; static capacitance, $C_0 = \varepsilon \cdot A/t$; piezoelectrically stiffened elastic stiffness, $c = \rho \cdot v^2$; square of the piezoelectric coefficient, $e^2 = k^2 \cdot \varepsilon \cdot c$; piezoelectric transformer turns ratio, $n = A \cdot e/t$; imaginary operator, j; circular frequency, $\omega = 2\pi f$, where f is the frequency variable; normalized frequency variable, $X = (\pi/2) \cdot (f/f_0)$; and mechanical loads at ports 1 and 2, Z_L & Z_R.

Elements of the symmetric impedance matrix of the ceramic are defined as follows: $z_{11} = z_{22} = Z_0/[j \cdot \tan(2X)]$; $z_{12} = Z_0/[j \cdot \sin(2X)]$; $z_{13} = z_{23} = n/[j \, \omega \, C_0]$; and $z_{33} = 1/[j \, \omega \, C_0]$. Elements of the symmetric load impedance matrix are defined as follows: $z_{11LOAD} = Z_L$; $z_{22LOAD} = Z_R$; $z_{33LOAD} = z_{12LOAD} = z_{13LOAD} = z_{23LOAD} = 0$. The total impedance matrix is the sum of these two matrices, and its reciprocal is the admittance matrix, [Y], of the ceramic resonator with mechanical loads at ports 1 and 2. All elements of the ceramic impedance matrix are functions of frequency; this will usually also be true for the mechanical load impedance matrix elements.

INCLUSION OF MATERIAL LOSSES

One includes the effects of loss within the material differently, depending upon the loss mechanism. Viscous loss, resulting, e.g., from various acoustic scattering mechanisms, is incorporated by adding an imaginary component to the acoustic stiffness: $c^* = c \cdot (1 + j \, \omega \, \eta)$. Conductive and dielectric losses are modeled similarly, by making the permittivity complex: $\varepsilon* = \varepsilon \cdot (1 + 1/[j\ \omega\ \tau_o])$; for simplicity, we will omit dielectric loss and take τ_o to be constant: $\tau_o = \varepsilon/\sigma$. These substitutions render v, Z_o, X, C_o, and k complex.

MECHANICAL LOADINGS [6-12]

When the resonator is mechanically loaded only by the electrode masses, Z_L and Z_R are purely reactive, and represent inertial effects [13,14]. The electrode masses are represented in normalized form as the quantities μ_L and μ_R. Mu is equal to the mass per unit area of an electrode, divided by the mass per unit area of a ceramic plate of half-thickness. The acoustic loads are given in this case by the relations: $Z_L = j\ \mu_L\ Z_o\ X$ and $Z_R = j\ \mu_R\ Z_o\ X$.

In the case where electrode loading is negligible, but the resonator is immersed in a fluid, then one must distinguish among the modes of vibration executed by the ceramic. If the resonator vibrates in an extensional mode, the fluid loading is represented, to a very good approximation, by purely resistive, and generally dispersionless loads: $Z_L = Z_R = A \cdot \rho_{fluid} \cdot v_{fluid}$. This resistance is due to the transport of acoustic energy outward from the active surfaces of the plate. If the vibrational mode is shear, the newtonian shear viscosity of the fluid produces a combination of viscous and inertial effects. The viscous part is due to the lossy drag of the evanescent shear 'wave' in the fluid, while the inertial portion arises from fluid entrainment. The complex load impedances are: $Z_L = Z_R = A \cdot (1 + j) \cdot \sqrt{[\omega \cdot \rho_{fluid} \cdot \eta_{fluid}\ /2]}$.

NETWORK FUNCTIONS

The admittance matrix, Y, describes the behavior with frequency of the boundary-loaded plate resonator driven by an excitation voltage at its electrical port 3. The array elements y_{13} and y_{23} give the particle velocities, per volt applied, at surfaces 1 and 2, respectively. Division by ω yields the desired surface displacements, U_L and U_R; these in general will be complex numbers. Element y_{33} is the input electrical admittance seen looking into port 3; its real and imaginary parts are the input conductance and susceptance.

RESONANCE FREQUENCY

The frequency $f_o = 250$ MHz is the fundamental open-circuited (or 'antiresonance') frequency of the plate. This differs from the fundamental 'resonance' (short-circuited) frequency, f_R, because of the presence of piezoelectricity. Whereas, in the case of no loss, $f_o = v/2t$, the resonance frequency is determined from the root of the transcendental equation $\tan X = X/k^2$. In the lossy case, f_R will be complex. In Table II is given the real part of f_R as the viscosity is varied via τ_1, with conductivity set equal to zero; Table

III gives the real part of f_R as the conductivity is varied via σ, with viscosity set equal to zero. When loss arises from viscosity, $Re[f_R]$ ceases to be defined when the figure of merit, M, is less than two: $M = (8 Q k^2)/\pi^2 < 2$.

TABLE II. Resonance frequency vs. viscosity; $\sigma = 0$

τ_1	$Re[f_R]$
(sec)	(MHz)
0	220.45902179
1.0 E-14	220.45902191
1.0 E-13	220.45903405
1.0 E-12	220.46024780
1.0 E-11	220.58218082
1.0 E-10	$M = Q/r \sim 1.34 < 2$; resonance nonexistent for real freqs.

TABLE III. Resonance f_R vs. conductivity; $\tau_1 = 0$

σ	$Re[f_R]$
(S/m)	(MHz)
0	220.45902179
1 E-7	220.45902179
1 E-6	220.45902179
1 E-5	220.45902179
1 E-4	220.45902190
1 E-3	220.45903345
1 E-2	220.46018820

DISPLACEMENTS
Static

In the limit where f approaches DC, the static displacement of the plate without boundary loadings, per volt applied, is $|U/V| = |e/c^E|$, where c^E is the elastic stiffness at constant electric field (the 'isagric' value, that is, the 'unstiffened' stiffness). The isagric stiffness is related to c by $c^E = c \cdot (1 - k^2)$. For our material we find the static displacement to be $|U/V| = |e/c^E| \cong 0.750$ angstroms per volt. This is the total displacement of the entire plate; the displacement of each side, reckoned from the center plane as origin, is just half this, or about 0.375 Å per volt. This displacement will be along the plate thickness (making the resonator become alternately thicker and thinner each half cycle) if the mode is compressional (longitudinal, extensional, dilatational), or will be in a lateral direction if the mode is shear (transverse, distortional, equivoluminal).

Dynamic

We normally expect that in the case of resonance the displacements should be enhanced by a multiplier equal to the quality factor, Q. In our case, the material quality factor is just $Q = 1/[2\pi f_o \cdot \tau_1] = 2000/\pi$. Because of the piezoelectric effect, however, the dynamic enhancement is not $Q \cdot |e/c^E|$, but rather $Q \cdot |e/c|$. The displacement of each side, in the absence of electrode mass, fluid loading, and bulk conductivity is thus expected to be about equal to $(2000/\pi)$ times (c^E/c), or ≈ 177 angstroms per volt; see Table IV for $\mu = 0$.

EFFECT OF ELECTRODE MASS LOADINGS

Table IV gives results for a plate with equal mass loadings lumped on each surface. The plate has only viscous loss, characterized by the motional time constant τ_1. The table gives, on a per volt basis, the maximum surface displacement, $|U|_{surf}$; frequency of maximum displacement, f_{max}; surface strain, S_{surf}; surface particle velocity, v_{surf}; acceleration at the surface in earth-gravity units, a_{surf}; surface stress, T_{surf}; and T_{surf} normalized to the acoustic stiffness, $(T/c)_{surf}$. This last quantity is a measure of breaking strength; material failure may take place when this value is between 0.1 and 1.0%. It is seen that the presence of the lumped masses has the effect of increasing the surface displacement, strain, and stress. The stress levels at the ceramic/electrode interface become quite large, and can lead to failure of the bond. The table has been extended, well beyond practical limits of μ, to illustrate the monotonic increase in $|U|$. Usual values of mass loading are 3%, or less; beyond about 10% one cannot treat the masses as lumped, but one has to take into account acoustic wave propagation within the electrode.

TABLE IV. Plate with equal mass loadings; viscous loss only; $\tau_1 = 1$ ps

μ	$|U|_{surf}$	f_{max}	S_{surf}	v_{surf}	a_{surf}	T_{surf}	$(T/c)_{surf}$
(%)	(Å/V)	(MHz)	(%/V)	(m/s•V)	(g/V)	(Pa/V)	(1/V)
0	178	220.46	0.405	24.7	3.49 E9	0	0
1	181	218.04	0.412	24.8	3.47 E9	1.13 E7	7.73 E-5
3	187	213.38	0.426	25.1	3.43 E9	3.34 E7	2.29 E-5
10	207	206.58	0.469	25.8	3.29 E9	1.07 E8	7.33 E-4
30	253	168.36	0.576	26.8	2.89 E9	2.81 E8	1.93 E-4
100	371	117.99	0.844	27.5	2.08 E9	6.75 E8	4.64 E-3
1 E3	1037	42.58	2.356	27.7	7.58 E8	2.46 E9	0.0169
1 E4	3230	13.67	7.342	27.7	2.43 E8	7.88 E9	0.0541
1 E5	1.020 E4	4.328	23.18	27.7	7.69 E7	2.50 E10	0.171
1 E6	3.225 E4	1.369	73.30	27.7	2.43 E7	7.90 E10	0.542

When all of the mass is aggregated on one side of the plate, then the motion is not symmetric. As the mass-loading increases, the displacement of the mass-loaded side at first increases, then decreases; the displacement at the free surface increases monotonically. These, and other features are seen from Table V and Table VI. Once again, the table has been extended well beyond practical limits of μ for the purposes of illustration.

TABLE V. Single-sided mass-loaded surface; viscous loss only; $\tau_1 = 1$ ps

μ	$|U|_{surf}$	f_{max}	S_{surf}	v_{surf}	a_{surf}	T_{surf}	$(T/c)_{surf}$
(%)	(Å/V)	(MHz)	(%/V)	(m/s•V)	(g/V)	(Pa/V)	(1/V)
0	178	220.46	0.405	24.7	3.49 E9	0	0
1	180	219.24	0.408	24.7	3.47 E9	1.12 E7	7.74 E-5
3	182	216.87	0.414	24.8	3.44 E9	3.35 E7	2.30 E-4
10	189	209.14	0.429	24.8	3.33 E9	1.08 E8	7.41 E-4
30	200	191.52	0.454	24.0	2.95 E9	2.87 E8	1.97 E-3
100	194	159.50	0.440	19.4	1.98 E9	6.43 E8	4.42 E-3
1 E3	66.3	118.36	0.151	4.93	3.74 E8	1.21 E9	8.33 E-3
1 E4	8.31	111.10	0.0189	0.580	4.13 E7	1.34 E9	9.20 E-3
1 E5	0.85	110.32	1.94 E-3	0.059	4.17 E6	1.35 E9	9.30 E-3
1 E6	0.085	110.24	1.94 E-4	5.9 E-3	4.17 E5	1.36 E9	9.31 E-3

TABLE VI. Single-sided mass loading; traction-free surface; viscous loss only; $\tau_1 = 1$ ps

μ	$|U|$	f	S	v_s	a
(%)	(Å/V)	(MHz)	(%/V)	(m/s•V)	(g/V)
0	178	220.46	0.405	24.7	3.49 E9
1	180	219.24	0.409	24.8	3.48 E9
3	184	216.87	0.417	25.0	3.47 E9
10	195	209.14	0.444	25.7	3.44 E9
30	227	191.52	0.517	27.4	3.36 E9
100	318	159.50	0.722	31.8	3.25 E9
1 E3	602	118.36	1.369	44.8	3.40 E9
1 E4	700	111.10	1.590	48.8	3.48 E9
1 E5	712	110.32	1.618	49.3	3.49 E9
1 E6	713	110.24	1.621	49.4	3.49 E9

EFFECT OF CONDUCTIVITY

The effect of setting the acoustic viscosity to zero, but retaining conductive loss is seen in Table VII. The plate is devoid of any loadings at its mechanical ports. The responses consist of a real component, symmetric with frequency about the maximum, and an imaginary component, antisymmetric about this same point. The conductive loss decreases the displacement monotonically with σ.

TABLE VII. Effect of conductive loss; no electrode or fluid loadings; $\tau_1 = 0$

σ	Re[U]	Im[U]	$|U|_{surf}$
(S/m)	(Å/V)	(Å/V)	(Å/V)
0	0	∞	∞
1 E-7	~2.6 E7	~±1.3 E7	~2.6 E7
1 E-6	~2.5 E6	~±1.3 E6	~2.5 E6
1 E-5	~2.6 E5	~±1.3 E5	~2.6 E5
1 E-4	~2.6 E4	~±1.3 E4	~2.6 E4
1 E-3	~2.5 E3	~±1.3 E3	~2.5 E3
1 E-2	~2.5 E2	~±1.3 E2	~2.5 E2

EFFECTS OF BOTH VISCOSITY AND CONDUCTIVITY
The results in Table VII are modified considerably when viscosity is added; the results are shown in Table VIII. For small conductivity values, the effect of the bulk viscosity predominates; as the conductivity gets large, it predominates, and produces a counterintuitive result: an increase in the amplitude with increasing conductivity.

TABLE VIII. Plate with no electrode or fluid loadings; $\tau_1 = 1ps$; effect of conductivity

σ	Re[U]	Im[U]	$|U|_{surf}$
(S/m)	(Å/V)	(Å/V)	(Å/V)
0	~-175	~±80	~178
1 E-7	~-175	~±80	~178
1 E-6	~-175	~±80	~178
1 E-5	~-175	~±80	~178
1 E-4	~-175	~±80	~180
1 E-3	~-190	~±95	~192
1 E-2	~-600	~±300	~595

Table IX provides, on a per volt basis, the maximum surface displacement, $|U|_{surf}$; frequency of maximum displacement, f_{max} ; surface strain, S_{surf}; surface particle velocity, v_{surf}; and acceleration at the surface in earth-gravity units, a_{surf}. These are given as functions of conductivity. It is

Dielectric Ceramic Materials 449

seen that σ has a negligible effect on these quantities until it becomes quite large, but, once again, $|U|_{surf}$ increases with increasing conductivity.

TABLE IX. Plate with no electrode or fluid loadings; $\tau_1 = 1ps$

σ	$|U|_{surf}$	f_{max}	S_{surf}	v_{surf}	a_{surf}
(S/m)	(Å/V)	(MHz)	(%/V)	(m/s•V)	(g/V)
0	178	220.46	0.405	24.7	3.49 E9
1 E-7	178	220.46	0.405	24.7	3.49 E9
1 E-5	178	220.46	0.405	24.7	3.49 E9
1 E-4	180	220.46	0.408	24.9	3.51 E9
1 E-3	192	220.46	0.436	26.6	3.75 E9
1 E-2	595	220.64	1.353	82.5	1.17 E10

EFFECT OF BULK VISCOSITY

When the conductivity is set equal to zero, and the plate is without mechanical loads, the effect of acoustic viscosity may be studied. Representative values are shown in Table X, where the viscosity is given in terms of the motional time constant: $\tau_1 = \eta/c$. Values of τ_1 for quartz average about ten femtoseconds (1 E-14 s); the value 1 E-12, used in [13,14], and here, represents a future goal for modern ferroelectric ceramics. Values of 1 E-10 are more common today. As the time constant increases, the shape of the resonance curve becomes increasingly blunted. When the viscosity increases to the point where the product $\omega_1 \cdot \tau_1 = 1$, the effect is severe; for $\tau_1 = 1$ E-9, it equals 1 at $f_1 = 160$ MHz, which is below the design resonance frequency of our posited resonator. The effect of loss is such then to reduce the quality factor, Q, to unity, and the figure of merit far below two.

TABLE X. Plate with no electrode or fluid loadings

τ_1	$|U|_{surf}$	f_{max}	S_{surf}	v_{surf}	a_{surf}
(s)	(Å/V)	(MHz)	(%/V)	(m/s•V)	(g/V)
1 E-13	1624.	220.46	3.690	224.9	3.18 E10
1 E-12	178.	220.46	0.405	24.7	3.49 E9
1 E-11	17.8	220.44	0.0405	2.47	3.49 E8
1 E-10	1.79	218.05	0.00406	0.245	3.42 E7

EFFECT OF FLUID LOADS

In the case of microbalances [6-12], the resonator is immersed in a fluid; in other situations, it may be used as a transducer in a fluid. Some resonators are operated in a gaseous environment. It is instructive to simulate the effects of fluid loading on our plate resonator. The effect differs, depending upon the mode of motion, as mentioned above. We take as input the data contained in Table XI [15-17]. The last two columns have the following definitions:

Shear ratio: $\sqrt{[\omega_0 \cdot \rho \cdot \eta/2]}/[\rho_{ceramic} \cdot shearvelocity_{ceramic}]$, and

Long ratio: $[\rho_{fluid} \cdot long.velocity_{fluid}]/[\rho_{ceramic} \cdot long.velocity_{ceramic}]$.

TABLE XI. Fluid properties.

Substance	Density	Velocity (long.)	Viscosity	Shear ratio+	Long. ratio+
	ρ, kg/m^3	v, m/s	η, Pa·s	$f_0 = 250$ E6	
air	1.2929	344	1.827 E-5	4.12 E-6	1.34 E-5
water	1000	1497	0.001	8.47 E-4	4.52 E-2
25% water*	1205	1738	0.046	6.31 E-3	6.33 E-2
20% water*	1217	1765	0.076	8.15 E-3	6.49 E-2
15% water*	1228	1798	0.13	1.07 E-2	6.67 E-2
10% water*	1239	1828	0.25	1.49 E-2	6.85 E-2
05% water*	1250	1870	0.58	2.28 E-2	7.06 E-2
00% water*	1260	1909	1.5	3.68 E-2	7.27 E-2
ceramic	7520	4400	0.146	--	--

+ Acoustic impedance, (shear or longitudinal), normalized to that of ceramic.
* Water/glycerol ($C_3H_8O_3$) volume percent mixture.

From these data, we calculate the elastic moduli, viscous time constants, and relaxation frequencies. These are given in Table XII. We define the relaxation frequency, f_1, as the frequency at which the magnitude of the fluid shear impedance equals the impedance arising from the longitudinal wave energy transport. Thus we find: $|(1+j) \sqrt{[\omega_1 \cdot \rho \cdot \eta/2]}| = \sqrt{[\omega_1 \cdot \rho \cdot \eta]} = \rho \cdot v$, so $\omega_1 \cdot \rho \cdot \eta = \rho^2 \cdot v^2$, and $\omega_1 \cdot \eta = \rho \cdot v^2 = c$, and, $\omega_1 = c/\eta = 1/\tau_1$, so $f_1 = 1/(2\pi \cdot \tau_1)$. It is seen that the relaxation frequencies are well above the frequencies of concern here.

When the resonator mechanical port is terminated with the appropriate fluid impedance, the surface displacements, etc., may be determined. We give two examples: termination by air, and by water. For each fluid, both shear and longitudinal motion are considered. The results are shown in Table XIII for air, and in Table XIV for water. As expected, the air damping of the motion is considerably smaller than that for water. Also, for either fluid, the effect on damping is more severe for longitudinal motion than for shear provided that one is below the relaxation frequency for the fluid; the situation is reversed above f_1.

CONCLUSION

The effects of various internal loss mechanisms on the amplitude of vibration of a ceramic plate have been calculated and simulated. In addition, the further influences of mechanical loads, consisting of lumped inertial masses, and fluid loads, have been characterized and described.

TABLE XII. Elastic moduli, viscous time constants and relaxation frequencies

Substance	$c = \rho \cdot v^2$	$\tau_1 = \eta/c$	$f_1 = 1/(2\pi \cdot \tau_1)$
	Pa	sec	GHz
Air	1.530 E5	1.194 E-10	1.33
Water	2.241 E9	4.462 E-13	356.7
25% water*	3.640 E9	1.264 E-11	12.59
20% water*	3.791 E9	2.005 E-11	7.939
15% water*	3.970 E9	3.275 E-11	4.860
10% water*	4.140 E9	6.038 E-11	2.636
05% water*	4.371 E9	1.327 E-10	1.199
00% water*	4.592 E9	3.267 E-10	0.4872
Ceramic	145.6 E9	1.000 E-12	159.2

* Water/glycerol ($C_3H_8O_3$) volume percent mixture.

TABLE XIII. Lossless ceramic plate with equal air loadings

mode	$|U|_{surf}$	f_{max}	S_{surf}	v_{surf}	a_{surf}
	(Å/V)	(MHz)	(%/V)	(m/s•V)	(g/V)
long.	1.490 E4	220.46	33.86	2.06 E3	2.91 E11
shear	5.181 E4	220.46	117.7	7.18 E3	1.01 E12

TABLE XIV. Lossless ceramic plate with equal water loadings.

mode	$|U|_{surf}$	f_{max}	S_{surf}	v_{surf}	a_{surf}
	(Å/V)	(MHz)	(%/V)	(m/s•V)	(g/V)
long.	4.43	220.17	0.0101	0.613	8.64 E7
shear	250	220.32	0.568	34.6	4.88 E9

Dielectric Ceramic Materials

REFERENCES

[1] J. Messina and A. Ballato, "Dependence of Strain Amplitude on Loss Mechanisms in Plate Resonators," IEEE Intl. Frequency Control Symp. Proc., Honolulu, HI, June 1996, 366-370.

[2] T. R. Meeker, "Thickness Mode Piezoelectric Transducers," Ultrasonics, Vol. 10, No. 1, January 1972, 26-36.

[3] A. Ballato, "Doubly Rotated Thickness Mode Plate Vibrators," in Physical Acoustics: Principles and Methods, (W. P. Mason and R. N. Thurston, eds.), Vol. 13, Academic Press, New York, 1977, Chap. 5, 115-181.

[4] T. R. Meeker, "Piezoelectric Resonators and Applications," in Encyclopedia of Applied Physics, G. Trigg, ed., Vol. 14, 147-169, VCH Publishers, New York, 1996.

[5] A. Ballato and J. R. Vig. "Piezoelectric Devices," in Encyclopedia of Applied Physics, G. Trigg, ed., Vol. 14, 129-145. VCH Publishers, New York, 1996.

[6] W. P. Mason, "Viscosity and Shear Elasticity Measurements of Liquids by Means of Shear Vibrating Crystals," J. Colloid Sci., Vol. 3, 1948, 147-162.

[7] H. Ohigashi, T. Itoh, K. Kimura, T. Nakanishi, and M. Suzuki, "Analysis of Frequency Response Characteristics of Polymer Ultrasonic Transducers," Japanese J. Appl. Physics, Vol. 27, No. 3, 1988, 354-360.

[8] G. Hayward, "Viscous Interaction with Oscillating Piezoelectric Quartz Crystals," Analytica Chimica Acta, Vol. 264, 1992, 23-30.

[9] G. L. Hayward and G. Z. Chu, "Simultaneous Measurement of Mass and Viscosity Using Piezoelectric Quartz Crystals in Liquid Media," Analytica Chimica Acta, Vol. 288, 1994, 179-185.

[10] S. J. Martin, G. C. Frye, and K. O. Wissendorf, "Sensing Liquid Properties with Thickness-Shear Mode Resonators," Sensors and Actuators, Vol. A44, 1994, 209-218.

[11] E. Benes, M. Gröschl, W. Burger, and M. Schmid, "Sensors Based on Piezoelectric Resonators," Sensors and Actuators, Vol. A48, 1995, 1-21.

[12] S. J. Martin, J. J. Spates, K. O. Wessendorf, and T. W. Schneider, "Resonator/Oscillator Response to Liquid Loading," Analytical Chemistry, Vol. 69, No. 11, June 1, 1997, 2050-2054.

[13] A. Ballato and J. Ballato, "Accurate Electrical Measurements of Modern Ferroelectrics," Ferroelectrics, Vol. 182, Nos. 1-4, 1996, 29-59.

[14] A. Ballato and J. Ballato, "Electrical Measurements of Ferroelectric Ceramic Resonators," Technical Report ARL-TR-436, U. S. Army Research Laboratory, Fort Monmouth, NJ, January 1996, 73 pp.

[15] F. T. Schulitz, Y. Lu, and H. N. G. Wadley, "Ultrasonic Propagation in Metal Powder-Viscous Liquid Suspensions," J. Acoust. Soc. Am., Vol. 103, No. 3, March 1998, 1361-1369.

[16] *Basic Laboratory and Industrial Chemicals, A CRC Quick Reference Handbook*, edited by D. R. Lide (CRC, Boca Raton, 1993).

[17] *CRC Handbook of Chemistry and Physics*, 77th ed. (CRC, Boca Raton, 1996), 15-16.

NEW TREND IN MULTILAYER CERAMIC ACTUATORS

Kenji Uchino
International Center for Actuators and Transducers
Materials Research Laboratory, The Pennsylvania State University
University Park, PA 16802, USA

and

Sadayuki Takahashi
Research & Development Group, NEC Corporation
Miyazaki, Miyamae, Kawasaki 216, Japan

ABSTRACT
During the past few years, piezoelectric and electrostrictive ceramic materials have become key components in smart actuator/sensor systems for use as precision positioners, miniature ultrasonic motors and adaptive mechanical dampers. In particular, multilayer structures have been intensively investigated in order to improve their reliability and to expand their applications. Recent developments in USA, Japan and Europe will be compared.

INTRODUCTION
    Piezoelectric actuators are forming a new field midway between electronic and structural ceramics [1-4]. Application fields are classified into three categories: positioners, motors and vibration suppressors. The manufacturing precision of optical instruments such as lasers and cameras, and the positioning accuracy for fabricating semiconductor chips, which are adjusted using solid-state actuators, is of the order of 0.1 μm. Regarding conventional electromagnetic motors, tiny motors smaller than 1 cm^3 are often required in office or factory automation equipment, and are rather difficult to produce with sufficient energy efficiency. Ultrasonic motors whose efficiency is insensitive to size are superior in the mini-motor area. Vibration suppression in space structures and military vehicles using piezoelectric actuators is also a promising technology.
    Multilayer structures are mainly used for practical applications, because of their low drive voltage, high energy density, quick response and long lifetime [5,6]. Figure 1 illustrates multilayer, multimorph and multilayer-moonie structures, which will be covered in this article. Recent investigations have focused on the improvement of reliability and durability in multilayer actuators.

This article reviews the investigations of device structures, reliability issues, and recent applications of multilayer actuators, comparing the developments in USA, Japan and Europe.

MULTILAYER STRUCTURES

Two preparation processes are possible for multilayer ceramic devices: one is a cut-and-bond method and the other is a tape-casting method. The tape-casting method requires expensive fabrication facilities and sophisticated techniques, but is suitable for the mass-production of thousands of pieces per month. Tape-casting also provides thin dielectric layers, leading to low drive voltages of 40 - 100 V [7,8].

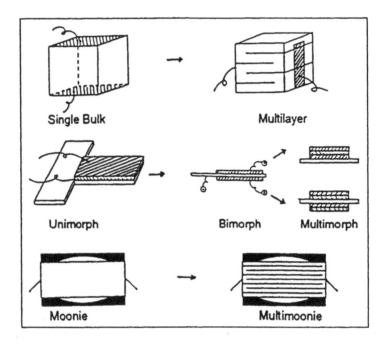

Fig.1 Examples of multilayer, multimorph and multilayer-moonie structures.

A multilayer actuator with interdigital internal electrodes has been developed by Tokin [9,10]. In contrast to the conventional electrode configuration in Fig. 1, line electrodes are printed on piezoelectric ceramic green sheets, and are stacked in such a way that alternating electrode lines are displaced by one-half pitch (see Fig. 2). This actuator generates motions at right angles to the stacking direction using the longitudinal piezoelectric effect. Long ceramic actuators up to 74 mm in length are manufactured.

A three-dimensional positioning actuator with a stacked structure has been proposed by PI Ceramic (Fig.3), in which shear strain is utilized to generate x and y displacements [11]. Composite actuator structures called "moonies" and "cymbals" have been developed at Penn State University to provide characteristics intermediate between the multilayer and bimorph actuators. These transducers exhibit an order of magnitude larger displacement than the multilayer, and much larger generative force with quicker response than the bimorph [12,13]. The device consists of a thin multilayer piezoelectric element and two metal plates with narrow moon-shaped cavities bonded together as shown in Fig. 1. A moonie 5 x 5 x 2.5 mm^3 in size can generate a 20μm displacement under 60V, eight times as large as the displacement of a multilayer of the same size [14]. This new compact actuator has been used to make a miniaturized laser beam scanner [14]. Moonie/cymbal characteristics have been investigated for various constituent materials and sizes [15,16].

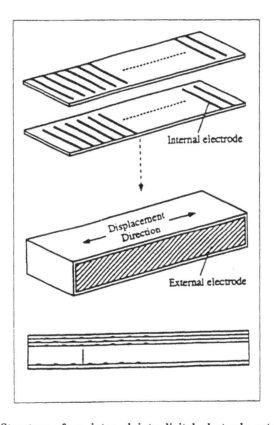

Fig.2 Structure of an internal interdigital electrode actuator.

RELIABILITY OF MULTILAYER ACTUATORS

As the application fields expand, the reliability and durability issues of multilayer actuators become increasingly important. The reliability of ceramic actuators depends on a number of complex factors, which can be divided into three major categories: reliability of the ceramic itself, reliability of the device design, and drive technique.

Compositional changes of actuator ceramics and the effect of doping are primary issues used in stabilizing the temperature and stress dependence of the induced strains. A multilayer piezo-actuator for use at high temperatures (150°C) has been developed by Hitachi Metal, using Sb_2O_3 doped $(Pb,Sr)(Zr,Ti)O_3$ ceramics [17]. Systematic data on uniaxial stress dependence of piezoelectric characteristics have been collected on various Navy PZT materials [18]. Grain size and porosity control of the ceramics are also important in controlling the reproducibility of actuators [19]. Aging phenomena, especially the degradation of strain response, are, in general, strongly dependent on the applied electric field as well as on temperature, humidity and mechanical bias stress [20].

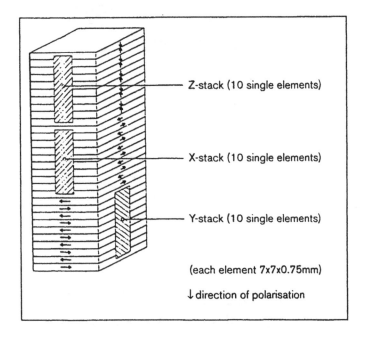

Fig.3 3-D controllable multilayer piezoelectric actuator.

The device design strongly affects its durability and lifetime. Silver electrode metal tends to migrate into the piezoceramic under a high electric field in high humidity. Silver:palladium alloys suppress this behavior effectively. Resistive coatings of the device should also be taken into account [21]. To overcome electrode delamination, improved adhesion can be realized by using a mesh-type electrode or an electrode material with mixed metal and ceramic (the matrix composition!) powders. Pure ceramic electrode materials have also been developed using semiconductive perovskite oxides (barium titanate-based PTCR ceramics) [22]. The lifetime characteristics of a multilayer actuator with applied DC or unipolar AC voltage at various temperatures [23,24] and at various humidities [25] were investigated by Nagata et al. The relationship between the logarithm of the lifetime and the reciprocal of absolute temperature showed linear characteristics similar to Arrhenius type. Nevertheless, the degradation mechanism remains a critical problem.

In multilayer actuators, reduction of the tensile stress concentration around the internal electrode edge of the conventional interdigital configuration is the central problem. Regarding the destruction mechanism of multilayer ceramic actuators, systematic data collection and analysis have led to considerable progress [26-33]. Two typical crack patterns are generated in a conventional interdigital multilayer device: one is a Y-shaped crack located on the edge of an internal electrode, and the other is a vertical crack located in a layer adjacent to the top or bottom inactive layer, connecting a pair of internal electrode.

To overcome this crack problem, three electrode configurations have been proposed as illustrated in Fig. 4: plate-through, interdigital and slit, and interdigital and float electrode types. The "float electrode" type is an especially promising design which can be fabricated using almost the same process as the conventional multilayer actuator, and lead to much longer lifetimes [34]. An empirical rule "the thinner the layer, the tougher the device" [27,28] is also very intriguing, and will be more theoretically investigated in the near future.

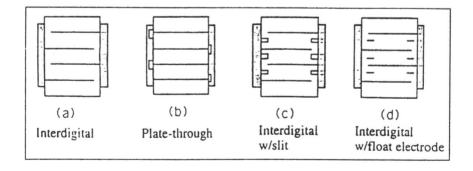

Fig.4 Various internal electrode configurations in multilayer actuators. (a) Interdigital, (b) Plate-through, (c) Slit-insert, and (d) Float electrode.

Failure detection or lifetime prediction methods are expected to remarkably increase the reliability of multilayer actuators. Acoustic emission and surface potential monitoring are promising methods [35]. Penn State has developed a modified multilayer actuator containing a strain gauge as an internal electrode [36]. This internal strain gauge electrode can detect the crack initiation sensitively and monitor the field-induced strain [37].

Regarding drive techniques for ceramic actuators, pulse drive and AC drive require special attention; the vibration overshoot associated with a sharp-rise step/pulse voltage causes a large tensile force, leading to delamination of the multi-stacked structure, while long-term application of AC voltage generates considerable heat. A special pulse drive technique using a mechanical bias stress are required in the first case, and heat generation can be suppressed by changing the device design. An analytical approach to the heat generation mechanism in multilayer actuators has been reported, indicating the importance of larger surface area [38]. Heat generation in piezoelectric ceramics is mainly attributed to the P-E hysteresis loss under large electric field drive [39,40]. For ultrasonic motors, antiresonance drive is preferable to resonance drive because of higher efficiency and lower heat generation for the same vibration level [41].

APPLICATIONS OF MULTILAYER ACTUATORS

Table I compares a variety of ceramic actuator developments in the USA, Japan and Europe. Additional details will be described in this section.

Table I   Ceramic actuator developments in the USA, Japan and Europe.

	US	Japan	Europe
TARGET	Military-oriented products	Mass-consumer products	Laboratory equipment
CATEGORY	Vibration suppressor	Mini-motor Positioner	Mini-motor Positioner Vibration suppressor
APPLICATION FIELD	Space structure Military vehicle	Office equipment Cameras Precision machines Automobiles	Lab stage-stepper Airplanes Automobiles Hydraulic systems
ACTUATOR SIZE	Up-sizing (30 cm)	Down-sizing (1 cm)	Intermediate size (10 cm)
MAJOR MANUFACTURER	AVX/Kyocera Morgan Matroc Itek Opt. Systems Burleigh AlliedSignal	Tokin Corp. NEC Hitachi Metal Mitsui-Sekka Canon Seiko Instruments	Philips Siemens Hoechst Ceram Tec. Ferroperm Physik Instrumente

USA

The principal target is military-oriented applications such as vibration suppression in space structures and military vehicles. Substantial up-sizing of the actuators is required for these purposes.

A typical example is found in the aircraft wing proposed by NASA [42]. A piezoelectric actuator was installed near the support of the wing, allowing immediate suppression of unwanted mechanical vibrations. Several papers have been reported on damper and noise cancellation applications [43,44].

Passive dampers constitute another important application of piezoelectrics, where mechanical noise vibration is radically suppressed by the converted electric energy dissipation through Joule heat when a suitable resistance, equal to an impedance of the piezoelectric element $1/\omega C$, is connected to the piezo-element [45].

A widely-publicized application took place with the repair of the Hubble telescope launched by the Space Shuttle. Multilayer PMN electrostrictive actuators corrected the image by adjusting the phase of the incident light wave (Fig.5) [46]. PMN electrostrictors provided superior adjustment of the telescope image because of negligible strain hysteresis.

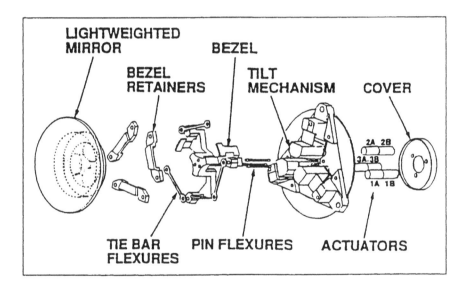

Fig.5 Articulating fold mirror using PMN multilayer actuators.

Dielectric Ceramic Materials   461

JAPAN

Japanese industries seek to develop mass-consumer products, especially mini-motors and micro positioners, aiming at applications such as office equipment and cameras/video cameras. Tiny actuators smaller than $1cm^3$ are the main focus in these products.

A dot matrix printer was the first widely-commercialized product using multilayer ceramic actuators [47]. Each character formed by such a printer was composed of a 24 x 24 dot matrix in which a printing ribbon was impacted by a multiwire array. The printing element was composed of a multilayer piezoelectric device with a sophisticated hinge lever magnification mechanism. The magnification by a factor of 30 resulted in an amplified displacement of 0.5 mm and an energy transfer efficiency greater than 50%. A modified impact printer head has been developed by Tokin, using new interdigital internal electrode type actuators, which have lowered production cost [48]. A color ink-jet printer has also been commercialized by Seiko Epson, using multilayer piezo-actuators [49].

Automotive applications by Toyota Motor have been accelerated recently. Multilayer actuators have been introduced to an electronically controlled suspension [50] and a pilot injection system for diesel engines [51].

Efforts have been made to develop high-power ultrasonic vibrators as replacements for conventional electromagnetic motors. The ultrasonic motor is characterized by "low speed and high torque," which is contrasted with the "high speed and low torque" of the electromagnetic motors. After the invention of - shaped linear motors [52], various modifications have been reported [53,54]. The Mitsui Petrochemical model is of particular interest because the motor body is composed of only one component prepared by a cofiring method as illustrated in Fig. 6 [54]. A maximum speed of 200 mm/s and a maximum thrust of 60 gf were reported for this motor.

Camera motors utilize a traveling elastic wave induced by a thin piezoelectric ring. A ring-type slider in contact with the "rippled" surface of the elastic body bonded onto the piezoelectric can be driven in both directions by exchanging the sine and cosine voltage inputs. Another advantage is its thin design, making it suitable for installation in cameras as an automatic focusing device. Nearly 80 % of the exchange lenses in Canon's "EOS" camera series have been replaced by ultrasonic motor mechanisms [55].

Intriguing research programs are underway in Japan on the vibration damping of earthquakes, using piezoelectric actuators [56,57]. Active damping of a multilayer piezo-actuator was tested using an actual size H-type steel girder, and was verified to be effective during earthquakes.

EUROPE

Ceramic actuator development has begun relatively recently in Europe with a wide range of research topics. A current focus of several major manufacturers is on laboratory equipment products such as laboratory stages and steppers with rather sophisticated structures.

Figure 7 shows a walking piezo motor with 4 multilayer actuators by Philips [58]. Two short actuators function as clampers while the longer two provide a proceeding distance in an inchworm mechanism. Physik Instrumente has developed more complicated two-leg type walkers [59].

Dielectric Ceramic Materials

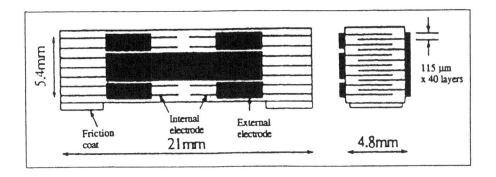

Fig.6 Monolithic multilayer piezoelectric linear motor.

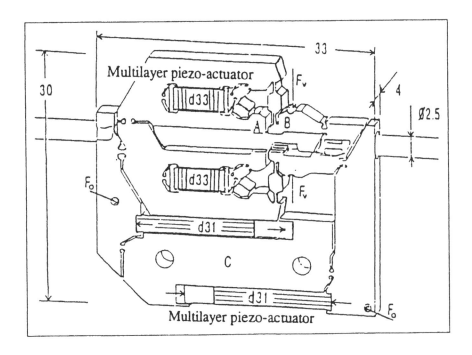

Fig.7 An inchworm using multilayer piezoelectric actuators.

Dielectric Ceramic Materials

463

CONCLUSIONS

Twenty years have passed since the intensive development of piezoelectric actuators began in Japan, then spread worldwide. Presently, the focus has been shifted to practical device applications. This article has reviewed several reliability issues of the multilayer ceramic actuators as well as new actuator structures, and compared the developments of recent applications among USA, Japan and Europe.

The markets in USA are chiefly limited to military and defense applications, and it is difficult to estimate the amount of sales. The current Navy needs include smart submarine skins, hydrophone actuators, and prop noise cancellation. Smart aircraft skins are an Air Force objective, while Army requires helicopter rotor twisting, aeroservoelastic control and cabin noise/seat vibration cancellation.

Meanwhile in Japan, piezoelectric shutters (Minolta Camera) and automatic focusing mechanisms in cameras (Canon), dot-matrix printers (NEC) and part-feeders (Sanki) are now commercialized and mass-produced by tens of thousands of pieces per month. During the commercialization, new designs and drive-control techniques of the ceramic actuators have been mainly developed over the past few years. A number of patent disclosures have been generated by NEC, TOTO Corporation, Matsushita Electric, Brother Industry, Toyota Motors, Tokin, Hitachi Metal and Toshiba.

If we estimate the annual sales in 2001 (neglecting the current economic recession in Japan), ceramic actuator units, camera-related devices and ultrasonic motors are expected to reach $500 million, $300 million and $150 million, respectively. Regarding the final actuator-related products, $10 billion is a realistic goal [55].

Future research trends can be divided in two ways: up-sizing in space structures and down-sizing in office equipment. Further down-sizing will also be required in medical diagnostic applications such as blood test kits and surgical catheters.

Key words for the future of multilayer ceramic actuators are "miniaturization" and "hybridization." Layers thinner than 10 µm, corresponding to current multilayer capacitor technology, will also be introduced in actuator devices replacing the present 100 µm sheets. Piezoelectric thin films compatible with silicon technology are a focus in micro-electromechanical systems. Ultrasonic rotary motors as small as 2 mm in diameter [60] and two-dimensional micro-optical-scanner [61], both of which were fabricated on a silicon membrane are good examples.

Non-uniform configurations or hetero-structures of different materials, layer thickness, or electrode patterns will be adopted for practical devices. Functionally gradient piezoelectric actuators now being prototyped indicate a new trend [62].

REFERENCES

1. Uchino K: Piezoelectric Actuators/Ultrasonic Motors. Kluwer Academic Pub., MA 1996.
2. Uchino K: Ceramic Actuators: Principles and Applications. MRS Bull. 1993, 18: 42 - 48.
3. Uchino K: Materials Update: Advances in Ceramic Actuator Materials. Mater. Lett. 1995, 22: 1 - 4.

4. Uchino K: New Piezoelectric Devices for Smart Actuator/Sensor Systems. Proc. 4th Int'l Conf. Electronic Ceramics & Appl. 1994, 179 -191.
5. Yoshikawa S, Shrout T: Multilayer Piezoelectric Actuators -- Structures and Reliability. AIAA/ASME/ASCE/AHS/ASC Struct. Struct. Dyn. Mater. Conf. 1993, 34: 3581-3586.
6. Uchino K: Manufacturing Technology of Multilayered Transducers. Proc. Amer. Ceram. Soc., Manufacture of Ceramic Components 1995, 81-93.
7. Dibbern U: Piezoelectric Actuators in Multilayer Technique. Proc. 4th Int'l Conf. New Actuators (Actuator '94), Germany 1995, 114 - 118.
8. Takada S, Inoue Y, Oya K, Inagawa M: 100 V DC Driving Type Multilayer Piezoelectric Actuator. NEC Giho 1994, 47: 98 -102.
9. Ohashi J, Fuda Y, Ohno T: Multilayer Piezoelectric Ceramic Actuator with Interdigital Internal Electrodes. Jpn. J. Appl. Phys. 1993, 32: 2412 - 2414.
10. Ohashi J, Fuda Y, Ohno T: Multilayer Piezoelectric Ceramic Actuator with Interdigital Internal Electrodes. Tokin Tech. Rev. 1993, 19: 55 - 60.
11. Bauer A, Moller F: Piezo Actuator Special Design. Proc. 4th Int'l Conf. New Actuators (Actuator '94), Germany 1995, 128 - 132.
12. Sugawara Y, Onitsuka K, Yoshikawa S, Xu QC, Newnham RE, Uchino K: Metal-Ceramic Composite Actuators. J. Amer. Ceram. Soc. 1992, 75: 996 - 998.
13. Onitsuka K, Dogan A, Tressler JF, Xu QC, Yoshikawa S, Newnham RE: Metal-Ceramic Composite Transducer, The "Moonie". J. Intelligent Mater. Systems and Struct. 1995, 6: 447 - 455.
14. Goto H, Imanaka K, Uchino K: Piezoelectric Actuators for Light Beam Scanners. Ultrasonic Techno 1992, 5: 48 -51.
15. Tressler JF, Xu QC, Yoshikawa S, Uchino K, Newnham RE: Composite Flextensional Transducers for Sensing and Actuating. Ferroelectrics 1994, 156 (Proc. 8th Int'l Mtg. Ferroelectricity): 67 - 72.
16. Dogan A, Fernandez JF, Uchino K, Newnham RE: New Piezoelectric Composite Actuator Designs for Displacement Amplification. Proc. 4th Euro Ceramics 1994, 5: 127 - 132.
17. Watanabe J, Sometsugu T, Watanabe Y, Johmura S. Kurihara K, Kazama Y: Multilayer Piezoelectric Ceramic Actuator for High Temperature Use. Hitachi Metal Giho 1993, 9: 59 - 64.
18. Zhang QM, Zhao JZ, Uchino K, Zheng JH: Change of the Weak Field Properties of Pb(Zr,Ti)O3 Piezoceramics with Compressive Uniaxial Stresses and Its Links to the Effect of Dopants on the Stability of the Polarizations in the Materials. J. Mater. Res. 1996, [in press].
19. Okada N, Ishikawa K, Murakami K, Nomura T, Ogino M: Improving Hysteresis of Piezoelectric PZT Actuator. Bull. Shizuoka Univ. Electron. Sci. Grad. School 1993, 14: 7 -13.
20. Sakai T, Ishikiriyama M, Terai Y, Shimazaki R: Improvement in the Durability of PZT Ceramics for an Actuator. Toyota Tech. Rev. 1992, 42: 52 -59.
21. Takada S: Quality Improvement of Multilayer Piezoelectric Actuator. NEC Giho 1992, 45: 109 -113.
22. Abe K, Uchino K, Nomura Late S: Barium Titanate-Based Actuator with Ceramic Internal Electrodes. Ferroelectrics 1986, 68: 215 - 223.

23. Nagata K, Kinoshita S: Life Time of Multilayer Ceramic Actuator at High Temperature. J. Powder and Metallurgy 1994, 41: 975 - 979.
24. Nagata K, Kinoshita S: Relationship between Lifetime of Multilayer Ceramic Actuator and Temperature. Jpn. J. Appl. Phys. 1995, 34: 5266 -5269.
25. Nagata K, Kinoshita S: Effect of Humidity on Life Time of Multilayer Ceramic Actuator. J. Powder and Metallurgy 1995, 42: 623 - 627.
26. Furuta A, Uchino K: Dynamic Observation of Crack Propagation in Piezoelectric Multilayer Actuators. J. Amer. Ceram. Soc. 1993, 76: 1615 - 1617.
27. Aburatani H, Harada S, Uchino K, Furuta A, Fuda Y: Destruction Mechanisms in Ceramic Multilayer Actuators. Jpn. J. Appl. Phys. 1994, 33: 3091 - 3094.
28. Furuta A, Uchino K: Destruction Mechanism of Multilayer Ceramic Actuators: Case of Antiferroelectrics. Ferroelectrics 1994, 160: 277 - 285.
29. Wang H, Singh RN: Electric Field Effects on the Crack Propagation in a Electro-strictive PMN-PT Ceramic. Ferroelectrics 1995, 168: 281 - 291.
30. Cao HC, Evans AG: Electric-Field-Induced Fatigue Crack Growth in Piezoelectrics. J. Amer. Ceram. Soc. 1994, 77: 1783 - 1786.
31. Schneider GA, Rostek A, Zickgraf B, Aldinger F: Crack Growth in Ferroelectric Ceramics under Mechanical and Electrical Loading. Proc. 4th Int'l Conf. Electronic Ceram. and Appl., Germany, 1994, 1211 - 1216.
32. Suo Z: Models for Breakdown Resistant Dielectric and Ferroelectric Ceramics. J. Mech. Phys. Solids 1993, 41: 1155 - 1176.
33 Hao TH, Gong X, Suo Z: Fracture Mechanics for the Design of Ceramic Multilayer Actuators. J. Mech. Phys. Solids 1996, [in press].
34. Aburatani H, Uchino K, Furuta A and Fuda Y: Destruction Mechanism and Destruction Detection Technique for Multilayer Ceramic Actuators. Proc. 9th Int'l Symp. Appl. Ferroelectrics 1995, 750 - 752.
35. Uchino K, Aburatani H: Destruction Detection Techniques for Safety Piezoelectric Actuator Systems. Proc. 2nd Int'l conf. Intelligent Materials 1994, 1248 - 1256.
36. Sano M, Ohya K, Hamada K, Inoue Y, Kajino Y: Multilayer Piezoelectric Actuator with Strain Gauge. NEC Giho 1993, 46 (10): 100 - 103.
37. Aburatani H, Uchino K: Stress and Fatigue Estimation in Multilayer Ceramic Actuators Using an Internal Strain Gauge. Abstract Annual Mtg. & Expo. Amer. Ceram. Soc., Int'l Symp. Solid-State Sensors and Actuators 1996, SXIX-37-96: 191.
38. Zheng JH, Takahashi S, Yoshikawa S, Uchino K: Heat Generation in Multilayer Piezoelectic Actuators. J. Amer. Ceram. Soc. 1996, [in press].
39. Hirose S, Takahashi S, Uchino K, Aoyagi M, Tomikawa Y: Measuring Methods for High-Power Characteristics of Piezoelectric Materials. Proc. Mater. Res. Soc., '94 Fall Mtg. 1995, 360: 15 - 20.
40. Takahashi S, Sakaki Y, Hirose S, Uchino K: Stability of $PbZrO_3$-$PbTiO_3$-$Pb(Mn_{1/3}Sb_{2/3})O_3$ Piezoelectric Ceramics under Vibration-Level Change. Jpn. J. Appl. Phys. 1995, 34: 5328 - 5331.
41. Hirose S, Aoyagi M, Tomikawa Y, Takahashi S, Uchino K: High-Power Characteristics at Antiresonance Frequency of Piezoelectric Transducers. Proc. Ultrasonic Int'l 1995, 1 - 4.

42. Heeg J: Analytical and Experimental Investigation of Flutter Suppression by Piezoelectric Actuation. NASA Tech. Pap., NASA-TP-3241 1993, 47.
43. Mather GP, Tran BN: Aircraft Cabin Noise Reduction Tests Using Active Structural Acoustic Control. Pap. Amer. Inst. Aeronaut Astronaut 1993, AIAA-93-4437: 7.
44. Agrawal BN, Bang H: Active Vibration Control of Flexible Space Structures by Using Piezoelectric Sensors and Actuators. Amer. Soc. Mech. Eng. Des. Eng. Div. 1993, 61:169 - 179.
45. Uchino K, Ishii T: Mechanical Damper Using Piezoelectric ceramics. J. Jpn. Ceram. Soc. 1988, 96: 863 - 867.
46. Wada B: Summary of Precision Actuators for Space Application. JPL Document D-10659, 1993.
47. Yano T, Fukui I, Sato E, Inui O, Miyazaki Y: Impact Dot-Matrix Printer Head Using Multilayer Piezoelectric Actuators. Proc. Electr.& Commun. Soc. (Spring) 1984, 1: 156.
48. Ono Y, Fuda Y: Multilayer Piezoelectric Ceramic Actuator with Interdigital Internal Electrodes for Impact Dot-Matrix Printers. Proc. Symp. on Electro-Magnetic Related Dynamics 1994, 6th: 135 - 138.
49. Yonekubo S: Color Inkjet Printer by Multi-Layer Piezoelectric Actuator. J. Electron. Photo Soc. 1995, 34: 226 -228.
50. Fukami A, Yano M, Tokuda H, Ohki M, Kizu R: Development of Piezo-electric Actuators and Sensors for Electronically Contorolled Suspension. Int. J. Veh. Des. 1994, 15: 348 - 357.
51. Abe S, Igashira T, Sakakibara Y, Kobayashi F: Development of Pilot Injection System Equipped with Piezoelectric Actuator for Diesel Engine. JSAE Rev. (Soc. Automot. Eng. Jpn.) 1994, 15: 201 - 208.
52. Tohda M, Ichikawa S, Uchino K, Kato K: Ultrasonic Linear Motor Using a Multilayered Piezoelectric Actuator. Ferroelectrics (Proc. ECAPD-1/ISAF '88) 1989, 93: 287 - 294.
53. Funakubo T, Tsubata T, Taniguchi Y, Kumei K, Fujumura T, Abe C: Ultrasonic Linear Motor Using Multilayer Piezoelectric Actuators. Jpn. J. Appl. Phys. 1995, 34: 2756 - 2759.
54. Saigoh H, Kawasaki M, Maruko N, Kanayama K: Multilayer Piezoelectric Motor Using the First Longitudinal and the Second Bending Vibrations. Jpn. J. Appl. Phys. 1995, 34: 2760 - 2764.
55. Uchino K: Piezoelectric Actuators/Ultrasonic Motors - Their Developments and Markets. Proc. 9th Int'l Symp. Appl. Ferroelectrics 1995, 319 - 324.
56. Shimoda H, Ohmata J, Okamoto F: Seismic Response Control of a Piping System by a Semiactive Damper with Piezoelectric Actuators. J. Precision Engr. Jpn. 1992, 58: 2111 - 2117.
57. Fujita T: Can Piezoelectric Actuators Control Vibrations of Buildings? New Ceramics 1994, 7 (12): 52 - 55.
58. Koster MP: A Walking Piezo Motor. Proc. 4th Int'l Conf. New Actuators, Germany 1994, 144 - 148.
59. Gross R: A High Resolution Piezo Walk Drive. Proc. 4th Int'l Conf. New Actuators, Germany 1994, 190 - 192.

60. Flynn AM, Tavrow LS, Bart SF, Brooks RA, Ehrlich DJ, Udayakumar KR, Cross LE: Piezoelectric Micromotors for Microrobots. IEEE J. Microelectromechanical Systems 1992, 1: 44 - 51.
61. Goto H: Miniature 2D Optical Scanner and Its Application to an Optical Sensor. J. Opt. Tech. Contact 1994, 32: 322 - 330.
62. Kim HS, Choi SC, Lee JK, Jung HJ: Fabrication and Piezoelectric Strain Characteristics of PLZT Functionally Gradient Piezoelectric Actuator by Doctor Blade Process. J. Korean Ceram. Soc. 1992, 29: 695 - 704.

INFLUENCE OF CERAMIC MICROSTRUCTURE ON VARISTOR ELECTRICAL PROPERTIES

F. A. Modine and L. A. Boatner
Solid State Division, Oak Ridge National Laboratory, Oak Ridge, TN 37831-6030

M. Bartkowiak and G. D. Mahan
Solid State Division, Oak Ridge National Laboratory, Oak Ridge, TN 37831-6030
Department of Physics and Astronomy, University of Tennessee, Knoxville, TN 37996-1200

H. Wang and R. B. Dinwiddie
Metals and Ceramics Division, Oak Ridge National Laboratory, Oak Ridge, TN 37831-6064

ABSTRACT
    Recent theoretical modeling together with experimental measurements using infrared imaging provide an improved understanding of the electrical behavior of zinc oxide varistors in terms of their ceramic microstructure. The ceramic microstructure of varistors forms a random electrical network of grain-boundary microjunctions that are of three distinct types: high nonlinearity "good", low nonlinearity "bad", or linear "ohmic". The mixture of these microjunction types determines the breakdown voltage, nonlinearity, and the sharpness of the switching characteristic of a varistor. Most varistors contain a high percentage of "bad" and "ohmic" microjunctions, but varistors with good properties have predominately "good" microjunctions. Moreover, it is possible to obtain even better characteristics by fabricating varistors with essentially all "good" junctions.
    The current flow in varistor ceramics is nonuniform. Either microstructural disorder or macroscopic inhomogeneities can combine with high electrical nonlinearity and cause the current to localize along narrow conduction paths. Microstructural disorder is the principal cause of current localization in small varistors, but processing-induced inhomogeneity on a macroscopic scale is the primary cause in large varistors. The current localization causes non-uniform Joule heating which promotes varistor failure. Although the current can localize on the scale of the grain size, the heat transfer on this microscopic scale is too fast to permit temperature differences sufficient to cause failure. Varistor failures are mostly caused by non-uniform heating on a macroscopic scale, and such failures are more prevalent in large blocks because it is difficult to obtain uniformity in large blocks by ceramic processing. Thermo-mechanical modeling of the temperature and stress in varistors due to localized currents identifies puncture, cracking, or thermal runaway as the dominant failure mode, depending on operating conditions.

INTRODUCTION

Zinc oxide varistors are ceramic materials that are widely used in surge arresters which protect electrical equipment and electronics from overvoltages caused by lightning and switching transients. Varistors have an extremely nonlinear electrical resistance that may change by 12 orders of magnitude or more due to a small voltage change. Essentially, varistors switch from a nonconducting to a conducting state in order to limit overvoltages and prevent electrical damage.

A basic understanding of the electrical nonlinearity of varistors has existed for some time.[1-5] The varistor electrical resistance is controlled by Schottky-like grain boundary barriers. The art of varistor manufacture is largely concerned with small amounts of additives (e.g., Co and Bi) that dope the grains or segregate to the ZnO grain boundaries to create the Schottky barriers during ceramic sintering.[6] When the applied voltage exceeds the barrier breakdown voltage, the current increase roughly conforms to the empirical power law $I \approx I_o V^\alpha$, where the nonlinearity coefficient α can be 60 or higher. This highly nonlinear behavior is understood, at the fundamental level, to occur when the electrical field is sufficient to cause impact ionization and generate minority carriers that become trapped in the grain boundaries and cause the Schottky barriers to collapse.[1,2]

Despite the fundamental understanding of individual grain-boundary barriers, an understanding of the electrical properties of varistor ceramics has been limited because of the difficulty of treating large random networks of nonlinear electrical elements. The grain-boundary barriers are microjunctions that are not identical, and a bulk varistor is a complex multijunction device composed of large numbers of both ohmic and nonlinear elements of different types that are randomly connected. The connection between the global properties of varistor ceramics and the electrical properties of the individual grain boundaries is obscured by this complexity, making it difficult to understand the changes in varistor characteristics that result from varying the composition and ceramic processing parameters. In the present paper, we describe the influence of the ceramic microstructure on the electrical characteristics and failure modes of bulk varistor materials.

STATISTICAL MODEL OF THE VARISTOR MICROSTRUCTURE

The random character of a zinc-oxide varistor ceramic is simply described if it is modeled as a two-dimensional Voronoi network.[7] The geometry and topology of Voronoi networks closely resemble polycrystalline ceramics with grain growth from random nucleation seeds. In the Voronoi model, the ZnO grains are represented by space-filling polygons of different sizes and shapes with a variable number of neighbors. The polygons are generated by a Wigner-Seitz construction on a disordered lattice. Each polygon represents a ZnO grain, and edges shared by neighboring cells correspond to grain boundaries. The disorder of a granular microstructure is obtained by introducing a controllable degree of randomness. Starting from a regular triangular lattice of "seed points" that correspond to a hexagonal cell structure, disorder is introduced by randomly displacing the individual seeds within a disk of radius d around their original positions, where d is defined in units of the distance between nearest neighbors in the original triangular lattice. The total number of cells is kept constant by imposing periodic boundary conditions. For small values of d, the hexagonal cells merely become deformed, but when the disorder increases above d ≈ 0.2, the lattice becomes topologically disordered.[8-10]

The Voronoi cell boundaries are randomly assigned the electrical characteristics of one or another of three basic types: (1) high nonlinearity (i.e., "good" barriers), (2) low nonlinearity (i.e., "bad" barriers), and (3) linear (i.e., ohmic barriers). The microjunctions referred to as "good" have high leakage resistance and a high coefficient of nonlinearity ($\alpha > 30$). Bad microjunctions have 2 to 3 orders of magnitude lower leakage resistance and a much lower coefficient of nonlinearity ($\alpha \approx 10$). Finally, ohmic microjunctions are nearly linear with a resistance 2 to 5 orders of magnitude lower than the leakage resistance of the "good" junctions. These three barrier types have been confirmed by measurements made on individual grain-boundaries,[11-14] and they are present in commercial varistors, even when their quality is good. The model characteristics assumed[10] for the three barrier types are shown in Fig. 1. The corresponding coefficients of nonlinearity are shown versus the grain-boundary voltage in the inset.

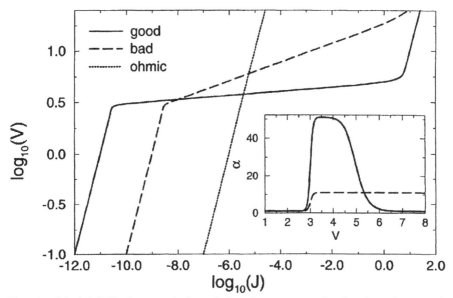

Fig. 1. Model I-V characteristics of the three types of microjunctions. The corresponding coefficients of nonlinearity α are shown in the inset.

A potential difference is assumed to be applied across opposite edges of the Voronoi network. The cells at the bottom and the top are given the fixed potentials, 0 and V, respectively. The potentials V_i of each of the other cells have to be calculated from the Kirchhoff equations:

$$\sum_j l_{ij} J(V_i - V_j) = 0, \tag{1}$$

where the summation index j runs over the cells that are neighbors of cell i, l_{ij} is the length of the edge shared by cells i and j, and the current density $J(V_i-V_j)$ is calculated depending on the type of microjunction which is present at the boundary between the grains i and j. This large set of nonlinear equations is solved numerically using an iterative method.[8-10] From the solution of the currents and voltages at all the boundaries in the random network, it is then easy to compute such global characteristics as the total current flowing through the network as a function of the applied potential and to determine thereby the I-V characteristic, as well as its nonlinearity coefficient $\alpha = d(\log I)/d(\log V)$.

The Voronoi network simulations yield clear explanations of the important features of the current-voltage characteristics of varistors. By varying the concentrations of good, bad, and ohmic barriers as well as the granular disorder, simulated I-V characteristics can be made to reproduce all the variations in breakdown field, nonlinearity, and shapes of switching characteristics that are experimentally measured.[8-10] The breakdown voltage of the network decreases rapidly as the grain-size variability increases. This explains the discrepancy between the breakdown voltage per grain boundary that is calculated from the average grain size and that which is measured as the single-barrier voltage (e.g., 2.5 compared to 3.3 V). However, the statistical dispersion in the grain sizes by itself has no significant effect on the shape of the global I-V characteristic.[9] Such features as double-inflected breakdown characteristics and local maxima of the nonlinearity coefficient in the prebreakdown region of small varistor samples[15] can be explained only by the presence of ohmic grain boundaries, and the reduction of the nonlinearity coefficient of bulk varistors relative to that of isolated grain boundaries can be explained only by the presence of "bad" barriers.[10]

In varistor fabrication, it is better to place emphasis on obtaining a better mixture of junctions rather than on obtaining universal improvements in the junction characteristics. It is control of the statistical mix of junction types rather than the uniformity of grain sizes or shape that is most important to varistor properties. In support of this understanding, Fig. 2 shows the electrical characteristics of two varistor materials that have similar breakdown fields–but quite different switching characteristics. One material has the less-abrupt switching characteristic typical of commercial varistors having a multiphase microstructure and a mixture of junctions. The improved varistor material with the sharper switching characteristic – the sharpest ever reported – has a two-phase microstructure that was achieved by control over composition and processing.[10] This two-phase material has a less-diverse mixture of microjunction characteristics, and it is well modeled by a Voronoi network with only "good" microjunctions. On the other hand, typical commercial materials can be modeled by a mixture that includes about 35% "bad" and 5% ohmic junctions.[10]

The Voronoi simulations also disclose that the current percolates preferentially over one or more of the different type barriers, depending upon the barrier-type concentrations, the disorder, and the applied electric field.[9-10] The topologically disordered networks (d > 0.2) exhibit a very interesting current localization phenomenon. In the ohmic prebreakdown region, the current is uniformly distributed over the Voronoi network; but as illustrated in Fig. 3, in the nonlinear region of the I-V characteristic, the current localizes along the narrow path with the fewest barriers between electrodes. At very high currents, however, the ohmic "upturn" region is entered, and the current is again uniformly distributed over the network. Clearly it is the combination of statistical disorder and nonlinearity that leads to current localization.

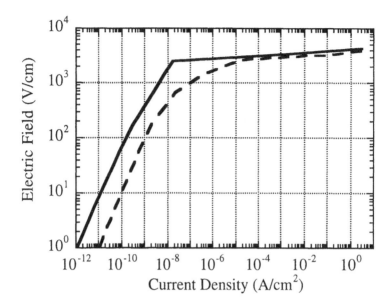

Fig. 2. Switching characteristics of a varistor with all "good" junctions (solid curve) compared to those of a commercial varistor with the same breakdown field.

OBSERVATION OF NON-UNIFORM VARISTOR CURRENTS

Because non-uniform electrical currents cause non-uniform Joule heating, they can be conveniently observed by use of high-speed, high-resolution infrared (IR) imaging of the varistor temperature.[16] Such imaging allows the varistor heating to be monitored on both a microscopic and a macroscopic scale during and after electrical transients. The IR imaging has additional relevance for varistors because heating is the primary cause of their failure.

The theoretical calculations based on the Voronoi model are well confirmed by infrared imaging of thin varistor slices. Low-voltage varistors were mounted on glass slides and polished down to about the 100 μm thickness of their grain size. Electrical contact was made through aluminum evaporated onto the ends of the slices so as to leave a 4 mm wide gap on the broad face. Thus, the electrical conduction was essentially confined to a single two-dimensional layer of grains, and the number of junctions between the electrodes was about 40—a geometry that roughly matches that of the Voronoi simulations. The area between the electrode gap was placed in front of the IR camera. When a sequence of images was taken, distinct conducting paths could be seen only in the first image, which was taken during the electrical pulse. The heat diffused away so rapidly (i.e., in less than 17 ms) that no clear conducting paths could be seen in subsequent images; but in the first image, the current paths exhibited localized "hot spots" occurring at the grain boundaries where the potential is dropped across the Schottky-type barriers and heat is generated. Local temperatures at the grain boundaries were found to rise well above that of the interiors of the grains during 1 ms current pulses of about 300 A/cm^2. These IR images represent the first experimental confirmation that varistor heating is localized at the grain boundaries during electrical breakdown. The image in Fig. 4 showing distinct current paths can be compared with Fig. 3, which was obtained from computer simulations. The similarity provides a striking experimental confirmation of the modeling predictions.

The IR images of grain-boundary heating were analyzed to understand the temperature increase at the grain boundaries as well as its spatial and temporal decay. The maximum temperature difference between the hot spots and the grain interiors was found to be about 35°C. This is essentially a steady-state temperature

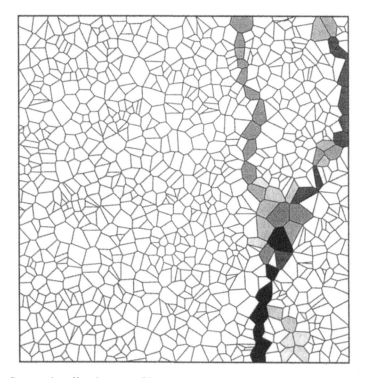

Fig. 3. Current localization on a Voronoi network. A dark shading indicates high current density.

difference, because a dynamic equilibrium is quickly established in which both the grain boundaries and grain interiors increase in temperature at nearly the same rate. In fact, the temperature in the interior of the grains increased by about 30°C during application of an electrical pulse. The temperature at a grain boundary increases at the same rate as that of a grain interior, while staying about 20°C higher to maintain the heat flow away from the grain boundaries. Because the thermal diffusivity is about 0.1 cm²/s, the heat does not remain localized on the scale of the grain size during a millisecond electrical pulse. The high spatial resolution of the IR camera allows the temperature profile along individual conducting paths to be plotted.

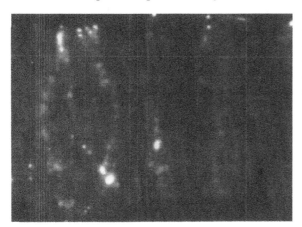

Figure 5(a) shows one such profile measured along the line through the centers of two distinct hot spots. The particular hot spots were chosen for the purpose of clarity, although the size of the grain between them (about 200 µm) was larger than the average grain size. The profile shows two clear peaks corresponding to grain boundaries, and the difference between the temperature at the grain boundaries and the lowest temperature in the grain interiors is about 15–20°C.

Fig. 4. Infrared image of a thin varistor slice taken during a current pulse. The lighter areas are hotter.

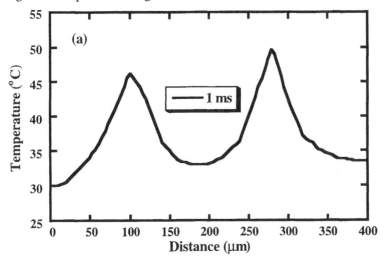

Fig. 5(a). Temperature along a line connecting two bright spots in Fig. 4.

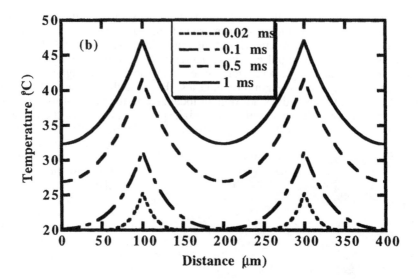

Fig. 5(b). Time-dependent temperature profiles obtained as solutions of the heat transfer equation.

The observed temperature profile can be interpreted by using a simple one-dimensional diffusion equation to compute the temperature between two grain boundaries.[16] This heat transport equation takes the form:

$$CD\frac{\partial^2 T(x,t)}{\partial x^2} - C\frac{\partial T(x,t)}{\partial t} + q\Theta(t)\delta(x) - HT(x,t) = 0 \quad , \tag{2}$$

where $C = 3.3$ J/(cm^{3}K) is the heat capacity, $D = 0.1$ cm^2/s is the thermal diffusivity of the material. The time-dependent temperature $T(x,t)$ at the distance x from the grain boundary is measured relative to the sample temperature at $t = 0$, taken as 20°C. Almost all of the voltage drop appears across the grain boundaries. So for simplicity, it is assumed that the heat is generated only at the grain boundaries. Since the grain-boundary region is very narrow (~1000 Å) compared to the grain size, it is reasonable to approximate the spatial dependence of the power input in Eq. (2) by the Dirac delta function. The time dependence of the power input is assumed to be a constant pulse of amplitude q that starts at $t = 0$. So $\Theta(t)$ is the unit step function. Heat diffuses not only along the line x, as explicitly described by Eq. (2), but it dissipates in directions perpendicular to the current flow to the grains adjacent to the conducting path, to the substrate beneath the varistor sample, and to some extent to the air above the sample. The term $-HT(x,t)$ describes this loss of heat from the conducting path under the assumption that it is proportional to the temperature rise $T(x,t)$. The parameter H is estimated at $H = 5000$ W/(cm^{3}K). The solutions of Eq. (2) are assumed to satisfy boundary conditions that require the temperature profile to be symmetric about the grain boundary and to have no

Dielectric Ceramic Materials

gradient along x (i.e., heat flow along x) at the center of a grain. The equation can be solved analytically using Laplace transforms.[16]

For a grain size of 200 μm the solution of Eq. (2) yields the temperature profiles for $t = 0.02$ ms, 0.1 ms, 0.5 ms, and 1 ms which are shown in Fig. 5(b). The final temperature profile at the end of a 1 ms current pulse, shown in Fig. 5(b) as the solid curve, is in excellent agreement with the measurement shown in Fig. 5(a). The calculations show that at the beginning of the current pulse the temperature rises rapidly only at the grain boundaries and in the immediate vicinity; but as the pulse continues and the heat diffuses into the grain interiors, the temperatures of the grain boundaries and grain interiors increase at nearly the same rate. This is consistent with the experimental observations and has significant implications with regard to varistor failures. Varistors exhibit a rather surprising increase in their energy handling capability at very high currents. This was first noticed by Sakshaug et al.,[17] who attributed this increase to a change on a microscopic scale (i.e., the scale of the grain size) in the temperature distribution that occurs at high current. In contrast to currents in the prebreakdown or the nonlinear region of the *I-V* characteristic, where the voltage drop appears almost exclusively across the grain boundaries, the current is so high when a varistor operates in the upturn region that the voltage drop inside the grains becomes significant, and the heating becomes more uniformly distributed. This, according to the interpretation of reference 17, should lead to a reduced temperature difference between grain interiors and grain boundaries, thereby reducing the thermal shock and leading to higher energy-handling capability. However, as discussed above, the difference between the temperature at the grain boundary and the grain interior quickly reaches an essentially steady-state value, and it never becomes very large—perhaps only about 35°C for large current pulses. This means that the heat transfer is too fast to permit temperature differences on the scale of the grain size that are great enough to cause failure by generating thermal stresses. Thus, varistor failures are not caused by temperature differences on the microscopic scale.

The Voronoi simulations reveal a tendency for the current to flow along those conducting paths that have the lowest breakdown voltages because, for example, they are comprised of the largest grains.[8-10] This understanding is also confirmed by IR imaging of non-uniform heating of small varistors. The image of Fig. 6 was taken immediately after a pulse of about 1 A and 1 ms duration was applied to a 100 volt varistor. The thickness of the varistor was about 1 mm and its width was 8 mm. The IR camera was aimed at a cut edge of the varistor. The temperature distribution of Fig. 6 shows that the current flows preferentially along several highly conducting paths. Since the number of ZnO grains between the electrodes is only about 40, a variation of 3 – 4 grains can change the current flow in a path by an order of magnitude relative to surrounding paths. Thus, conducting paths with low breakdown voltages carry most of the current and become hotter.

Although the Voronoi model gives some remarkable insights into varistors, it has been used only in relatively small-scale simulations (e.g., 1000 ZnO grains) which are appropriate to the small, low-voltage varistors used in electronic applications, but not to the large varistors used in power applications. The non-uniform current conduction in small varistors is mostly of a statistical origin. Statistical fluctuations are in general more significant in small systems than in large systems. Moreover, macroscopic inhomogeneities are less prevalent because it is easier to fabricate ceramic disks with uniform properties when the disks are small. Large varistors are a different matter. Statistical variations have less influence and ceramic processing has more influence on the properties of large varistors.

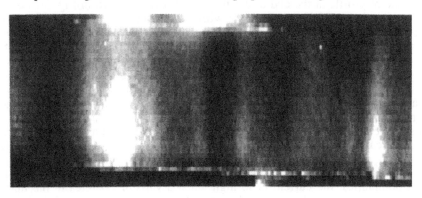

Fig. 6. Infrared image of the temperature distribution on the edge of a varistor. The lighter color shows localized heating along highly conducting current paths.

If the number of grains between the varistor electrodes is sufficient to use the statistics of large numbers, it is easy to explain the difference in the degree of current localization due to statistical disorder in large and small varistors.[16] A grain-size distribution characterized by a mean diameter $\bar{d}$ and a variance σ_d leads to a differing number of grain boundaries along the various conducting paths that is approximately Gaussian distributed with a mean $\bar{N}$ and a variance σ_N given by

$$\bar{N} = L / \bar{d} \quad \text{and} \quad \sigma_N = \bar{N}^{1/2} \sigma_d / \bar{d} , \tag{3}$$

where L is the path length between the electrodes. If it is assumed that the breakdown voltage of a current path is just the sum of the breakdown voltage V_b of N identical grain boundaries, then the distribution of the breakdown voltage of the various paths is described by a mean $\bar{V}$ and a variance σ_V given by:

$$\bar{V} = \bar{N} V_b \quad \text{and} \quad \sigma_v = \bar{N}^{1/2} V_b \sigma_d / \bar{d} . \tag{4}$$

While the voltage increases in proportion to the number of grain boundaries, the variance of the voltage increases only as the square root of the number of boundaries because smaller grains tend to compensate for larger grains. (In reality, grain boundaries are not identical, and the breakdown voltage V_b also varies.[9,10,18] However, variations in V_b contribute a dependence to $\bar{V}$ and σ_v which is similar to that of σ_d, and it can be included by just assuming a larger value for $\sigma_d / \bar{d}$.) If a varistor is well into the breakdown region where the current I has a nonlinear dependence on the voltage V that can be described by a power-law exponent α, then the distribution of the current among the possible paths is described by a mean $\bar{I}$ and a variance σ_I. In the simplest approximation,

$$\bar{I} = I_0 \bar{V}^\alpha \quad \text{and} \quad \sigma_I = \alpha \bar{I} \bar{N}^{-1/2} \sigma_d / \bar{d} . \tag{5}$$

The above classification of the current will be accurate only if the distributions are relatively narrow and approximately Gaussian, but the results agree with the predictions of more realistic modeling, at least in a rough way.[16] Clearly, current localization on highly conducting paths will be negligible if the variance of the current distribution is small relative to its mean, or when $\sigma_I / \bar{I} \ll 1$. This condition, together with Eq. (5) above leads to a criterion for the number of grain boundaries required in a varistor in which current localization due to statistical variations in grain size will be unimportant:

$$\bar{N} \gg \alpha^2 (\sigma_d / \bar{d})^2 . \tag{6}$$

If it is assumed, for example, that $\sigma_d / \bar{d} = 0.1$ and $\alpha = 50$, then $\bar{N} \gg 25$ should result in minimal current localization due to grain size variations. (Note the strong influence of the nonlinearity exponent.) Large arrester blocks contain thousands of barriers in each path, so current localization due to statistical disorder of the grain sizes is insignificant. Based on the results of small-scale computer simulation of Joule heating in disordered networks, puncture failures sometimes have been attributed to microstructural disorder.[19] However, this conclusion is not valid for large (i.e., high-voltage) varistor blocks. In large varistors, microstructural disorder is of little consequence, and failures are the result of macroscopic nonuniformities due to processing (e.g., compaction and sintering).[20-22] As discussed above, small varistors for electronic applications are another matter. Because they may have fewer than 10 barriers in a current path, current localization due to grain-size variations can be severe, but the heat does not localize along microscopic paths.

For small values of α or $\sigma_V / \bar{V}$, the distribution $P(I)$ becomes approximately Gaussian. In this simple case, the fractional number of hot paths f_{hp} in a varistor sample is approximately given by a complementary error function that describes the number of paths in the high current tail of the distribution:

$$f_{hp} = \frac{1}{2} \text{cerf} \left[(n-1)\sqrt{z} \right] = \frac{1}{2} \left[1 - \frac{1}{\sqrt{\pi}} \int_{-(n-1)\sqrt{z}}^{(n-1)\sqrt{z}} \exp\left(-x^2\right) dx \right] , \tag{7}$$

where $z = (\bar{N} / 2\alpha^2)(\bar{d} / \sigma_d)^2$ and the parameter n is a rather arbitrary number between about 2 and 5 that depends upon the criterion used for "hot" and is defined by $I_{hp} > n\langle I \rangle \approx n\bar{I}$, where $\langle I \rangle$ is the true mean and $\bar{I}$ is defined by Eq. (5).

In general, however, f_{hp} for a given factor n should be calculated from the non-Gaussian current paths distribution.[16] The result of a more general calculation is shown in Fig. 7 for n = 2 and 3, and for $\alpha = 10$ and 50. It is found, for example, that for varistors with $z > 5$, only one in about 10^4 paths conducts current 3 times ($n = 3$) greater than the average, which is still not the case for a very "hot" path. The choice of z as an independent variable in Eq. (7) and in Fig. 7 leads to an approximate scaling of the results. But this does not mean that f_{hp} is not sensitive to the value of α. Since z is a strongly varying function of α, so is f_{hp}.

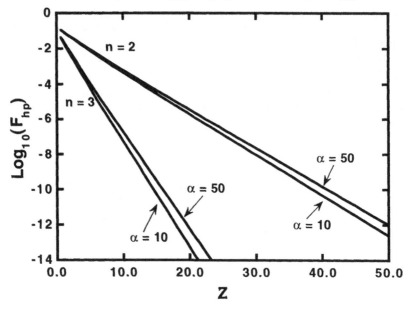

Fig. 7. Fractional number of hot paths as a function of $z = (\overline{N} / 2\alpha^2)(\overline{d} / \sigma_d)^2$. Since z depends on α, the fractional number of hot paths is a strong function of α. The slightly different curves for $\alpha = 10$ and $\alpha = 50$, but with the same n, show a functional dependence on z that is not single-valued.

The predictions of Fig. 7 for small and large varistors contrast remarkably. A typical varistor for electronic applications is a small disk of about 1 cm diameter and 0.1 cm thickness. Typically, the grains of such a varistor are described by $\overline{d} = 100$ μm and $\sigma_d / \overline{d} = 0.1$, and the nonlinearity coefficient might be $\alpha = 25$. If the hot paths are assumed to be those paths that carry a current $I > 2\langle I \rangle$ (i.e., $n = 2$), then the parameter z in Fig. 7 is 0.8, and $f_{hp} \approx 0.1$. Hence, current localization is predicted, and the amount of current localization roughly agrees with the observations of Fig. 6. Commercial varistors for electronic applications have rather low nonlinearity coefficients, although it is not difficult to produce small varistors having an α of 50 or more. However, increasing α, would lead to smaller values of z and [by Eq. (7) and Fig. 7] to an even higher probability of hot paths.

On the other hand, a distribution-class arrester block is typically a disk of about 4 cm diameter and 4 cm thickness. The smaller grains of such a varistor block can be described by $\overline{d} = 20$ μm and $\sigma_d / \overline{d} = 0.1$, and the nonlinearity coefficient of such a varistor might be $\alpha = 50$. Again, if the hot paths are assumed to be those paths carrying a current $I > 2\langle I\rangle$ (which is a very conservative definition), then the parameter z of Fig. 7 is about 40, and $f_{hp} \approx 10^{-10}$. So there will be no hot paths due to statistical disorder in such blocks, even though the total number of available current paths exceeds a million. Current localization resulting from statistical disorder is only important in small varistors.

On a macroscopic scale, the IR imaging confirms that nonuniformity in larger varistors is due to ceramic processing. Figure 8 is a thermal image of the surface temperature of a 4.2 cm arrester block that was heated by a 10 A current pulse with a 1 ms duration. The circumference of the block represents the hottest region, and the temperature distribution is roughly radially symmetric. On average, the block temperature increases by about 0.5°C, but the temperature at the outside of the disk increases by more than a degree—while the temperature change at the center of the block is much less. The results shown in Fig. 8 were obtained

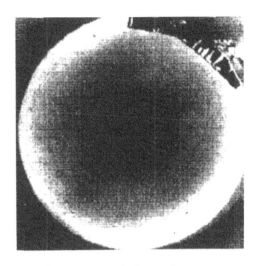

Fig. 8. Image of the surface-temperature generated in a varistor block fabricated at a low aspect ratio. The strip at the upper right is an electrical contact to the surface.

with a block fabricated with a 1 cm thickness, and a symmetrical pattern of localized heating was observed. However, when 1 cm thick pieces were cut from commercial blocks made with the same diameter but with a 4.6 cm length, asymmetric heating was observed[16] which was similar to that reported by Mizukoshi et al.[23] These results are consistent with the greater difficulty of fabricating large blocks with uniform properties, particularly when the blocks have a high aspect ratio.

Thermal images of 4.2 cm blocks were analyzed to determine the temperature distribution along a disk diameter before, during, and after the current pulse. Temperature profiles obtained at different times are shown in Fig. 9(a). (The same block was used in Fig. 8.) The background temperature is flat (i.e., the temperature before the pulse, or at time zero), except that the edge of the block is slightly warmer because of a much earlier pulse. An initial temperature profile was taken from an image made during the current pulse. Because the camera was synchronized with the electrical pulse, and because the detector integration time and width of the current were both 1 ms, the first profile best represents the temperature of the block at a time half-way through the pulse, or at 0.5 ms. During the current pulse, the temperature was rising very fast, particularly at the periphery of the block. But after the pulse, the temperature changed relatively slowly on the centimeter-size scale of the block. (The thermal diffusivity of the material is about 0.1 cm^2/s.) Hence, the temperature at the periphery of the block at 17.5 ms after the current pulse has decreased only slightly from the value attained immediately after the pulse, and it is almost twice the temperature of the first profile because twice as much energy has been absorbed. As expected, the temperature profiles taken from later images (e.g., 817.5 and 1650 ms) show the temperature dropping at the block periphery and increasing at the center in response to heat flow.

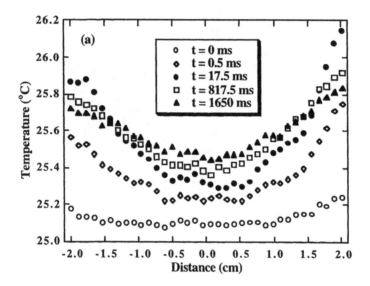

Fig. 9(a). Measured temperature profiles along a varistor diameter before, during, and after a current pulse.

The temperature rises most at the periphery of the block because most of the current flow is taking place there since the breakdown voltage is lower. This inference was confirmed by the method of placing pairs of small electrodes at various positions on the block.[24] A 3 to 4% decrease in the breakdown voltage from the center to the edge was measured. The values of the breakdown voltage were used to describe the current as an approximate analytical function of position, and this enabled the thermal behavior of the block to be simulated by solving numerically the heat transfer equation for the block:

$$D\nabla^2 T - \frac{\partial T}{\partial t} = -\frac{JF}{C\rho} \quad . \tag{8}$$

Close to room temperature: the heat capacity $C = 0.6$ J/(gK), the diffusivity $D = 0.1$ cm^2/s, and the density $\rho = 5.5$ g/cm^3. The heat input is JF, and it was assumed that the disk can dissipate heat only radially through the sidewall.[16,20]

For the case of a 1 ms, 10 A current pulse, the temperature profiles of the disk were calculated for the times corresponding to those in the measurements. The results are presented in Fig. 9(b). They are in very good agreement with the experimental data in Fig. 9(a). Although the thermal parameters of zinc oxide varistor materials are well known, they could also have been found from the IR measurements. In the present case, checking the thermal parameters implied by the IR measurements provides both calibration and consistency checks.

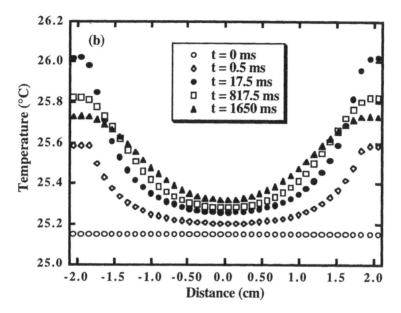

Fig. 9(b). Computer simulations of the temperatures profiles along a varistor diameter before, during, and after a current pulse.

Because the temperature distribution shown in Figs. 8 and 9(a) has the radial symmetry of the block, it is logically attributed to the fabrication process. The temperature increases more at the edge of the block because there is a higher current that results from a lower breakdown voltage that can be attributed to a larger grain size. The implied grain size increases monotonically from the center to the edge of the block. Friction in the die used for powder compaction causes density variations that can result in such a systematic variation. Moreover, grain growth during sintering can be more rapid in the outer part of the block than in the center of the block. Less symmetry is found with longer blocks. Obtaining uniformity is especially important in varistor manufacture because any nonuniformity is amplified in the conduction process by the electrical nonlinearity. Thus, the production of large varistors requires particularly close control of pressing and sintering conditions. As discussed below, improving the uniformity of varistor ceramics would improve their energy-handling capability and reduce failure rates.

VARISTOR FAILURES

Although varistors have a large capacity to absorb energy (e.g., 500 J/cm^3), their energy-handling capability is limited. The energy-handling capability of varistors is defined as the amount of energy that a varistor can absorb before failure. There are three main failure modes of varistors: thermal runaway, puncture, and cracking, and all are typically initiated by non-uniform Joule heating. Thermal runaway can occur because of a varistor's negative temperature coefficient of resistivity. The leakage current of a varistor, and consequently the Joule heating, increase with temperature. Thus, if the temperature rises above a thermal-stability temperature T_S, the power input may exceed heat dissipation and thermal runaway occurs. Puncture occurs at spots with lower breakdown voltage and higher current conduction than their surroundings. In puncture, a small hole typically appears where the current and heat are concentrated and the material melts.[24] Nonuniform heating can also cause thermal stresses higher than the failure stress of the ceramic and lead to cracking.[24,25]

In the nonlinear region of the varistor I-V characteristics, the electrical current tends to concentrate at electrically weak spots that originate with imperfect processing or statistical disorder. In principle, non-uniform heating on the microscopic size scale of statistical disorder can lead to failure, but, as discussed above, heat transfer on the scale of the grain size is too fast to allow temperature differences sufficient to cause failure. Moreover, current localization on a microstructural scale is significant only in small varistors. Most varistor failures originate with non-uniform heating on a macroscopic scale, and these failures are mainly a problem in large blocks where uniform properties are difficult to achieve by ceramic processing. Thus, the emphasis here is on the failure of large varistor blocks.

Significant understanding of the energy-handling capability of varistors has emerged from earlier studies.[17,24,26] It is shown that puncture can occur when electrically weak spots reach about 800°C and local melting of the Bi$_2$O$_3$ phase results.[24] However, competing influences have been overlooked: (1) the influence of the upturn in the I-V characteristics which causes the current to be uniform at high current and suppresses puncture, and (2) thermal cracking which may occur before puncture takes place. Electrical testing has established that the energy-handling capability initially decreases with increasing current, but it then increases if the current becomes very high and the pulse duration becomes very short.[17] Recent results show that the energy absorption capability of varistors used in station-class arresters increases almost 4 times as the peak current level increases from 0.8 to 35 kA.[26] This complex variation of the energy-handling capability with pulse magnitude and duration has proven difficult to interpret.

The failure modes of large varistor blocks have recently been predicted by thermo-mechanical modeling.[20-22] Computer simulations of varistor failures have included the influences of varistor size and the degree of nonuniformity, as well as variations in the I-V characteristic, to achieve results in excellent agreement with experimental testing. By solving the heat transfer equations for a varistor disk with non-uniform electrical properties, the time dependence of the temperature profile (together with the resulting distribution of thermal stresses) were established and the points of failure were determined. The energy-handling capability was

Dielectric Ceramic Materials

established by analyzing all three common failure modes to determine if, or when, they occur for current surges corresponding to experimental testing.

The coupled equations of current conduction and heat diffusion are simplified by assuming cylindrical symmetry, and the diffusion equation can be expressed as:

$$D(T')\left(\frac{\partial^2 T'}{\partial r^2} + \frac{1}{r}\frac{\partial T'}{\partial r} + \frac{JF}{K_0}\right) = \frac{\partial T'}{\partial t}, \tag{9}$$

where $D(T')$ is the thermal diffusivity and K_0 is the thermal conductivity at 20°C. The strong temperature dependence of the thermal conductivity K is included in the equation without undue complication by use of a scaled temperature T' that varies with the actual temperature T in proportion to the thermal conductivity. The only boundary condition is the assumption that a disk dissipates heat only radially through a sidewall by convection to air at 20°C. Although this condition takes no account of the arrester design and environment, puncture and cracking are insensitive to this boundary condition. An I-V characteristic typical of high-voltage varistors was assumed. At T = 293 K, the breakdown field (i.e., the field at 1 mA cm^2) is $F_b \approx 1870$ V/cm, the prebreakdown resistivity $\rho_{pb} = 5 \times 10^{11}$ ohm cm, and the resistivity in the upturn region (i.e., the resistivity of the grains) is $\rho_{up} = 1$ ohm cm. The coefficient of nonlinearity has a maximal value $\alpha = 50$. The temperature dependence of the I-V characteristic is included by assuming that it is exponentially activated with temperature so as to conform with experiments.[20-22]

The electrical nonuniformity in varistor disks is simulated by assuming a hot spot of diameter $2R_{hs}$ that extends axially through a block at its center; so the heat diffusion problem remains cylindrically symmetric. The breakdown voltage at the hot spot F_{bhs} is assumed to be a factor of p lower than that for the rest of the varistor block, i.e., $p = (F_b - F_{bhs})/F_b$. The quantity p (expressed in percent) is called the hot spot intensity. Typical varistor disks have hot spots[24] with p = 5, but the intensity of hot spots in highly nonuniform disks can be as high as 10. The small difference between the electrical properties of the hot spot and those of its surroundings causes high nonuniformity in the spatial distribution of the current and, thereby, in the input power density due to Joule heating. Depending upon the applied current and the temperature, the current flowing through the hot spot may be more than a 100 times that in the rest of the disk.

Once the temperature profile for the disk and its time evolution are known, they are used to calculate the distribution of thermal stresses and to identify the cracking failure mode. The stresses caused by the nonuniform temperature T(r) are:[27]

$$S_{rr}(r) = \frac{\gamma Y}{1-v}\left(\frac{1}{R^2}\int_0^R r'T(r')dr' - \frac{1}{r^2}\int_0^r r'T(r')dr'\right) \tag{10}$$

$$S_{\theta\theta}(r) = \frac{\gamma Y}{1-v}\left(\frac{1}{R^2}\int_0^R r'T(r')dr' + \frac{1}{r^2}\int_0^r r'T(r')dr' - T(r)\right) \tag{11}$$

$$S_{zz}(r) = S_{rr}(r) + S_{\theta\theta}(r) . \tag{12}$$

Stresses in the radial, tangential, and axial direction are denoted by S_{rr}, $S_{\theta\theta}$, and S_{zz}, respectively. The thermal expansion coefficient,[28] Young's modulus, and Poisson's ratio are denoted as γ, Y, and v. Varistor disks in station-class arresters typically have an aspect ratio lower than 1. In this case, the thermal stresses in the axial direction are negligible and v can be set to zero in determining radial and tangential stresses.[27] To determine the maximum energy that can be absorbed by a varistor disk before it cracks, it is necessary to know the strength of the material. The value for varistors strongly depends on the details of the ceramic processing used, on the presence of pre-existing flaws, and on other factors that may vary considerably for individual disks. The simple assumption that the tensile strength of the varistor ceramic is $S_{ft} = 1.4 \times 10^8$ nt/m^2 is used.[25] Two types of failures are considered: (1) cracking in tension, when the thermal tensile stress exceeds S_{ft}; and (2) puncture, when the temperature at the hot spot exceeds 800°C.[23]

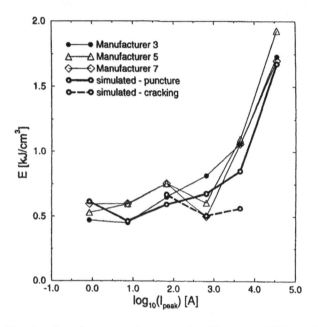

Fig. 10. Simulated and measured energy-handling capabilities of station-class-arrester disks as a function of the peak applied current.

The predictions of the thermo-mechanical modeling can be compared with measurements[26] of the energy absorption capability of varistor disks used in station-class arresters. A typical disk, 23 mm high and 63 mm diameter, is assumed to have a 1.2 cm diameter hot spot ($R_{hs} = 0.6$ cm). Computed energy-handling capabilities as functions of the peak test current causing puncture and cracking are shown in Fig. 10 together with experimental mean values of total energy needed to destroy the varistor disks of three manufactures. (The symbols and code numbers for manufactures are those used in reference 26.) At low

Dielectric Ceramic Materials

currents corresponding to 0.8 A and 7 A, the only predicted failure mode (besides a thermal runaway) is puncture. This is consistent with the experimental results, which at these currents exhibited failure resulting in a single hole through the bulk ceramic, but no cracking or fragmentation failures. At a peak current of 600 A, the tensile tangential stresses around the hot spot become higher than the strength of the varistor ceramic before the temperature at the hot spot reaches 800°C. Therefore, the disks become likely to crack along lines branching out from the hot spot. Indeed, the most significant external damage was found for peak currents of 600 A.[26] The simulations indicate that cracking remains the most likely failure mode at 4 kA peak. However, for very high peak current pulses of 35 kA, the simulated thermal stresses do not become high enough to cause cracking, and the disks fail either due to a puncture or to overheating. This is again consistent with the experimental data.[26]

Varistors from manufacturers 5 and 7 exhibit a dip in energy absorption capability at 600 A peak that agrees surprisingly well with the predicted energy handling characteristic for cracking. On the other hand, the energy handling curve for puncture in Fig. 10 is in good agreement with the data on varistors from manufacturer 3. Apparently, the latter varistor disks have higher mechanical strength and are more likely to exhibit puncture than cracking. The predicted cracking patterns, as well as the transition from puncture failure at low current densities to cracking at high currents, are also in agreement with the observations of Eda.[24]

The assumption that a typical disk can be modeled as having a 1.2 cm diameter hot spot with intensity $p = 5$ is supported by the excellent agreement with mean values obtained from statistical analyses of experimental data.[26] However, varistor disks are not identical, and results of individual measurements vary considerably. In particular, disks have different hot spots. Therefore, the thermo-mechanical behavior of the disks with hot spots 1 cm in diameter was simulated for three different intensities: $p = 2.5$, 5, and 7.5. The energy-handling characteristics of disks with such hot-spot characteristics are shown in the left panels of Fig. 11.

The disks with hot spots of higher intensity have lower energy handling capability. The minimum energy handling decreases from about 1 kJ/cm³ for p = 2.5, to 450 J/cm³ for $p = 5$, and 200 J/cm³ for $p = 7.5$. The current densities for which disks exhibit puncture extends to lower currents as the hot-spot intensity increases. Simulations for the case of $p = 2.5$ show that the thermal stresses never become high enough to cause cracking, independent of the magnitude of the current. Thus, disks with hot spots of low intensity do not crack, but fail only by puncture. Cracking caused by tensile thermal stresses in the tangential direction is the dominating failure mode at high current densities for $p = 5$ and $p = 7.5$. Moreover, for larger p, disks are more likely to crack at lower current densities.

The size of hot spots can also significantly deviate from a typical value of 1 cm. Hot spots with $p = 5$, and three different diameters: 0.1, 1.0, and 2.0 cm were simulated to evaluate the influence of hot spot size on disks of station-class size. The energy handling characteristics are shown in the right panels of Fig. 11. The minimum energy handling only weakly depends on the size of the hot spot, and varies from about 650 J/cm³ for the disk with a 1 mm hot spot, through 450 J/cm³ when $2R_{hs} = 1$ cm, to 480 J cm³ for the disk with a 2 cm diameter hot spot. The minimum energy handling is not a monotonic function of R_{hs}, but has a minimum for hot spots of intermediate size. The minimum energy corresponds to puncture, and it moves toward lower current densities as the size of the hot spot increases. As can be seen from Fig. 11, disks are more prone to cracking failures when they have large hot spots. A disk with a 0.1 cm hot spot never cracks, whereas cracking is the dominating failure mode even at 1 A/cm^3 for a disk with $2R_{hs} = 2$ cm.

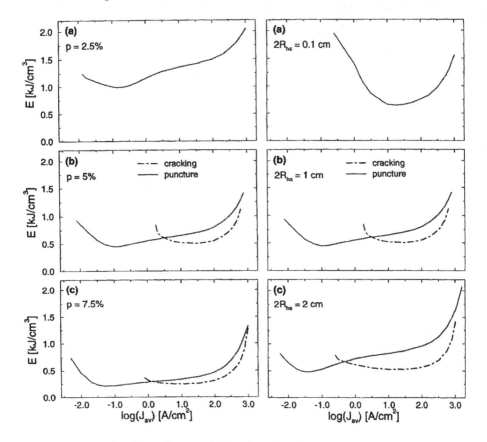

Fig. 11. Energy handling characteristics of station-class varistor disks having 1 cm diameter hot spots with different intensities (left panels); and having hot spots with intensity p = 5% and three different sizes (right panels).

In summary, a simple thermo-mechanical model can identify energy-handling limitations of ZnO varistors imposed by three different failure modes: puncture, thermal runaway, and cracking. Each of these failure modes can be limiting, depending on the disk shape, its electrical uniformity, and the applied current. The model predictions conform to available failure data and explain the observation of Sakshaug et al.[17] that energy handling improves at high current densities. The model confirms that cracking and puncture are caused by current localization which causes local heating leading to nonuniform thermal expansion and thermal stresses. Puncture is most likely in varistor disks with a low geometrical aspect ratio and when the current density has intermediate values. Puncture is the dominant failure mode for slightly nonuniform disks, but cracking becomes more likely as the degree of nonuniformity increases. Cracking dominates at high current densities and for disks with high aspect ratio. Puncture and cracking do not occur when the current is small because the heating is too slow to cause a large temperature gradient. Puncture and cracking do not occur at large currents corresponding to the upturn region of the *I-V* characteristic, since the current becomes uniformly distributed. Hence, for low and very high current densities the most likely failure mode is thermal runaway.

ACKNOWLEDGMENTS

This work was supported by the U.S. Department of Energy through the Office of Basic Energy Sciences and by the Office of Transportation Technologies, as part of the High Temperature Materials Laboratory User Program under contract DE-AC05-96OR22464, managed by Lockheed Martin Energy Research Corporation.

REFERENCES

[1]G. D. Mahan, L. M. Levinson, and H. R. Philipp, "Theory of Conduction in ZnO Varistors," *Journal of Applied Physics* **50**, 2799–2812 (1979).

[2]G. E. Pike, "Electronic Properties of ZnO Varistors: A New Model," *Materials Research Society Symposia Proceedings,* **5**, 369–379 (1982).

[3]G. E. Pike, "Semiconductor Grain-Boundary Admittance: Theory," *Physical Review B* **30**, 795–802 (1984).

[4]G. Blatter and F. Greuter, "Carrier Transport Through Grain Boundaries in Semiconductors," *Physical Review B* **33**, 3952–3966 (1986).

[5]G. Blatter and F. Greuter, "Electrical Breakdown at Semiconductor Grain Boundaries," *Physical Review B* **34**, 8555–8572 (1986).

[6]M. Matsuoka, "Nonohmic Properties of Zinc Oxide Varistor Ceramics," *Japanese Journal of Applied Physics* **10**, 736–746 (1971).

[7]D. Weaire and N. Rivier, "Soaps, Cells, and Statistics - Random Patterns in Two Dimensions," *Contemporary Physics* **25**, 59–99 (1984).

[8]M. Bartkowiak and G. D. Mahan, "Nonlinear Currents in Voronoi Networks," *Physical Review B* **51**, 10825–10832 (1995).

[9]M. Bartkowiak, G. D. Mahan, F. A. Modine, and M. A. Alim, "Influence of Ohmic Grain Boundaries in ZnO Varistors," *Journal of Applied Physics* **79**, 273–281 (1996).

[10]M. Bartkowiak, G. D. Mahan, F. A. Modine, M. A. Alim, R. J. Lauf, and A. D. McMillan, "Voronoi Network Model of ZnO Varistors with Different Types of Grain Boundaries," *Journal of Applied Physics* **80**, 6516–6522 (1996).

[11]M. Tao, A. Bui, O. Dorlanne, and A. Loubiere, "Different 'Single Grain' Junctions Within a ZnO Varistor," *Journal of Applied Physics* **61**, 1562–1567 (1987).

[12]E. Olsson and G. L. Dunlop, "Characterization of Individual Interfacial Barriers in a ZnO Varistor Material," *Journal of Applied Physics* **66**, 3666–3675 (1989).

[13]Z.-C. Cao, R.-J. Wu, and R.-S. Song, "Ineffective Grain Boundaries and Breakdown Threshold of Zinc Oxide Varistors," *Materials Science and Engineering B* **22**, 261–266 (1994).

[14]H. Wang, W. Li, and J. F. Cordaro, "Single Junctions in ZnO Varistors Studied by Current-Voltage Characteristics and Deep Level Transient Spectroscopy," *Japanese Journal of Applied Physics* **34**, 1765–1771 (1995).

[15]H. Wang, W. A. Schulze, and J. F. Cordaro, "Averaging Effect on Current-Voltage Characteristics of ZnO Varistors," *Japanese Journal of Applied Physics* **34**, 2352–2358 (1995).

[16]H. Wang, M. Bartkowiak, F. A. Modine, R. B. Dinwiddie, L. A. Boatner, and G. D. Mahan, "Non-uniform Heating in Zinc Oxide Varistors Studied by Infrared Imaging and Computer Simulation," *The Journal of the American Ceramic Society*. Accepted for publication.

[17]E. C. Sakshaug, J. J. Burke, and J. S. Kresge, "Metal Oxide Arresters on Distribution Systems. Fundamental Considerations," *IEEE Transactions on Power Delivery* **4**, 2076–2089 (1989).

[18]C.-W. Nan and D. Clarke, "Effect of Variations in Grain Size and Grain Boundary Barrier Heights on the Current-Voltage Characteristics of ZnO Varistors," *The Journal of the American Ceramic Society* **79**, 3185–3192 (1996).

[19]A. Vojta and D. R. Clarke, "Microstructural Origin of Current Localization and "Puncture" Failure in Varistor Ceramics," *Journal of Applied Physics* **81**, 985–991 (1997).

[20]M. Bartkowiak, M. G. Comber, and G. D. Mahan, "Energy Handling Capability of ZnO Varistors," *Journal of Applied Physics* **79**, 8629–8633 (1996).

[21]M. Bartkowiak, M. G. Comber, and G. D. Mahan, "Failure Modes and Energy Absorption Capability of ZnO Varistors," IEEE Preprint PE-135-PWRD-0-12-1997.

[22]M. Bartkowiak, M. G. Comber and G. D. Mahan, "Influence of Nonuniformity of ZnO Varistors on Their Energy Absorption Capability," Submitted for publication in the *IEEE Transactions on Power Delivery*.

[23]A. Mizukoshi, J. Ozawa, S. Shirakawa, and K. Nakano, "Influence of Uniformity on Energy Absorption Capabilities of Zinc Oxide Elements as Applied in Arresters," *IEEE Transactions on Power Apparatus and Systems* **102**, 1384–1390 (1983).

[24]K. Eda, "Destruction Mechanism of ZnO Varistors Due to High Currents," *Journal of Applied Physics* **56**, 2948–2955 (1984).

[25]H. F. Ellis, R. M. Reckard, H. R. Philipp, and H. F. Nied, *Fundamental Research on Metal Oxide Varistor Technology*, EPRI Report EL-6960 (Electric Power Research Institute, Palo Alto, CA, 1990).

[26]K. G. Ringler, P. Kirkby, C. C. Erven, M. V. Lat, and T. A. Malkiewicz, "The Energy Absorption Capability and Time-to-Failure of Varistors Used in Station-Class Metal-Oxide Surge Arresters," *IEEE Transactions on Power Delivery* **12**, 203–208 (1997).

[27]B. A. Boley and J. H. Weiner, *Theory of Thermal Stresses.* John Wiley and Sons, New York, 1960.

[28]Y. S. Touloukian, R. W. Powell, C. Y. Ho, and P. Klemens, *Thermophysical Properties of Materials.* Plenum, New York - Washington, 1970.

ELECTRIC CHARACTERISTICS OF
LOW TEMPERATURE SINTERED ZnO-VARISTORS

Atsushi Iga, Kei Miyamoto* and Hiroki Miyamoto*
Zinctopia Laboratory,
Daiwa 1-14-11, Takatsuki, Osaka 569-1047, Japan
*Technology Research Institute of Osaka Prefecture,
2-7-1, Ayumino, Izumi, Osaka 594-1157, Japan

ABSTRACT

The present paper shows example of the electric characteristics of low temperature sintered ZnO varistors [1]. Some varistors were sintered at 750°C by modifying the amount of B_2O_3 addition. Effect of annealing on electric characteristics were insignificant for the varistors sintered at low temperatures.

1   Introduction

ZnO varistors are ceramic semiconductor devices with compositions of ZnO +M.O. (Metal Oxides) and are used as surge protectors for electrical circuits[2].

In the phase diagram of $ZnO-Bi_2O_3$ system, there is an eutectic point at 740°C in the Bi_2O_3-rich side[3,4]. Therefore ZnO varistor with basic composition $(ZnO + Bi_2O_3 + CoO + MnO)$ can be sintered at low temperatures (800~900°C). Sb_2O_3 is added to the basic composition for producing high voltage varistors. These high voltage varistors, however, should be sintered at higher temperatures (1200°C~1300°C)[2]. The sublimation of Sb_2O_3 which takes place at low temperatures caused the formation of thin films which hinder the effective contact between ZnO−ZnO grains and induce retardation of sintering.

It has been found recently that $ZnO + Bi_2O_3 + Sb_2O_3$ can be sintered at 900°C or higher temperature when a mixture of $Bi_2O_3 + Sb_2O_3$ is pre-fired in advance and added to ZnO[5]. In the same way, $ZnO + Bi_2O_3 + Sb_2O_3 + CoO + MnO + NiO$ can be sintered at a low temperature when the mixture of $Bi_2O_3 + Sb_2O_3$ is preveously pre-fired and added to ZnO with other metal oxides[1]. Thus, it became clear that ZnO varistors can be sintered at lower temperatures (850°C∼950°C).

2 Experimental Method

[1] Various ZnO varistors.

The following ceramics were sintered by varying the X concentration. The electric characteristics of obtained varistors are discussed.

$$Z n O + (Co, Mn, Ni) + Al_2O_3 + X$$

    1 mol   1.75 g   10 ppm

In the present paper, (Co, Mn, Ni) means that a mixed powder of composition CoO(0. 954 g) + MnO_2(0. 414 g) + NiO (0. 383 g) was prepared in advance and [A+B] means that the powders A+B were mixed and pre-fired at 555°C for 5 h 55 min in advance.

Al_2O_3 was doped by using aluminium nitrate aqueous solution. Compact pellets were pressed under a pressure of 500 kg/cm². Heating and cooling rates were ± 50°C/h in the sintering process.

The prepared specimens are as follows.

(1) X = $[Bi_2O_3 + Cr_2O_3]$, $[Bi_2O_3 + B_2O_3]$,

(2) X = $[Bi_2O_3 + Sb_2O_3]$,
        $[Bi_2O_3 + Sb_2O_3] + [Bi_2O_3 + B_2O_3]$,$[Bi_2O_3 + Cr_2O_3]$

(3) X = $[Bi_2O_3 + M.O.]$,

(4) X = $[Bi_2O_3 + Sb_2O_3] + [Bi_2O_3 + M.O.]$,

(5) X = $[Bi_2O_3 + M.O. + Sb_2O_3]$,
        $[Bi_2O_3 + M.O. + Sb_2O_3] + [Bi_2O_3 + B_2O_3]$, $[Bi_2O_3 + Cr_2O_3]$.

[2] Effect of annealing.

High and low temperature sintered ZnO varistors were annealed and electrical characteristics were studdied.

3 Results and Discussions

[1] Electric characteristics of various varistors.

Example 1 Abnormal grain growth and retardation of grain growth.

Composition   (Sintering : 950°C*13 h)

☆3316 X =[0. 8 Bi_2O_3 + 0. 5TiO_2 + 0. 025Sb_2O_3] (3. 45 g)
            + [2Bi_2O_3 + B_2O_3] (0.345 g)

☆3317 X =[0. 8 Bi_2O_3 + 0. 5TiO_2 + 0. 05Sb_2O_3] (3. 45 g)
            + [2Bi_2O_3 + B_2O_3] (0.345 g)

☆3318 X =[0. 8 Bi_2O_3 + 0. 5TiO_2 + 0. 15Sb_2O_3] (3. 45 g)
            + [2Bi_2O_3 + B_2O_3] (0.345 g)

Electric characteristics (Electrodes : In-Ga)   Similar results to

specimen	$_{1mA}\alpha_{10mA}$	$V_{1mA/mm}$
☆3316	23	10. 7
☆3317	26	37.2
☆3318	64	189

Example 1 were obtained
in X =
[Bi_2O_3 + TiO_2 + Sb_2O_3]
+ [Bi_2O_3 + Cr_2O_3] system.

Example 2   Low voltage varistor (Sintering : 960°C*6 h, 100°C/h up,
             150°C/h down)

☆3862 X = [0. 75 Bi_2O_3 + 0. 5TiO_2 + 0. 05Sb_2O_3] (3. 81 g)
            + [2Bi_2O_3 + B_2O_3] (0.19 g)

Pulse characteristics,   (Ag-electrodes, t = 0.98 mm)

V_{1mA}/mm	V_{10A}/V_{1mA}	$\triangle V_{1mA}$ (%) after repeated surges (8/20 μs)	
		1000 A*2times	1000 A*30times
26. 51	1. 92	(+) +2. 4	(+) +2. 2
		(−) -0. 0	(−) -1. 4

Example 3   High voltage varistor − 1   (Sintering : 950°C*13 h),

☆3408   X =[0. 8 Bi_2O_3 + 0. 15Sb_2O_3] (3. 0 g),

Initial characteristics (Electrodes : In-Ga), $V_{1mA/mm}$=203 , $_{1mA}\alpha_{10mA}$ =62

Pulse characteristics , (Electrodes : Ag-paste, t = 1.64 mm)

V_{1mA}/mm	V_{50A}/V_{1mA}	$\triangle V_{1mA}$ (%) after repeated surges (8/20 μs)		
		2500 A*2times	2500 A*10times	2500 A*30times
183. 1	1. 39	(+) +0. 6	(+) +1. 1	(+) -21. 1
		(−) 0	(−) -1. 7	(−) -22. 5

Example 4 High voltage varistor— 2 (Sintering : 950°C*13 h),

☆3427 X =[0. 8 Bi_2O_3+0. 5TiO_2 +0. 15Sb_2O_3] (4. 0 g)

+ [2Bi_2O_3+B_2O_3] (0.2 g) + [Bi_2O_3+Cr_2O_3] (0.8 g)

Initial characteristics, (Electrodes : In-Ga) $V_{1mA/mm}$=147 , $_{1mA}\alpha_{10mA}$ =54

Pulse characteristics, (Electrodes : Ag-paste , t = 1.59 mm)

V_{1mA} /mm	V_{50A}/V_{1mA}	$\triangle V_{1mA}$ (%) after repeated surges (8/20 μs)		
		2500 A*2times	2500 A*10times	2500 A*30times
108. 2	1. 77	(+) +1. 0	(+) +2. 0	(+) +2. 8
		(−) -0. 2	(−) -0. 0	(−) +0. 2

Example 5 High voltage varistor (Sb-less) (Sintering : 950°C*13 h),

☆3409 X =[0. 8 Bi_2O_3+0. 2SnO_2] (3. 0 g)

Initial characteristics,(Electrodes : In-Ga), $V_{1mA/mm}$=200 , $_{1mA}\alpha_{10mA}$ =66

Pulse characteristics. (Electrodes : Ag-paste , t = 1. 59 mm)

V_{1mA}/mm	V_{50A}/V_{1mA}	$\triangle V_{1mA}$ (%) after repeated surges (8/20 μs)		
		2500 A*2times	2500 A*10times	2500 A*30times
176. 8	1. 42	(+) +0. 9	(+) -2. 2	(+) -27. 7
		(−) 0	(−) -4. 4	(−) -28. 9

Example 6 High voltage varistors sintered at 750 °C

☆3405 X =[0. 8 Bi_2O_3+0. 5TiO_2+0. 15Sb_2O_3] (2. 92 g)

+ [2Bi_2O_3+B_2O_3] (0. 877 g)

☆3406 X =[0. 8 Bi_2O_3+0. 5TiO_2+0. 15Sb_2O_3] (2. 53 g)

+ [2Bi_2O_3+B_2O_3] (1.267 g)

☆3407 X =[0. 8 Bi_2O_3+0. 5TiO_2+0. 15Sb_2O_3] (2. 17 g)

+ [2Bi_2O_3+B_2O_3] (1.629 g)

Sintering : 750°C*21 h , ± 50°C up and down , t = 1. 5 mm
Electric characteristics (Electrodes : In-Ga)

specimen	$V_{10\mu A}$	α	$V_{100\mu A}$	α	V_{1mA}	α	V_{10mA}		V_{1mA}/mm
☆3405	689. 3		747. 9		-		-		
☆3406	567. 3	33	608. 0	74	627. 2	62	650. 9		~400V/mm
☆3407	584. 6	33	627. 4	74	647. 0	62	671. 4		~410V/mm

[2]   Effect of Annealing
   Both high (1200°C) and low temperature (800°C~950°C) sintered varistors were annealed at 700°C for 1 h and the effect of annealing on the electric characteristics were studied.
Composition of specimens
  (1) ☆3362 X = [0. 8Bi$_2$O$_3$ +0. 5TiO$_2$+0. 15Sb$_2$O$_3$] (3.17 g)
       + [2Bi$_2$O$_3$+B$_2$O$_3$] (0.32 g) + [Bi$_2$O$_3$+Cr$_2$O$_3$] (0.32 g)
  (2) ☆3363 X = [0. 8Bi$_2$O$_3$ +0. 5TiO$_2$+0. 15Sb$_2$O$_3$] (3.17 g)
       + [2Bi$_2$O$_3$+B$_2$O$_3$] (0.63 g)
  (3) ☆3408 X = [0. 8Bi$_2$O$_3$ +0. 15Sb$_2$O$_3$] (3.0 g)
  (4) ☆3409 X = [0. 8Bi$_2$O$_3$ +0. 15SnO$_2$] (3.0 g)
  (5) ☆3423 X = [0. 8Bi$_2$O$_3$ +0. 15Sb$_2$O$_3$] (1.9 g)
       +[0. 8Bi$_2$O$_3$ +0. 15SnO$_2$] (1.9 g)

   Annealing at 700°C decreased the α-value in ZnO varistors sintered at 1200°C. Meanwhile the α-value changed slightly for low temperature sintered varistors. Those effects are found remarkable. On the other hand, all Bi$_2$O$_3$

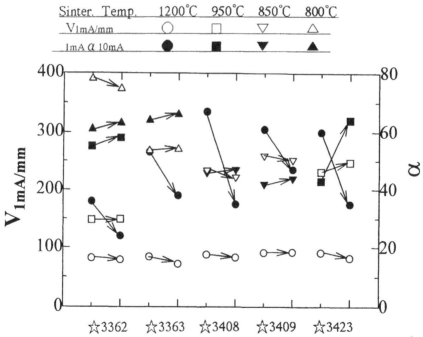

Fig. 1 The change on electric characteristics of varistors by annealing at 700°C.

phases transferred to γ-Bi_2O_3 phase by the annealing at $700^{\circ}C$.

Annealing at $700^{\circ}C$ will cause migration of interstitial Zn^+ in the accumulation layer to the grain boundary, connection of Zn^+ to O^- at the grain boundary and decrease of α-value [6]. It is considered that the density of interstitial Zn is less in the low temperature sintered ZnO ceramics[7] and the conduction electrons mainly come from doped Al in the lattice. As a result α-value did not decrease by annealing in the low temperature sintered varistors where the density of interstitial Zn is less.

5 Conclusions

[1] Various varistors from low to high voltage were obtained by sintering at low temperature.

[2] Some varistors were sintered at $750^{\circ}C$.

[3] Effect of annealing on the electric characteristics were insignificant for low temperature sintered varistors.

References

1 A.Iga,"Mechanism of Low Temperature Sintering of ZnO-Varistor," Proc. American Ceramic Society's 99th Annual Meeting & Exposition, SXIII-021-97 (1997).

2 M.Matsuoka, "Nonohmic Properties of Zinc Oxide Ceramics," Jpn. J. Appl. Phys., 10 [6] 736-46 (1971).

3 G.M.Safranov, V.N.Batog, T.V.Stepanyuk, and P.M.Fedorov, "Equilibrium Diagram of the Bismuth Oxide- Zinc Oxide System," Russ.J. Inorg. Chem.(Engl.Transl), 16 [3] 460-61 (1971).

4 К о с о в А.В.,К у т в и ц и й В.А., С к о р и к о в В.М., У с т а л о в а О.Н., К о р я г и н а Т.И.,"Н е о р г а н-и ч е с к и е М а т е р и а л ы,Т о м 12, 466 (1976).

5 M.Ito, M.Tanahashi, M.Uehara and A.Iga,"The Sb_2O_3 Addition Effect on Sintering ZnO and $ZnO+Bi_2O_3$,"Jpn. J. Appl. Phys.36 pp.L1460-L1463 (1997).

6 T.K.Gupta, "Appliction of Zinc Oxide Varistors, "J.Am.Ceram.Soc.,73 [7] 1817-1840 (1990).

7 D.G.Thomas, "Interstitial Zinc in Zinc Oxide," J. Phys. Chem. Solids, 3, 229-237. (1957).

Miniaturization Techniques of Microwave Components
for Mobile Communication Systems
– using low loss dielectrics –

Kikuo WAKINO
Murata Manufacturing Company Limited
2-26-10, Tenjin, Nagaokakyo-Shi, Kyoto, 617-8555 Japan

Abstract

The extremely low loss and high dielectric constant ceramics are playing the important role in the field of microwave communication systems: as a key material of the temperature stable high Q resonator filters in the satellite transponder, the satellite broadcasting receiver and many other microwave communication equipment; for the miniaturized components of cellular radio system both for a base station and a terminal. such as dielectric resonators, duplexer, multilayer substrates for MCM, antenna element and others. In this report a history and recent topics will be reviewed of the low loss and temperature stable dielectric ceramics with respect to the composition and characteristics, the basis for a design of the resonator and filter, and other application. The explanations of the recent progress are also described of the multilayer modules and SAW devices.

I. INTRODUCTION

After the utilization of the microwave has become popular as a communication media, the miniaturization of the devices was pointed out as the important subject to be improved. The key board and display of a portable telephone terminal are requested a proper size for the ease of operations as a man-machine interface, but other devices or components are expected to be as small as possible. Battery, antenna, antenna filter and earphone are typical bulky devices in the terminal. Great deals of effort of the reduction in size and weight of these devices have been continued to design lightweight pocket size telephone terminals. The antenna filter of base station for micro cell system is also the urgent subject to be miniaturized.

Meanwhile, in the low frequency application area, a temperature compensating ceramic capacitor was developed in Germany in the 1930s, and widely used for the temperature stabilization of electronic circuits such as LC resonance circuit, CR timing circuit, and so on. Fortunately, the temperature compensating ceramic capacitor (so called Class I Ceramic Capacitor) showed enough high Q value (2000 - 10000) compared with that of inductor (200 - 700, 1000 at most). By this reason, a special attention to be improved the Q value of Class I Ceramic Capacitor was not considered until the end of 1960s, when the application to microwave equipment started.

The research and development work of low loss and temperature stable dielectric ceramics for the microwave application started at the beginning of 1970s. At present, the Q values of these materials have been remarkably improved. Extremely low loss dielectric materials at microwave frequency are now available with specific dielectric constant from 20 to 95.

The advantage of using the dielectric materials is to reduce the size of the microwave devices. It based on the phenomenon that the wavelength of electromagnetic wave is reduced to $1/\sqrt{K}$ in the dielectrics compared with that in the free space, where K is the relative dielectric constant.

The word "dielectric resonator" was described on the paper by Richtmyer in 1939 [1], where he showed theoretically that the ring shaped dielectrics could work as a resonator.

In the 1960s, pioneers researched the behavior of dielectrics at microwave frequencies and tried to apply them to microwave devices. For example, the dielectric loss at microwave frequency of $SrTiO_3$ crystal was measured and its mechanism was discussed by Silverman et.al. [2], and the far infrared dispersion was studied by Spitzer et.al. [3].

A.Okaya and L.F.Barash measured the dielectric constant and the Q value of TiO_2 and $SrTiO_3$ single crystals at from room temperature down to 50 K in the GHz range, using the commensurate transmission line method in 1962 [4]. The simple and accurate method for dielectric measurement at microwave frequency was developed by Hakki and Coleman using a pair of parallel metal plates sample holder [5]. This method was improved by Y.Kobayashi [6]. This invention accelerated the progress of material research.

The first microwave filter using TiO_2 ceramics was designed by S.B.Cohn in 1968 [7]. But this filter was not put into practical use because it's large temperature variation of pass band frequency of approximately 450 ppm/$^{\circ}$C. Y.Konishi [8] and J.K.Plourde [9] developed a stacked resonator using two opposite signed temperature characteristic dielectric disks. But this composite resonator was not used either in practice, because the too precise and careful controls of material handling, processing, and machining, were required to achieve reproducible mass production conditions.

We have developed the low loss and temperature stable dielectric ceramics for microwave application referring the composition of Class I Ceramic Capacitor materials in 1974, and reported of the temperature stable 6.8 GHz microwave filter loaded with this dielectric resonator in 1975 on the MTT-S meeting at Palo Alto [10].

Along with the development of the improved microwave materials, new designing and manufacturing techniques of microwave passive devices have been proposed and applied to realize the miniaturization of the dielectric filters, hybrid circuits oscillators and others.

On the other hand, in the elastic vibration and sound wave technology area, the new materials, the new design and new manufacturing technique of the low attenuation and temperature stable resonator have made a great progress using piezoelectric crystals and ceramics. Filter utilizing mechanical vibration started at the low frequency from a few kHz to a few MHz, and its operating frequency gradually extended up to 100 MHz. Remarkable extension of operating frequency range was triggered by the innovation of SAW technology. The current upper limit attained over 2 GHz of operating frequency of SAW device.

Meanwhile, multilayer LC microwave modules are now coming back in some areas supported by the progress of the multilayer ceramic capacitor technology. The great feature of LC circuit is spurious response free characteristics, small size and low cost.

These newly coming in devices are indispensable for the miniaturization of microwave equipment today as well as multilayer ceramic capacitor, chip resistor, and chip coil, and are contributing in the reduction of size, weight and cost of advanced electronics equipment.

II. HISTORICAL WORKS of DIELECTRIC RESONATORS and FILTERS

Fig. 1 shows the scheme of dielectric resonator proposed by R.D.Richtmyer in 1939 [1]. Donut shaped dielectric ring resonates when the average length in the circumference is equal to $n\lambda$, where λ is the wavelength of electromagnetic wave along with the circle in the dielectric ring and n is integer. This scheme is quite analogous to the de Brois's electron orbital image.

Fig. 2 shows the microwave filter composed with a dielectric resonator that developed by S.B.Cohn in 1968 [7]. The dielectric resonators are placed in the cylindrical waveguide. Therefore, the electromagnetic wave with frequency lower than cut-off frequency cannot propagate through this simple guide. But, the cut off frequency, in the region where dielectric resonator is placed, is lowered and resonates with the specified frequency defined by the geometry and dielectric properties of the resonator. The electromagnetic wave with this frequency can propagate and store the energy proportionally to the Q value in this region. The DR filter, proposed by S.B.Cohn, showed the fairly large temperature deviation of pass band frequency due to the temperature behavior of TiO_2 ceramics.

III DIELECTRIC MATERIALS for MICROWAVE

3.1. Typical materials for microwave

Since 1940s, the compositions of Class I Ceramic Capacitor, $MgTiO_3$-$CaTiO_3$ and BaO-TiO_2 systems, have been well known as same as TiO_2 ceramics as mentioned before. These compositions could be referred as the starting material for microwave applications.

Beside these materials, many investigators developed several similar, modified and different concept materials during the last two decades.

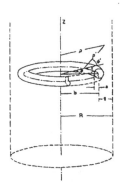

Fig. 1. Dielectric resonator proposed by Richtmyer [1]

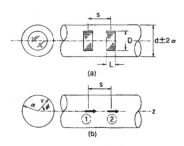

Axially oriented dielevtric disks in a circular tube.
(a)Coupled resonators. (b)Equivalent magnetic dipoles.

Fig. 2. Dielectric resonator filter proposed by S.B.Cohn Axially oriented dielectric disks in a circular wave-guide. (a) Coupled resonator. (b) Equivalent magnetic dipoles [4]

Dielectric Ceramic Materials

Table 1 shows the microwave characteristics of the typical dielectric materials currently used. In the table, τ_f is the temperature coefficient of resonant frequency, which is given by the following equation:

$$\tau_f = -\frac{1}{2}\tau_K - \alpha \qquad (1)$$

where,

τ_K : temperature coefficient of the dielectric constant,
α : linear thermal expansion coefficient of the dielectrics.

3.2. Dielectric properties at microwave

In general, the theory of solid state physics tells us that the materials, which have the higher melting temperature, show higher hardness, higher elastic constant, higher Q value, and so on.

According to the classical dispersion theory, the complex dielectric constant at a frequency ω, $\varepsilon^*(\omega)$, is expressed by the following equation, [3], [17]

$$\varepsilon^*(\omega) - \varepsilon(\infty) = \sum_i \frac{\omega_i^2 \cdot 4\pi\rho_i}{\omega_i^2 - \omega^2 - j\gamma_i\omega} \qquad (2)$$

where $\varepsilon(\infty)$ is the dielectric constant at very high frequency. ω_i, ρ_i and γ_i are the eigen-frequency, the intensity, and the damping factor of the i-th harmonic oscillator in the unit cell, respectively.

This equation is separated into real and imaginary parts as follows:

$$\varepsilon'(\omega) = \omega(\infty) + \sum_i \frac{4\pi\rho_i \cdot \omega_i^2 \cdot (\omega_i^2 - \omega^2)}{(\omega_i^2 - \omega^2)^2 + (\gamma_i \cdot \omega)^2} \qquad (3)$$

$$\varepsilon''(\omega) = \sum_i \frac{4\pi\rho_i \cdot \omega_i^2 \cdot (\gamma_i \cdot \omega)^2}{(\omega_i^2 - \omega^2)^2 + (\gamma_i \cdot \omega)^2} \qquad (4).$$

In the eq. (1) all the absorption edges, $\omega_i's$, which correspond to the i-th optical resonance modes of lattice vibration, are in the frequency ranges of infrared or far infrared. Consequently, $\omega_i's$ are approximately two order higher than the microwave frequency. Accordingly, we can assume that $(\omega/\omega_i)^2 \ll 1$, and simplify the eq. (2) and (3) as follows:

$$\varepsilon'(\omega) - \varepsilon(\infty) = \sum_i 4\pi\rho_i \qquad (5)$$

$$\tan\delta = \frac{\varepsilon''}{\varepsilon'} = \frac{\sum_i (4\pi\rho_i/\omega_i)}{\omega(\infty) + \sum_i 4\pi\rho_i} \cdot \omega \qquad (6)$$

where ε' and ε'' are the real and imaginary part of the dielectric constant.

Considering the eq. (5) and (6), it is concluded that the dielectric constant is unchanging and the

Table 1. Typical Materials for Microwave Application

Materials	K	Q · f GHz	τ_f ppm/°C	reference
Sapphire	9.4	1215000	−51	
MgTiO₃ - CaTiO₃	21	55,000	+10 ~ −10	[11]
Ba(Sn,Mg,Ta)O₃	25	200,000	+5 ~ −5	[12]
Ba(Zn,Ta)O₃	30	168,000	+5 ~ −5	[13]
Ba(Zr,Zn,Ta)O₃	30	100,000	+5 ~ −5	[14]
(Zr,Sn)TiO₄	38	50,000	+5 ~ −5	[15]
Ba₂Ti₉O₂₀	40	32,000	+10 ~ +2	[16]
BaO-PbO-Nd₂O₃-TiO₂	90	5,000	+10 ~ −10	[15]

Dielectric Ceramic Materials

501

dielectric loss increases proportionately to frequency at microwave. Therefore the product $(Q \cdot f)$ is inherent for each material as shown in Fig. 3.

IV. DIELECTRIC RESONATOR

The advantages of using a dielectric resonator are to miniaturize its size without a significant degradation of unloaded Q, and to stabilize the temperature deviation of resonant frequency without using expensive gold plated invar.

Dielectric resonator resonates with multiple higher order frequencies, at which boundary conditions satisfy the requirement of the Maxwell's equation, similar to the air cavity, although the mode chart for dielectric resonator is quite complicated compared with that of the simple air cavity.

Intensive works were investigated of the mode analysis and the preparation of mode chart on the cylindrical resonator. [18]-[21] An appropriate mode can be selected during the designing process of particular application: (a) higher Q is required, (b) smaller in size is required, and (c) average in size and Q are accepted.

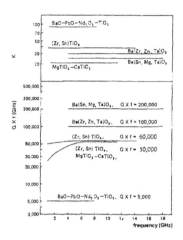

Fig. 3. Frequency dependence of K and Q value of typical resonator materials

4-1. Single mode resonator

4.1.1. Fundamental mode resonator

The resonators, which are most commonly used, are the following three fundamental mode resonators. Fig. 4 shows the electric and magnetic distribution of three fundamental modes, $TE_{01\delta}$, TM_{010} and TEM.

Table 3 summarizes the distinctive characteristics among these three modes.

The unloaded Q of the dielectric resonator, shielded with metal cavity, is given by the following equation:

$$\frac{1}{Q_0} = \frac{1}{Q_d} + \frac{1}{Q_c} \qquad (10)$$

where

Q_0 : Q unloaded

Q_d : Q due to the dielectric loss

Q_c : Q due to the conductor loss.

Here the loss due to the radiation is neglected.

TEM mode has the largest size reduction effect (more than 1/20 in volume). However, it has the disadvantage of largest degradation effect on the Q value, because the electric current density, in the shielding metal for TEM mode resonator, is the largest among these three. On the contrary, $TE_{01\delta}$ mode design is not effective in reducing the size of a resonator (1/3 - 1/5 of conventional air cavity), but it realizes the highest Q value among them.

Table 3. Comparison of character between modes

Mode	distinctive characteristics
$TE_{01\delta}$	highest Q, largest size
TM_{010}	medium Q, medium size
TEM	lowest Q, smallest size

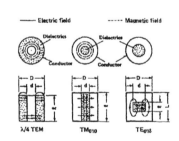

Fig. 4. The three fundamental modes of commonly used dielectric resonator

Dielectric Ceramic Materials

V. FILTER and MULTIPLEXER

Microwave filters using $TE_{01\delta}$ and TEM mode dielectric resonators have been developed since the middle of 1970s. Recently, other modes, such as TM mode or HE mode, have become popular in filter design [22]. Since the magnetic field intensities for TE and TM mode are far less than that for conventional air cavity, the influences of metallic shield (such as deformation, temperature changes, current loss, etc.) are suppressed, and then the expensive invar and gold-plating are not required. Moreover, since the value of τ_f is adjustable by changing the composition of material, the dielectric resonator has great advantage, which can compensate over all temperature drift of resonant frequency being adaptable to all surrounding circuits' performance.

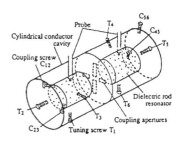

Fig. 6. Structure of hybrid mode multiplexed filter

5.1 Small signal filter

5.1.1 $TE_{01\delta}$ mode filter

Remarkable advantage of this mode is that the resonator shows the lowest degradation of the unloaded Q, because the metal shield does not contact directly to the resonator as mentioned above.

The practical, applicable DR loaded microwave filter was reported by Murata MFG Co., Ltd. on MTT-S meeting in 1975, at Palo Alto [10]. Since then, many researchers developed wide varieties of microwave filters and multiplexers for the microwave networks and the satellite communication systems.

5.1.2 TM mode filter

Since the unloaded Q of this mode resonator is slightly worse compared with $TE_{01\delta}$ mode one, simple TM mode resonators have not been commonly used, even

they are smaller in size than $TE_{01\delta}$ mode ones. This situation has now being reconsidered by making a compromise between performance and size. A filter with coupled two TM mode resonators was developed for the power-combining bank in a cellular base station [23].

5.1.3 Hybrid mode filter

The multimode filter design technique, using the coupling between several desired higher modes in the one piece of resonator, was developed and put in practical use by Fiedziuszko [24]. This design concept is able to reduce the number of resonators, and is effective to the reduction of weight and the miniaturization of satellite transponders.. Fig. 6 shows the structure of the 6 pole multiplexed filter developed by Y.Kobayashi [25]. Each section is consisting of a triple multiplexed resonator

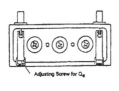

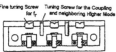

Fig. 5. 6.4 GHz TE01δ mode dielectric resonator filter.

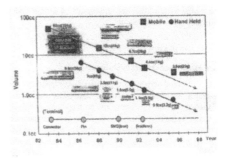

Fig. 7. Size trend of filter and duplexer for terminal

Dielectric Ceramic Materials

503

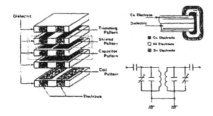

Fig. 8. Construction of the semi-lumped LC filter

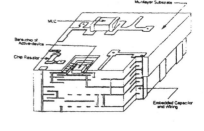

Fig. 9. Schematic illustration of MCM

with one $TE_{01\delta}$ and two $HE_{11\delta}$ modes. Two sections are coupled through the coupling slots punched on the separating metal wall.

5.1.4 TEM mode filter

Distinctive feature of this mode is the tremendous miniaturization capability on a resonator. The resonator carries the disadvantage in fairly high degradation of Q. The Q value of this resonator is approximately proportional to $V^{1/3}$, where V is the volume of resonator. Hence a designer should decide an adequate size, compromising with the contradictory relation between the size and the unloaded Q for his/her application.

5.1.4.1 Duplexer in cellular phone

Cellular mobile radiophones at 800 and 1800 MHz band have become popular in many countries. We have reported several times on band pass filters using coaxial TEM mode resonators. [26]-[29] The size and weight of antenna duplexer for mobile telephone terminal were reduced to about 1/10 in the last 7 years. Fig. 5 shows the trend of the size and weight of a filter and a duplexer in a cellular phone.

VI. OTHER APPLICATIONS

6. 1 Multilayer circuit module

Several types of microwave components and modules have been developed combining the planer circuit, the ceramic substrate and package, and multilayer ceramic capacitor technologies. The low temperature sintering ceramic technology firable under reduced atmosphere enables us to design a Cu wire embedded multilayer substrate. The technology of cofirable multilayer ceramics with different kinds of materials, (insulator, dielectric, ferrite, and so on) provides the more flexibility on various designs for sophisticated circuit modules.

Fig. 8 shows the structure and characteristics of semi lumped circuit LC filter for mobile telephone. The size and volume are 5.7 mm × 5.0 mm ×2.5 mm and 0.07 cm^3 respectively. The insertion loss is less than 4 dB. Fig. 9 shows the schematic structure of a multi chip module (MCM) substrate with/without a package. MCM is one of the most promising devices for the future electronics in cooperation with the Si- and GaAs-devices.

6.2 Surface acoustic wave (SAW) device

Although SAW is not a simple electromagnetic phenomenon, several SAW devices are becoming popular and are contributing for the miniaturization in the field of microwave components. SAW device is capable to miniaturize RF devices, due to the low propagation velocity comparing with that of electromagnetic wave. Rayleigh or pseudo Rayleigh waves easily exited along the surface of electrostrictive or piezoelectric material using interdigital electrodes. $LiNbO_3$, $K(Ta,Nb)O_3$, quartz, ZnO and PZT ceramics are very popular as a substrate material. With the progress of LSI technologies, the recent technologies in fine pattern lithography have been expanding the operating frequencies of SAW devices.

The SAW device is a particularly promising candidate for wrist watch size electronic equipment, because of large potentiality of miniaturization that comes from its principle and structure.

6.3 Dielectric antenna

Several types of distinctive antennas have been designed using adequate dielectric materials depending upon the operation frequency and application.

Fig. 10 shows the examples of a recent dielectric block antenna in a mobile telephone set. This compact block antenna has the potentiality to replace a clumsy whip antenna.

Dielectric Ceramic Materials

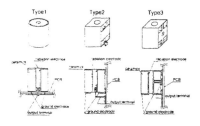

Fig. 10. Schematic draws of dielectric antenna and its directivity data.

VII. CONCLUSION

The idea of a dielectric resonator and microwave device has been proposed since the 1930's by several pioneers but has not realized before the 1970's, until the wide use of microwave became popular as a communication media.

Extremely low loss, temperature stable ceramics for microwave have been aggressively investigated and made remarkable progresses during the past 15 years. Using these improved dielectric ceramics, a small size, temperature stable and low loss microwave filter, an antenna duplexer, a power combiner, a fixed or variable frequency oscillator and other devices were developed.

These miniaturized microwave devices greatly contribute for popularization on mobile and portable communication equipment.

The unloaded Q of a cavity resonator filled up with dielectric material is ideally proportional to the cubic root of its volume. Compromising between size reduction and sharp response characteristics of resonator is unavoidable. TEM mode filter designing is the good example of this trade off problem.

The design technique of multimode resonator is effective to reduce the necessary number of resonators.

Microwave devices such as dielectric antennas or oscillators are expected to contribute in the field of satellite communication systems, GPS system, wireless LAN system, and so forth.

SAW devices should be the key devices for super miniaturization to realize wrist watch size equipment.

ACKNOWLEDGMENT

The research and development program of microwave ceramics and device in Murata MFG Co., Ltd. had been triggered by the sample request from Professor Y.Kobayashi at Saitama University with his suggestions.

Regarding the initiation of design technique on filters, we owe to Dr. Y. Konishi. These works were the fruits of all the related engineers' cooperation in Murata MFG Co., Ltd.. Dr. Toshio Nishikawa. Dr. Youhei Ishikawa. Dr. Haruo Matsumoto. Dr. Hiroshi Tamura, Dr. Kenji Tanaka. and others.

The research and development of SAW devices in the company of Murata were advised by Professor Akira Kawabata and Dr. Tadashi Shiosaki at Kyoto University, and were proceeded by the group of Dr. Satoru Fujishima, Mr. Seiichi Arai. Dr. Michio Kadota. Mr. Eiji Ieki. Mr. Tohru Kasnami. Mr. Yukio Yoshino and others.

I would like to express my hearty thanks to all of these gentlemen.

References

[1] R.D.Richtmyer, "Dielectric Resonators," J. Appl. Phys., vol., 15. pp., 391-398 (1939)

[2] B.D.Silverman, "Microwave absorption in Cubic Strontium Titanate," Phys. Rev., vol., 125, pp., 1921-30 (1962)

[3] W.G.Spitzer, R.C.Miller, D.A.Kleinman, and L.E.Howarth, "Far Infrared Dielectric Dispersion in BaTiO₃, SrTiO₃, and TiO₂," Phys. Rev., vol., 126, pp., 1710-21 (1962)

[4] A.Okaya and L.F.Barash, "The Dielectric Microwave Resonator." Proc. IRE, vol., 50, pp., 2081-2092, October (1962)

[5] B.W.Hakki and P.D.Coleman, " A Dielectric Resonator Method of Measuring Inductive Capacitance in the Millimeter Range," IRE Trans. on MTT, MTT-8, pp., 402-410 (1960)

[6] Y.Kobayashi and M.Katoh, "Microwave Measurement of Dielectric Properties of Low-Loss Materials by the Dielectric Resonator Method," IEEE Trans. on MTT, MTT-33, pp., 586-92 (1985)

[7] S.B.Cohn, "Microwave Filters containing High-Q Dielectric Resonators." G-MTT Symposium Digest, pp., 49-53. (1965) ; "Microwave Bandpass Filters containing High-Q Dielectric Resonators," IEEE Trans. on MTT, MTT-16, n-227 (1968)

[8] Y.Konishi, "Microwave Dielectric Resonator," Tech. Report of NHK (Nippon Hohsoh Kyoukai), pp., 111-117 (1971) (in Japanese)

[9] J.K.Plourde, "Temperature Stable Microwave Dielectric Resonators Utilizing Ferroelectrics," IEEE MTT-S Digest, pp.,202-205(1973)

[10] K.Wakino, T.Nishikawa, S.Tamura and Y.Ishikawa, "Microwave Bandpass Filters Containing Dielectric Resonator with Improved Temperature Stability and Spurious Response," IEEE MTT-S. Int, Microwave Symp. Dig., pp., 63-65 (1975)

[11] K.Wakino, M.Katsube, H.Tamura, T.Nishikawa, and Y.Ishikawa, "Microwave Dielectric Materials," Joint Convention Record for Four Institute of Electrical Engineers Japan, No., 235 (1977) (in Japanese)

[12] H.Tamura, D.A.Sagala, and K.Wakino, "High-Q Dielectric Resonator Material for Millimeter-Wave Frequency," Proc. of the 3rd US-Japan Seminar on Dielectric and Piezoelectric Ceramics,

Dielectric Ceramic Materials

pp., 69-72 (1986)

[13] S.Kawashima, M.Nishida, I.Ueda, and H.Ouchi, "Ba(Zn,Ta)O₃ Ceramic with Low Dielectric Loss," J. Am. Ceram. Soc., vol., pp., 66, 421-23 (1983)

[14] H.Tamura, T.Konoike, and K.Wakino, "Improved High-Q Dielectric Resonator with Complex Perovskite Structure," J. Am. Ceram. Soc., vol., 67, C-59-61 (1984)

[15] K.Wakino, K.Minai, and H.Tamura, "Microwave Characteristics of (Zr,Sn)TiO₄ and BaO-PbO-Nd₂O₃-TiO₂ Dielectric Resonator," J. Am. Ceram. Soc., vol., pp., 67, 278-81 (1984)

[16] H.M.O'Bryan,Jr., J.Thomson,Jr., and J.K.Plourde, "A New BaO-TiO₂ Compound with Temperature-Stable High Permittivity and Low Microwave Loss," J. Am. Ceram. Soc., vol., 57, pp., 450-53 (1974)

[17] K.Wakino, D.A.Sagala, and H.Tamura, "Far Infrared Reflection Spectra of Ba(Zn,Ta)O₃-BaZrO₃ Dielectric Resonator Material," Proc. of 6th IMF: Jpn. J. Appl. Phys., vol., 24 Suppl., 24-2, pp., 1042-44 (1985) ; H.Tamura, D.A.Sagala, and K.Wakino, "Lattice Vibration of Ba(Zn₁/₃Ta₂/₃)O₃ Crystal with Ordered Perovskite Structure," Jpn. J. Appl. Phys., vol., 25 [6] pp., 787-91 (1986)

[18] D.L.Rebsch, D.C.Webb, R.A.Moore and J.D.Cowlishaw, "A Mode Chart for accurate Design of Cylindrical Dielectric resonators," IEEE Trans., MTT-S, MTT-13, pp. 468-469, (1965)

[19] W.E.Courtney, "Analysis and Evaluation of a Method of Measuring the Complex Permittivity and permeability of Microwave Insulators," IEEE Trans., MTT-S, MTT-18, pp. 476-485 (1970)

[20] Y.Kobayashi and S.Tanaka, "Resonant Mode of a Dielectric Rod Resonator short-circuited at both Ends by parallel conducting Plates," IEEE Trans. vol., MTT-28, 10, pp., 1077-1085 (1980)

[21] K.A.Zaki and A.E.Atia, "Modes in Dielectric-loaded Waveguides and Resonators," IEEE Trans., vol., MTT-31, 12, pp., 1039-1045 (1983); K.A.Zaki and C.Chen, "New Results in Dielectric-loaded Resonators," IEEE Trans., vol., MTT-34, 7, pp., 815-824 (1986)

[22] A.E.Williams and A.E.Atia, "Dual-mode Canonical Waveguide Filters," IEEE Trans., vol., MTT-25, No. 12 (1977)

[23] Y.Ishikawa, J.Hattori, M.Andoh and T.Nishikawa, "800 MHz high Power Duplexer using TM Dual-Mode Dielectric Resonators," IEEE MTT-S Dig., II-3, pp. 11617-1620 (1992); T.Nishikawa, K.Wakino, H.Wada and Y.Ishikawa, "800 MHz Band Dielectric channel dropping Filter using TM₁₁₀ triple mode Resonator," IEEE MTT-S Dig., K-5, pp. 289-292 (1985); Y.Ishikawa, J.Hattori, M.Andoh and T.Nishikawa, "800 MHz high power Bandpass Filter using TM dual Mode Dielectric Resonators," 21st Europ. Microwave Conf. Proc., vol. 2 pp. 1047-1052 (1991)

[24] S.J.Fiedziuszko, "Dual-mode Dielectric Resonator loaded Cavity Filters" IEEE Trans., col., MTT-30, pp., 1311-1316 (1982)

[25] S.Komatsu and Y.Kobayashi, "Design of Bandpass Filters using Triple-Mode Dielectric Rod Resonators," IECIE Japan, C-1, pp. 96-103 (1995)

[26] K.Wakino, T.Nishikawa, H.Matsumoto and Y.Ishikawa, "Miniaturized Bandpass Filters using Half Wave Dielectric Resonators with Improved Spurious Response," IEEE MTT-S, Cat.

No. 78CH1355-7, pp. 230-232 (1978)

[27] K.Wakino, T.Nishikawa, H.Matsumoto and Y.Ishikawa, "Quarter Wave Dielectric Transmission Line Duplexer for Land Mobile Communications," IEEE MTT-S, Cat. No. 79CH1439, pp. 278-280 (1979)

[28] K.Wakino, T.Nishikawa and Y.Ishikawa, "Miniaturized Duplexer for Land Mobile Communication High Dielectric Ceramics," IEEE MTT-S, Cat. No. 0149-645X81, pp. 185-187 (1981)

[29] K.Wakino, "High Frequency Dielectrics and Their Applications," Proc. IEEE Intnl. Symp. on Application of Ferroelectrics, pp. 97-106 (1986); H.Tamura, H.Matsumoto, and K.Wakino, "Low Temperature Properties of Microwave Materials," Jpn. J. Appl. Phys., vol., 28, Suppl., 28-2, pp., 21-23 (1989); K.Wakino, "Recent Development of Dielectric Resonator Materials and Filters in Japan," Proc. of The Second Sendai Int. Conference, YAGI Symposium on Advanced Technology Bridging the Gap between Light and Microwaves, pp. 187-196 (1990)

$M_{(1-x)}Ca_xMnO_3$ (M=La, Y) THIN FILMS FOR HIGH TEMPERATURE THIN FILM THERMOCOUPLE APPLICATIONS

Ramakrishna Vedula, Hemanshu D. Bhatt and Seshu B. Desu
Materials Science and Engineering Department,
Virginia Polytechnic Institute and State University, Blacksburg, VA 24061-0237

Gustave C. Fralick,
NASA Lewis Research Center, 21000 Brookpark Road, Cleveland, OH 44135

ABSTRACT
    In this study, rare earth manganates have been investigated for high temperature thin film thermocouple applications. Thin films of $La_{(1-x)}Ca_xMnO_3$ and $Y_{(1-x)}Ca_xMnO_3$ were deposited on to sapphire substrates using pulsed laser ablation technique and annealed at temperatures ranging from 1173 K to 1573 K in air. Their thermal stability was studied in terms of phase, microstructure and composition using x-ray diffraction, atomic force microscopy and electron spectroscopy for chemical analysis respectively. It was observed that manganate thin films were stable up to 1373 K temperatures above which, breakdown in chemical composition occurs. XRD studies also indicate that these materials exist as a single phase perovskite structure up to 1373 K beyond which mixed phases of individual oxides were detected. Electrical conductivity and seebeck coefficients were measured in-situ as a function of temperature up to 1023 K. Electrical conductivity results indicated that these manganate thin films conduct using a thermally activated small polaron hopping conduction mechanism. Stable and fairly high seebeck coefficients along with high electrical conductivity indicate that these materials offer significant improvements over conventional Pt/PtRh thermocouples and hence, hold excellent promise for high temperature thin film thermocouple applications.

INTRODUCTION
    The thermocouple systems used in high temperature applications such as advanced aerospace propulsion systems have to function in rapidly fluctuating and extremely high heat fluxes. Also, they have to withstand severe environments corresponding to high temperatures (in excess of 1400 K) and high speed flows[1]. Another important constraint lies in the fact that the thermocouple materials must be compatible with ceramic (such as sapphire) and/or metal matrix composites and other nonmetallic materials in most cases, and operate under extremely hostile environments with respect to temperature, pressure, flow and cycling. Thin film thermocouples have been demonstrated to possess several advantages arising

mainly due to their small size (1 micron thick and less than 1 mm wide). They offer excellent spatial resolution, low cost in large volume production, extremely fast response and minimal disturbance to the field [2-5]. However, the bottleneck lies in finding suitable thin film thermocouple materials which can withstand the above mentioned temperature/heat flux/pressure fields, and at the same time easy to fabricate in the thin film form. Traditionally, Platinum/Platinum-Rhodium based thin film thermocouples have been used for surface temperature measurements. However, recent studies indicated several problems associated with these thermocouples at temperatures exceeding 1000 K, some of which include poor adhesion to the substrate, rhodium oxidation, reaction with the substrate at high temperatures and coalescing of the films due to electromigration resulting in electrical discontinuity at higher temperatures[6].

Several nonmetallic materials were investigated for this purpose, some of which are carbides, borides, silicides and nitrides. These refractory ceramics have high melting points and also possess excellent electrical conductivities and Seebeck coefficients at high temperatures. Most of these materials however, exhibit poor oxidation resistance at high temperatures even in vacuum. For instance, recent studies from our group indicated that carbide (TiC and TaC) thin films oxidize completely at 1200K, even under pressures less than 10^{-4} torr resulting in the thermocouple breakdown[7]. Silicide thin films ($ReSi_2$ and WSi_2) have also been found to oxidize completely at temperatures above 1100 K[8].

The important criteria of selection of candidate materials are chemical stability and phase stability, stable and fairly large electrical conductivity and seebeck coefficient, resistance to oxidation in the entire temperature range of operation and easy processibility in the thin film form. Previously mentioned candidate materials have demonstrated the absence of at least one of the above mentioned properties. Earlier studies for different applications showed that some of the mixed oxide families (perovskite type) have good chemical and phase stability in both oxidizing and reducing atmospheres, good thin film processibilities, and fairly large electrical conductivities. Some cobaltites, manganates and chromites have demonstrated large seebeck coefficients and electrical conductivities in the bulk form throughout the temperature range of interest.

$La_{(1-x)}M_xMnO_3$ (M=Ca, Sr, Ba) type compounds have been studied with renewed interest ever since the discovery of giant magnetoresistance effects in these perovskite type oxides. These compounds have shown unusually large negative magnetoresistance at low temperatures, commonly referred to as giant magnetoresistance. In all these perovskite oxide systems a mixed valence ion state (+3/+4) in the Mn ion is created for charge compensation resulting in the creation of a "polaron". The transport of polarons has been extensively studied in the investigation of giant or colossal magnetoresistance in these compounds[10,11].

Apart from the giant magnetoresistive effects in these perovskite oxides, $La_{(1-x)}Ca_xMnO_3$ and $Y_{(1-x)}Ca_xMnO_3$ have also been studied as candidate cathode materials for solid oxide fuel cells (SOFCs)[12,13]. Their oxidation-reduction behaviour, electrical resistivity and thermoelectric properties have been studied in the bulk form both at low temperatures as well as at high temperatures by several researchers for the above mentioned applications. Although substantial amount of information on these perovskite oxides is available in the bulk form, there have been few reports in thin film form.

The purpose of this study is to investigate the high temperature thermal stability, electrical resistivity and seebeck coefficient as a function of temperature for the rare earth metal doped $CaMnO_3$ systems. An attempt is made to identify the optimum composition from each of the Y and La doped $CaMnO_3$ systems from the point of view of high temperature thin film thermocouple applications. A systematic study of stability of phase, structure, composition and morphology of these perovskite oxides is presented. The temperature dependence of electrical resistivity as well as seebeck coefficients is also reported.

EXPERIMENTAL PROCEDURE
Thin Film Deposition
Powders of La_2O_3, $CaCO_3$ and $MnCO_3$ (99.99% purity) in the required stoichiometric ratios were mixed thoroughly in ethanol solution and the resulting mixture dried overnight at 90 °C. After drying, the powder was then calcined at 900 °C for 3 hours. A bulk sample was made by compacting the powder at 5 Mpa, and then sintering at 1350 °C for 3 hours in air. This bulk sample was used as the target in PLD. A KrF_2 (248 nm) excimer laser was used to ablate the target. The substrate (sapphire) was heated to 650 °C during deposition. The laser beam was collimated and focused onto the target. Both the target and the substrate were rotated to ensure uniform film deposition on the substrate. The energy of the laser used was 700 mJ, at a reprate frequency of 30 Hz. the deposition time was 25 mins. Deposition was done in oxygen ambience and the pressure inside the chamber was maintained at 400 mtorr.

Characterization
After deposition, the thickness of the films was measured by a Dektak profilometer. Then, X-ray diffraction (XRD) studies were conducted on all the films using a Scintag diffractometer. After the phase was studied, composition was analyzed using Electron spectroscopy for chemical analysis (ESCA). Then, the grain size, microstructure and surface roughness were characterized by Atomic force microscopy (AFM). Electrical resistivity and seebeck coefficient were measured in-situ in air as a function of temperature.

Electrical Resistivity And Seebeck Coefficient Measurement

From the samples deposited as described earlier, resistors of width 4 to 4.5 mm were cut. These resistors were pasted on electronic grade alumina samples. The height of the samples was about 12 mm and Platinum/Platinum-10 %(by weight) Rhodium thermocouples were used to measure the thermoemf of the film. Each thermocouple was inserted in an alumina tube having two isolated holes for the two thermocouple legs. The two thermocouples were placed one on top of the other 8 mm distant from each other. They were fixed to a stainless steel holder which could be slid length wise. The whole system was spring loaded to ensure contact with the resistor. Fig.1 shows the principle of measurement of thin film seebeck coefficient of manganates.

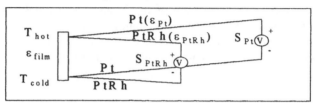

Fig. 1 Principle of measurement of thin film seebeck coefficients

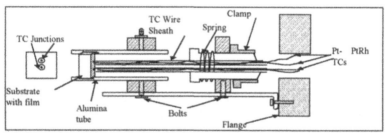

Fig. 2 Experimental setup for high temperature thermoelectric and electrical conductivity measurement system.

The resistance of the sample was measured by a Keithley 2000 digital multimeter. The top thermocouple was insulated during Seebeck coefficient measurement using fiber glass to incorporate a temperature difference of about 10-15 °C. The voltages generated between the Pt and Pt-Rh legs of the thermocouple and the temperatures of the hot and cold junctions were measured using a Tempbook 7-channel thermocouple readout system. The whole system was inserted in a furnace which has a built in temperature controller system. From the data collected, electrical resistivity and Seebeck coefficients were plotted as a function of temperature. The equations employed to evaluate the Seebeck coefficient are listed below

$$S_{Pt} = -\varepsilon_{Pt}\left(T_h - T_c\right) + \varepsilon_{film}\left(T_h - T_c\right)$$

$$S_{PtRh} = -\varepsilon_{PtRh}\left(T_h - T_c\right) + \varepsilon_{film}\left(T_h - T_c\right)$$

$$\varepsilon_{film} = \frac{S_{Pt}}{\left(T_h - T_c\right)} + \varepsilon_{Pt} = \frac{S_{PtRh}}{\left(T_h - T_c\right)} + \varepsilon_{film}$$

$$\Delta T_{theoretical} = \left(T_h - T_c\right) = \frac{S_{Pt} - S_{PtRh}}{\varepsilon_{PtRh} - \varepsilon_{Pt}}$$

where, S_{Pt} and S_{PtRh} correspond to the voltage drops in the Pt and PtRh legs of the two thermocouples respectively, ε_{film}, ε_{Pt} and ε_{PtRh} are the Seebeck coefficients of the film, Pt and PtRh respectively and T_h and T_c are the temperatures of the hot and cold junctions respectively.

RESULTS AND DISCUSSION
I. Phase, Composition And Microstructure

The thickness of the deposited films as measured using a Dektak profilometer was 6000 $^{\circ}$A.

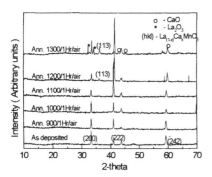 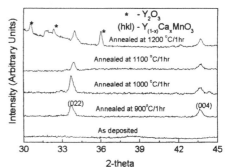

Fig.3: XRD of $La_{0.5}Ca_{0.5}MnO_3$ Fig.4: XRD of $Y_{0.5}Ca_{0.5}MnO_3$

The thickness of the films after annealing at various temperatures remains unchanged as observed from Auger depth profiles. The phase of the films was studied by X-ray diffraction (XRD) using Cu-Kα radiation as a function of annealing temperature to observe the phase stability of the thin films. All the films were annealed in air in conventional annealing furnaces. Figs 3 and 4 show the XRD patterns of $La_{0.5}Ca_{0.5}MnO_3$ and $Y_{0.5}Ca_{0.5}MnO_3$ thin films as a function of annealing temperature.

Earlier studies on $Y_{(1-x)}Ca_xMnO_3$ in the bulk form by several researchers have indicated that, $YMnO_3$ has two polymorphs viz., orthorhombic perovskite and hexagonal nonperovskite structures. The former is found to be stable at high pressures and high temperatures whereas at lower temperatures and normal pressures, the nonperovskite phase is dominant[14, 15]. Stevenson et al explained the nonstability of $YMnO_3$ on the basis of the ionic radius and the coordination number of the Y 3+ ion in the two structures[16]. In the hexagonal structure, the coordination number of Y is smaller than in the case of orthorhombic structure, thereby increasing its stability. They also observed the presence of a miscibility gap in the $YMnO_3$ and $CaMnO_3$ systems for the compositions of $Y_{(1-x)}Ca_xMnO_3$ 0<x<0.3. In our case, as the composition of the films was beyond the miscibility gap, it was observed that the films both as-deposited as well as annealed at temperatures below 1100 °C has a stable orthorhombic perovskite structure. XRD data from the films annealed 1200 °C and 1300 °C however, showed extraneous peaks which were identified as individual oxides such as La_2O_3, CaO and Y_2O_3. This indicates that the perovskite phase is not stable beyond 1100 °C and hence these materials cannot be used for thin film thermocouple applications involving temperatures greater than 1100 °C. Studies on the thin film composition using electron spectroscopy for chemical analysis (ESCA) also confirmed that these materials are stable upto 1100 °C, temperatures above which, there was an observable loss of manganese from the system. Microstructural studies on the films using atomic force microscopy (AFM) showed a progressive increase in the grain size upto 1100 °C (grain sizes ranging from 30nm to 75nm).

One of the primary objectives of this study was to identify the optimum composition in each of the $Y_{(1-x)}Ca_xMnO_3$ and $La_{(1-x)}Ca_xMnO_3$ systems. From the point of view of thermal stability, it was observed that in the $La_{(1-x)}Ca_xMnO_3$ as well as $Y_{(1-x)}Ca_xMnO_3$ systems, compositions with x=0.5 possesses better phase stability than those with x=0.7 as observed from the XRD patterns and results from composition analyses. There was little difference in the microstructure of the films. It can be concluded that the above mentioned compositions of $Y_{(1-x)}Ca_xMnO_3$ and $La_{(1-x)}Ca_xMnO_3$ systems show remarkable stability in phase, structure, compoaition and microstructure up to 1100 °C and hence show great promise for thin film thermoelectric device applications.

II. Electrical Conductivity

The as deposited thin films were annealed at 1000 °C for 1 hour in air prior to the measurement of electrical resistivities to avoid any grain growth effects. Electrical resistivity of the films was measured using both Pt as well as the PtRh legs of the thermocouples in the experimental setup explained earlier. From the resistivities, electrical conductivites were calculated using the specimen geometry and plotted as a function of temperature as shown in Fig. 5. All the films show fairly large electrical conductivities and the linearity of the plots of $Ln(\sigma T)$ vs 1/T indicates that electrical conduction in these materials occurs through a thermally activated small polaron conduction mechanism consistent with the results from literature. Although increase in the dopant concentration should result in an increase in the electrical conductivity, experimental results indicate little or no difference in the electrical conductivities of the samples. At this point the reason for such behaviour is unclear, however, it could be speculated that the films differ in the oxygen stoichiometry resulting in a difference in the oxygen vacancy concentrations, although such a difference in the oxygen concentration in the films could not be detected from ESCA studies. A difference in the concentrations of the oxygen vacancies results in a change in the charge carrier concentration thereby accounting for a change in the electrical conductivity of the films. Several researchers have postulated that for materials in which conduction occurs through small polaron hopping mechanism, it electrical conductivity (σ) follows the relation[17]:

$$\sigma = \frac{(1 - c)}{kT} n e^2 a_0 \gamma \exp\left(\frac{-E_a}{kT}\right)$$

where, c-fraction of sites occupied by polarons, Ea-activation energy of hopping, n-small polaron concentration, e-electron charge, a₀-distance between sites, γ-optical phonon frequency, z-number of nearest neighbors and k-Boltzmann constant.

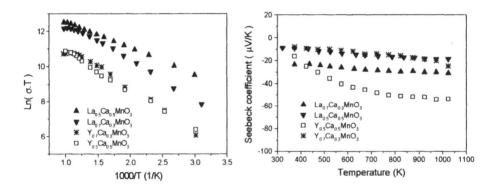

Fig.5 Electrical conductivities and Fig.6 Seebeck coefficients of manganate thin films

From fig. 5 it is clear that the charge transport occurs through a thermally activated small polaron hopping mechanism consistent with other reports in literature. The slope of the plot gives the activation energy of hopping (E_a). Table .1 shows the activation energy of hopping as a function of amount of Y or La doping. These values are consistent with earlier reports of Stevenson et. al[16] of 0.19 eV and 0.1 eV for $Y_{0.7}Ca_{0.3}MnO_3$ and $Y_{0.5}Ca_{0.5}MnO_3$ respectively and several other researchers[18, 19]. This also confirms that electrical conduction occurs predominantly through thermally activated small polaron hopping conduction mechanism.

Table. 1: The Activation energy of hopping and the fraction of available sites for hopping that are occupied for $M_{(1-x)}Ca_xMnO_3$ (M = La,Y), x= 0.5, 0.3.

Material	E_a (eV)	C
$La_{0.7}Ca_{0.3}MnO_3$	0.18	0.55
$La_{0.5}Ca_{0.5}MnO_3$	0.128	0.589
$Y_{0.7}Ca_{0.3}MnO_3$	0.196	0.554
$Y_{0.5}Ca_{0.5}MnO_3$	0.164	0.651

In addition to activation energy, table. 1 also shows the calculated values of fraction of available hopping sites that are occupied (C). The details of calculation of C are available in literature and hence, not reported here[17]. From the table, it is clear that as the Y doping increases from 0.5 to 0.7, the fraction of available sites for hopping that are occupied decreases.III Seebeck Coefficient

Therefore, a linear plot of Ln(σT) vs 1/T indicates that conduction occurs predominantly through small polaron hopping as is the case with our films. An important aspect to note, from the data is that there is no observable deviation in the linearity of the curves, indicating no change in the mechanism of conduction throughout the temperature range of measurement. This is a significant factor in choosing candidate materials for thin film thermocouple applications indicating that these materials are excellent candidates for the thin film thermoelectric device applications.

The seebeck coefficient of the films was measured in-situ as a function of temperature simultaneously with the electrical conductivity measurement. The temperature differential between the hot and cold junctions during the measurement was maintained at around 10-15 °C. The seebeck coefficient was calculated from the thermoemf based on the temperature differential measured. Fig. 6 shows the seebeck coefficients of $Y_{(1-x)}Ca_xMnO_3$ and $La_{(1-x)}Ca_xMnO_3$ systems as a function of temperature. It can be seen that all the films showed a negative seebeck coefficient and there is little dependence of the seebeck coefficients of the films with temperature. It is well established in literature that the sign of the seebeck coefficient indicates the nature of the charge carriers (electrons or holes). In our case, electrons appear to be the predominant charge carriers. This can be explained as follows: $CaMnO_3$ is the host perovskite and when the A site in the perovskite (Ca +2 valence) is doped with La or Y (+3 valence state donors), the valence of the Mn ion (+4 in $CaMnO_3$) changes to +3 for charge compensation. This results in the formation of a localised charge carrier, an electron, commonly referred to in such cases as a polaron.

Several researchers have identified the temperature dependence of seebeck coefficient(α) of materials with predominantly small polaron hopping conduction mechanism as follows[17].

$$\alpha = \left(\frac{k}{e}\right) \ln \left[\frac{2(1-c)}{c}\right] + \left(\frac{k}{e}\right) \frac{zJ^2 k}{E_B^3} T$$

where, c-fraction of sites occupied by polarons, Ea-activation energy of hopping, e-electron charge, z-number of nearest neighbors, J-intersite transfer energy during hopping process, E_B-binding energy of polaron and k-Boltzmann constant.

Fig. 6 shows that these materials have stable and fairly large seebeck coefficients in thin film form throughout the temperature range of measurement and hence show excellent promise for thin film thermoelectric device applications. Also noteworthy from Fig. 6 is the fact that among the two compositions in the systems $M_{(1-x)}Ca_xMnO_3$ (M=La, Y) studied, x=0.5 shows a higher seebeck coefficient value compared to x = 0.3. This is important because, the absolute value of the seebeck coefficient determines the output of the thin film thermocouple fabricated from these materials and the higher the seebeck coefficient, the greater the thermocouple output and the higher the sensitivity.

CONCLUSIONS

From this study, the following conclusions can be drawn.

(1): Both $Y_{(1-x)}Ca_xMnO_3$ and $La_{(1-x)}Ca_xMnO_3$ thin films crystallize in a single phase orthorhombic perovskite structure. This phase is stable up to annealing temperatures of 1100 °C, above which, the perovskite structure is no longer stable. Loss of manganese from the systems results in the formation of individual oxides of La, Y and Ca. This limits the temperature range of operation of thermoelectric devices based on these materials up to 1100 °C.

(2): Electrical conductivity and seebeck coefficient studies indicate that these materials have fairly large and stable conductivities and seebeck coefficients throughout the temperature range of measurement. The linearity of the Ln(σT) vs 1/T plot indicates that small polaron hopping conduction mechanism is the predominant conduction mechanism in these perovskite oxides.

(3) From the phase, composition analyses, as well as from the seebeck coefficient and electrical conductivity measurements, it is clear that for both Y(1-

$_{x)}Ca_XMnO_3$ and $La_{(1-x)}Ca_XMnO_3$ systems, $x = 0.5$ is the optimum composition.

Both these materials exhibit significant improvements over Pt/PtRh thermocouples in terms of thermal stability and substrate compatibility and hence, show excellent potential for use in high temperature thin film thermocouple applications.

REFERENCES

1: Terrel J.P, Hager J. M, Onishi S and Diller T. E., NASA Conf. Pub., 69-80 (1992).

2: E. A. Johnson and L. Harris, Phys. Rev., 44 (1933), 944-945.

3: R. Marshall, L. Atlas and T. Putner., J. Sci. Instrum., 43 (1966), 144-149.

4: T. C. Kuo, J. Flattery, P. K. Ghosh and P. K. Kornreich, J. Vac. Sci. Tech., A, 6 (1988), 1150-1152.

5: K. G. Kreider, J. Vac. Sci. Tech., A, 11 (4), (1993), 1401-1405.

6: A. S. Darling and G. L. Selman, *Some Effects of Environment on the Performance of Noble Metal Thermocouples in Temperature*, Soc. America, PA., (1972) Vol.   4., 1633.

7: R. Vedula, H. D. Bhatt, S. B. Desu and G. C. Fralick, "*TiC/TaC Thinfilm Thermocouples*" submitted to Thin Solid Films.

8: K. G. Kreider, Mat. Res. Sym. Proc. Vol. 322., (1994) 285.

9: H. D. Bhatt, R. Vedula, S. B. Desu and G. C. Fralick, "*$La_{(1-x)}Sr_xCoO_3$ for Thinfilm Thermocouple Applications*" submitted to Thin Solid Films.

10: B. Chen, C. Uher, D. T. Morelli, J. V. Mantese, A. M. Mance and A. L. Micheli, Physical Review B., Vol. 53, No. 9 (1996),5094.

11: M. Jaime, M. B. Salamon, K. Pettit, M. Rubinstein, R. E. Treece, J. S. Horwitz and D. B. Chrisey, Appl. Phys. Lett., 68(11), 1996, 1576.

12: B. C. H. Steele, MRS Bulletin, 1989, p. 19.

13: A. Hammouche, E. J. L. Schouler and M. Henault, Solid State Ionics, 28-30, 1205 (1988).

14: H. L. Yakel, W. C. Koehler, E. F. Bertaut and E. F. Forrat, Acta Crystallogr., 16 (1963), 957.

15: V. Wood, A. Austin, E. Collings and K. Brog, J. Phy. Chem. Solids 34, (1973), 859.

16: J. W. Stevenson, M. M. Nasrallah, H. U. Anderson and D. M. Sparlin, J. Solid State Chem., 102 (1993), 175.

17: F. T. J. Smith, J. Appl. Phys.,41 (10), 4227.

18: R. Raffaelle, H. U. Anderson, D. M. Sparlin and P. E. Parris, Phys. Rev. B 43(1991), 7991.

19: D. P. Karim and A. T. Aldred, Phys. Rev. B 20(1979),2255.

BOROSILICATE GLASS-COATED STAINLESS STEEL ELECTROSTATIC CHUCKS FOR WAFER HANDLING SYSTEM

Jesus Noel Calata, Sihua Wen and Guo-Quan Lu

Department of Materials Science and Engineering
Virginia Polytechnic Institute and State University
Blacksburg, Virginia 24061, USA

ABSTRACT

Electrostatic chucks (ESC) are important components in semiconductor wafer processing equipment. Using electrostatic force to hold wafers during processing, ESCs reduce wafer bowing and contamination problems which are common in mechanical wafer-handling systems. This study demonstrated the feasibility of using tape casting method followed by sintering to fabricate borosilicate glass-coated stainless steel chucks for low-temperature processes. The glass coatings on the substrates ranged from 100 μm to 150 μm thick. The adhesion of the coating was also excellent such that it was able to withstand moderate drop tests and temperature cycling to over 300°C without spalling. The electrostatic clamping force generally followed the theoretical voltage-squared curve except at elevated temperatures when a deviations were observed to occur at higher applied voltages. Based on these results, we believe that we have a viable technique for manufacturing low-cost electrostatic wafer chucks.

INTRODUCTION

Electrostatic wafer clamping represents an improvement over mechanical clamping devices used in semiconductor processing equipment. A mechanical clamp holds the wafer in place through a clamping ring around the wafer periphery[1,2]. This allows nonreactive gas to be introduced under pressure between the wafer and susceptor for efficient cooling. However, mechanical clamps produce uneven force distribution that causes the wafer to bow, leading to non-uniform temperature distribution and lower heat dissipation. Electrostatic chucks (ESC) do not suffer from this limitation since the electrostatic force is distributed uniformly over the chuck area covered by the electrode, which for practical purposes may be considered to be the entire chuck surface.

Electrostatic chucks have design advantage over mechanical systems mainly because of the absence of moving parts. In its simplest form, It consists of an electrically insulating layer covering an electrically conductive substrate. The elimination of mechanical parts (e.g., clamping ring) minimizes particle contamination and increases the available wafer area. The applied clamping force is sufficiently high to prevent bowing of the wafer and can even be used to flatten bowed wafers. Enhanced contact with the chuck surface also provides a more uniform heat transfer over the entire wafer area. Variations in the design implementation and type of materials can be found in the literature and patents but electrostatic chucks can be classified into three basic configurations, namely : the monopolar, bipolar, and Johnsen-Rahbek (J-R)[1,3]. The monopolar configuration is the simplest and is very much like a parallel plate capacitor with the wafer acting as the second electrode. The plasma serves as a conductor to complete the electrical circuit. Theoretically, it also produces the largest clamping force among the three types. For other applications, variants of the bipolar design are better suited because they do not require the presence of the plasma in order to function. Because of its simple design, the monopolar configuration was used as the basis for fabricating the chucks used in this study.

This research was undertaken to demonstrate the feasibility of using glass coatings on metal substrates for electrostatic wafer clamping. They have lower firing temperatures compared to currently favored materials such as alumina and silica. The fabrication method borrows heavily from the tape casting process that is used in the production of microelectronics packaging substrates and multilayer ceramic capacitors. [4-6].

EXPERIMENTAL PROCEDURE
Materials and Processing
The insulating layer of the electrostatic chucks consisted of a sodium borosilicate glass (BSG) obtained from Sem-Com Co., Inc. (P.O. Box 8428, Toledo, OH 43623). It has a softening point of $710°C$. The substrates were fabricated from an austenitic stainless steel stock. The glass has a lower coefficient of thermal expansion (CTE), $5.6 \times 10^{-6}/°C$, than the steel (about $11 \times 10^{-6}/°C$), such that it will be under compression at any temperature below the firing temperature. The slurry was prepared from the glass powder using a two-stage milling procedure[4,5]. The powder was mixed with fish oil and solvents and milled for 24 hours. The polyvinyl butyral (PVB) binder and plasticizer were then added and the mixture was milled for another 24 hours. The deaired slurry was cast on the substrates using a doctor blade to form dried green films roughly 150 to 250 μm thick. The films were allowed to dry in air for 24 hours before the coated chucks were trimmed. The dried coatings had an initial relative density of 60% of the

Dielectric Ceramic Materials

theoretical density of 2.32. The stainless steel substrates had to be pre-oxidized above 500°C before being coated with the glass slurry to improve adhesion. The BSG-coated substrates were fired in air at 710°C for 3 hours after binder burnout at 500°C. The chucks were subsequently polished with diamond paste to improve the flatness of the chuck and eliminate much of the surface roughness. The insulating layer of the test chuck had a thickness of 125 μm after final polishing with 1 μm diamond paste.

Clamping Force Measurements

Clamping force measurements were performed using a custom-built vacuum chamber enclosing a disk heater and a load cell attached to a motor-driven micrometer screw as shown in Fig. 1. The glass-coated substrate is mounted directly on the disk heater with a hollow screw. A thermocouple is inserted into the substrate through the screw hole such that the tip is as close as possible to the dielectric layer. The metal substrate serves as the bottom electrode and a rigid alumina plate electroless-plated with a thin layer of copper acts as the top electrode. A machinable ceramic rod attached to the load cell is linked to the top electrode through rotatable joint to enable the electrode to be seated properly on the chuck surface. The voltage is applied to the top electrode as shown schematically. Data acquisition was done automatically by a personal computer equipped with a data acquisition card. In a typical run, the electrode is lowered onto the chuck and charged at a set voltage for 5 minutes. After charging, the electrode is slowly lifted until it breaks free from the electrostatic force developed in the chuck.

RESULTS AND DISCUSSION
Coating-to-substrate Adhesion

Good adhesion of the glass coating on the substrate is essential to the reliability and life of the electrostatic chuck and it is for this reason that the stainless steel substrate was pre-oxidized. A good discussion of the mechanism involved is described elsewhere[7]. In the case of stainless steel, the native oxide layer provides a chemical bond that maintains the continuity of the electronic and atomic structure across the interface. This approach is not desirable if the oxide layer is not strongly bonded to the base metal. In stainless steel, the chromium on the surface is readily oxidized although an added pre-oxidation seemed to enhance adhesion. Excessive oxidation is not desirable either, as was observed along the edges of the coating, which resulted in weaker adhesion. This can be remedied by switching to a non-oxidizing atmosphere during the sintering stage.

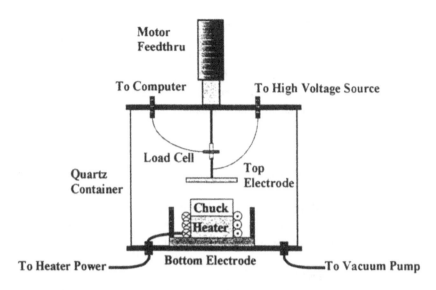

Figure 1. Schematic diagram of the apparatus for clamping force measurement.

Clamping Force

Clamping force measurements were conducted at room temperature, 100°C and 150°C at applied voltages up to 550°C. The results, plotted as clamping pressure values, are shown in Fig. 2. The superimposed dashed lines are non-linear least square fits based on a more realistic model[8,9] that includes an air gap between the electrode and the insulating layer. The model can be represented by an equation of the form

$$P = \frac{\varepsilon_o V^2}{2\left(\dfrac{h_d}{k} + h_{gap}\right)^2}$$

(1)

where P is the clamping pressure and V is the applied voltage. ε_o is the permittivity of a vacuum; k is the relative dielectric constant of the coating; h_d and h_{gap} are the coating thickness and airgap, respectively. For the room temperature measurements, the clamping pressure closely followed the V^2 relationship as predicted by Eq. 1. The average airgap obtained by curve-fitting ranged from about 1.3 μm at room temperature to 2.5 at 150°C. The increase in the airgap with increasing temperature, roughly 1 μm/100°C, was caused to the relatively large CTE mismatch between the glass and the substrate. Measurements obtained at higher temperatures revealed some interesting results. At 100°C, the clamping

Dielectric Ceramic Materials

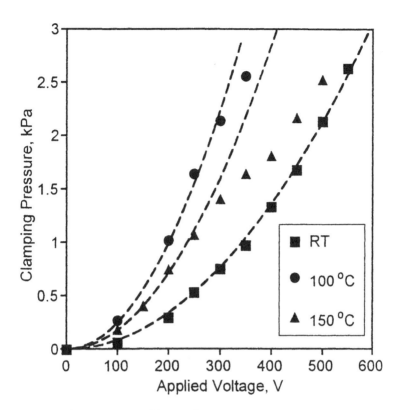

Figure 2. Clamping pressure of borosilicate glass/stainless steel chuck at various
temperatures.

pressure was consistently higher than at room temperature for all voltages. This
trend is similar to the results obtained from glass-ceramic coatings on metal
substrates[10] which was attributed to increased flatness of the dielectric and
higher dielectric constant at elevated temperature. However, in this instance, the
insulating layer was already polished flat such that the increase can only be
accounted for in Eq. 1 by an increase in the dielectric constant. The data also
followed the pressure-voltage relationship although it showed a hint of deviation
beginning at 300 volts. At 150°C, the trend was reversed and the clamping
pressure was lower than at 100°C, though still clearly higher than at room
temperature. The deviation was also distinct above 250 volts and moved quickly
towards the room temperature curve. We think that the deviation is due to
increased leakage current at higher temperature and applied voltage as the
electrical resistivity drops. Insufficient charging time, although not very likely, is
also a possibility under such conditions. The resistivity of the glass drops below

10^{10} ohm-cm at 150°C and below 10^{8} ohm-cm at 200°C, as shown in Fig. 3, such that it's useful operating temperature should be no higher than 200°C. Replacement of the glass with a high-expansion and zero or low-alkali composition should increase the useful temperature range.

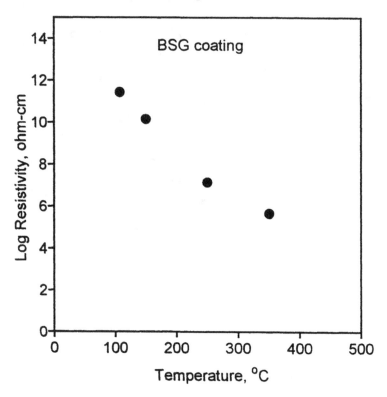

Figure 3. Electrical resistivity of the borosilicate glass coating.

CONCLUSIONS

Electrostatic chucks were successfully produced by casting glass slurry on stainless steel substrates followed by sintering at a temperature well under 1000°C. The borosilicate glass/stainless steel chuck has a maximum operating temperature of around 200°C making it a suitable alternative for low temperature chucking applications. Measurements showed an overall increase in clamping pressure above room temperature that may be the effect of increased surface flatness and dielectric constant. Deviations from the clamping pressure-applied voltage relationship also occurred above room temperature at higher applied voltages. We think that this is a result of increased leakage current due to decreasing resistivity of the insulating layer.

ACKNOWLEDGMENT

This project was funded by Triad Investors Corp.and was also partially supported by the National Science Foundation, under the Faculty Early Career Development (CAREER) program, DMR-9502326.

REFERENCES

[1]J. Field, "Electrostatic wafer clamping for next-generation manufacturing," *Solid State Technology*, 91-98 Sept. 1994.

[2]J.-F. Daviet, L. Peccoud and F. Mondon, "Electrostatic Clamping Applied to Semiconductor Plasma Process I. Theoretical Modeling," *J. Electrochem. Soc.* **140**, 3245-3255 (1993).

[3]L. D. Hartsough, "Electrostatic Wafer Holding," *Solid State Technology*, 87-90, Jan. 1993.

[4]J S. Reed, *Principles of Ceramic Processing*, 2nd ed. Wiley, New York, 1978.

[5]G. Y. Onoda and L. L. Hench (eds.), *Ceramic Processing Before Firing*, Wiley, New York, 1978.

[6]R. R. Tummala, Ceramic Packaging, in *Microelectronics Packaging Handbook*, (R. R. Tummala and E. J. Rymaszewski, eds.), Van Nostrand Reinhold, New York 1989.

[7]G. Partridge, C. A. Elyard, and M. I. Budd, Glass-ceramics in substrate applications, in *Glasses and Glass-Ceramics*, edited by M. H. Lewis (Chapman and Hall, London 1989) Chapter 7.

[8]M. Nakasuji and H. Shimizu, "Low voltage and high speed operating electrostatic wafer chuck," *J. Vac. Sci. Technol.* A **10** [6] 3573-3578 (1992).

[9]Sematech, *Electrostatic Chuck Tutorial, Technology Transfer Number 93031533A-GEN*, Sematech, Inc., 1993.

[10]G-Q. Lu, J. N. Calata, J. Bang, T. Kuhr, A. Amith, *Final Project Report to Triad Investors Corp.*, 1997.

STRONTIUM-LEAD TITANATE BASED COMPOSITE THERMISTOR MATERIALS

Zhilun Gui, Dejun Wang, Jun Qiu, and Longtu Li
Dept. of Mater. Sci.&Eng., State Key Lab. of New Ceramics & Fine Processing
Tsinghua University, Beijing 100084, CHINA

Yttrium-doped $(Sr_{0.45}Pb_{0.55})TiO_3$ ceramics were prepared by liquid chemical method and the conduction mechanism was studied by the complex impedance analysis. The materials exhibited a significantly large negative temperature coefficient of resistivity below Tc in addition to the ordinary PTC characteristics above Tc. The minimum resistivity was in the order of $10^2 \Omega \cdot cm$ and the negative temperature coefficient of resistivity was better than $-3\%^\circ C^{-1}$. Complex impedance analysis indicated that the strong NTC effect did not originate from the deep energy level of donor (bulk behavior), but from the electrical behavior of the grain boundary. Furthermore, the NTC behavior in the materials was dependent on the composition and microstructure of the grain boundary. The NTC-PTC ceramics was a grain boundary controlled material.

I INTRODUCTION

It is well known that perovskite PTC materials consist of semiconducting barium titanate and its solid solutions with strontium titanate and lead titanate,[1] which exhibit a rapidly increasing resistivity when heated above Tc and an almost constant resistivity below Tc. $(Sr,Pb)TiO_3$ ceramics is a new perovskite ferroelectric semiconductor with NTC–PTC (V-type PTC) characteristics,[2,3] which can fulfill multiple functions such as temperature control and overcurrent protection, etc. Therefore this research is of great practical significance.[4] Meanwhile, it has aroused great interest in the interpretation of the strong NTC effect in $(Sr,Pb)TiO_3$ materials.

In the previous literature,[2] it was suggested that the NTC effect was related to the depth of donor energy level and the strong NTC effect was induced by the deep donor energy level, although there is not enough experimental evidence. In this paper, the bulk and grain boundary resistances of the V-type PTC material have been investigated by complex impedance analysis, which is an efficient method to study the grain boundary conduction mechanism of the polycrystalline ceramics.

II EXPERIMENT PROCEDURE

1. Sample Preparation

The samples with formulation of $(Sr_{0.45}Pb_{0.55})TiO_3$ were prepared from strontium nitrate (>99.5%), lead (II) nitrate (>99%), yttrium nitrate (>99%), and tetrabutyl titanate $(Ti(OC_4H_9)_4,>98\%)$. The weighed nitrates and the tetrabutyl titanate were dissolved in a $CH_3COOH-C_2H_5OH-H_2O$ system. The coprecipitates were obtained by adding an excess amount of oxalic acid dihydrate (>99.8%) into the solution. The chemical reaction is described by the following equation:[5]

$$0.45Sr(NO_3)_2 + 0.55Pb(NO_3)_2 + Ti(OC_4H_9)_4 + 2H_2C_2O_4 \cdot 2H_2O + H_2O \rightarrow$$
$$(Sr_{0.45}Pb_{0.55})TiO(C_2O_4)_2 \cdot 4H_2O \downarrow + 4C_4H_9OH + 2HNO_3 \qquad (1)$$

The precipitates were washed, dried, and calcined at 700°C for 45 min in air to obtain $(Sr_{0.45}Pb_{0.55})TiO_3$ powders.

Small amount of ultrafine SiO_2 particles were coated on the surface of $(Sr_{0.45}Pb_{0.55})TiO_3$ powders by surface modification method[6] in order to obtain $SiO_2-(Sr_{0.45}Pb_{0.55})TiO_3$ composite powders.

The $SiO_2-(Sr_{0.45}Pb_{0.55})TiO_3$ composite powders were then pressed into pellets (10mm in diameter and 1mm in thickness) and sintered at 1080 °C~1240°C for 10min~180min to prepare the glass-ceramics composite materials.

2. Measuring Procedure

The surfaces of the $(Sr,Pb)TiO_3$ specimen were coated with In-Ga alloy and the resistance-temperature property was measured with a DC resistance-temperature measuring system, while the applied voltage was generally about 0.5V. The temperature range was 15°C~450°C, the temperature interval of the measurement was 1°C, and the heating rate was 100°C/h.

Complex impedance was measured with a LF impedance analyzer (HP4192A). The frequency range was 5Hz~13MHz, the temperature range was 20°C ~250°C, and the heating rate was 1°C/min.

The microstructure and composition of the materials were analyzed by H9000NAR transmission electron microscope and EDAX/9100 system, respectively.

Thermo-mechanical analysis was performed by SETARAM92-24 TMA analyzer (SETARAM, France). The atmosphere was oxygen and the heating rate was 10°C/min.

III RESULTS AND DISCUSSION
1 Resistance-temperature Characteristics

A series of (Sr,Pb)TiO₃ materials with various Curie temperatures can be obtained by modifying the Sr/Pb ratio. The relation between Sr/Pb ratio and Curie temperature is shown in Fig.1. It can be seen that due to the Pb loss in the sintering process, the experimental results are slightly lower than the theoretical values.

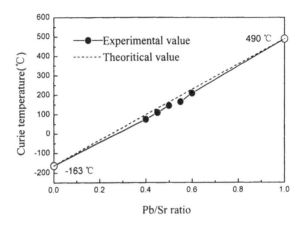

Fig.1 The dependence of curie temperature on Pb/Sr ratio

In the composition design of (Sr,Pb)TiO₃ materials, it is found that both the dopant Y and the additive SiO₂ have remarkable effects on the resistance-temperature characteristics of (Sr,Pb)TiO₃ ceramics. It is an essential condition in the fabrication of the V-PTCR materials that the composition of Si and Y must be higher than those of the typical PTCR material.

By modifying the process parameters, the typical R-T curve of V-PTCR materials is obtained, as shown in Fig.2 as solid line. The room temperature resistance was 7.8kΩ, the lowest resistance was 83.95Ω and the highest resistance was 3.83MΩ. In the PTC range, the resistance jump was 4.7 orders of magnitude and the temperature coefficient of resistance was +6.61%/°C. In the NTC range, the temperature coefficient of resistance was -3.54%/°C. Besides the strong PTC effect above Tc, the sample also presented strong NTC effect below Tc.

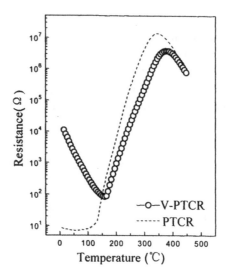

Fig. 2 Resistance-temperature curves of (Sr,Pb)TiO$_3$ ceramics.

2 Bulk and Grain Boundary Resistance

For V-PTCR elements, the simplified equivalent circuit and the Cole-Cole plot are shown in Fig.3 and the corresponding impedance equation (2) is

$$Z' = R_b + \frac{R_{gb}}{1+\omega^2\tau_{gb}^2} \quad ; \quad Z'' = \frac{\omega\tau_{gb}R_{gb}}{1+\omega^2\tau_{gb}^2} \qquad (2)$$

where Z', Z" represent the real and imaginary part of the impedance, respectively; R_b and R_{gb} represent bulk and grain boundary resistance, respectively; ω represents angular frequency; τ_{gb} represents the relaxation time of the grain boundary. When $\omega \to 0$, $Z' = R_b + R_{gb}$; $\omega \to \infty$, $Z' = R_b$, so we can obtain R_b and R_{gb} from the intercept of Z' of the Cole-Cole plot .

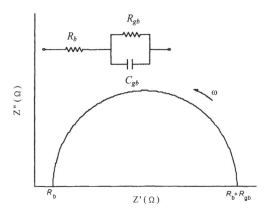

Fig. 3 Simplified equivalent circuit and its Cole-Cole plot.

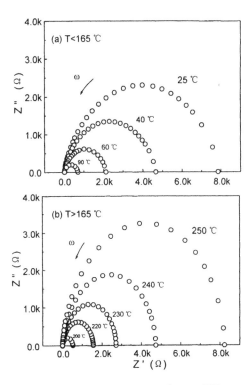

Fig.4 Cole-Cole patterns of V-PTCR ceramics at different temperature.

The Cole-Cole plots of the sample at different temperatures are shown in Fig.4. It can be seen that the variation of the left intercept Rb at the real axis Z' of the plot was small, and the variation of the right intercept $R_{gb}+R_b$ was large, which indicated the large variation of R_{gb}. The right intercept $R_{gb}+R_b$ decreased below Tc (Tc=165°C) and increased rapidly above Tc as the temperature increased. According to the impedance diagrams of the specimen at different temperatures, the variation of grain resistance R_b and grain boundary resistance R_{gb} with the temperature were obtained in Fig.5. For V-PTC materials, grain boundary resistance was large and presented V-type PTC characteristics with the variation of the temperature. When the temperature was below Tc, grain boundary resistance decreased as the temperature increased and the temperature coefficient is -3.54%/°C; when the temperature was above Tc, grain boundary resistance elevated rapidly as the temperature increased and the temperature coefficient is +6.61%/°C. It is known that the overall resistance consists of grain resistance and grain boundary resistance. Because grain resistance was small, the overall resistance R was approximately equal to grain boundary resistance R_{gb}. Therefore, It was concluded that V-PTC effect originated from the electrical behavior of grain boundary.

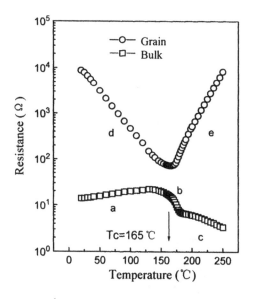

Fig.5 Dependence of Rb and Rgb on temperature

3 Microstructure and Composition

TEM micrographs clearly indicated that the glass phase existed in the grain boundary, which is shown in Fig.6. EDAX analysis of the bulk and grain boundary phase, shown in Fig.7, also indicated that element Si and Y were rich in glass phase. Therefore, The NTC behavior was closely related to the composition of the grain boundary phase.

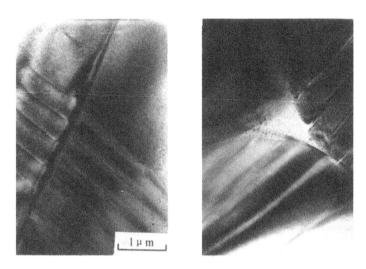

Fig.6 TEM microscopies of (Sr,Pb)TiO$_3$ composite thermistor material

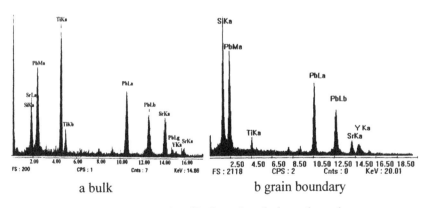

a bulk   b grain boundary

Fig.7 EDAX analysis of bulk and grain boundary phase

4 Thermo-Mechanical Analysis

Thermo-mechanical analysis also provided an implemental evidence for the above therory. Fig.8 are the TMA analysis results of $(Sr,Pb)TiO_3$ materials with SiO_2 additive of 10mol% and 0.3mol% respectively. Unlike the material with single phase structure, V-PTCR material expand in the final stage of the sintering process above 1160°C due to the existence of the glass phase, meanwhile the previous work indicates that the sintering temperature for the V-PTCR material was above 1170°C. Therefore it can be concluded that V-PTCR characteristics were closely related to the formation of the glass phase in the materials.

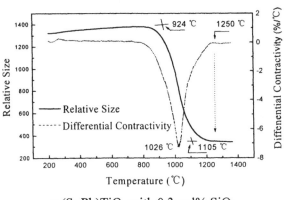

a $(Sr,Pb)TiO_3$ with 0.3mol% SiO_2

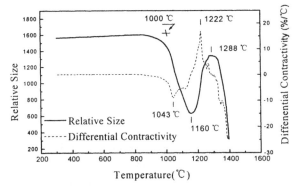

b $(Sr,Pb)TiO_3$ with 10mol% SiO_2

Fig.8 Thermo-mechanical analysis of $(Sr,Pb)TiO_3$ material

V. CONCLUSIONS

In summary, our study indicates that V-PTC effect was a grain boundary effect, which contradict the theoretical model of deep donor energy level in the previous literature. The NTC-PTC ceramics was a grain boundary controlled materials.

REFERENCES

[1]B.M. Kulwicki, "PTC Materials Technology, 1955-1980"; p.138-54 in Advances in Ceramics, Vol.1, Grain Boundary Phenomena in Electronic Ceramics. Edited by L.M. Levinson. American Ceramic Society, Columbus, OH, 1981.

[2]S. Iwaya, H. Masumura, H. Taguchi, and M. Hamada, "V-type Characteristics of PTC Materials," Electron. Ceram. Jpn., 19, 33 (1988).

[3]C. Lee, I-N Lin, and C.-T. Hu, "Evolution of Microstructure and V-Shaped Positive Temperature Coefficient of Resistivity of $(Pb_{0.6}Sr_{0.4})TiO_3$ Materials," J. Am. Ceram. Soc., 77, 1340 (1994).

[4]Longtu Li, Shan Wan, Shiping Zhou, and Zhilun Gui, "Preparation and Characteristics of $(Sr,Pb)TiO_3$ Ultrafine Particles," Chin. J. Mater. Res., Suppl., 148 (1994).

[5]Dejun Wang, J. Zhou, Z. Gui, and L. Li, "Preparation and Characteristics of $(Sr,Pb)TiO_3$ Ultrafine Particles," Wuji Cailiao Xuebao / Journal of Inorganic Materials, 12, 231 (1997).

[6]Y. Azuma and K. Nogami, "Coating of Ferric Oxide Particles with Silica by Hydrolysis of TEOS," J. Ceram. Soc. Jpn., 10, 646 (1992)

MORPHOTROPIC PHASE BOUNDARY SYSTEMS IN FERROELECTRIC TUNGSTEN BRONZE

RUYAN GUO

Materials Research Laboratory
The Pennsylvania State University
University Park, PA 16802 U.S.A

ABSTRACT

This paper reports studies to enhance the present database of information on the interrelation among the property, the macro-/micro-structure, and the chemical compositions in the ferroelectric tungsten bronze (TB) structural families that contain morphotropic phase boundary (MPB) regions. Several binary, ternary, and quaternary solid solution systems, such as $(Ba_{1-x}Sr_x)_2Na_{1-y}K_yNb_5O_{15}$, $(Pb_{1-x}K_x)_2Na_{1-y}Sr_yNb_5O_{15}$, and $(Pb,Ba)Nb_2O_6$ have been systematically investigated. The effect of cation substitution on the structural phase transitions through site-occupancies and orderings, are reviewed. Characteristics and properties found in near morphotropic phase boundary (MPB) compositions are summarized.

INTRODUCTION AND SCIENTIFIC BACKGROUND

In oxygen octahedron type ferroelectric compositions, tungsten bronze oxides constitute the second largest family. Similar as the perovskites, the ferroelectric tungsten bronzes typically have large spontaneous polarization and high dielectric constants. The structural flexibility and the chemistry versatility of this family make it more attractive in many applications than the ferroelectric perovskites. Excellent pyroelectric and electrooptic properties found in this family, e.g., in

$Sr_{5-x}Ba_xNb_{10}O_{30}$ (SBN)[1,2] and $K_{6-x-y}Li_{4+x}Nb_{10+y}O_{30}$ (KLN),[3,4] have made the way for the materials in applications such as IR detector, electric & electrooptic nonvolatile memory, and photorefractive holography. Besides ferroelectric, pyroelectric, piezoelectric and electrooptic applications, TB materials have also been used in composite form for high energy density capacitor applications utilizing their high dielectric constant and high electric strength.[5] Recent interests on material development of TB solid solutions also showed great potential of TB dielectrics in microwave applications with high dielectric constant (>150) and ultra-low dielectric loss ($<10^{-5}$).[6]

Additionally, an unique type of morphotropic phase boundary (MPB) that separates two ferroelectric phases with mutually **perpendicular** polarization directions, distinctive from a MPB that separates ferroelectric tetragonal and rhombohedral phases found in some perovskites (e.g., MPB in $PbZr_{1-x}Ti_xO_3$), has been identified so far only in TB meta-niobate solid solution systems (e.g., in $Pb_{1-x}Ba_xNb_2O_6$, PBN). The most significant advantage of the MPB composition for device applications is the enhancement of numerous physical properties over a wide temperature region. Unlike a ferroelectric phase transition, which is a function of temperature and the physical properties such as dielectric permittivity and polarization change drastically with temperature, a MPB structural transition is ideally independent of temperature and hence high dielectric constant is preserved in broad temperature range in compositions near a MPB. Such feature is considered very useful in applications for which temperature independent high polarization and high dielectric constants are required.

Along with many optimized properties reported, it was demonstrated earlier by this author and coworkers that in PBN, ferroelectric phases can be controlled/switched by applying an external electric field on a crystal of near the MPB composition.[7] It was also shown by precision optical refractive indices measurement that crystals grown with proper compositions could encounter MPB in temperature span and still maintain good crystal quality and homogeneity.[8] Those encouraging results indeed pointed to exciting application directions such as optical bi-stable and electromechanical switching devices using these materials.

Evidently there are numerous designing possibilities for different MPB systems (not limited to the orthorhombic:tetragonal MPB as found in PBN) within the ferroelectric TB structural family. Optimization of various device properties, such as the maximization of electrooptic coefficients, dielectric constants, shear mode piezoelectric coefficients, and spontaneous polarization, was also possible to obtain in non-polarization-switching type MPB compositions, e.g., in $Ba_{2-x}Sr_xK_{1-y}Na_yNb_5O_{15}$ (BSKNN) (possibly pseudo-tetragonal and tetragonal MPB) and $Sr_{2-x}Ca_xNaNb_5O_{15}$ (SCNN) (possibly MPB between two different pseudo-tetragonal phases).[9]

Although TB is technically an extremely important family and there are considerable data available in literature for individual TB compositions (e.g., ref.[9,10 11,12]), systematic experimental studies for MPB systems in TB structure family are scarce. An overall understanding of the material science that correlate structure-property-chemistry of this family is critically lacking.

This paper reviews recent study carried out at the Materials Research Lab, Penn State University, on systematic exploration, synthesizing, and characterization of selected TB solid solution compositions that contain MPB. Ceramic and single crystal samples of a range of compositions belonging to several different types of MPB in TB family were synthesized and characterized. Compositions that corresponding to a MPB have been identified. In depth studies on their crystallographic structure and property dependence have been also in progress.

The Ferroelectric Tungsten Bronze Structure

The tetragonal TB prototype (point group 4/mmm) structure is shown in Fig. 1 as projected on to the (001) plane. The chemical formula in an extended sense has the general form:

$$(A1)_2^{XII} (A2)_4^{XV} (C)_4^{IX} (B1)_2^{VI} (B2)_8^{VI} O_{30}^{VI}$$

in which A1 (2 identical sites), A2 (4 identical sites), and C (4 identical sites) are the 12-, 15-, and 9-fold coordinated sites in the crystal structure surrounded by oxygen anions.[13] The unit cell in the prototype symmetry (4/mmm) is only one octahedron high (~0.4 nm) in the c-direction with an a=b dimension of typically 1.25 nm. The BO_6 oxygen octahedra are corner linked to form three (A1:square, A2:pentagonal, and C:triangular) different types of tunnels running through the structure parallel to the c-axis.[14] These three types of cation sites can accommodate a large variety

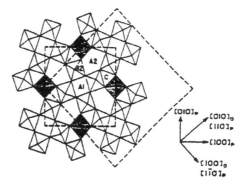

Fig. 1. Tungsten Bronze Type Structure Projected on (001) Plane.

of metal atoms. A1 and A2 sites can be occupied by large mono-valent, divalent, or trivalent ions, the small C-site may be vacant or occupied by very small ions such as Li^{1+}. Two different B-sites, B1 (shaded dark) and B2, are usually occupied by smaller, highly charged cations such as Nb, Ta, W, or Ti.[15,16,17]

Morphotropic Phase Boundary

The term "Morphotropic phase boundary" (MPB) is used to denote an abrupt structural change within a solid solution with variation in composition but nearly independent of temperature.[18] In practice, a MPB composition in polycrystalline materials usually refers to a specific composition that is a two-phase zone where the two phases are considered present in equal quantity. Depending on recent understanding of the MPB, the two phase zone in a complete solid solution necessarily has certain width that is inversely proportional to the volume of element such as grain size of the polycrystalline ceramics.[19] In single crystals, therefore, the width of this two-phase zone will approach theoretical limit of minimum at MPB composition $x=x_{MPB}$. However, all MPB identified has finite temperature dependencies therefore even in single crystals of optic quality, it is found that MPB is slanted therefore x_{MPB} is a weak function of temperature or $x_{MPB}(T)$. So far the interests have been focused on the characteristic properties of such a boundary, few detailed studies for the crystallographic nature that governs the property of such a boundary is available.

Various types of morphotropic phase boundaries are possible and can be classified primarily depending on their macroscopic space groups and additionally by their microscopic structural elements such as order-disorder and short-range or long-range order.[20] Consequently, there are MPB systems that separate phases with various properties such as ferroelectric (including relaxor)-ferroelectric, ferroelectric-antiferroelectric, antiferroelectric-antiferroelectric, etc., in different detail according to one's requirement. To avoid ambiguous description, the following discussion of the MPB is limited to the MPB between two ferroelectric phases with different crystallographic space groups.

Although more than a hundred individual ferroelectric complex compounds of the TB structure and innumerous possible solid solutions between those end members are known (a compilation of the known ferroelectric compounds and some solid solutions can be found in Landolt-Börnstein[21]), only two ferroelectric forms, orthorhombic form ($m2m$ $P_S||<010>_O$ or pseudo-tetragonal $mm2$ $P_S||<001>_T=<001>_O$) and tetragonal form ($4mm$ $P_S||<001>_T$), have been reported. Among those, several different types of ferroelectric subspecies, $P4bm$ as in SBN, and four of the same point group (*mm2*) but different space group symmetry (e.g., *Ccm2₁* as in $Ba_{4+x}Na_{2-2x}Nb_{10}O_{30}$ (BNN), *Cm2m* as in $Pb_2KNb_5O_{15}$ (PKN),[22,23] *Bb2₁m* as in $PbNb_2O_6$, and *Bbm2* as in $Sr_2NaNb_5O_{15}$) have been reported. Correspondences of the polarization orientation with crystallographic axis of the prototype structure in these two ferroelectric species are shown pictorially in Fig. 2 (the pseudo-tetragonal mm2 is not shown).

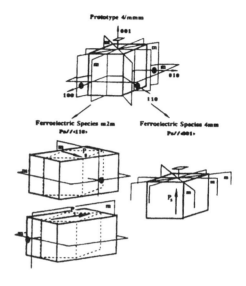

Fig. 2. Schematic representation of two (orthorhombic m2m and tetragonal 4mm) ferroelectric sub-phases for prototype point group 4/mmm.

The appearance of a MPB can usually be related to the instability of one ferroelectric phase with respect to another upon critical composition change. It is logical to expect that the two phases separated by the MPB differ slightly in composition are energetically similar. The mechanical restrains to preserve one phase against the other may very well be relaxed, or softened, because of the incipient structural change. Hence, many physical properties will be either greatly enhanced or suppressed in compositions near the MPB. Spontaneous polarization P_S, for example, may increase either due to the increase in magnitude of ionic displacement or in number of possible dipole vector directions. Elastic stiffness c_{ijkl}, as another example, may decrease because the crystallographic structure is softened due to impending phase change. The linear electrooptic effect in a ferroelectric crystal (at $T<T_c$) of a centric prototype symmetry (in this case, 4/mmm), is fundamentally a quadratic effect biased by P_S, the spontaneous polarization.[24] Large and temperature independent values of electrooptic

coefficients therefore can be anticipated. Though the individual magnitude of physical properties will depend on the symmetry and the chemistry of each phase next to the MPB, the basic trend of the behavior mentioned above will be similar.

Types of MPB in Tungsten Bronze Family

Depending on the symmetry classifications, five different phases of a lower symmetry, namely, *4mm, mm2, m(A4), m(A2)* and *1*, are possible derivations of the prototype phase 4/mmm,[25] where *A4* or *A2* indicating the polarization orientation is arbitrary with respect to the 4- or 2- axis.

For structural phase transitions in crystal, however, it is the space group and the irreducible representation form a continuum therefore must be considered when we discuss the possible morphotropic phase boundaries. Depending on the combination of different space groups and the implicated polarization orientations, using the following nomenclature:

Orthorhombic phases (*Cm2m, Bb2₁m* etc.) $P_S \perp$c-axis	Orthorhombic (O, OI; OII, etc.)
Orthorhombic phases (*Cbm2, Ccm2₁* etc.) P_S//c-axis	Pseudo-Tetragonal (PT, PTI; PTII, etc.)
Tetragonal phase (*P4bm*) P_S//c-axis	Tetragonal (T)

Total five different MPB systems therefore can be derived:

(a) Pseudo-Tetragonal and tetragonal system (MPB: PT-T)

　　　Example: Solid solution $Ba_{2-x}Sr_xK_{1-y}Na_yNb_5O_{15}$,[26]

(b) Orthorhombic and Pseudo-Tetragonal System (MPB: O-PT)

　　　Example: $(1-x)Pb_4K_2Nb_{10}O_{30}-(x)Ba_{4+x}Na_{2-2x}Nb_{10}O_{30}$,[9]

(c) Orthorhombic and Tetragonal System (MPB: O-T)

　　　Example: $Pb_{1-x}Ba_xNb_2O_6$,[27,11]

(d) Pseudo-Tetragonal I and Pseudo-Tetragonal II System (MPB: PTI-PTII)

　　　Example:$(1-x)(Ba,Na)Nb_2O_6-(x)(Sr,Na)Nb_2O_6$ (Ccm2₁ for BNN, Bbm2 for SNN[28,29]).

(e) Orthorhombic I and Orthorhombic II System (MPB: OI-OII)

　　　Example: $PbNb_2O_6-PbTa_2O_6$,[27]

EXPERIMENTAL RESULTS AND DISCUSSION

Morphotropic Phase Boundary System Selection

Five different systems were selected as described in a previous section. Three composition systems:

Pseudo-Tetragonal and tetragonal system (MPB: PT-T)

(a) $Ba_{2-x}Sr_xK_{1-y}Na_yNb_5O_{15}$ (BSKNN)

Orthorhombic and Pseudo-Tetragonal System (MPB: O-PT):

(b) $Pb_{2-x}Sr_xK_{1-y}Na_yNb_5O_{15}$ (PSKNN), and

Orthorhombic and Tetragonal System (MPB: O-T)

(c) $Pb_{1-x}Ba_xNb_2O_6$ (PBN)

are addressed in detail in this paper. The three composition families are A-site cation solid solutions representing A-site filled Pb-free (a), A-site filled Pb-containing (b), and A-site vacant Pb-containing MPB systems (c). Morphotropic types (d) and (e), as addressed in the previous page, are the topic of separate publications.

High Density Ceramic Sample Preparation

Ceramic samples used in this study were synthesized by conventional solid solution technique. Differential thermal analysis (DTA) and XRD were extensively used to optimize the processing parameters. Chemical compositions are adjusted and refined, depending on the trend of the structure parameters and property variations, to obtain the composition near a MPB.

Single Crystal Fiber Growth

Single crystals of desired composition were grown in one dimensional fiber form ($\approx$1mm in diameter) using a laser heated pedestal growth (LHPG) technique. The LHPG technique is an extremely powerful and versatile technique in crystal growth. The LHPG capability of growing single crystal fibers for materials characterization is remarkably beneficial that allow one to obtain single crystals (small yet sufficient in size for characterization) of different compositions in significantly less time than any other conventional crystal growth techniques.

Figs. 3 (a)-(d) show photographs of single crystal samples as grown by the LHPG technique. Five different compositions (Fig. 3(a)) of BSKNN family are grown that allowed systematic studies of the A-site filled lead-free TB MPB system. High quality SCNN and SBN:Ce doped crystals were also grown using LHPG technique (Figs. 3(b) and 3(c)). As it is known that large single crystals of BSKNN and SCNN have been previously grown successfully using Czochralski technique,[30] they are of high potential, being available, for device applications. The successful growth of Pb-containing ferroelectric single crystals (Fig. 3 (d) PSKNN) demonstrated the capability of the LHPG technique for growing crystals with unusual characters (e.g., high volatility). It enhances the research capabilities to study a family of MPB systems and also potentially results in

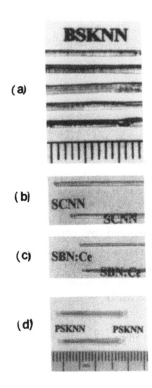

(a)

(b)

(c)

(d)

Fig. 3. Examples of LHPG
grown single crystals of
ferroelectric TBs..

developing an economical way of producing crystallographically oriented high quality high performance ferroelectric single crystals suitable for various applications where longer length, lower applied voltage and reproducible quantities of small-diameter (several mm) crystals are desirable.

Property and Characteristics of near MPB Compositions

All samples after synthesis and growth, were characterized for their crystallographic structures and were measured for their ferroelectric, pyroelectric, piezoelectric and thermal expansion properties. Electrooptic properties have also been measured for PBN, SCNN, KLN and SBN crystals. Their corresponding MPB compositions were identified along with relevant characteristics near the transition temperature. General characteristics for compositions near a MPB are found to include: average dielectric permittivities in polycrystalline samples approach maximum near the MPB; the phase transition temperatures approach minimum near the MPB; and the thermal expansion coefficients also approach minimum near the MPB. Sharp unit cell volume change in near MPB composition was also found for O-T or O-PT type MPBs where polarization vector-direction changes.

(A) Pseudo-Tetragonal and tetragonal system (MPB: PT-T): $Ba_{2-x}Sr_xK_{1-y}Na_yNb_5O_{15}$ (BSKNN)

The particular interests of this study focus on the composition range where a completely filled tungsten bronze structure is formed by solid solution of 2:1 ratio between tungsten bronze and perovskite end members, $2(A1_{1-x}A2_x)Nb_2O_6:(A3_{1-y}A4_y)NbO_3$, or $(A1_{1-x}A2_x)_2(A3_{1-y}A4_y)Nb_5O_{15}$.

Compositions studied in this research are 2SBN75:KNN50, 2SBN75:KNN40, 2SBN75:KNN25, 2SBN60:KNN40, and 9SBN61:2KNN50, and abbreviated as B2, SK, B3, K4, and C2, respectively (as seen in Table 1). All of them but one (C2) are filled TB-type compositions while C2 is chosen to compare the role of vacancy in this series.

The quaternary phase diagram $(BaNb_2O_6\text{-}SrNb_2O_6\text{-}KNbO_3\text{-}NaNbO_3)$ depicting the related solid solutions is shown in Fig. 4. As can be seen in Fig. 4, the filled $A_6B_{10}O_{30}$ compositions (e.g., known as BSKNN-1 through BSKNN-5 by the authors[31]) are clustered near the 2:1 (2SBN:KNN) ratio (projected on a solid line).[31]

Powder X-ray diffraction for all the compositions prepared were obtained, analyzed and refined using least-square fitting approach. All ceramic samples processed are essentially single phase and crystallized in tungsten bronze structure. Tetragonal or pseudo-tetragonal cell parameters are found to give complete indexing to the obtained diffraction patterns, which are shown in Fig. 5.

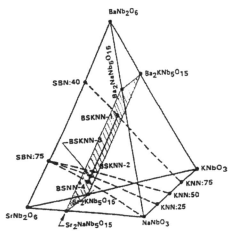

Fig. 4. $BaNb_2O_6\text{-}SrNb_2O_6\text{-}KNbO_3\text{-}$ $NaNbO_3$ Quaternary Phase Diagram (Ref.31). Shaded area is added by the current authors to indicate the 2:1 ratio plane.

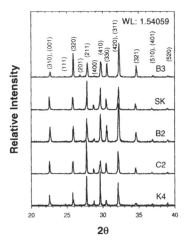

Fig. 5. X-ray diffraction patterns of the compositions studied.

The lattice parameters found for these compositions in tetragonal symmetry are listed in Table 2. As expected, compositions with higher fraction of Sr or Na, give smaller cell dimensions and cell volume. Dielectric and thermal expansion properties of the ceramic samples are summarized in Table 3. Dielectric properties obtained from single crystal samples give full account for the highly anisotropic property of these samples. Example of a single crystal dielectric permittivity and dielectric loss as function of temperature and frequency is given in Fig. 6 for single crystal SK (2SBN75:KNN40) that was grown for the first time. The dielectric permittivity κ_{33} in [001] direction is ~20 times higher than that of κ_{11}. Dielectric loss in [100] direction is noticeably low ($<2\times10^{-3}$) over a broad temperature range (-50~100°C). Low temperature relaxation behavior in all compositions were also noted, similar as reported earlier for SBN family.[32]

Table 1. Compositions studied in this work

Composition $\alpha(Ba_{1-x}Sr_x)Nb_2O_6:\beta(Na_{1-y}K_y)NbO_3$	Abbreviation	Crystal Growth Methods
$\alpha:\beta = 2:1$, $x = 0.75$, $y = 0.50$	B2	Cz.[a,b], LHPG[c]
$\alpha:\beta = 2:1$, $x = 0.75$, $y = 0.40$	SK	LHPG[c]
$\alpha:\beta = 2:1$, $x = 0.75$, $y = 0.25$	B3	Cz.[d,e], LHPG[c]
$\alpha:\beta = 2:1$, $x = 0.60$, $y = 0.40$	K4	LHPG[c]
$\alpha:\beta = 9:2$, $x = 0.61$, $y = 0.50$	C2	Cz.[a], LHPG[c]

[a] Ref. 26; [b] Ref. 33; [c] current study; [d] Ref.34; [e] Ref. 35

Table 2. Lattice parameters of BSKNN compositions by XRD analysis from ceramic samples (in tetragonal symmetry).

Composition	Volume (A^3)	Lattice Parameters (A)		
$(SrBa)Nb_5O_{15}$- $(K_{0.5}Na_{0.5})Nb_5O_{15}$	614.68	a=12.4653	c=3.9559	JCPD 38-1255
C2	613.42	a=12.4603	c=3.9509	*
K4	612.55	a=12.4463	c=3.9543	*
$(SrBa)Nb_5O_{15}$- $NaNb_5O_{15}$	609.97	a=12.4309	c=3.9473	JCPD 38-1256
B2	609.52	a=12.4344	c=3.9422	*
SK	606.02	a=12.4137	c=3.9327	*
B3	604.51	a=12.4072	c=3.9269	*

* current study.

Table 3. Dielectric and thermal expansions in ceramic samples

Composition	$T\kappa_{max}$, heating, 1kHz	$T\kappa_{max}$, cooling, 1kHz	$\Delta T\kappa_{max}$ (°C)	T_α (°C)	$T_{(Burns)}$ (°C)	κ_{max} (1 kHz)
K4	258.0	251.8	6.2	223	~253	5,000
B2	227.5	216.0	11.5	194	~500	20,600
SK	229.3	221.2	8.1	208	528	20,900
B3	229.8	221.1	8.7	208	511	19,500
C2	233.0	222.2	10.8	176	345	25,500

T_α: Temperature at minima of the thermal expansion coefficient
$T_{(Burns)}$: Temperature at thermal expansion deviating from HT linear behavior

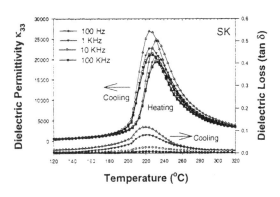

Fig. 6(a). Temperature and frequency dependence of dielectric permittivity and loss factor for LHPG grown single SK crystal along [001] direction.

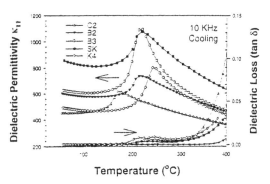

Fig. 6(b). Temperature and frequency dependence of dielectric permittivity and loss factor for LHPG grown single BSKNN crystals along [100] direction.

The tungsten bronze phase has been observed in a wide range of the $A_5Nb_{10}O_{30}$ type compositions, though the end components for A=Ba or Sr are not of the TB structure. Introduction of Na or K into the A-site has been known to enhance the TB stability and cause the compositional variation from $A_5Nb_{10}O_{30}$ (unfilled TB niobate structure: partially empty 15- and 12-fold coordinated sites with a vacancy of up to 1/6) to $A_6Nb_{10}O_{30}$ (filled TB niobate structure where both 15- and 12-fold coordinated sites are completely occupied) type.[36] The ferroelectric relaxor characteristics known to exist in typically unfilled TB niobate structures, e.g., in SBN60, are expected to maintain since the fundamental randomness in this structure due to various atoms occupying both the two A1 (12-coordinated) and the four A2 (15-coordinated) sites, is sustained. In addition, due to the fact that A-site is vacancy-free in BSKNN family, the aging behavior, the dielectric loss, and the easiness of growth, may be anticipated. Large BSKNN crystal grown by Czochralski technique,[37] high polarizability[38] and stability against temperature quenching[39] in some of the BSKNN compositions have been reported earlier.

On the basis of ionic radii of constituent ions, it seems clear that the radius of Ba (r_{Ba}^{2+}=0.150nm) ion is too large but that of Sr (r_{Sr}^{2+}=0.136nm) is too small to form the simple TB compound. That solid solution of $(Ba_{1-x}Sr_xNb_2O_6)$ for $0.20<x<0.80$ crystallized in tetragonal TB structure[40] indicates the tolerance range for the mixture of these two ionic species to form the TB structure. Introduction of Na (r_{Na}^{1+}=0.116nm) or K (r_K^{1+}=0.152 nm) increases the stability to form TB structures but of orthorhombic (BNN, SNN) or tetragonal (BKN, SKN) symmetry, respectively. It was found that in BNN Na occupies ONLY the A1-site (12-coordinated) while Ba primarily occupies the A2-site (15-coordinated).[41] It is presumed that K and Ba/Sr can occupy both the A1- and the A2-sites (though not necessarily randomly) in BKN and SKN however there is no detailed structure determination available for the BKN or SKN compositions. Following the trend of point symmetry in relation to the preference of A1-, A2-site occupancies, the morphotropic phase boundary between the tetragonal and the orthorhombic ferroelectric phases may exist near a critical ratio of Na^{1+} in the A1-site (rich in Na for a stable orthorhombic phase or when ordered arrangement destroyed, the structure becomes tetragonal).

Several indications may diagnose the change in critical Na population for the structure transforms from orthorhombic to tetragonal, e.g., cell dimension reduces (cell parameter a decreases); relaxor behavior starts to dominate; lower ionic conductivity and lower low frequency dielectric loss; higher polarizability, etc..

Utilizing the experimental results obtained and combined with studies reported in literature, the morphotropic phase boundary is proposed to lie in the plane near a composition of 30-40% of BNN, as schematically shown in Fig. 7. MPB composition in BNN-BKN system has been studied to be near 40% of BNN[31] and a possible MPB in BNN-SNN was reported to be near 25% of BNN.[42] The proposed phase boundary is in agreement with the experimental results obtained in this study that compositions with low BNN (B2, B3, and SK) are ferroelectric tetragonal with characteristics of relaxors and high polarizability; compositions with higher BNN ratio (K4 and C2) are of pseudo-tetragonal symmetry with lower polarizability at room temperature.

This hypothesis also predicts there should be two MPB compositions in the solid solution of SNN-BKN. The results obtained by Kim et al.[43] appear to support such possibility as they found Tc and the dielectric properties do not change monotonically and two minima were shown near 40 and 60% BKN.

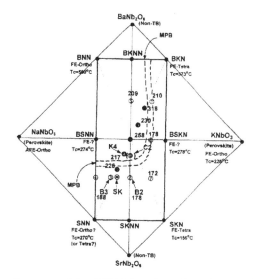

Fig. 7. Phase diagram of the 2SBN:KNN section plane, where proposed morphotropic phase boundary near the 30-40% BNN is shown. The numbers shown in figures are T_C in °C.

(B) ORTHORHOMBIC AND PSEUDO-TETRAGONAL SYSTEM (MPB: O-PT): $Pb_{2-x}Sr_xK_{1-y}Na_yNb_5O_{15}$ (PSKNN)

Quaternary phase diagram of the solid solution system is shown in Figure 8. Four filled ternary compounds (PKN, SKN, PNN, SNN) have been reported to form ferroelectric tungsten bronze structure.[21] Of these compounds, PKN has been studied in greater detail. $2PbNb_2O_6$:$KNbO_3$ (PKN) has orthorhombic (m2m) symmetry at room temperature with polarization along b-axis of the orthorhombic unit cell ($T_c \sim 460°C$)[9] while $2SrNb_2O_6:NaNbO_3$ (SNN) has pseudo-tetragonal symmetry or orthorhombic mm2, with very weak orthorhombic distortion along a- and b-axes ($T_c \sim 266°C$) at room temperature with polarization along c-axis.[21,9] As the spontaneous polarization is parallel to the 2-fold axis ($P_s//[010]_o$) in PKN, large transverse electro-optic and piezoelectric coefficients ($d_{15} \sim 470$ pC/N) have been reported.[44,45,46] SNN, however, is likely to have large longitudinal coefficients (values not available in the literature). Therefore, there exists a possibility of a MPB with perpendicular polarization directions in PKN-SNN solid solution. As an A-site filed Pb-containing composition, this family of materials may have potential applications in SAW devices, shear-mode transducers and electro-optic devices.[44,45,46] Table 4 gives the crystal parameters of PKN and SNN (in pseudo-tetragonal symmetry). Earlier study by Oliver et al.[9] indicates the possibility of a MPB around 25-30 mol % PKN in SNN but supporting evidence of its existence has not been given. The systematic study carried out on ceramic samples of PKN-SNN solid solution system has identified the morphotropic phase boundary composition and associated characteristics.

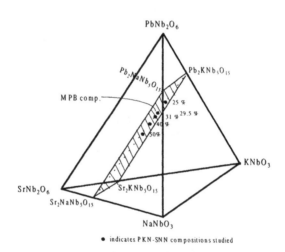

Figure 8. $PbNb_2O_6$-$SrNb_2O_6$-$KNbO_3$-$NaNbO_3$ quaternary phase diagram. Shaded area and hairline indicate 2:1 ratio plane and studied composition range respectively.

Table 4. Lattice parameters of the PKN[45] and SNN[47] unit cell.

	$Pb_2KNb_5O_{15}$ (PKN)	$Sr_2NaNb_5O_{15}$ (SNN)
Space Group	Cm2m (P_s // b-axis)	Bbm2 (P_s // c-axis)
a (Å)	17.754	12.3562
b (Å)	18.014	12.3562
c (Å)	7.830	3.8979
Unit Cell volume (Å^3)	2504.2	595.1

The solid solution of end members PKN and SNN were prepared to investigate the existence of MPB in PKN-SNN system. Eight compositions (25, 29, 29.5, 30, 30.5, 31, 40 & 50 mol % PKN in SNN) were prepared, chosen based on available literature,[9] in ceramic form by solid state reaction, and several of them were grown as single crystals using the LHPG technique.

Powder XRD patterns (Figure 9) indicate a definite trend in transition from pseudo-tetragonal phase to orthorhombic phase with equal intensity of peaks of both the phases at 29.5 mol % PKN. Least square refinement of the XRD data to the appropriate crystal system indicates tetragonal phase (tetragonal peaks observed due to the very weak orthorhombic distortion in SNN unit cell) for Sr-rich compositions and orthorhombic fit for Pb-rich compositions. Figure 10 indicates the trend in unit cell volume with varying composition. Normalized unit cell volume is used to compare the volume across all the compositions. A sharp decrease in unit cell volume at about 29.5 mol % PKN composition can be attributed to a large structural instability induced near the MPB. This structural instability may be mainly due to larger cations Pb^{2+} and K^+ being substituted for relatively smaller cations Sr^{2+} and Na^+ in A sites.

Variations of the peak dielectric permittivity and corresponding transition temperature with the composition are shown in Figure 11. These trends also indicate existence of MPB for 29.5 mol % PKN as permittivity reaches a maximum and transition temperature reaches a minimum at this composition. This behavior is likely to be due to the structural instability and related 'lattice softening' causing high polarizability.

Thermal expansion measurements were carried out from $-160^{\circ}C$ to $650^{\circ}C$ in order to study the phase transition characteristics of the PKN-SNN ceramics. Figure 12 shows the transition from ferroelectric to paraelectric phase due to difference in thermal expansion coefficient of these phases in the case of 29.5 mol

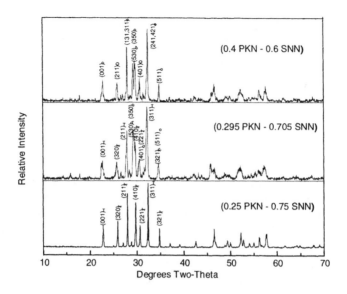

Figure 9. XRD patterns of calcined powders of different PKN-SNN compositions.

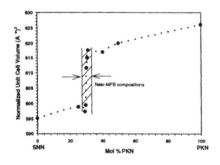

Figure 10. Unit cell volume vs. PKN mol % in PKN-SNN system.

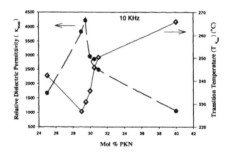

Figure 11. Dielectric constant (κ_{max}) and transition temperature vs. PKN mol%.

Dielectric Ceramic Materials

% PKN composition. MPB composition has a lower thermal expansion (shown in Fig. 13) as compared to compositions with single pseudo-tetragonal or orthorhombic ferroelectric phase. For MPB compositions, each grain has micro-domains with agility to align to either pseudo-tetragonal or orthorhombic phase as compared to macro-domains of single-phase material and as a result, higher level of compensation for thermal expansion of each domain may be resulting in a net lower expansivity.[9]

Phase transition from pseudo-tetragonal to orthorhombic phase is accompanied by a maximum in dielectric permittivity along with a minimum in the transition temperature and the coefficient of thermal expansion for near MPB composition. These characteristics can be of considerable importance for piezoelectric devices that require both high polarizability and low noise level (low thermal fluctuation).[48] Detailed study on the tensor properties such as piezoelectric, dielectric and electro-optic properties has been carried out on single crystals of these compositions in order to evaluate their potential for application in various devices.[49]

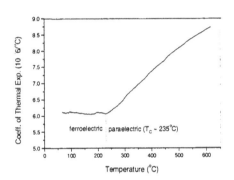

Figure 12. Coefficient of thermal expansion for 29.5 mol % PKN composition vs. temperature.

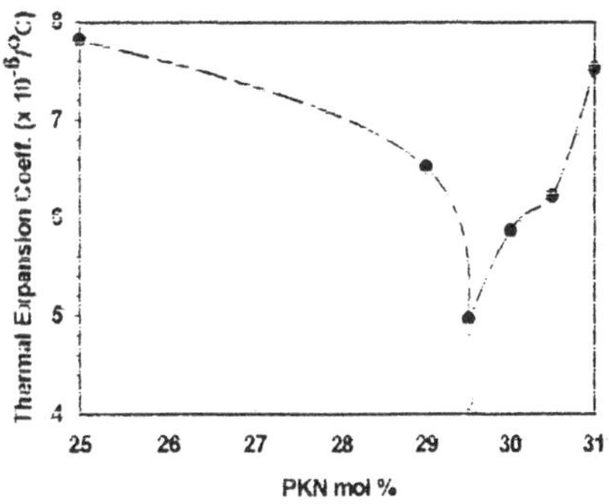

Figure 13. Coefficient of thermal expansion vs. PKN mol % at 100°C.

(C) Orthorhombic and Tetragonal System (MPB: O-T):$Pb_{1-x}Ba_xNb_2O_6$

The morphotropic phase boundary in this important solid solution family has been studied extensively. Detailed dielectric and ferroelectric properties have also been conducted on PBN single crystals. However, in this unique system where perpendicular polarization-switch takes place in compositions near the MPB, a precise determination of its crystallographic structure has not been available. This author and co-workers have been carrying out a structure analysis by single crystal x-ray diffraction of this family for compositions near the MPB in tetragonal symmetry and on-going studies for compositions of the orthorhombic symmetry.

The structure refinement results indicated that there are strong tendencies for cation preference for site-occupancy. There is, essentially, no barium atoms in the A1site when the Ba:Sr ratio is lower than 60:40. When an attempt was made to place barium atoms in the A1-site, the population for this site is refined to nearly zero.

The lattice parameters refined for the tetragonal single crystal, $Pb_{0.596}Ba_{0.404}Nb_{2.037}O_6$, (unannealed state) are a=b=12.4970 Å, c=3.9603 Å, and the population of Pb in the A1-site is p=0.551. The lattice parameters found for the powder x-ray diffraction of the annealed powder, however, are a=b=12.5019 Å, c=3.9605 Å, and the population of Pb in the A1-site was found to be somewhat higher than p=0.551 as in the case of unannealed crystal. It is likely that the A1-site in the annealed state is larger than that in the unannealed state (or poled state), presumably due to the relaxation of oxygen octahedron distortion and redistribution of vacancies after annealing that makes the A1-site more accommodate to the Pb^{2+} ions. This observation is in agreement with Trubelja et al.[50] that an increase in Sr occupancy in the A1-site was observed in $Sr_{0.50}Ba_{0.50}Nb_2O_6$ upon annealing. It is a further support for the vacancy non-random distribution hypothesis proposed earlier, through an annealing and quenching effect study in tungsten bronze $Sr_{1-x}Ba_xNb_2O_6$ and $Ba_{2-x}Sr_xK_{1-y}Na_yNb_5O_{15}$ crystals.[39] This non-random distribution of A-site vacancies appears to be an important factor in interpreting the structural changes that occur at the MPB and their contribution to the polarization of the materials at temperature below the Curie temperature range.

Oxygen anisotropic vibration has been difficult to refine at room temperature for PBN crystals however, preliminary results showed significance in the oxygen vibration magnitude and orientation. An oblique view of the structure (Figure 14.) shows that the oxygen atoms are highly compressed normal to the Nb-O bonds as would be expected. There is however no clear splitting in oxygen positions at the room temperature, different that the O5 half-atom model used by Jamieson et al.[41] in other tungsten bronze crystals such as SBN.

The direction of cation displacements can be related directly to the orientation of the macroscopic ferroelectric polarization as illustrated in the Figure 15. Note that the Nb(2) position was selected for the z=0 plane. It is known that for lead-free oxygen octahedral compounds, the displacement of Nb^{5+} from the oxygen plane is the major contribution to the macroscopic polarization. Substantial polarization contributions in this crystal, however. can be attributed to the Pb^{2+} and Ba^{2+} displacements as they are of similar magnitude as that of the Nb^{5+}.

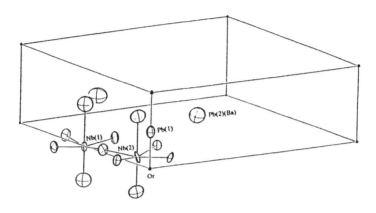

Fig. 14. PBN60 single crystal refined structure showing oxygen anisotropy.

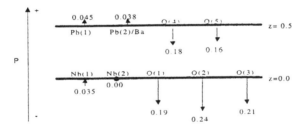

Fig. 15. Magnitude and sense of atomic displacements (in angstroms) relative to the c-axis in PBN60. Macroscopic polarization is indicated by the large arrow.

The apparent valences (V_i) of the cations for PBN60 single crystal using r_os' listed in Brown and Altermatt[51] were calculated for the refined structure. The results are summarized in Table 5.

Table 5. Apparent valences for cations in $Pb_3Ba_2Nb_{10}O_{30}$.

Atom	Site	Coordination	V_i
Nb(1)	B1	6	5.498
Nb(2)	B2	6	4.999
Pb(1)	A1	12	2.045
Ba	A2	15	2.088
Pb(2)	A2	15	1.307

As it is clear from the Table 5 that the Pb^{2+} in the A2-site (15-coordination) are severely underbonded. High polarizability in the unique c-axis direction in this tetragonal structure therefore evidently includes the contribution from the largely rattling Pb^{2+} ions in the A2-site. Together with the high temperature anisotropy, the cation fluctuation in the A2-site could give dynamic polarization components perpendicular to the mirror plane, in the <110> direction. With the increase of the Pb:Ba ratio, it is inevitable that more Pb^{2+} will occupy A2-site, that may aggravate the rattling status of the Pb^{2+} in the open cage until the orthorhombic structure became energetically more favored in the morphotropic phase boundary composition (Pb:~63%). Slight overbonding is found for Pb^{2+} in the A1-site and Ba^{2+} in the A2-site. Nb(1) in B1-site is also slightly overbonded and is known to form less distorted oxygen octahedra compared to the more deformed Nb(2) in B2-site.

Currently on-going study on orthorhombic PBN single crystal reveals the detail of the structure manifesting itself to develop spontaneous polarization in the <010> direction, upon critical Pb:Ba ratio change. The tilting of the oxygen octahedra is evident that results in the additional puckering of the oxygen plane as well as unit-cell double in <001> direction. Forthcoming results will be interesting in shedding light on the nature of MPB and the local polarization status in this family.

SUMMARY

Several distinctive types of A-site solid solution MPB families are systematically studied and reviewed in this paper, as summarized in Table 6. These three solid solutions were chosen as they represent MPB-types of ferroelectric orthorhombic: tetragonal (O:T), pseudo-tetragonal: tetragonal (PT:T), orthorhombic: pseudo-tetragonal (O:PT) combinations.

Table 6. Summary of Different TB MPB Composition Families

MPB Type	Solid Solution System	Synthesis Studies	MPB
PT-T.	$\alpha(Ba_{1-x}Sr_x)Nb_2O_6\text{-}\beta(Na_{1-y}K_y)Nb_2O_6$ $\alpha:\beta=2:1; x=0.75, 0.60, 0.50$ $y=0.50, 0.40, 0.25$	Ceramics of five different compositions along 2:1 (SBN:KNN) tie line synthesized; Single crystals of five different compositions grown by LHPG.	30~40 mol% BNN
O-T	$Pb_{1-x}Ba_xNb_2O_6$	Full range of ceramic composition studied; Single crystals were grown by Cz technique	1-x~0.63
O-PT	$(Pb_{1-x}Sr_x)_2K_{1-x}Na_xNb_5O_{15}$	Ceramic samples with x=0.5-0.75 synthesized; single crystal were grown by LHPG up to Pb~25mol%.	1-x~0.295

With the growing potential for tungsten bronze materials in fields such as high speed communications, high performance memories, infrared imaging, and electrooptics, etc., there are demanding needs to gain basic knowledge and understandings about this family, the polarization mechanisms and the MPB systems it possesses. This paper reviews some aspects of a systematic study on ferroelectric TB families with various types of morphotropic phase boundaries. Further reports and ongoing studies is anticipated to add a clearer understanding of polarization and dielectric properties in compositions near a MPB in tungsten bronze family, and of conditions needed for optimization for device applications.

ACKNOWLEDGMENT

This research was supported partially by the National Science Foundation under Grant No. DMR-9510299.

REFERENCES

1. R.R. Neurgaonkar, W.K. Cory, and J.R. Oliver, *SPIE 739, Conjugation; Beam Combining and Diagnostics*, 91 (1987).
2. L.F. Jelsma, M.E. Lowry, R.R. Neurgaonkar, W.K. Cory, and H.K. Welch, *SPIE 836, Optoelectronic Materials* (1987).
3. W.A. Bonner, W.H. Grodkiewicz, and L.G. Van Uitert, *J. Cryst. Growth* **1**, 318 (1967).
4. L.G. Van Uitert, H.J. Levinstein, J.J. Rubin, C.D. Capio, E.F. Dearborn, and W.A. Bonner, *Mat. Res. Bull.* **3**, 47 (1968).
5 e.g., R. Koontz, G. Blokhina, S. Gold, A. Krasnykh, *1998 Annual Report*

Conference on Electrical Insulation and Dielectric Phenomena, v.1, 23-26 IEEE by Omnipress (1998)

6   X.M. Chen, J.S. Yang and J. Wang, (SXII-55) presented at 100th ACerS Annual Meeting, May 3-6, Cincinnati (1998).

7.   R. Guo, A.S. Bhalla, and L.E. Cross, *Applied Optics,* **29**, 904 (1990).

8.   G. Burns, F.H. Dacol, R. Guo, and A.S. Bhalla, *Appl. Phys. Letters,* **57**, 543 (1990).

9   J.R. Oliver, R.R. Neurgaonkar and L.E. Cross, *J. Am. Ceram. Soc.* **72**(2), 202 (1989).

10   T. R. Shrout, H.C. Chen and L.E. Cross, *Ferroelectrics,* **74** 317 (1987).

11. R. Guo, *Ferroelectric Properties of Lead Barium Niobate Compositions near the Morphotropic Phase Boundary,* Publication No. 9117682 (U.M.I. Ann Arbor, MI, 1991).

12   Yuhuan Xu, "Ferroelectric Materiasl and Their Applications," Chapter 6 and references, North-Holland (1991).

13. G. Hagg and A. Magneli, *Rev. Pure Appl. Chem. (Australia)* **4**, 235 (1954).

14. M.E. Lines and A.M. Glass, *Principles and Applications of Ferroelectrics and Related Materials,* Oxford University Press, Oxford (1977).

15   L.G. Van Uitert, H.J. Levinstein, J.J. Rubin, C.D. Capio, E.F. Dearborn, and W.A. Bonner, *Mat. Res. Bull.* **3**, 47 (1968).

16. S.C. Abrahams, P.B. Jamieson and J.L. Bernstin, *J. Chem. Phys.* **54**, 2355 (1971).

17. B.A. Scott, E.A. Giess, B.L. Olson, G. Burns, A.W. Smith and D.F. O'Kane, *Mater. Res. Bull.* **5**, 47 (1970).

18. B. Jaffe, W.R. Cook, Jr., and H. Jaffe, *Piezoelectric Ceramics* (Academic Press, London and New York, 1971).

19. W. Cao and L.E. Cross, *J. Appl. Phys.* **73**(7), 3250-55 (1993).

20. A.S. Bhalla, IUMRS meeting (1992) Tokyo, Japan; ONR Workshop on Smart Materials (1993) U. of Maryland, MD.

21. Landolt-Börnstain, *Ferroelectrics and Related Substances,* Springer-Verlag, Berlin, New York (1981)

22. J. Nakano and T. Yamada, *J. Appl. Phys.* **46**(6) 2361-65 (1975).

23. K. Nagata, T. Yamazaki and K. Okazaki, *Proc. 2nd Meeting on Ferroelectric Matl's and their Applications,* No. F-10 (1979).

24. M. DiDomenico, Jr. and S.H. Wemple, *J. Appl. Phys.* **40**(2), 720 (1969).

25. L.A. Shuvalov, *J. Phys. Soc. Jpn.,* **28S**, 38 (1970).

26. Y. Xu, H. Chen and L.E. Cross, *Ferroelectrics* **54**, 123-26 (1984).

27. E.C. Subbarao, G. Shirane, and F. Jona, *Acta Crystallogr.* **13**, 226 (1960).

28. J.C. Toledano, *Phys. Rev. B.* **12**(3), 943-50 (1975).
29. J. Ravez, A. Perron-Simon and P. Hagenmuller, *Ann. Chim. T.* **I**, 251-68 (1976).
30. R.R. Neurgaonkar, W.K. Cory, and J.R. Oliver, *Ferroelectrics*, **142**, 167-88 (1993).
31 J.R. Oliver and R.R. Neurgaonkar, *J. Am. Ceram. Soc.* **72**(2) 202-11 (1989).
32 J.M. Povoa, Ruyan Guo, and A.S. Bhalla, *Ferroelectrics*, **158** 283 (1994).
33 R. R. Neurgaonkar, W.K. Cory, J.R. Oliver, W.W. Clark, III, G.L. Wood, M. J. Miller, and E.J. Sharp, *J. Cryst. Growth* **84**, 623-37 (1987).
34 R. R. Neurgaonkar, W.K. Cory, J.R. Oliver, M. Khoshnevisan, and E.J. Sharp, *Ferroelectrics*, **102**, 3-14 (1990).
35 R. R. Neurgaonkar, W.K. Cory, J.R. Oliver, E.J. Sharp, G.L. Wood, and G.J. Salamo, *Ferroelectrics* **142**, 167-88 (1993).
36 T. Ikeda, K. Uno, K. Oyamada, A. Sagara, J. Kato, S. Takano, H. Sato, *Jap. J. Appl. Phys.* **17**(2) 341 (1978).
37 R.R. Neurgaonkar, W.K. Cory, J.R. Oliver, W.W. Clark III, G.L. Wood, M.J. Miller, E.J. Sharp, *J. Crystal Growth*, **84** 629 (1987).
38 A.S. Bhalla, R. Guo, L.E. Cross, G. Burns, F.H. Dacol, R.R. Neurgaonkar, *J. Appl. Phys.* **71**(11) 5591-95 (1992).
39 R. Guo, A. S. Bhalla, G. Burns, F.H. Dacol, *Ferroelectrics* **93** 397 (1989).
40 A.A. Ballman and H. Brown, *J. Cryst. Growth,* **1**, 311 (1967); M.H. Francombe, *Acta Crystallog.* **13**, 131 (1960)
41 P.B. Jamieson, S.C. Abrahams, J.L. Bernstein, *J. Chemical Physics* **50**(10), 4352 (1969).
42 L.G. Van Uitert, J.J. Rubin, W.H. Grodkiewicz, W.A. Bonner, *Mat. Res. Bull.* **4**, 63 (1969).
43 Y. Kim, H. Joo, S. Bu, and G. Park, *Ferroelectrics* **157**, 287-292 (1994).
44 T. Yamada, *J. Appl. Phys.*, **46**, 2894 (1975).
45 J. Nakano and T. Yamada, *J. Appl. Phys.* **46**, 2361 (1975).
46 H. Yamauchi, *Appl. Phys. Lett.*, **32**, 599 (1978).
47 M. Pouchard, J. P. Chaminade, A. Perron, J. Ravez, P. Hagenmuller, *J. Solid State Chemistry*, **14**, 274 (1975).
48 Kumar Pandya, R. Guo, and A. S. Bhalla, *Ferroelectrics Lett.* **25** (3/4) (1998).
49 R. Guo, Y. Jiang, K. Pandya, A.S. Bhalla, *J. Materials Sci.* (to be published 1999).
50 M.P. Trubelja, E. Ryba, and D.K. Smith, *J. Mat. Sci.*, **31**, 1435-1443 (1996).
51 I.D. Brown and D. Altermatt, *Acta Cryst.* **B41**, 244-247 (1985).

KEYWORD AND AUTHOR INDEX

560

Printed and bound by CPI Group (UK) Ltd, Croydon, CR0 4YY

16/04/2025

14658450-0002